Sustainable Development Goals Series

The **Sustainable Development Goals Series** is Springer Nature's inaugural cross-imprint book series that addresses and supports the United Nations' seventeen Sustainable Development Goals. The series fosters comprehensive research focused on these global targets and endeavours to address some of society's greatest grand challenges. The SDGs are inherently multidisciplinary, and they bring people working across different fields together and working towards a common goal. In this spirit, the Sustainable Development Goals series is the first at Springer Nature to publish books under both the Springer and Palgrave Macmillan imprints, bringing the strengths of our imprints together.

The Sustainable Development Goals Series is organized into eighteen subseries: one subseries based around each of the seventeen respective Sustainable Development Goals, and an eighteenth subseries, "Connecting the Goals," which serves as a home for volumes addressing multiple goals or studying the SDGs as a whole. Each subseries is guided by an expert Subseries Advisor with years or decades of experience studying and addressing core components of their respective Goal.

The SDG Series has a remit as broad as the SDGs themselves, and contributions are welcome from scientists, academics, policymakers, and researchers working in fields related to any of the seventeen goals. If you are interested in contributing a monograph or curated volume to the series, please contact the Publishers: Zachary Romano [Springer; zachary.romano@springer.com] and Rachael Ballard [Palgrave Macmillan; rachael.ballard@palgrave.com].

Michael Odei Erdiaw-Kwasie ·
G. M. Monirul Alam
Editors

Circular Economy Strategies and the UN Sustainable Development Goals

palgrave
macmillan

Editors
Michael Odei Erdiaw-Kwasie
Business and Accounting Discipline
Faculty of Arts and Society
Charles Darwin University
Darwin, NT, Australia

G. M. Monirul Alam
Department of Agribusiness under the Faculty of Agricultural Economics and Rural Development
Bangabandhu Sheikh Mujibur Rahman Agricultural University
Gazipur, Bangladesh

ISSN 2523-3084 ISSN 2523-3092 (electronic)
Sustainable Development Goals Series
ISBN 978-981-99-3082-1 ISBN 978-981-99-3083-8 (eBook)
https://doi.org/10.1007/978-981-99-3083-8

Cover illustration: Reed Kaestner/gettyimages

This Palgrave Macmillan imprint is published by the registered company Springer Nature Singapore Pte Ltd.
The registered company address is: 152 Beach Road, #21-01/04 Gateway East, Singapore 189721, Singapore

To all the authors who provided novel insights into the book's topic. Your relentless efforts made this book project a reality.

Preface

In this book, circularity is introduced and discussed as a driver of the UN Sustainable Development Goals (SDGs) and its progress in emerging and developing countries. As we move forward, we want to capture the realities of the circular economy concept as much as possible by moving quickly to the fascinating theoretical and practical progress across various sectors in these economies. The circular economy approach is inherently accompanied by a wide range of other issues, which one must address while exploring its progress path: early in the book, a state-of-the-art review provides a theoretical framework for mapping the subtle characteristics, scope, and progress of emerging and developing economies, while the diverse case studies provide insight into the real stories—the progress and challenges.

This book was written to help students at universities, and other institutions of higher education find empirical case studies on circular economy from emerging and developing countries. It is also intended for people pursuing a professional career in sustainability and readers with a general interest in circular economy and the SDGs. At an advanced level, this text can also be used as a handbook, providing an overview of current theoretical and empirical debates and controversies regarding circular strategies and SDGs. As well as providing a non-technical entry point towards circular economy strategies in emerging and developing economies, the book provides a broader perspective on circular economy as an emerging field.

This indispensable reference is written by a team of international scholars from a variety of disciplines, including development, education, business, ecology, geography, and planning, and presents the current state of circular economy research within emerging and developing economies.

Darwin, Australia — Michael Odei Erdiaw-Kwasie
Gazipur, Bangladesh — G. M. Monirul Alam

Introduction

Developing and emerging economies rely heavily on the availability of bountiful and inexpensive natural resources, which are key drivers of a linear economy. In most cases, raw natural resources are taken (TAKE), transformed into products (MAKE), and get disposed of (WASTE). A linear economy is characterized by two unsustainable processes—resource scarcity, and excessive pollution load, which are occurring at an unprecedented pace in many developing and emerging countries.[1] Complicating matters further, many of these economies are already set to become the global drivers of consumption. Recent sustainability calls have emphasized a shift from linear-based strategies to circular ones,[2] although such a transition is far from straightforward. The ambition to understand the processes and pathways for a linear-to-circular transition has mostly remained in the background, particularly in the developing economy context. This book aims to contribute to emerging discussions around understanding and theorizing the pathways, mechanisms, diversity, determinants, impacts, and interactions of circular business strategies towards meeting the SDGs.

[1] Schröder, P., Anantharaman, M., Anggraeni, K., & Foxon, T. J. (Eds.). (2019). *The circular economy and the Global South: Sustainable lifestyles and green industrial development*. Routledge.

[2] Wu, Y., Zhu, Q., & Zhu, B. (2018). Comparisons of decoupling trends of global economic growth and energy consumption between developed and developing countries. *Energy Policy, 116*, 30–38.

The book captures contemporary themes, including circular food systems, product-to-service design, waste management, net-zero emissions, eco-innovativeness, circular operations model, and many others. The book contains theoretical, empirical, experimental, and case study research contributions that address key themes of the book. The book is categorized broadly into three Parts: Part I contains contributions that offer an extensive review of the extant knowledge on the topic. The contributions offer insights into key theoretical gaps in the current literature stream, particularly in the emerging and developing countries' context. Chapters captured in Part I aim to offer an understanding of the existing theoretical debates relevant to the book's topic and set the pace for further conceptual and empirical discussions. Part II of the book presents chapters extending the review's knowledge to offer conceptual insights about the topic. The chapters offer conceptual models that open new discussion points and themes on how circularity impacts SDGs in developing economies. Part III of the book presents intriguing case studies across various developing regions worldwide. Empirical studies captured in this part touch on circular issues across developing and emerging countries such as Ghana, Nigeria, Iran, Mauritius, Uganda, Malaysia, Turkey, and among others, highlighting the lessons learned from the experiences of all case countries. The empirical chapters help to validate some of the theoretical arguments and multiple hypotheses levelled in the previous parts using various robust methodological approaches to increase evidence and knowledge on the topic. The multi-case study approach taken in this book will allow a wider discovery of theoretical evolution and arguments to increase the reliability and validity of the findings presented in this book.

Conclusion

This book project aims to look beyond the linear economy model and to change the longstanding take-make-dispose approaches that have characterized many economies in the Global South. The book is categorized broadly into three parts, and to achieve this goal, the book captures review, conceptual and empirical chapters that address each aspect of the book's overall goal. Part I contains contributions that offer an extensive review of the extant knowledge on the topic. Part II of the book presents chapters extending the review's knowledge to offer conceptual insights about the topic. Part III of the book presents intriguing case studies across various developing regions worldwide. The findings from each part in this book demonstrate and address the book's core themes, and the discussions illustrate how the respective parts complement each other in addressing the book's overall goal.

Praise for *Circular Economy Strategies and the UN Sustainable Development Goals*

"This is a useful book with a significant collection of articles written by forty-eight researchers and experts on circular economy and the SDGs. The book presents appropriate policies, cases and theoretical frameworks which give a thorough idea of the paradigm shift from the EPCD2 (Extract-Produce-Consume-Dispose-Deplete) concept of the linear economy to the regenerative circular economy for sustainable futures. The collection is comprehensive in its coverage of various issues, focusing on the transitional requirements for countries. This book will definitely make a significant difference to the understanding of the readers about the circular economy, globalisation, and the SDGs".

—Prof. Sadhan Kumar Ghosh, *Research Director, SDCERC, ISWMAW, India*

Contents

Notes on Contributors

Mohammed Abubakari is an interdisciplinary Ph.D. student at the University of Saskatchewan, Canada working on the intersections of environmental knowledge systems through migration. He has an interest in development planning and environmental sustainability research, and aims to become a scholar-practitioner in sustainability science.

Matthew Abunyewah holds a Ph.D. from Newcastle University and is a research-focused Lecturer at the Australasian Centre for Resilience Implementation for Sustainable Communities within the College of Health and Human Sciences and an Adjunct Senior Lecturer at the University of Newcastle, Australia. He leverages participatory and interdisciplinary research approaches to inform government policies on disaster management and resilience, circular economy and sustainability, community development, industry and business development, and climate change and variability. Before joining Charles Darwin University, he worked as a community and industry development practitioner researching workforce and training development, public policy development and disaster resilience across Africa and Australia for over eight years. He has also published his research outputs in world-class reputable academic outlets and translated findings from most of his research into simple and plain language, accessible to industry practitioners in outlets such as the conversation.

Ebenezer Afum holds a Master's degree in Logistics Engineering and Management from Dalian Maritime University, China. He also has a Bachelor's degree in Management Studies from the University of Cape Coast, Ghana. His research expertise spans circular economy, green supply chain management, zero waste practices, green human resource management, green innovation management, lean operations, and quality management practices. His articles have been published in top-tier international journals, including the International Journal of Logistics Management, Journal of Cleaner Production, Supply Chain Management: An International Journal, Sustainable Production and Consumption, Journal of Manufacturing Technology Management, and International Journal of Lean Six Sigma.

Yaw Agyabeng-Mensah is a Ph.D. student in Curtin University. He holds a Master's degree in Logistics Engineering and Management from Dalian Maritime University, China, and a Bachelor's degree in Management Studies from the University of Cape Coast, Ghana. His research outputs have been published in reputable journals, such as *Supply Chain Management: An International Journal*, *Journal of Cleaner Production*, *International Journal of Physical Distribution and Logistics Management*, *International Journal of Logistics Management*, *Journal of Business and Industrial Marketing*, and *Sustainable Production and Consumption*.

Kofi Dua Agyei holds a Master's degree in Climate Change and Sustainable Development from the University of Ghana. His research focuses on waste management, environmental sustainability, climate change adaptation, and water resources management. He is interested in exploring and understanding sustainability transitions in developing countries. His ultimate goal is to conduct research that creates awareness and informs change among local authorities and various waste management stakeholders to understand and apply local adaptive strategies and innovative methods to manage waste sustainably.

Douglas Opoku Agyeman is a Ph.D. candidate at the School of Management Operations, and Marketing, University of Wollongong, Australia. *He holds a Master of Philosophy in Marketing from the University of Ghana and a first-class Bachelor's degree in Management Studies from the University of Cape Coast. He has held teaching and research positions at the University of Ghana Business School, the University of Cape Coast Distance Education, and Knutsford University College, Accra. He consults* *for and has research interests in branding, electronic business, digital and social media marketing, and general consumer/firm-level studies. He has published in internationally peer-reviewed journals.*

Dr. G. M. Monirul Alam is currently a Professor at the Department of Agribusiness, Bangabandhu Sheikh Mujibur Rahman Agricultural University (BSMRAU), Bangladesh. He also holds the position of Adjunct Research Fellow at USQ, Australia. He is an award-winning researcher whose current research domains are climate change adaptation, poverty, food security, ICT, circular economy, and impact analysis.

Desmond Alugnoa is the Africa Climate and Zero Waste programme Manager for the Global Alliance for Incinerator Alternative. He co-founded the Green Africa Youth Organization (GAYO) which he uses as a mechanism to provide youth and gender-based leadership and support frontline communities to adapt to climate change. He has served as an advisor to the Global Center on Adaptation (GCA) on youth related issues since 2021. He has led and supported communities to develop and implement climate and circular economy models in Ghana including Water for Adaptation, Ghana's Zero Waste Strategy, and the Sustainable Community Project. He is a public speaker and a strategic campaigner who has authored many publications on crucial topics like Period Poverty and Climate Change, Representing the Voiceless and Zero Waste as a Climate Solution.

Dr. Rashid Ameer is a Lecturer in the postgraduate and degree programmes at IPU New Zealand. His teaching areas are auditing and assurance, financial reporting, postgraduate research methods, and corporate social responsibility. His research interests include business ethics, environmental management, corporate fraud, and sustainability.

Ivy Agyeiwaa Amponsah holds a Master's degree in International Studies and Diplomacy and also pursuing a second M.A. degree in International Administration and Policy. As a Green House Gas (GHG) and Sustainable Development Goal (SDG) analyst, her current work focuses on ensuring integrity in voluntary carbon offset projects. She aims to be an expert in sustainable land-use practices in Africa.

Joshua Amponsem is a Ghanaian climate activist and the Climate Lead at the Office of the UN Secretary General's Envoy on Youth. He has over 8 years of experience working with young people on Climate Action

Disaster Risk, and Resilience Building. He founded Green Africa Youth Organization (GAYO), serves as a member of the IRENA Global Council on Enabling Youth Action for SDG 7, and has been an Adaptation Fellow at the Global Center on Adaptation(GCA). In the past years, he has raised over $500,000 to support grassroots youth-led organizations and is continuously engaged in the advocacy to shift climate philanthropy to youth and locally led organizations. He is the lead author of the first-ever background paper on youth and climate change adaptation which provided a pathway for youth engagement in global adaptation efforts. He has initiated locally led projects like the Water for Adaptation and Sustainable Communities Project in sub-Saharan Africa which is creating jobs for over 100 people and is championing the establishment of Youth Climate Councils across the Global South.

Michael Afful Andoh holds a Ph.D. from RMIT University and is currently a Research Fellow and a part-time Lecturer in Aerospace and Aviation Department, School of Engineering, RMIT University, Australia. He obtained his Ph.D. in Space Science from RMIT University, Meng Electronics engineering with major in Satellite Systems Engineering from Stellenbosch University, South Africa, M.Sc. in Computational Science from Ecole Normale Superieure de Lyon, France and La Sapienza University in Rome, Italy, B.Sc. (Hons) in Space Science and Astronomy, University of Cape Town, South Africa and B.Sc. in Engineering Physics, University of Cape Coast, Ghana. He serves as a technical committee member on the Space Education and Outreach Committee (SEOC) of the International Astronautical Federation (IAF) and a member of the International Academy of Astronautics (IAA) Commission 6—Space and Society, Culture and Education. He worked in the South African National Space Agency helping develop the first South African CubeSat (Zacube-1) as a research engineer. His interests are Space Systems, Space Sustainability, AI4Space, Distributed Satellite Systems, Trusted Autonomy for Space platforms and Space-Based Intelligence Systems, Space Domain Awareness (SSA), Debris Mitigation, Surveillance and Reconnaissance (ISR), and Space Education.

Samuelson Appau is an Assistant Professor of Marketing at the Melbourne Business School. His multidisciplinary research examines how transitions and transformations in people's lives, society, markets, and the environment affect consumer behaviour and well-being.

Patrick Arhin is a graduate student pursuing a 2-year joint Master's Programme in Spatial Planning and Regional Development at the Technical University of Dortmund University, Germany, and the Federal University of ABC, Brazil. His research and professional experience spans across project planning and management, natural resources governance and livelihood, circular economy and sustainability, climate change and adaptation, and urban and regional planning.

Dennis Asare is a Research Consultant at Imani Centre for Policy and Education, Accra-Ghana. His research focuses on energy policy, climate change policy and finance, public policy and anti-corruption, and political economy analysis.

Zehra Binnur Avunduk holds a Ph.D. in Business Administration from the Faculty of Economics, Istanbul University, Istanbul, Turkey. She is currently a Lecturer with the Vocational School of Social Sciences, Istanbul University-Cerrahpaşa. Her research interests include innovation strategies, open innovation, sustainability, technology, and production management. She has published book chapters in well-known international publishers and articles in national and international peer-reviewed journals, served as an ad-hoc reviewer in several prestigious journals, and presented many papers at scientific conferences. She has received the "Best Paper Award" from the 2018 International EMI Entrepreneurship and Social Sciences Congress. She also has served as a reviewer for the Academy of Management annual meetings. She has been a Researcher in projects, one of which is funded by the Istanbul Development Agency, and the second is an R&D project in Technopark of Istanbul University.

Charles Baah is a Ph.D. candidate at the Asia Pacific College of Business and Law, Charles Darwin University (CDU), Australia. He received his B.Sc. degree in Management Studies and Master's degree in Engineering from the University of Cape Coast and Dalian Maritime University in 2015 and 2020, respectively. His research interests include sustainable management, circular economy, green manufacturing, green logistics, sustainable leadership, green entrepreneurship, and innovation. He has published in Q1 and A journals, such as the *Journal of Cleaner Production*, *Sustainable Production and Consumption*, *International Journal of Physical Distribution and Logistics Management*, and *Journal of Business and Industrial Marketing*.

Kingsley Tetteh Baako is a Lecturer in Property at RMIT University in Melbourne, Australia. His research uses mixed methods to study multiple themes in applied research in the areas of property markets and policy impacts, and technological disruptions in developing markets.

Festival Godwin Boateng is a political Economist. Most of his highly interdisciplinary work focuses on political pathways to low emissions, sustainable, and equitable infrastructure and land use. He is currently a postdoctoral Scholar at the Centre for Sustainable Urban Development of Columbia University's Climate School/Earth Institute.

Betty Osei Bonsu is a Ghanaian circular economy enthusiast and climate activist serving as the Uganda Country Manager for the Green Africa Youth Organization (GAYO). She has over 5 years' experience leading the implementation of sustainable community projects and youth in climate policy engagements. This is providing green jobs for rural communities, diverting waste from landfills, and amplifying the voices of young people in climate change actions. She led the first background paper on waste to energy that saw the conversion of palm kernel shells to charcoal briquette in Ghana. She has consulted for the World University Services of Canada where she provided Green Jobs training to women in mining extractive areas. She is currently pursuing her master's with the United Nations University studying Environmental Risk and Human Security. Following her passion in public speaking and advocacy, she founded the platform B.ISA (B. Inspired with Stories from Africa) where she is bringing stories on community values, passion, and environment to life.

King Carl Tornam Duho (ACMA CGMA CA) is a pracademic with experience in the private sector, academia, and public policy space. He is the Technical Advisor of the Dataking Research Lab, an Honorary Fellow with the VIT TVET CPE and an IMF Youth Fellow. He is an independent Researcher for a University of Chicago research project. He has published over 40 policy papers and peer-reviewed journal articles.

Kirk Bright Enu is presently a doctoral student at the Technical University of Munich studying the effective conservation and management of water-related ecosystems for mitigating flood risk. His research focuses on achieving resilience in urban areas concentrating primarily on planning, waste management, and climate-related risk reduction using nature-based solutions and how these issues interface with young people and the poor.

His thesis leverages on these systems to improve the quality of life in cities.

Angela Erdiaw-Kwasie is currently the Senior Research Scientist at PearlDiver Technologies, USA. Angela is a data scientist and analyst who holds Bachelor's degree in Actuarial Science from KNUST in Ghana, and a Master of Statistics from American University, Washington DC. Prior to joining PearlDiver Technologies, she worked for American Society for Engineering Education (AEES) as a Senior Data Analyst on different projects on health, sustainability, education, and population, where she applied her skills of displaying data with cutting-edge interactive dashboards. She has outstanding expertise in handling large and confidential data by applying machine learning algorithms, statistical methods in data science. She is specialized in data analytics using statistical programming languages, such as R software, STATA, SPSS, and KNIME for data processing, analysis, and management.

Dr. Michael Odei Erdiaw-Kwasie is an award-winning author and Lecturer in the Business and Accounting Discipline at Charles Darwin University. Dr. Erdiaw-Kwasie holds a PhD in Sustainability Management from the University of Southern Queensland. His research interests focus on sustainability transitions, circular economy, climate risks & business, digital futures, and global development.

Abebe Hailemariam is a Research Fellow with Bankwest Curtin Economics Centre at Curtin Business School, Curtin University. He holds a Ph.D. in Economics from Monash University. His research has focused on energy and climate change, tourism, gender, and inequality.

Farkhondeh Mortaz Hejri recently Defended her Ph.D. thesis in IT management at Islamic Azad University, South Tehran Branch, under supervision of Dr. Valmohammadi. Currently, she works as a consultant in an Iranian IT company. Also, she cooperates as a sessional lecturer with a private Iranian higher education institution.

Kris Ivanovski is a Lecturer in the Monash Business School, Monash University, Australia. He holds a Ph.D. in Economics and is an applied economist with interests in economic history and long-term time-varying relationships between various socioeconomic factors. His research has focused on energy economics, financial economics, and other issues related to applied economics. He has published widely on these topics in a variety of outlets.

Halim Kazan is a Professor and the Head of the Department of Business Administration, Faculty of Economics, at Istanbul University where he teaches Advanced Research Techniques, Production Management, Research Techniques, Advanced Research Applications, Quantitative Decision-Making Techniques in Businesses at undergraduate and graduate levels. He has articles and books published in prestigious journals in the field of production management and supply chain management. He carried out many projects and consultancy activities in public and private sector organizations.

Naiyer Fatema Khanom is a Ph.D. candidate at the University of Queensland. Her research area is STEAM education and gender equality in decision-making roles. She has worked in the Asia region for access to finance and women's economic empowerment projects. Her volunteer involvement includes working with Transparency International, Grameen Bank, GovHack Australia, and other organizations to promote quality education for girls, youth employment, entrepreneurship ecosystems, ICT literacy, green financing, and green jobs for sustainable development. She has also worked as a director for the Adelaide chapter of the Founder Institute and is a co-founder of Wisecars, a technology-based start-up.

Andrews Darfour Kwakye holds an M.B.A. in Management Information Systems, an M.Sc. in Data Analytics, and is pursuing a Ph.D. in Data Analytics in Aircraft Health Management. His research covers applying Data Analytics to improve efficiencies in early Fault Detection and Isolation (FDI) in aircraft health management.

Mudasiru Mahama is an Assistant Lecturer and a Ph.D. candidate at the University of Energy and Natural Resources (UENR), Ghana. His areas of research interest include electricity market designs for regional integration, energy/renewable energy policies, and development planning.

Bob Offei Manteaw is currently a Senior Research Fellow at the Centre for Climate Change and Sustainability Studies at the University of Ghana. His Ph.D. in Sustainable Development Thinking and Practice was from the Washington State University in Pullman, USA. He specializes in climate change and sustainable development governance where his research and teaching interests explore critical convergence issues around environment, sustainability, and climate transitions. He is particularly interested in catalytic and transformational learning processes for climate-resilient sustainable development. His most recent research works have focused on climate-informed decision-making in data-poor environments, climate and health impacts, planetary health framing, and post-pandemic imaginations of global health futures in a changing climate.

Dorothea Afriyie Opoku Mensah is pursuing her second Master's degree in M.Sc. Agribusiness & Applied Economics. Her interests are in production economics, efficiency, agricultural risk assessment, and management. Her research employs statistical, machine learning, and operations research techniques to understand policy issues related to production efficiency and risk in the agricultural industry in North Dakota State.

Joshua Mugambwa holds a Ph.D. and is a Senior Lecturer and the Dean of the Faculty of Management and Public Policy at Makerere University Business School (MUBS), Kampala, Uganda. He has been lecturing at universities and other tertiary institutions for over 10 years. His research interests are in public policy, circular economy, Africa-China relations, development, poverty reduction, urban management, business and work ethics, gender, leadership, and governance. He has authored over ten publications including peer-reviewed journal articles, book chapters, and edited books.

Annet K. Nabatanzi-Muyimba holds a Ph.D. and is a Senior Lecturer and the Director of the Knowledge for Development Centre at Makerere University Business School (MUBS), Kampala, Uganda. She has over 20 years of lecturing, research, consultancy, and managerial experience. She has special research interests in enterprise competitiveness, commercialization of innovations, new products and technologies, CSR management, gender mainstreaming, governance, and public policy implementation. She is a member of the International Business Conference (IBC), a best research paper awardee (South Africa International Conference on Entrepreneurship 2009), a reviewer for several journals, and an external examiner for many universities in and outside Uganda.

Andy Bonaventure Nyamekye holds a Ph.D. from Wageningen University in The Netherlands and is a Vice President of Digital Project Management at Citigroup. His research interests border on technology, innovation, and sustainability. He is interested in investigating how emerging economies can leverage technology and innovation in solving societal problems. As a multidisciplinary Researcher, his recent works employ multiple lenses, such as responsible innovation, adaptive decision-making, adaptive governance, and informational governance.

Adunola Okupe holds a Doctor of Philosophy in Strategic Leadership in the Tourism Industry from the University of Surrey and is currently a Lecturer of sustainability and strategy at Lagos Business School, Pan-Atlantic University, Lagos, Nigeria. Her research and teaching interests are in the areas of behavioural economics to understand decision-making as regards leadership for societal change, socio-cultural leadership, tourism, and leisure studies, and the impacts these have on the sustainable development agenda. She is also a Senior Advisor with Red Clay, a tourism advisory practice. She is a fellow of the Higher Education Academy and a member of the Academy of International Business. She has been a reviewer for *Journal of Cleaner Production*. She is on the Board of Open House Lagos, Kadara Heritage Foundation and a member of Legacy 1995 and the Nigeria Field Society.

Radiah Othman is an Associate Professor at the School of Accountancy, Massey University at Manawatu campus, Palmerston North, New Zealand. Her teaching areas include auditing, forensic accounting, and accounting information systems. She has a research interest in sustainability, forensic and fraud auditing, and public sector accountability. She is an Editorial Board Advisory member with the *Journal of Public Budgeting, Accounting & Financial Management*, *Journal of Accounting in Emerging Economies*, and *Journal of Financial Reporting and Accounting*.

Kofi Kusi Owusu-Ansah is a graduate student currently pursuing his M.A in International Studies at the School of Political Science, Public Affairs, and International Studies at the University of Wyoming. He also serves as a Graduate Research Assistant for the Wallop Civic Engagement programme, where he assists in conducting research related to the development of civic education and community engagement. He holds a Bachelor's degree in Economics from Kwame Nkrumah University of Science and Technology, Ghana. During his undergraduate studies, he conducted extensive research on macroeconomic factors that influence economic growth and savings in Ghana. His research interests primarily focus on circular economy, environmental politics, and strategies for sustainable and global development in developing economies.

Yaw Ndori Queku is a Senior Lecturer in the Department of Accounting and Finance (DAF) at Cape Coast Technical University (CCTU), Ghana. He is currently the Head of the Department and Vice-Dean of the School of Business. He holds a Ph.D. in Business Administration from Open University Malaysia (OUM), a Master of Commerce (MCOM), and a Bachelor of Commerce (BCOM) from the University of Cape Coast (UCC), Ghana. He is a qualified accountant from the Institute of Chartered Accountants Ghana (ICAG). His current research interest is in the areas of accounting standards, financial reporting, taxation, circularity accounting, and environmental accounting.

Samira Sanni is a passionate young woman with a strong interest in inclusive sustainable development and politics, public policy analysis, and gender relations. She has an interest in academic scholarship that focuses on inclusivity to promote the well-being of people and the environment through a collaborative research approach. She looks forward to making co-produced knowledge relevant for policymaking and implementation for change. Currently, she is pursuing a two-year Master's Programme in Global Development Studies at the Norwegian University of Life Sciences, Norway.

Emmanuel Junior Tenakwah recently completed his Ph.D. in Business at the School of Business, Western Sydney University. He holds a Master of Philosophy degree and a Bachelor of Science degree, all in Accounting, from the University of Ghana. He is currently a Lecturer at the Swinburne University of Technology and Excelsia College, Sydney. His research interests are Business Sustainability, Corporate Social Responsibility, Environmental Reporting, Corporate strategy and governance, Auditing, and governance. He has over 10 journal articles, books, and conference papers.

Emmanuel Senior Tenakwah is a Lecturer at the Asia Pacific College of Business and Law, Charles Darwin University, Australia. He received his Ph.D. from the University of Newcastle. He also holds a Master's and Bachelor's degree in Public Administration from the University of Ghana. He writes on people management, sustainability, and circular and associated practices and teaches courses in HRM, ethics, and management. He has produced journal articles, books, chapters, and conference papers.

Kwaku Owusu Twum works as a professional Geo-Data Scientist and a multidisciplinary Researcher with a focus on resource innovations and governance; data-decision systems and prediction science; sustainability modelling and regional development. He is currently the Head of Data and IT for Koa Impact Switzerland and Ghana leading the radical change in data-driven innovations within the cocoa beverage industry.

Amila Kasun Sampath Udage Kankanamge is a Master By Research student at Charles Darwin University, Northern Territory, Australia. His research interest includes sustainability transition, circular economy solutions, waste management, and urban mining. He holds a Master of Business Administration from the Postgraduate Institute of Management (PIM) of the University of Sri Jayawardenepura, Sri Lanka and a Bachelor of Business Management (Special) Degree in Marketing from the University of Kelaniya, Sri Lanka, with first-class honours. He acquired Postgraduate Diploma in Professional Marketing at the Chartered Institute of Marketing (CIM) in the UK with merit. Before joining Charles Darwin University, he worked in the service industry for over five years in Sri Lanka.

Adil Ünal works as a Research Assistant Doctor in the Department of Business Administration, Faculty of Political Sciences, Istanbul University. His main areas of research are procurement, sustainability, supply chain management, and process/performance management. He is also co-authored a book titled "Logistics Center Management". He teaches Business Intelligence at the undergraduate level and Supply Chain Strategies at the graduate level.

Changiz Valmohammadi is Associate Professor at Islamic Azad University, South Tehran Branch. His areas of interest are quality management, knowledge management, operations management, strategic management, and application of IT in business operations. For more than 25 years, he has taught and supervised undergraduate, graduate students at both national and international universities, such as Taylor's university and Universidad de las Américas Puebla (UDLAP) and offered consulting services to the Iranian public and private organizations. He has published research papers in journals, such as *Information & Management*, *International Journal of Production Economics*, *The TQM Journal*, *International Journal of Productivity and Performance*, *Resources Policy*, *Total Quality Management & Business Excellence*, and *Journal of Enterprise Information*

Management. He sits on editorial board of journals such as Management Decision, Industrial, and Commercial Training, just to name a few.

Thomas Yeboah is a Research Fellow/Lecturer with the Bureau of Integrated Rural Development (BIRD), Kwame Nkrumah University of Science and Technology. He holds a PhD in Development Studies from the University of Cambridge, UK. He has nearly a decade of experience working on issues around household relations, gender, and generational relations in rural development and migration studies, especially children and youth's labor migration and their engagement with rural and urban economies. Prior to joining BIRD, he worked as a Postdoctoral Research Fellow at the Centre for Trust, Peace and Social Relations, Coventry University, UK, where he was part of an interdisciplinary team of engineers and social science researchers that worked on an EPSRC-funded Humanitarian Engineering and Energy for Displacement (HEED) project led by Coventry University in partnership with Practical Action and Scene.

Nuran Çakır Yıldız is an Asst. Professor at the IU Vocational School of Social Sciences and the Head of the Marketing and Advertising Department in Istanbul University. She graduated from Istanbul University (IU) Faculty of Economics in 1993. She completed her master's and doctorate studies between 1994 and 2002 at the Faculty of Economics of IU Institute of Social Sciences. She completed her doctorate in the field of "Inflation and Nominal Anchor policies" at the Department of Economic Policies of IU Faculty of Economics. She worked as an editor in the IU Journal of Social Sciences in 2014–2017 and in the Journal of Applied Social Sciences between 2017 and 2019. She has been a member of the editorial board of the International Journal of Water Management and Diplomacy from 2020 to 2022. In the field of economy, she works especially on inflation, the economic dimension of water, and circular economy. She lectures on economics, research methods and techniques, statistics, and marketing research. Currently, she serves as a faculty member at Istanbul University Cerrahpaşa Vocational School of Social Sciences.

Melek Yurdakul graduated from Istanbul University International Business Administration Master's Programme in 2013 and from Istanbul University Institute of Social Sciences, Business Administration Doctorate Programme in 2018. She worked as a research assistant between 2011 and 2018 and has worked as an Ass. Professor since 2018. She is staff at Istinye University, Faculty of Economics, Administrative and Social Sciences, Department of Business Administration. Her research focuses on sustainability management at the business level. In this context, she has researched circular economy and eco-innovation topics, which are seen as solutions in the sustainability transformation of companies. Through her research areas, she teaches courses and manages theses at undergraduate, graduate, and doctoral programme levels.

Abbreviations

CE	Circular Economy
EU	European Union
FICF	Finland's Independence Celebration Fund
SMEs	Small and Medium-sized Enterprises
UK	United Kingdom
WBES	World Bank Enterprise Survey

List of Figures

STEAM Education and Innovation Learning Towards Circular Strategies

How Can Ghana Transition from a Linear to a Circular Economy of Waste Management? A Conceptual Analysis of Policy Approaches

Linear Economy Wolf in Circular Economy Sheep Clothing: Ride-Hailing in Ghana

A Transition Towards a Circular Food Economy in Ghana: An Institutional and Policy Analysis

Macroeconomic Performance and Progress Towards a Circular Economy in Developing Countries

Circular Economy Principles and Responsible Manufacturing: Assessing Implications for Resource Conservation, Emission Reduction, Cost Performance, and Environmental Legitimacy

Human Capital Transformation for Circular Economy and Sustainable Development: A Government-Linked Company Experience

Applying Circular Economy Thinking to Food Systems in the Hospitality Industry in Nigeria

Circular Economy in Turkish Manufacturing Sector: The Roles of Green Manufacturing and Innovation

Corporate Sustainability and Circular Economy in Turkish Service and Industrial Businesses

Circular Value Chains: Circular Strategies and Managerial Perceptions of Supply Chain Professionals from Turkey

Exploring Factors Affecting the Adoption of Green Process Management Model in the Software Industry: Progress Towards Sustainability and Circularity

Capital Maintenance and Depreciation Accounting Convergence: A New Model for Sustainability and Circularity in Capital Maintenance

Pathways Towards a Circular Economy in Ghana: The Contribution of Waste Transfer Stations and the Informal Waste Collectors in Solid Waste Management

GAYO's Sustainable Community Waste Management Model: Impacts and Lessons For Circular Futures

List of Tables

A Transition Towards a Circular Food Economy in Ghana: An Institutional and Policy Analysis

Macroeconomic Performance and Progress Towards a Circular Economy in Developing Countries

Circular Economy Practices in Mauritius: Examining the Determinants

Circular Economy Principles and Responsible Manufacturing: Assessing Implications for Resource Conservation, Emission Reduction, Cost Performance, and Environmental Legitimacy

Capital Maintenance and Depreciation Accounting Convergence: A New Model for Sustainability and Circularity in Capital Maintenance

Institutionalization and Enforcement of Circular Fisheries Practices in Uganda: The Military's Role in Implementing the National Fisheries Policy

GAYO's Sustainable Community Waste Management Model: Impacts and Lessons For Circular Futures

Literature Review of Circular Strategies and SDGs

Circularity Challenges in SDGs Implementation: A Review in Context

Michael Odei Erdiaw-Kwasie

1 Introduction

Transforming our world: the 2030 agenda for sustainable development was adopted by the UN General Assembly in September 2015. Integrated around the themes of people, planet, prosperity, peace, and partnership, the 2030 Agenda consists of 17 Sustainable Development Goals (SDGs), which incorporate all three dimensions of sustainable development (economic, social, and environmental). SDGs recognize the interconnectedness of eradicating poverty and inequality, creating inclusive economic growth, and protecting the environment. It is imperative that the 2030 Agenda is implemented in a manner consistent with circularity principles to achieve the vision of an equitable and inclusive world, as well as sustainable use of resources (Bringezu et al., 2016; Cobbinah et al., 2015; Mensah, 2019; Zhang et al., 2016). The new agenda is already being incorporated into the plans, strategies, and visions of several

M. O. Erdiaw-Kwasie (✉)
Business and Accounting Discipline, Faculty of Arts and Society, Charles Darwin University, Casuarina, NT 0810, Australia
e-mail: michael.erdiaw-kwasie@cdu.edu.au

M. O. Erdiaw-Kwasie and G. M. M. Alam (eds.), *Circular Economy Strategies and the UN Sustainable Development Goals*, Sustainable Development Goals Series,
https://doi.org/10.1007/978-981-99-3083-8_1

governments, institutions, and organizations. Despite these challenges, emerging and developing countries face a variety of challenges as they implement global goals and targets (Cobbinah et al., 2021; Griggs et al., 2017; Leal Filho et al., 2019).

In the 2030 Agenda, circularity occupies a central position and is addressed through SDG 12: Responsible Consumption and Production. Consequently, the 2030 Agenda has significant implications, and its implementation will require coherent, integrated strategies, and multisectoral initiatives (Castro et al., 2021; Weitz et al., 2018). At all levels, including the global, regional, and national, new approaches will be required to achieve each of these broad objectives. As part of SDG 12 there are 11 targets and 13 indicators related to the management of toxic materials that are harmful to the environment and that cover all priorities and international agreements. To achieve the SDGs and associated targets, an integrated approach will have to be adopted that promotes the circularity of resources across all sectors, which received much less attention during the millennium development goals period (MDGs) (Liu et al., 2016; Sachs et al., 2019). To maintain momentum and identify areas that require more effort, tracking progress was vital to achieving the MDGs. In light of the Sustainable Development Goals, it is no less important. In the 2030 Agenda, systematic follow-up and review of SDG-related implementation are heavily emphasized at the national, regional, and global levels (Janoušková et al., 2018; Organization, 2016). It is important to assess how well the world is progressing towards Sustainable Development Goals and to set targets aligned with circularity principles.

Throughout this article, we analyse and assess the circularity within the context of the SDG implementation—challenges as well as future research and policy directions. Specifically, we demonstrate how SDG implementation can be seen as a system solution framework in which circularity features prominently. A discussion of circularity in SDG implementation in emerging and developing countries is presented in this study, as well as some possible research avenues for the future.

2 Circularity Within the SDGs Framework

Despite the market-based approaches that have dominated sustainability efforts in the past (Biermann et al., 2017; Khurshid et al., 2022), the circularity dimension of SDG implementation is less taken into account but essential. As the twenty-first-century challenges grow more complex, circular economies have emerged as viable alternatives to unsustainable linear economies (MacArthur, 2013; Velenturf & Purnell, 2021). Circular economies maintain the maximum functionality of resources by not consuming them but reintroducing them to a continuous and long-lasting system. Recycling, repairing, remanufacturing, or reusing devices instead of destroying their value after use retains their value. In lieu of destroying value after use, a cycle of reusing, repairing, remanufacturing, or recycling is used to preserve value. Therefore, we need to develop new business models and innovative product designs that use non-toxic materials that can be endlessly recycled to achieve this goal (Erdiaw-Kwasie et al., 2023; Haupt & Hellweg, 2019). As we switch from our current consumption methods to a circular economy, wealth and prosperity move from one system to another (Prieto-Sandoval et al., 2018; Sauvé et al., 2016). By design, the system is regenerative, meeting all citizens' needs within the earth's natural capacity.

The circular economy model offers enormous potential compared to the traditional linear economy model of "take-make-consume-dispose". Reducing the requirement for virgin resources also re-thinks the entire process of handling both resources and wastes, re-designs the product so that it becomes cost-effective, creates jobs, facilitates the development of new and innovative technologies, and ultimately results in an environmentally friendly process. By utilizing material components to be reused, shared, repaired, refurbished, remanufactured, and recycled, circularity seeks to create a closed-loop system while reducing the use of natural resources (Brennan et al., 2015; Lüdeke-Freund et al., 2019; Reike et al., 2018). As part of the circularity concept, waste is brought back into the production stream, at least in part. Taking a circular approach to the goals and targets in the United Nations 2030 agenda for SDGs is discussed.

Circularity brings to light the damaging effects of economic activity on human and natural systems and then designs solutions to mitigate them. A number of factors contribute to this, including the release of greenhouse gases and hazardous substances, the pollution of the air, land, and water, as well as structural waste, such as traffic congestion (Giama & Papadopoulos, 2020; Hailemariam & Erdiaw-Kwasie, 2022; Zhang et al., 2022). In emerging and developing countries, climate change, biodiversity loss, waste, and pollution pose challenges that prevent them from achieving their Sustainable Development Goals (Al Mamun et al., 2022; Alam et al., 2021; Cobbinah et al., 2021). While climate change is a global phenomenon, and industrialized countries are the primary sources of greenhouse gases, the negative effects will most severely affect developing and emerging countries. As a result of circularity, reducing carbon emissions, minimizing raw resource extraction, and lessening our ecological footprint are possible by designing products and systems that encourage and facilitate circular behaviours, such as sharing, reusing, and recycling products and their components. Taking circularity actions can help achieve Sustainable Cities and Communities (Goal 11), Responsible Production and Consumption (Goal 12), Climate Action (Goal 13), Life Below Waters (Goal 14), and Life on Lands (Goal 15) by reducing production, consumption, and waste, and encouraging community cooperation.

Moreover, a circular food system should not only prevent waste, but also be regenerative, resilient, and healthy (Jurgilevich et al., 2016; Schreefel et al., 2020). Throughout the value chain, food is wasted and lost. Consumers mostly cause waste in high-income countries. Emerging and developing countries experience loss at the beginning of their production chains (Calvo-Porral et al., 2017; Parfitt et al., 2010). In most cases, waste is a result of financial, technical, and organizational limitations (Hodges et al., 2011; Thi et al., 2015). A growing number of circular initiatives create jobs, improve livelihood, and stimulate the economy. As an example, in emerging and developing countries, organizations work with waste pickers to build better employment opportunities (Fergutz et al., 2011; Gutberlet, 2021). It is also important to include gender concerns in circular development programmes to ensure a just transition (Schröder, 2020; Schroeder & Barrie, 2022). There is a direct impact of circular initiatives on multiple Sustainable Development Goals (SDGs), including No Poverty (Goal 1), Zero Hunger (Goal 2), and Reduced (gender) Inequalities (Goals 5 and 10).

Sustainable ecological growth is incompatible with long-term economic growth (Hepburn & Bowen, 2013). Currently, much production and work **are** environmentally destructive and not geared towards promoting the welfare of humans or the environment—but rather to earn a profit. For circularity to succeed, new systems, materials, and services must be developed. It is also imperative to maintain a safe operating space for humanity by transforming work, introducing new forms of work, and terminating work since work is a medium between society and the environment (UN, 2015). In certain sectors, such as mining, fracking, deforestation, aviation, shipping, and animal agriculture, there is a need to limit their work, whereas rewilding, nature restoration, and autonomous and solidarity activities need more focus (Salleh, 2016). To support economic development and human well-being, SDG 9 calls for quality, reliable, sustainable, and resilient infrastructure, including regional and transborder infrastructure, emphasizing affordable and equitable access. "Eco-LooBox" has been introduced as part of the UN Global Compact Sustainable Infrastructures for Belt and Road Initiative to accelerate the SDGs' Action Platform. Eco-LooBoxes are bio-toilet systems constructed from recycled cargo containers. Using a circular economy, we can create more meaningful jobs supporting Decent Work and Economic Growth (Goal 8) and Industry, Innovation, and Infrastructure (Goal 9).

Using this reasoning, we propose several contextual circularity conditions that will enhance the chances of the SDGs becoming a reality, particularly in developing and emerging countries.

3 Circularity Challenges in SDGs Implementation

The Sustainable Development Goals yield many positive results in reducing poverty and hunger, improving people's well-being and education, ensuring access to clean water and sanitation, and other objectives promoting and protecting the environment. Despite this, several challenges remain (Kumar et al., 2021; Sharma et al., 2021). As emerging and developing countries struggle to implement the SDGs, this section emphasizes key circularity challenges.

3.1 Under-Developed Circularity Indicators

The SDGs demonstrate circularity through national, regional, and subnational indicators, based on national priorities. As part of the 2030 Agenda, considerable emphasis is placed on country follow-up and review processes. Many emerging and developing countries need more data when reviewing progress towards SDG targets (Movilla-Pateiro et al., 2021; Workie et al., 2020). Consequently, it should be a priority to strengthen the country's resource circularity measurement system. In order to track progress effectively, it is argued that effective circularity indicators should be promoted. Analysing the indicators globally and regionally can only be done with comparable data. In many emerging and developing countries, circularity indicators and methods of collection, analysis, and use of data are underdeveloped or unproven (Bao & Lu, 2020; Ngan et al.,

2019; Pavlović et al., 2020). The circularity indicators inherited from the MDG era remain inadequate, and other metrics used by UN bodies to measure progress are still underdeveloped (Hillerbrand, 2018; Nijenhuis & Leung, 2017; Stuckler et al., 2010). Major investments will be required to improve the situation regarding estimates to strengthen the interactions between United Nations agencies and countries, emphasizing developing and emerging countries.

3.2 COVID-19 Shocks

Several nations have developed national policies to support sustainable development and signed international agreements to protect the environment. The European Union and 79 countries reported having implemented at least one national policy instrument contributing to implementing the 10-year Framework of Programmes on Sustainable Consumption and Production (Naumann et al., 2011). Evidence suggests that progress had stalled or been reversed in other areas prior to COVID-19's emergence at the end of 2019 and WHO's declaration of a global pandemic on 11 March 2020. The COVID-19 outbreak has exacerbated existing vulnerabilities to sustainability goals and is threatening every one of the 17 Sustainable Development Goals (Davila et al., 2021; Fagbemi, 2021). As a result of parallel threats to health, economic, and social crises, the global community is facing unprecedented financial hardships as countries attempt to contain the epidemic and provide immediate financial assistance to many people suffering from the macroeconomic downturns associated with it. As an example, according to SDG12, consumption and production patterns are not sustainable (Cohen, 2020; Kumar et al., 2020); oceans experienced an unsustainable depletion and environmental degradation (Aktar et al., 2021; Silva et al., 2021); forests continued to decline at alarming rates (Amador-Jiménez et al., 2020; Vittor et al., 2021); and species remained threatened by extinction (Ceballos et al., 2020; Khetan, 2020; Pinder et al., 2020). Although the pandemic has struck every part of the world, the poorest and most vulnerable have been disproportionally affected. There is a hypothesis that the pandemic abruptly halted the implementation of many of the SDGs, sometimes reversing years of progress.

3.3 *Limited Progress Tracking Systems*

As a starting point, material circularity is the new impetus for the 2030 agenda. Close monitoring of progress towards the SDGs is necessary since they cover a range of economic, environmental, and social factors. In circularity monitoring, resource-related targets from the Sustainable Development Goals must be taken into account, as well as the risks associated with sustainable material flow, which fall under the domain of other sectors (such as air, water, and plastics) (Dantas et al., 2021; dos Santos et al., 2022; Oberle et al., 2019). It is also important to consider factors that determine material circularity, such as private investments in recyclable materials, recycling of electronic waste, and other municipal waste. It will be necessary to have a more integrated and collaborative approach to monitor these determinants since many emerging and developing countries will often be responsible for tracking them (Geng et al., 2012; Ghisellini et al., 2016; Heshmati, 2017). The monitoring of material circularity should also monitor cross-sectoral actions. A clear picture of what is happening can be obtained by examining qualitative information as well as quantitative process indicators outside the SDG indicator framework.

3.4 *Data Gaps*

According to an analysis of the available global SDG Indicators, less than half of the countries have comparable data for 4 of the 17 Goals (Goals 5, 12, 13, and 14) (Hogan et al., 2018; Kraak et al., 2018; Lafortune et al., 2018). In particular, the lack of country-level data on Goal 12 (ensure sustainable consumption and production patterns) is worrisome (Gasper et al., 2019; Rodić & Wilson, 2017). Observed from the perspective of emerging and developing countries, the statistics get worse. Even countries with available data have only a small number of observations over time, which makes identifying trends and monitoring progress difficult. Most countries are missing data for global material footprint-related indicators (Goal 12) earlier than the 2000s. Also, indicators measuring gender equality (Goal 5), sustainable cities (Goal 11), and peace, justice, and strong institutions (Goal 16) demonstrate similar results. To ensure operational continuity, statistics groups have been working at warp speed around the globe to adapt and innovate methods and processes for data production. To effectively accelerate the SDGs over the next decade, we need to scale up these initiatives. For material circularity efforts and recovery strategies to continue, the national statistical systems urgently need financial and technical support. As we accelerate the SDGs' implementation, we must mobilize international and domestic resources to increase investment in national data and statistical systems.

3.5 Ineffective Policy Integration

The SDGs must, therefore, be effectively translated between global and national aspirations to succeed. A few studies have been conducted to contextualize the SDGs in national settings (Biermann et al., 2017; Kanie & Biermann, 2017; Wastl et al., 2020). Furthermore, some developing countries have taken steps forward, especially those with national mechanisms in place to handle MDGs, so they can now adopt SDGs (Allen et al., 2021; Horn & Grugel, 2018). The SDGs, however, cannot only be achieved by the government. A key question will be whether the 17 Sustainable Development Goals will be adequate to guide behaviours in the right direction without an integrated vision and principle for long-term sustainable development. As a result, effective implementation of the SDGs often requires systems to solve issues that transcend existing frameworks and institutions. The interlinkages between the SDGs should be kept in mind when implementing the goals to avoid negative trade-offs and create synergies. Leadership, foresight, innovation, finance, and collaboration at all levels will be key to reducing the risks and achieving SDG circularity indicators.

4 Outlook for Future Research and Policy Actions

As a result of the SDGs, academic research and policy analysis are presented with new questions.

As a matter of fact, circularity promotion and its communication to different stakeholders have increasingly become objectives and priorities as they help accelerate SDG implementation in many countries. Tools for measuring, tracking, and communicating progress are essential (Valls-Val et al., 2022). There is a need for tools that identify current conditions and create short-term, medium-term, and long-term objectives and goals, as well as monitor their progress. However, the level of circularity in national settings needs to be measured using better tools. It is imperative that a circular economy be established to achieve SDG targets since many countries now see it as a perfect channel to accomplish these goals. To achieve this, governments will need tools that allow them to evaluate and communicate (clearly and unequivocally) their progress.

A second challenge is an access to the right data, which is essential for businesses, governments, and all stakeholders to create effective circular strategies to achieve Sustainable Development Goals. It is still relatively new to measure circularity, and more studies are needed in the areas that have yet to be measured traditionally, particularly from the point of view of the Sustainable Development Goals. Regarding the country's transition to a circular economy and its achievement of its Sustainable Development Goals, having available metrics measuring a company's circular economy performance can have several benefits for both companies themselves and the country. By providing stakeholders with data on the circular economy's performance, stakeholders can make more informed decisions about their sustainability investments, which will speed up the transition to a circular economy.

Lastly, circular economy solutions in emerging and developing countries remain severely underfunded, which means they cannot contribute to the SDGs. As a result, the circularity approach to SDGs raises important questions about how to embed and integrate new financial and investment models to promote circular actions, especially in developing and emerging economies. Investing in projects that contribute to the development of the circular economy is essential for addressing sustainability concerns because circular economy initiatives reduce overconsumption, design out waste, and restore and regenerate ecosystems and natural capital. It was noted earlier, for instance, that circular economy investments can help achieve several of the Sustainable Development Goals, in particular SDG 12 (sustainable consumption and production). In light of the growing attention being paid to this issue, more insights are needed into how circularity financing can impact the SDGs, especially among emerging and developing economies.

5 Conclusion

In the context of implementing SDGs, circularity offers the opportunity to examine and model socio-environmental contexts. SDG implementation should consider using non-toxic, natural materials and processes that protect the health of water, air, and soil—thereby promoting the circularity approach. In this approach, raw materials are kept in production cycles longer and reused repeatedly, resulting in a much less wasteful production and consumption model. It is argued that the circularity approach relates to a number of different Sustainable Development Goals (SDGs) and can contribute to them. In a circular economy, responsible consumption and production are at the core. A circularity-based SDG implementation approach will be presented in this paper as a way to advocate for circular economy knowledge and cycle contexts in SDG implementation, particularly in emerging and developing economies. A dilemma regarding the balance between resource cycling costs and sustainability benefits cannot be resolved through techno-economic analysis; it is a circularity issue. Emerging and developing economies can significantly impact SDG implementation by using reliable data, promoting circularity measurement and monitoring, strengthening circularity financing, and integrating circularity policies effectively. In developing and emerging regions, circularity is crucial for achieving SDGs and offers alternatives to traditional techno-economic approaches.

References

Aktar, M. A., Alam, M. M., & Al-Amin, A. Q. (2021). Global economic crisis, energy use, CO_2 emissions, and policy roadmap amid COVID-19. *Sustainable Production and Consumption, 26*, 770–781.

Al Mamun, A., Islam, A. R. M. T., Alam, G. M., Sarker, M. N. I., Erdiaw-Kwasie, M. O., Bhandari, H., & Mallick, J. (2022). Livelihood vulnerability of char land communities to climate change and natural hazards in Bangladesh: an application of livelihood vulnerability index. *Natural Hazards*, 1–27.

Alam, G. M., Erdiaw-Kwasie, M. O., Nagy, G. J., & Leal Filho, W. (Eds.). (2021). *Climate vulnerability and resilience in the Global South: Human adaptations for sustainable futures*. Springer Nature.

Allen, C., Metternicht, G., & Wiedmann, T. (2021). Priorities for science to support national implementation of the sustainable development goals: A review of progress and gaps. *Sustainable Development, 29*(4), 635–652.

Amador-Jiménez, M., Millner, N., Palmer, C., Pennington, R. T., & Sileci, L. (2020). The unintended impact of Colombia's COVID-19 lockdown on forest fires. *Environmental and Resource Economics, 76*(4), 1081–1105.

Bao, Z., & Lu, W. (2020). Developing efficient circularity for construction and demolition waste management in fast emerging economies: Lessons learned from Shenzhen China. *Science of the Total Environment, 724*, 138264.

Biermann, F., Kanie, N., & Kim, R. E. (2017). Global governance by goal-setting: The novel approach of the UN Sustainable Development Goals. *Current Opinion in Environmental Sustainability, 26*, 26–31.

Brennan, G., Tennant, M., & Blomsma, F. (2015). Business and production solutions: Closing loops and the circular economy.

Bringezu, S., Potočnik, J., Schandl, H., Lu, Y., Ramaswami, A., Swilling, M., & Suh, S. (2016). Multi-scale governance of sustainable natural resource use—challenges and opportunities for monitoring and institutional development at the national and global level. *Sustainability, 8*(8), 778.

Calvo-Porral, C., Medín, A. F., & Losada-López, C. (2017). Can marketing help in tackling food waste?: Proposals in developed countries. *Journal of Food Products Marketing, 23*(1), 42–60.

Castro, G. D. R., Fernandez, M. C. G., & Colsa, Á. U. (2021). Unleashing the convergence amid digitalization and sustainability towards pursuing the Sustainable Development Goals (SDGs): A holistic review. *Journal of Cleaner Production, 280*, 122204.

Ceballos, G., Ehrlich, P. R., & Raven, P. H. (2020). Vertebrates on the brink as indicators of biological annihilation and the sixth mass extinction. *Proceedings of the National Academy of Sciences, 117*(24), 13596–13602.

Cobbinah, P. B., Erdiaw-Kwasie, M. O., & Amoateng, P. (2015). Rethinking sustainable development within the framework of poverty and urbanisation in developing countries. *Environmental Development, 13*, 18–32.

Cobbinah, P. B., Erdiaw-Kwasie, M. O., & Basson, M. (2021). Deconstructing Africa's urban space: Sustainable development and spatial planning challenge. In P. B. Cobbinah & M. Addaney (Eds.), *Sustainable urban futures in Africa* (pp. 21–41). Routledge.

Cohen, M. J. (2020). Does the COVID-19 outbreak mark the onset of a sustainable consumption transition? *Sustainability: Science Practice and Policy, 16*(1), 1–3.

Dantas, T. E., De-Souza, E., Destro, I., Hammes, G., Rodriguez, C., & Soares, S. (2021). How the combination of Circular Economy and Industry 4.0 can contribute towards achieving the Sustainable Development Goals. *Sustainable Production and Consumption, 26*, 213–227.

Davila, F., Bourke, R. M., McWilliam, A., Crimp, S., Robins, L., Van Wensveen, M., Alders, R. G., & Butler, J. R. (2021). COVID-19 and food systems in Pacific Island Countries, Papua New Guinea, and Timor-Leste: Opportunities for actions towards the sustainable development goals. *Agricultural Systems, 191*, 103137.

dos Santos, L. C. T., Giannetti, B. F., Agostinho, F., & Almeida, C. M. (2022). Using the five sectors sustainability model to verify the relationship between circularity and sustainability. *Journal of Cleaner Production, 366*, 132890.

Erdiaw-Kwasie, M. O., Abunyewah, M., Yusif, S., & Erdiaw-Kwasie, A. (2023). Does circular economy knowledge matter in sustainable service provision? A moderation analysis. *Journal of Cleaner production, 383*, 135429.

Fagbemi, F. (2021). COVID-19 and sustainable development goals (SDGs): An appraisal of the emanating effects in Nigeria. *Research in Globalization, 3*, 100047.

Fergutz, O., Dias, S., & Mitlin, D. (2011). Developing urban waste management in Brazil with waste picker organizations. *Environment and Urbanization, 23*(2), 597–608.

Gasper, D., Shah, A., & Tankha, S. (2019). The framing of sustainable consumption and production in SDG 12. *Global Policy, 10*, 83–95.

Geng, Y., Fu, J., Sarkis, J., & Xue, B. (2012). Towards a national circular economy indicator system in China: An evaluation and critical analysis. *Journal of Cleaner Production, 23*(1), 216–224.

Ghisellini, P., Cialani, C., & Ulgiati, S. (2016). A review on circular economy: The expected transition to a balanced interplay of environmental and economic systems. *Journal of Cleaner Production, 114*, 11–32.

Giama, E., & Papadopoulos, A. (2020). Benchmarking carbon footprint and circularity in production processes: The case of stonewool and extruded polysterene. *Journal of Cleaner Production, 257*, 120559.

Griggs, D., Nilsson, M., Stevance, A., & McCollum, D. (2017). *A guide to SDG interactions: From science to implementation*. International Council for Science.

Gutberlet, J. (2021). Grassroots waste picker organizations addressing the UN sustainable development goals. *World Development, 138*, 105195.

Hailemariam, A., & Erdiaw-Kwasie, M. O. (2022). Towards a circular economy: Implications for emission reduction and environmental sustainability. *Business Strategy and the Environment, 32*(3977).

Haupt, M., & Hellweg, S. (2019). Measuring the environmental sustainability of a circular economy. *Environmental and Sustainability Indicators, 1*, 100005.

Hepburn, C., & Bowen, A. (2013). *Prosperity with growth: Economic growth, climate change and environmental limits*. Edward Elgar Publishing.

Heshmati, A. (2017). A review of the circular economy and its implementation. *International Journal of Green Economics, 11*(3–4), 251–288.

Hillerbrand, R. (2018). Why affordable clean energy is not enough. A capability perspective on the sustainable development goals. *Sustainability, 10*(7), 2485.

Hodges, R. J., Buzby, J. C., & Bennett, B. (2011). Postharvest losses and waste in developed and less developed countries: Opportunities to improve resource use. *The Journal of Agricultural Science, 149*(S1), 37–45.

Hogan, D. R., Stevens, G. A., Hosseinpoor, A. R., & Boerma, T. (2018). Monitoring universal health coverage within the Sustainable Development Goals:

Development and baseline data for an index of essential health services. *The Lancet Global Health, 6*(2), e152–e168.

Horn, P., & Grugel, J. (2018). The SDGs in middle-income countries: Setting or serving domestic development agendas? Evidence from Ecuador. *World Development, 109*, 73–84.

Janoušková, S., Hák, T., & Moldan, B. (2018). Global SDGs assessments: Helping or confusing indicators? *Sustainability, 10*(5), 1540.

Jurgilevich, A., Birge, T., Kentala-Lehtonen, J., Korhonen-Kurki, K., Pietikäinen, J., Saikku, L., & Schösler, H. (2016). Transition towards circular economy in the food system. *Sustainability, 8*(1), 69.

Kanie, N., & Biermann, F. (2017). *Governing through goals: Sustainable development goals as governance innovation*. MIT Press.

Khetan, A. K. (2020). COVID-19: Why declining biodiversity puts us at greater risk for emerging infectious diseases, and what we can do. *Journal of General Internal Medicine, 35*(9), 2746–2747.

Khurshid, A., Qayyum, S., Calin, A. C., Saleem, S. F., & Nazir, N. (2022). The role of pricing strategies, clean technologies, and ecological regulation on the objectives of the UN 2030 Agenda. *Environmental Science and Pollution Research, 29*(21), 31943–31956.

Kraak, M. J., Ricker, B., & Engelhardt, Y. (2018). Challenges of mapping sustainable development goals indicators data. *ISPRS International Journal of Geo-Information, 7*(12), 482.

Kumar, A., Luthra, S., Mangla, S. K., & Kazançoğlu, Y. (2020). COVID-19 impact on sustainable production and operations management. *Sustainable Operations and Computers, 1*, 1–7.

Kumar, R., Verma, A., Shome, A., Sinha, R., Sinha, S., Jha, P. K., Kumar, R., Kumar, P., Das, S., & Sharma, P. (2021). Impacts of plastic pollution on ecosystem services, sustainable development goals, and need to focus on circular economy and policy interventions. *Sustainability, 13*(17), 9963.

Lafortune, G., Fuller, G., Moreno, J., Schmidt-Traub, G., & Kroll, C. (2018). SDG index and dashboards detailed methodological paper. *Sustainable Development Solutions Network.*

Leal Filho, W., Tripathi, S. K., Andrade Guerra, J., Giné-Garriga, R., Orlovic Lovren, V., & Willats, J. (2019). Using the sustainable development goals towards a better understanding of sustainability challenges. *International Journal of Sustainable Development & World Ecology, 26*(2), 179–190.

Liu, L., Oza, S., Hogan, D., Chu, Y., Perin, J., Zhu, J., Lawn, J. E., Cousens, S., Mathers, C., & Black, R. E. (2016). Global, regional, and national causes of under-5 mortality in 2000–15: An updated systematic analysis with implications for the Sustainable Development Goals. *The Lancet, 388*(10063), 3027–3035.

Lüdeke-Freund, F., Gold, S., & Bocken, N. M. (2019). A review and typology of circular economy business model patterns. *Journal of Industrial Ecology, 23*(1), 36–61.

MacArthur, E. (2013). Towards the circular economy. *Journal of Industrial Ecology, 2*(1), 23–44.

Mensah, J. (2019). Sustainable development: Meaning, history, principles, pillars, and implications for human action: Literature review. *Cogent Social Sciences, 5*(1), 1653531.

Movilla-Pateiro, L., Mahou-Lago, X., Doval, M., & Simal-Gandara, J. (2021). Toward a sustainable metric and indicators for the goal of sustainability in agricultural and food production. *Critical Reviews in Food Science and Nutrition, 61*(7), 1108–1129.

Naumann, S., Davis, M., Kaphengst, T., Pieterse, M., & Rayment, M. (2011). Design, implementation and cost elements of Green Infrastructure projects. *Final report, European Commission, Brussels, 138.*

Ngan, S. L., How, B. S., Teng, S. Y., Promentilla, M. A. B., Yatim, P., Er, A. C., & Lam, H. L. (2019). Prioritization of sustainability indicators for promoting the circular economy: The case of developing countries. *Renewable and Sustainable Energy Reviews, 111*, 314–331.

Nijenhuis, G., & Leung, M. (2017). Rethinking migration in the 2030 agenda: Towards a de-territorialized conceptualization of development. *Forum for Development Studies, 44*(1), 51–68.

Oberle, B., Bringezu, S., Hatfield-Dodds, S., Hellweg, S., Schandl, H., Clement, J., Cabernard, L., Che, N., Chen, D., & Droz-Georget, H. (2019). Global resources outlook 2019: natural resources for the future we want.

Organization, W. H. (2016). *World health statistics 2016: Monitoring health for the SDGs sustainable development goals*. World Health Organization.

Parfitt, J., Barthel, M., & Macnaughton, S. (2010). Food waste within food supply chains: Quantification and potential for change to 2050. *Philosophical Transactions of the Royal Society b: Biological Sciences, 365*(1554), 3065–3081.

Pavlović, A., Bošković, G., Jovičić, N., Nestić, S., Sliusar, N., & Stanisavljević, N. (2020). Determination of circularity indicators: A case study of MB International Company. *Recycling and Sustainable Development, 13*(1), 9–21.

Pinder, A. C., Raghavan, R., Britton, J. R., & Cooke, S. (2020). COVID-19 and biodiversity: The paradox of cleaner rivers and elevated extinction risk to iconic fish species. *Aquatic Conservation: Marine and Freshwater Ecosystems, 30*(6), 1061–1062.

Prieto-Sandoval, V., Jaca, C., & Ormazabal, M. (2018). Towards a consensus on the circular economy. *Journal of Cleaner Production, 179*, 605–615.

Reike, D., Vermeulen, W. J., & Witjes, S. (2018). The circular economy: New or refurbished as CE 3.0?—Exploring controversies in the conceptualization of the circular economy through a focus on history and resource value retention options. *Resources, Conservation and Recycling, 135*, 246–264.

Rodić, L., & Wilson, D. C. (2017). Resolving governance issues to achieve priority sustainable development goals related to solid waste management in developing countries. *Sustainability, 9*(3), 404.

Sachs, J. D., Schmidt-Traub, G., Mazzucato, M., Messner, D., Nakicenovic, N., & Rockström, J. (2019). Six transformations to achieve the sustainable development goals. *Nature Sustainability, 2*(9), 805–814.

Salleh, A. (2016). Climate, water, and livelihood skills: A post-development reading of the SDGs. *Globalizations, 13*(6), 952–959.

Sauvé, S., Bernard, S., & Sloan, P. (2016). Environmental sciences, sustainable development and circular economy: Alternative concepts for trans-disciplinary research. *Environmental Development, 17*, 48–56.

Schreefel, L., Schulte, R., De Boer, I., Schrijver, A. P., & Van Zanten, H. (2020). Regenerative agriculture–the soil is the base. *Global Food Security, 26*, 100404.

Schröder, P. (2020). *Promoting a just transition to an inclusive circular economy*. Royal Institute of International Affairs.

Schroeder, P., & Barrie, J. (2022). Is going circular just? Environmental justice and just transition–key elements for an inclusive circular economy. *Field Actions Science Reports. The Journal of Field Actions*(Special Issue 24), 20–25.

Sharma, H. B., Vanapalli, K. R., Samal, B., Cheela, V. S., Dubey, B. K., & Bhattacharya, J. (2021). Circular economy approach in solid waste management system to achieve UN-SDGs: Solutions for post-COVID recovery. *Science of the Total Environment, 800*, 149605.

Silva, A. L. P., Prata, J. C., Walker, T. R., Duarte, A. C., Ouyang, W., Barcelò, D., & Rocha-Santos, T. (2021). Increased plastic pollution due to COVID-19 pandemic: Challenges and recommendations. *Chemical Engineering Journal, 405*, 126683.

Stuckler, D., Basu, S., & McKee, M. (2010). Drivers of inequality in Millennium Development Goal progress: A statistical analysis. *PLoS Medicine, 7*(3), e1000241.

Thi, N. B. D., Kumar, G., & Lin, C.-Y. (2015). An overview of food waste management in developing countries: Current status and future perspective. *Journal of Environmental Management, 157*, 220–229.

United Nations. (2015). *Resolution adopted by the General Assembly on 25 September 2015*. UN.

Valls-Val, K., Ibáñez-Forés, V., & Bovea, M. D. (2022). How can organisations measure their level of circularity? A review of available tools. *Journal of Cleaner production, 354*, 131679.

Velenturf, A. P., & Purnell, P. (2021). Principles for a sustainable circular economy. *Sustainable Production and Consumption, 27*, 1437–1457.

Vittor, A. Y., Laporta, G. Z., Sallum, M. A. M., & Walker, R. T. (2021). The COVID-19 crisis and Amazonia's indigenous people: Implications for conservation and global health. *World Development, 145*, 105533.

Wastl, J., Porter, S., Draux, H., Fane, B., & Hook, D. (2020). Contextualizing sustainable development research. *Digit. Sci.*

Weitz, N., Carlsen, H., Nilsson, M., & Skånberg, K. (2018). Towards systemic and contextual priority setting for implementing the 2030 Agenda. *Sustainability Science, 13*(2), 531–548.

Workie, E., Mackolil, J., Nyika, J., & Ramadas, S. (2020). Deciphering the impact of COVID-19 pandemic on food security, agriculture, and livelihoods: A review of the evidence from developing countries. *Current Research in Environmental Sustainability, 2*, 100014.

Zhang, C., Hu, M., Di Maio, F., Sprecher, B., Yang, X., & Tukker, A. (2022). An overview of the waste hierarchy framework for analyzing the circularity in construction and demolition waste management in Europe. *Science of the Total Environment, 803*, 149892.

Zhang, Q., Prouty, C., Zimmerman, J. B., & Mihelcic, J. R. (2016). More than target 6.3: A systems approach to rethinking sustainable development goals in a resource-scarce world. *Engineering, 2*(4), 481–489.

Sharing Instead of Owning: A Synthesized Review on Collaborative Consumption in Sub-Saharan Africa

Kwaku Owusu Twum, Patrick Arhin, Samira Sanni, and Mohammed Abubakari

1 Introduction

The concept of ownership is frequently discussed in the business and sustainability literature (Bardhi & Eckhardt, 2012). The discussions have largely focused on the fluid nature of the ownership and its nexus with possession towards access-based consumption in the context of circular economy and sustainability (Belk, 2014b; Watkins et al., 2016). The traditional conceptualization of ownership presents it as a sole control

K. O. Twum (✉)
Koa Switzerland and Ghana, Assin Akrofuom, Ghana
e-mail: kay4blez@gmail.com

P. Arhin
Department of Spatial Planning, TU Dortmund University, Dortmund, Germany

S. Sanni
Department of International Environmental and Development Studies (NORAGRIC), Norwegian University of Life Sciences, Oslo, Norway

M. O. Erdiaw-Kwasie and G. M. M. Alam (eds.), *Circular Economy Strategies and the UN Sustainable Development Goals*, Sustainable Development Goals Series,
https://doi.org/10.1007/978-981-99-3083-8_2

and total exclusionary and right transfer one has over a product and its utilization. However, contemporary research has argued that ownership is fragmented such that different people can own resources through diverse ownership configurations (Gaus, 2012; Haase & Kleinaltenkamp, 2011). The latter understanding of ownership is based on multiple consumers' emergent sharing of products instead of a full liberal (permanent) ownership of products (Bardhi & Eckhardt, 2012; Belk, 2014b). As a result, there is a rise in temporal forms of 'ownership' where people manifest 'ownership' simply via access-based consumption. In access-based consumption, people' temporarily own' products and consume them through both market (renting) and non-market-mediated (sharing, gifting, borrowing) means (Bardhi & Eckhardt, 2012; Jenkins et al., 2014; Ozanne & Ballantine, 2010). In this way, ownership becomes more of a contractual arrangement between different actors who get the chance to use a particular product in different ways.

In the current dispensation, different ownership configurations manifest due to increasing attention to circular economy and sustainability. The circular economy, primarily based on material reduction, reuse, and recycling, has significantly driven ownership interest based on sharing (and exchange), a sustainable approach to using products (Baumeister et al., 2015). As a result, consumers are becoming more interested in accessing resources based on user experiences rather than buying and owning them permanently (Bardhi & Eckhardt, 2012; Baumeister et al., 2015; Erdiaw-Kwasie et al., 2023; Watkins et al., 2016). Therefore, the traditional ownership approach, which considers ownership *in toto*, is a setback to global environmental sustainability efforts (Hoang, 2014; Schandl & West, 2010). Because first, such ownership contributes to environmental destruction through excessive waste generation and the emission of carbon (Martin, 2016). Second, it becomes a hurdle for consumers as they must be responsible for caring for and maintaining the products without any cost-sharing mechanisms (Philip et al., 2015). And third, it creates inequality and fuels power imbalances between the 'haves and the have-nots' (Hawlitschek et al., 2016). The result is that an access-based consumption labelled 'collaborative consumption' has increasingly

---

M. Abubakari
School of Environment and Sustainability, University of Saskatchewan, Saskatoon, SK, Canada

become a common means of 'owning and consuming' products in the context of the global circular economy.

In developed countries, there have been extant scholarly works on circular economy and collaborative consumption (CC) regarding the scale of adoption, the mechanism for its practices, its advantages and disadvantages, and its relationship with sustainability (Matzler et al., 2015; Möhlmann, 2015). On the contrary, in developing countries, circularity and CC-oriented studies are few (Hailemariam & Erdiaw-Kwasie, 2022; Mair & Reischauer, 2017). For instance, in sub-Saharan Africa (SSA), such studies are a few and disjointed. As a result, little is known about how CC manifests in different geographical contexts in the SSA, despite studies (Preston & Lehne, 2017; Wright et al., 2019) arguing that circularity is more entrenched in the economies of developing countries, including the SSA than those of developed countries. Moreover, the circular activities in the SSA and the developing world take place in an informal context and are spearheaded through poverty and unemployment with very limited environmental considerations (Wright et al., 2019). A strategy for 'domesticating' circularity is thus a requisite to catalyse sustainability in such countries by propelling industrialization whilst addressing environmental concerns. This also calls for evidence-based studies (Preston & Lehne, 2017), which are, unfortunately, scanty (Mair & Reischauer, 2017).

This study contributes to academic and policy literature on circularity and CC by 'homogenizing' the few SSA disjointed studies to provide a 'springboard' for future country-specific policy research and case studies for an appropriate 'domestication' of circularity, CC, and sustainable use of resources in the SSA. Since CC is based on 'sharing', the study uses the Social Exchange Theory (SET) (see the classical work of Emerson in 1962) as useful theoretical premises for synthesizing CC practices in the SSA. The SET argues that 'sharing' (and exchange) can occur through an established relationship between different sets of consumers. On this basis, understanding the nature of relationships amongst CC actors is significant to unravel how sharing (and exchange) of products takes place and how policy can create an enabling environment for sharing and exchange relationships to happen.

With an overview of the study provided, the remaining sections are organized as follows: Sect. 2 defines key concepts (circular economy, collaborative consumption, and sustainability) and their linkages. Section 3 talks about the theoretical framework of the study, focusing on the Social Exchange Theory (SET); Sect. 4 presents insight into the literature review approach; Sect. 5 reviews CC in developed countries and draws out useful lessons for the SSA; Sect. 6 presents the relationships between CC and sustainability with implications for the SSA; Sect. 7 reviews CC and sustainable resource use practices in the SSA; Sect. 8 discusses and synthesizes the findings, and Sect. 9 concludes the study with policy pathways concerning CC in the SSA.

2 Circular Economy, Collaborative Consumption, and Sustainability: Meaning and Linkages.

Circular economy (CE) is one of the most trending but slippery sustainability-imbibed concepts: its definitions are many but blurred with no universal acceptance (Gladek, 2017; Lieder & Rashid, 2016). On this basis, Kirchherr et al. (2017) conceptualized CE based on an in-depth analysis of 114 definitions to bring forth a more comprehensive definition of the circular economy. Whilst this definition may still suffer from universal acceptance, it is somewhat robust and provides good leverage for studies on CE to reduce blurriness in the CE literature. Also, as this study draws linkages between CE, collaborative consumption (CC, the key facet of the study), and sustainability, Kirchherr et al.'s definition establishes such linkages directly (***bolded and underlined in the adopted definition***) that can simplify and facilitate a better understanding of the study. CE is '... an economic system that is based on business models which replace the "end-of-life" concept with reducing, alternatively reusing, recycling and recovering materials in production/ distribution and **consumption process**, thus operating at the micro level (**products, companies, consumers**), meso level (eco-industrial parks) and macro level (city, region, nation, and beyond), to accomplish **sustainable development**, which implies creating environmental quality, economic prosperity and social equity, to the benefit of current and future generations' (Kirchhler et al., 2017, pp. 224–225). CE practices are driven by the R-framework, which is usually considered as the principles of CE: Reduce, Recycle, Replace, Repair, and Reuse, amongst others (Preston et al., 2019).

The definition emphasizes replacing the end-of-life of a product with 'reusing, recycling and recovering' in the consumption process ties in with CC. CC is defined as an economic model based on sharing, swapping, trading, or renting products and services where people coordinate the acquisition and distribution of resources for a fee or other compensation (Belk, 2014a). CC mechanisms of 'sharing, swapping, trading, or renting products' is a pathway to 'reusing, recycling, and recovering', making CE a broad arena that supposedly enhances CC practices. Through CC, instead of the linear economic model of 'buy-use-dispose', consumers can rather 'share' their 'idle capacity' and 'underused' products with others needing the product (i.e. the circularity of products) to impede outright disposal. The increasing scholarly works on CC present it as trans-dimensional, encompassing collaborative online (e.g. sharing networks and Wikipedia); social commerce, a form of commerce mediated by social media (e.g. Facebook market); sharing online, a form of commerce facilitated by IT-based e-commerce for digital and physical products (e.g. eBay); and of the more recent, ideological underpinnings facilitated by the advances in technology used by consumers for organized actions (e.g. awareness for ethical consumption) (Bardhi & Eckhardt, 2012; Botsman & Rogers, 2011; Godelnik, 2017; Hamari et al., 2016). Botsman and Rogers (2010), who have written extensively on CC, categorized it into product/service system (a product traditionally sold in a discrete transaction, but now offered as a service by an entity (e.g. Zipcar), redistribution markets (a 'second life' marketplace for unwanted products offered at an affordable price, and may (not) be contingent on the use of the internet (e.g. flea market, eBay), and collaborative lifestyles (a 'peer-to-peer' network where peers deliver products and services to other peers in an IT-driven network [e.g. Uber]) (see also, Garrett et al., 2016).

Also, Kirchhler et al.'s definition provides that the mission of CE is 'to accomplish sustainable development', a development that provides for the needs of the present generation and makes room for the future generation to provide for themselves with emphasis on the environment, economy, and society (World Commission for Environment and Development [WCED], 1987). Also, as a key component of CE, CC is thought to contribute to the sustainable use of resources (Belk, 2014a). These revelations offer areas of synergy between CE, CC, and sustainability. The literature on sustainability has presented it as outrightly vague and a 'theoretical dream' that cannot be practically realized (Engelman, 2013; Naudé, 2011). CE (and its associated practices, such as CC) is seen as one of the operational 'arms' of sustainability (Murray et al., 2017) where it can be practised and promoted.

3 Theorizing CC Practices Using the Social Exchange Theory

CC is practically manifested based on continuous consumer interactions instead of a simple one-time, linear buyer–seller relationship that usually occurs with conventional transactions (World Economic Forum, 2017). CC is, thus, driven by relationships that potentially influence and shape the behaviours of individuals involved. Because CC aims to promote sustainability, its practices are influenced by sustainable consumer behaviour (Heinrichs, 2013). Sustainable consumer behaviour is needed to ensure effective and efficient use of idle and underused resources in CC spaces, such as Facebook market, Airbnb, and Uber (Cohen & Munoz, 2016; Sigala, 2014). Wang et al. (2019) argue that such a behaviour thrives on a two-sided interactive relationship between individuals based on trust, reputation/goodwill, and reciprocity (social variables). The emphasis on 'relationship' in CC practices and sustainable consumer behaviour validates the thinking around the Social Exchange Theory (SET) that grounds this study.

The SET argues that relationships are built based on perceived cost and benefits, so individuals can only engage in resource exchanges (such as CC) when benefits exceed the cost (Emerson, 1962). The cost and benefits are not just economic (e.g. profit and value for money) but also social (trust, reciprocity, reputation, perceived support, friendship, sense of belongingness) that can lead to sustainable consumer behaviour (Flaherty & Pappas, 2009; Jiang & Kim, 2015; Tsai et al., 2011; Wang et al., 2019; Wetzel et al., 2014). Thus, whilst individuals engaging in CC would expect to gain economically, their behaviour can be influenced by social factors, for example, trust and reciprocity, i.e. they are committed to treating others the same way others have treated them (Cropanzano & Mitchell, 2005). CC is known to have the potential to drive extensive social relationships and interactions, and as such, the actions of others can be influenced by other actors. On this basis, the use of SET can help to investigate how economic factors (e.g. price of a product) and social influence (e.g. reciprocity, trust, friendship) can drive sustainable consumer behaviour in CC practices, as well as the implications for advancing CC focusing on the SSA.

4 Literature Review Approach

The study reviewed relevant CE, CC, and sustainability studies with key attention to the SSA regional literature. Four main steps were followed for the review process as advised by Moher et al. (2009) for careful and replicable reviews—(i) identification of relevant literature, (ii) screening of the literature initially gathered, (iii) eligibility for inclusion and exclusion, and (iv) inclusion and utilization of eligible documents.

A search criterion was designed to identify relevant literature by combining the study's keywords (collaborative consumption, rethinking consumer ownership; sustainable resource use; sub-Saharan Africa). Also 'collaborative consumption and rethinking consumer ownership in sub-Saharan Africa'; 'collaborative consumption and sustainable resource use in sub-Saharan Africa'; 'definitions and meanings of collaborative consumption'; 'relevant theories for collaborative consumption'; 'collaborative consumption models and practices in sub-Saharan Africa', was considered. Since many of the studies on CC and its related concept of CE are spearheaded by policy (and other non-academic) actors (Schut et al., 2015), the search for materials was done in both academic and non-academic outlets: Scopus and Web of Science as the primary academic outlets; Google as a non-academic outlet; and Google Scholar and ResearchGate as outlets with both academic and non-academic materials.

Through the initial search, some noticeable trends were identified. First, other terms such as Circular Economy, Sharing Economy, Sustainability, and Sustainable Development were dominant themes. They provided an understanding of CC embeddedness and the need to obtain materials around these terms and their relationship with CC. Second, in the SSA context, some countries were commonly identified in the literature, whilst others were hardly mentioned. South Africa, in particular, stood out in the literature, but also as Kenya, Nigeria, Tanzania, and Ghana in sequential order in terms of the magnitude of studies identified to be related to the SSA. This gave a fair idea at the early stages of the literature search in terms of paper structuring, some SSA countries to be selected and visualized, and the manner for synthesizing the SSA literature in the study.

The second, third, and fourth steps (screening, eligibility, and inclusion) were done concurrently with the focus of them based on the following factors: (i) the publication year of the materials, (ii) materials written in English, and (iii) relevant materials based on authors' discretion. The authors were more interested in post-2014 materials to tie in with adopting the 17 sustainability-driven goals. Therefore, efforts were made to eliminate materials that did not meet such a requirement; however, some time-ineligible materials (not more than 15) were still considered due to their great relevance to the aim of the study (for instance, Emerson, 1962 for the Social Exchange Theory [SET]). Also, since all the authors are only proficient in English as an international language, all literature materials prepared in other languages, for example, French were excluded. Each author had assigned sections, and their own discretion informed the papers considered for the sections they worked on.

Based on the final materials considered, the authors became interested in the representativeness and potential generalization of the study to fit the entire SSA. As a result, the authors became conscious of studies from each of the SSA sub-regions (Central Africa, East Africa, Southern Africa, and West Africa), which also led to specific web searches focusing on websites designed for CC practices (e.g. Uber, Airbnb, Facebook market, etc.). However, whilst some countries (five of them) stood out in terms of works on CC in the SSA (South Africa, Kenya, Nigeria, Tanzania, and Ghana), the others had very scanty materials, probably because of the search for materials was conducted in English.

The authors preferred only English-speaking countries in sampling countries in the SSA sub-regions for the study due to their limitation in other international languages spoken in the SSA region (e.g. French, Portuguese). Based on the literature search, the five countries that stood out were automatically considered for their respective sub-regions. Of the five countries, four were already from East Africa (Kenya, Tanzania) and West Africa (Ghana, Nigeria), with only one from Southern Africa and none from Central Africa. As a result, the focus was to get an additional one from Southern Africa and two from Central Africa to equate those of East Africa and West Africa. An 'intentional' literature search focused on English-related materials from Southern Africa and Central Africa. Botswana stood out from Southern Africa regarding the degree of availability of materials compared to the others from the sub-region, whilst the Democratic Republic of Congo and Rwanda were the ones identified from Central Africa. Whilst Rwanda speaks French and English, the Democratic Republic of Congo speaks French but was considered because some relevant materials for conceivable analysis were obtained concerning

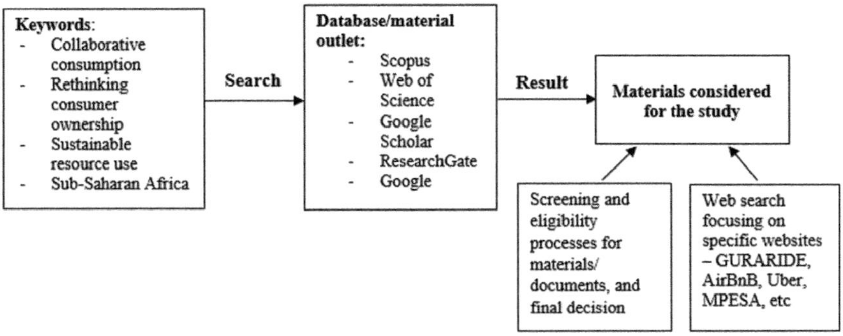

Fig. 1 Literature review process

such a country compared to the others in the Central Africa sub-region (making it the only exempted country from the authors' preference in favour of English-speaking countries). Figure 1 summarizes the literature review approach of the study.

5 Circular Economy and Collaborative Consumption in the Developed World: Any Lessons for Sub-Saharan Africa?

5.1 Circular Practices in Developed Countries

Circular economy (CE) is a sustainable development strategy proposed to address the pressing problems of environmental degradation and resource depletion. The circular economy is based on the R principle—material *reduction*, *reuse*, and *recycling*. The principle considers a circulation system in which materials are recycled, with energy coming from renewable sources such that activities support and rebuild ecosystems, human and societal health. Resources are used to create value (Heshmati, 2015). The R principle has become a common phenomenon and has received wide acceptance in the developed world, from which collaborative consumption (CC) has been birthed. A study by Vaughan and Daverio (2016) confirmed the presence of circularity driven by CC in Europe, specifically countries including Austria, Germany, Czech Republic, Belgium, Luxembourg, Romania, Poland, France, United Kingdom, Netherlands, Italy, and Spain, which are seen as active players in the collaborative economy and have introduced radical initiatives towards sustainable management of resources, especially non-renewable resources.

Delving into CC in the developed world, Barreiro-Gen (2017) highlighted why CC is predominating the economies of developed countries, notably those in Europe—saving money and resources, economic gains, good for the environment, and helping to build communities—and these benefits should encourage developing countries to appreciate and intensify its CC initiatives. The practice of CC catalyses the efforts to reduce carbon footprints (e.g. reducing the demand to purchase new vehicles). In addition, it stimulates joint economic growth by creating financial and employment opportunities (Schor, 2016). CC in developed countries represents a wide range of activities and actions, including co-creation, production, distribution, trade, and consumption of goods and services between different individuals and organizations that can contribute to the economic learnings of SSA countries. In the developed world, the significant penetration of information technology into everyday life has created scalable sharing economies, gradually narrowing the need for conventional product ownership (Matzler et al., 2015). The scalable sharing economies of developed countries allow for significant cost savings and social benefits that motivate consumers to participate in sharing (Guttentag, 2015). These benefits have been enabled by digital technology that acts as a mediator to reduce transaction costs and increase seamless connectivity between providers who offer services and/or goods for sharing and those who intend to receive them. However, this is quite unusual in the case of developing countries, particularly in the SSA regions, where consumption patterns are more conventional, with single-use and non-sharing patterns more pronounced.

Figures 2 and 3 display the level of prominence of CC—Fig. 2 presents a geographical display where four (4) countries have been highlighted to show the degree of prominence of CC (with colour shades), and Fig. 3 provides a holistic picture (in a chart) to show the proportion of CC practices (in per cent terms) and the rankings of countries. They also give an impression of the progressive nature of CC and circular economy in the context of the developed world (particularly Europe, which is leading such a change towards sustainability).

The prominence of CC in Europe has come about due to the supportive role of the Europe Union other factors (e.g. information technology, strong economies) that have allowed for building a resource-saving and environment-friendly society in such countries.

Fig. 2 The prominence of CC in Europe—a geographical display (*Source* Authors' construct with insight from Barreiro-Gen [2017] and Eurostat data [2014])

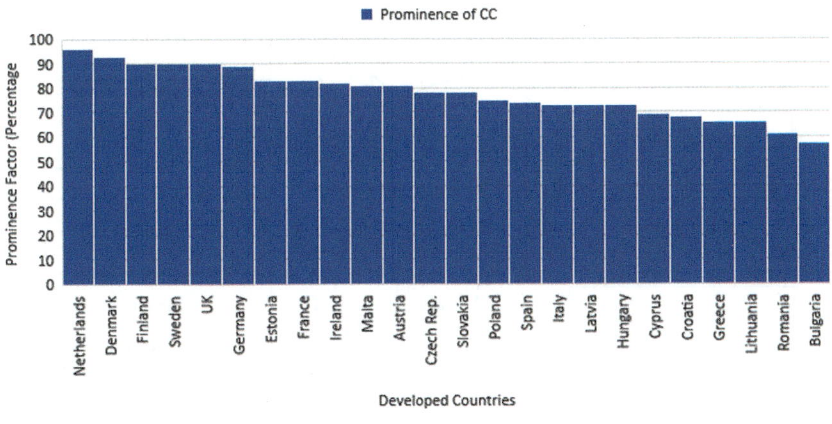

Fig. 3 Degree of CC in Europe—per cent value and ranking of countries (*Source* Authors' construct with insight from Barreiro-Gen [2017])

5.2 *Potential Lessons for Sub-Saharan Africa*

Circularity exists in developing countries, including those in sub-Saharan Africa (SSA). In strengthening circularity in the SSA, the circular practices in developed countries can provide useful lessons that can inform an effective practice of a comprehensive circular system. The conceptual framework (Fig. 4) highlights the demand and supply cycle through the circular economy principles and framework CC. On this basis, the goal of the SSA region for CC and circularity should be towards reordering supply chain optimization through the principles of reducing, recycling, replacing, repairing, and reusing across the entire value chain of products (Preston et al., 2019). Particularly for CC, consumption aims to replace products with idle capacity ones of another consumer instead of purchasing new ones, repairing durable products for use by another consumer, and reusing products by another consumer instead of outright disposal.

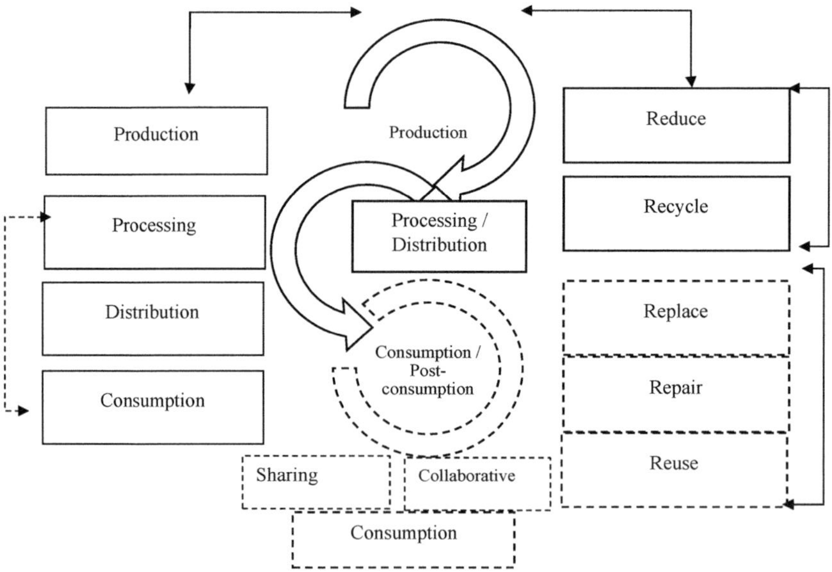

Fig. 4 A Conceptual framework for CC (*Source* Authors' construct, insight from Preston et al. [2019]. *Note* Lines/boxes with dashes indicate areas in which CC takes [or can take] place in the value chain)

In developing countries, including the SSA, Cao and Zhang (2011) have argued that successful collaboration could create a joint competitive advantage which could lead to value creation, which could create a collaborative advantage for all partners. Furthermore, the challenges faced by companies in these countries (including the SSA), such as the difficulty in delivering consumer goods, lack of advanced technology, resource scarcity, and logistics barriers, can be overcome by collaboration (Amores Salvado, 2013). In the SSA, sustainability knowledge is evolving rapidly, with companies gradually becoming sensitive and committed to the concept of CE through the efficient use of resources (Singh & Singh, 2018). Although people in the SSA region are commonly connected to (natural) resources, with the gradual penetration of information technology, CE is gaining prominence, where organizations are drawing towards CC models and supply chain innovation (Lemille, 2016). Therefore, the need for stakeholder collaboration in SSA cannot be overemphasized. The State is seen as having an important play creating an enabling environment (e.g. policy framework and legislation for circularity) for CC practices and circularity to 'grow'.

6 The Nexus Between CC and Sustainability in the SSA

CC is a circular economy-related concept that contributes to sustainability. It significantly impacts the delivery of goods and services by lowering transactional costs and boosting the rate at which goods and services can effectively and efficiently be used (Goudin, 2016; Schor, 2016; Welsum, 2016). The traditional consumption model ('buy and own') through face-to-face exchanges remains common in developing countries, including the SSA (Perren & Grauerholz, 2015), and has been described to be dyadic—fostering a relationship between buyers and sellers (Perren et al., 2014). This seems to create a fear that CC, usually happening digitally, may erode dyadic relationships that enhance social relationships and development in the SSA. This may hold valid depending on the CC model, especially the frequently fee-paying ones and involves access to goods or assets that individuals often use for economic gain (see: Belk, 2014a; Eckhardt & Bardhi, 2015; Hamari et al., 2016). Nevertheless, CC has generally been considered more dyadic as it is flexible and less institutionalized than traditional consumption, which presents a more rigid and bureaucratic form of consumption (Perren et al., 2014). In developing economies like the SSA countries, the practice of CC can help people access resources, allowing those deprived to improve the quality of their life (Perren & Grauerholz, 2015).

Under Agenda 21, the United Nations declares that sustainability is pursuing economic and social development in harmony with the natural environment to enhance all living standards (United Nations General Assembly, 2015). Many scholars have argued that the practice of CC can contribute to sustainability by addressing and mitigating over-consumption issues that have positive implications for the environment, economy, and society (Huang & Rust, 2011). For instance, Botsman and Rogers (2010) contend that the practice of CC can result in a more sustainable consumption behaviour that significantly benefits consumers, businesses, and the community. Affirming Botsman and Rogers' perspective, Frenken and Schor (2017) add that, unlike the traditional consumption methods known to depreciate the environment significantly, CC promotes the conservation of resources and contributes to less waste. At the individual level, CC serves as an avenue to practice their political and personal beliefs associated with sustainability (Perren & Grauerholz, 2015). The ability for individuals to manipulate political consumption under CC enables them to exchange goods and services and fulfil their needs at a lower cost (Parsons, 2014).

As a social-driven model, CC allows for the temporary possessions of certain goods and services by the less privileged in society, which the traditional consumption model cannot and will not allow (Tussyadiah, 2015). Also, the economic impact of CC is related to its ability to contribute to affordable and effective consumer access to goods and services (Binninger et al., 2015). In addition, CC produces self-employed "microentrepreneurs", allowing people to earn income for a living. In furtherance, CC contributes to resource conservation and saves costs (Belk, 2014a; Dall Pizzol et al., 2017). Finally, CC brings together consumers to balance the demand and supply of products and services effectively and efficiently (Vasirani & Ossowski, 2012; Xiaoli & Yang, 2015).

Whilst CC comes with benefits for sustainability, it has setbacks that may be worrisome in practising it in the SSA if policy safeguards are not put in place. Perren and Grauerholz (2015) have indicated that CC can be illegal as it can easily lack effective regulation and oversight of informal transactions. This has implications for the SSA, where informal transactions are common, and legal measures and regulations remain weak, putting some consumers at risk. Again, CC depends on sharing and exchanging resources using modern technology (e.g. phones and the internet). This may impede the illiterates from benefiting from CC (Dillahunt & Malone, 2015; Taeihagh, 2017). This aspect of CC may not allow greater use of CC models in the SSA, where many people need to be ICT literate. Morozov (2015) describes CC as a 'platform capitalism' allowing large (international) corporations (e.g. Uber) to 'parasite' products they do not produce through a (digital) rearrangement whilst making an avalanche of profits which mostly swerve the formal tax systems. In developing countries such as the SSA, CC can possibly become an avenue for the monetization of existing socio-economic relations through westernized neoliberalism and exploitations by large corporations, negatively affecting the economic well-being of ordinary members.

7 CC Models and Practices in the SSA

Some CC models are used in SSA, and they have usually taken the form whereby consumers use mobile apps to share resources for commercial purposes (Dreyer et al., 2017). However, the SSA is known to have significant infrastructure deficiencies, for example, poor access to the internet characterized by the high cost of mobile data, disorganized and difficult-to-identify building addresses, poor product delivery services, online payment concerns, and the lack of trust in online vendors (Holmberg, 2016; Meltzer, 2014). This has implications for sustainable consumer behaviour amongst consumers in the SSA based on trust and reciprocity and creates concerns about whether CC is practised with a sustainability mindset. In addition, the SSA also has issues concerning limited education and knowledge about the use of Information and Community Technology (ICT) (World Economic Forum, 2015), poor electricity coverage, and frequent power cuts (Africa Progress Panel, 2015). All these affect CC practices negatively.

CC is somewhat practised in the SSA because of its adaptive dynamic and resilient nature (Hall et al., 2012) and the motivation of people to gain additional income from their under-utilized and idle resources. The downturn in the global economy in 2008 (Botsman & Rogers, 2011; Heinrichs, 2013) has been hard-hitting to the SSA, making living conditions difficult. The Covid-19 pandemic has also worsened economic downturns, possiblintensifyingfy the interest in CC (after Clark, 2020). Some elite groups, especially those living in the most urbanized cities in the SSA, practice 'western' CC as an extra income source. Car sharing (e.g. Uber, Bolt) and house renting (e.g. Airbnb) are the most common types of CC practices in the SSA (Dreyer et al., 2017; Dzisi et al., 2020). Table 1 summarizes the specific practices in the selected SSA countries.

In South Africa in Southern Africa, ride-sharing has gained prominence in their urban transport system for the past few years (Henama & Sifolo, 2017). This began in 2013 when Uber launched their operation in Johannesburg (Giddy, 2020). Since its inception in 2013, the Uber application has created employment for over 4000 drivers in South Africa and serves as major means of transport in major cities, like Johannesburg, Cape Town, Durban, and Port Elizabeth (Henama & Sifolo, 2017; Maziriri et al., 2020). Similarly, in 2015, Bolt (formerly known as Taxify) also launched their operation in Johannesburg and now operates in 34 cities and towns across South Africa (Carmody & Fortuin, 2019). In 2019, a Russian-based ride-sharing company, inDriver, also launched its services in Cape Town. Aside from international ride-sharing companies, local sharing platforms such as Yookoo Rides and Hailer have also emerged in South Africa (Papadopoulos & Eck, 2021). In addition to the e-hailing services, Uber in South Africa has launched a subsidiary business called UberEats that allows people to order food online from a restaurant of their choice and subsequently deliver it to a stated delivery address (Manavhela & Henama, 2019). Another CC model that has gained ground in South Africa for the past few years is Airbnb (Visser et al., 2017). For the past five years, more than 2 million people have used Airbnb to secure accommodation in South Africa (Guttentag, 2016). In Botswana, another Southern African country, the most common CC practices are ride-sharing and e-commerce services. Available ride-sharing platforms include Uber, Bolt, and Hello cabs. Like Uber and Bolt, the Hello cabs application allows users to hail taxis via their mobile phones. Available e-commerce platforms in Botswana include Skymartbw, Apex Mart, DHL Africa eShop, and Botswana Post E-Services (Dzimiri et al., 2018; Jaiyeoba et al., 2018).

In Kenya, a country in East Africa, some CC practices have been incorporated into its economy (Katende-Magezi, 2017). These CC practices include ride-sharing, house renting (Airbnb), and e-commerce. The ride-sharing sector in Kenya is largely dominated by Uber and Bolt (Okedi & Yano, 2020). However, in addition to Uber and Bolt, there are other locally owned ride-sharing companies, such as Beba Beba, Safaricom's Little Cabs, Mondo Ride, and Maramoja. The Beba Beba, for instance,

Table 1 Commonly used CC models and practices in selected SSA countries

No.	*SSA countries*	*Common (and new) models and practices of CC*	*Selected references*
Southern Africa			
1.	South Africa	Ride-sharing (Uber, Bolt, inDriver Yookoo rides, Hailer, etc.), home-sharing and rentals (Airbnb), and e-delivery systems for food (UberEats, etc.)	Dreyer et al. (2017), Giddy (2020), and Henama (2021)
2.	Botswana	Peer-to-peer accommodation/vacation rent (Airbnb); transport/ride-sharing (carpool, uber, etc.); online transport advice for tourists; e-commerce, e-banking, and peer-to-peer delivery; social media networks (Facebook market, etc.)	Dzimiri et al. (2018) and Jaiyeoba et al. (2018)
East Africa			
3.	Kenya	Second-hand products and informal swapping of durable products, online purchases (including second-hand products) through electronic transfers (MPESA, etc.), online delivery, and payment for second-hand products usually from developed countries, home-sharing, and rentals (Airbnb), ride-sharing (Uber, Bolt, Beba Beba, Safaricom's Little Cabs, Mondo Ride, and Maramoja, etc.)	Katende-Magezi (2017), Daily Nation Kenya (2018), and Africa Insight Report (2017)
4.	Tanzania	Informal sales and swapping of second-hand (durable) products, E-commerce facilitated by e-banking (MPESA, etc.), home-sharing and rentals (Airbnb), ride-sharing (Carpool-Tanzania, Moovn Tanzania, Little Ride and Twende, etc.)	Mvungi (2019), Katende-Magezi (2017), and Mtaho and Mselle (2014)

(continued)

Table 1 (continued)

No.	*SSA countries*	*Common (and new) models and practices of CC*	*Selected references*
West Africa			
5.	Ghana	Ride-sharing (Uber, Bolt, Yango, Yenko Ghana, etc.), home-sharing and rentals (Airbnb), informal swapping of (durable) second-hand products, collaborative online purchases (use of mobile banking, etc.), and delivery (DHL, EMS, etc.), flea market for second-hand products, informal sharing and gifting, and social media purchases (Facebook),	Dzisi et al. (2020), Simmons et al. (2019), Oreva (2016), and Kufuor (2018)
6.	Nigeria	Ride-sharing (Uber, etc.); peer-to-peer sharing of products; informal product swapping and gifting; ride-sharing; E-commerce for second-hand products and delivery support services (Amazon, DHL, Jumia, Konga); home-sharing and rentals (Airbnb)	Folorunso and Momoh (2015), Oreva (2016), and Olotewo (2017)
Central Africa			
7.	The Democratic Republic of Congo	Online purchases (including second-hand products) through mobile transfers (MPESA, PayGate), and delivery services that may (not) involve CC (e.g. the use of Amazon), peer-to-peer sharing of products, informal product swapping, and gifting, ride-sharing (Comfort Car, Blue Congo), home-sharing and rentals (Airbnb), second-hand products usually from foreign countries, the use of social media platforms (Facebook)	Bose (2019), Audu (2018), Mulligan (2013), and GSMA (undated)
8.	Rwanda	E-commerce for second-hand products and delivery support services (Amazon, DHL), bike-sharing (GuraRide), ride-sharing (Uber), home-sharing and rentals (Airbnb), second-hand products and informal swapping, social media commercial groups, and sharing (Facebook marketplace)	Galuszka et al. (2021), GuraRide (undated), and Katende-Magezi (2017)

NB Some sources were obtained from the website of corporations that foster CC practices

is owned by the Drivers and Partners Association of Kenya and has been in operation since 2018 (Daily Nation Kenya, 2018). In addition to ride-sharing, house renting (Airbnb) is increasingly becoming popular in Kenya. Available statistics indicate that the country has 5900 active listings on the Airbnb platform and welcomed 39,500 guests in 2016 (Airbnb, Africa Insight Report, 2017). Along with ride-sharing and Airbnb, Kenya's e-commerce sector is gaining ground (Murage et al., 2020) together with online payment platforms like M-Pesa. In Tanzania, another East African country, the ride-sharing operation has penetrated its economy: Uber and Bolt launched their operations in 2016 and 2017, respectively (Laizer, 2020). Since then, ride-sharing companies such as Carpool-Tanzania, Moovn Tanzania, Little Ride, and Twende have joined the ride-sharing market (Mvungi, 2019). Relatedly, online payment platforms like M-Pesa allow for purchasing goods and services online (Mtaho & Mselle, 2014).

In West Africa, focusing on Nigeria, e-commerce and ride-sharing are the most predominant CC practices. For example, the following e-commerce sites create specific services: Cheki Nigeria for cars, Wakanow for travel, slot.ng for travel, showroom.ng and furnish.ng for furniture, private property and property24 for real estate, Olx and jiji.ng for the marketplace, Fashpa and Traclist for fashion; Jumia, Konga, Deaddey, Yudala, PayPorte, Vconnect, and Kara for general merchandise (Anunike & Onuegbu, 2020; Folasade & Ogungbe, 2020; Olasanmi, 2019; Olotewo, 2017). Along with e-commerce, ride-sharing is also common such that in Lagos alone, there are 14 e-ride-sharing companies, which by far is the highest compared to other cities in Africa (Oreva, 2016). Also, in Ghana, another West African country, common CC practices are ride-sharing (largely dominated by Uber and Bolt), house renting (Airbnb), and e-commerce (Agyemang, 2020; Degadjor, 2018). Along with Uber and Bolt, several relatively smaller ride-sharing companies have emerged, including Yango, Yenko Ghana, Urughana, Accra Cab, and Poki Cab (Dzisi et al., 2020; Kufuor, 2018). As of 2018, Uber alone had about 180,000 active riders (Melia, 2020). Another emerging CC practice in Ghana is house renting through the Airbnb sharing platform, with over 300 host listings across the country's regions (Airbnb, 2017). Common e-commerce services in Ghana include DHL, EMS, Ghana Post, and FedEx.

In Central Africa, the Democratic Republic of Congo has also incorporated some CC practices into its economy. Among these practices are ride-sharing, peer-to-peer accommodation, and e-commerce. Ride-sharing applications for CC practices include Blue Car, and Comfort Car (Mulligan, 2013), with the most common retail eCommerce sites comprising lezando.com, you.cd, zephone.com, and coolprix.com (Bose, 2019). Online payment platforms like M-Pesa are also available and allow users to transact business online, supplemented with social media platforms like Facebook (Mulligan, 2013). Rwanda, another Central African country, has online platforms such as TripAdvisor and Facebook marketplace, which are rapidly increasing (Foster & Graham, 2015). In addition, several accommodation providers and tour operators have incorporated online-based CC practices through their websites and social media pages (Foster & Graham, 2015). Ride and bike-sharing applications such as GuraRide and Uber are also common, as well as Airbnb, the country's key home-sharing rental service (Galuszka et al., 2021).

8 Synthesis and Discussions: Implications for Sustainable Resource Use

8.1 Conceptual and Theoretical Discussions

As ascertained in the literature, the fundamental premise of CC is its potential to trigger relationships and shape the behaviours of consumers due to the continuous 'collaborative' interaction nature it provides for them (World Economic Forum, 2017). CC is also known to occur in private exchange at physical meetings (e.g. flea markets), the sales of second-hand items in the public domain, and online platforms for peer-to-peer exchanges. This implies that whilst information technology (e.g. the internet) has become the mainstream for practising CC through peer-to-peer social networks, especially in the developed world (Gruszka, 2017), the non-digital possibilities for practising CC provide an avenue for SSA countries (even with their infrastructural deficiencies) to promote CC based on sustainable consumption. We found different drivers of CC practices, encompassing social (belongingness, trust, reciprocity, reputation, perceived quality, etc.), environmental (ecologically sustainable consumption), and economic (cost savings, profits, internet technology, etc.) benefits; all are important for possibly driving sustainability in the SSA. This, however, calls for sustainable consumer behaviour in the practice of CC in the SSA for sustainable resource use. Individuals can engage in CC in a manner that may not advance the ideals of sustainability—profit exploitation, illegal transactions due to poor regulation systems, etc.—in the SSA.

Conceptually speaking, sustainable consumer behaviour in CC is predominantly driven by social drivers (Wang et al., 2019), as such drivers can influence and shape people's behaviour through constant exchanges and interactions. On this basis, if consumers understand that the social benefits outweigh the social cost of CC, they are likely to engage in it and learn and conduct themselves in tandem with how other CC actors treat and behave towards them (as argued by the Social Exchange Theory (SET)). This brings in key concerns and attention to behaviours of not only consumers (and 'consumer-sellers'), but also service providers and companies that provide (digital) platforms for such arrangements and interactions to take place (e.g. Uber, Airbnb) in the SSA. The implications for sustainable resource use imply that behaviours in CC practices should discourage excessive/over-consumption and promote ethical consumption (ecologically sustainable consumption) (Schuitema & De Groot, 2014). The extent to which the behaviours of all these actors promote sustainable behaviour for sustainable consumption in CC practices (and not just for huge profits and economic gains) should be part of policy discourses in the SSA.

In Fig. 5, we have developed a conceptually grounded context framework (looking at the various dimensions of sustainability as suggested by Joung et al., 2013) for which CC practices can contribute to sustainable consumer behaviour for sustainable resource use in the SSA. The dimensions of sustainability (social, environmental, and economic) are interconnected, but social drivers are identified to underpin and drive sustainable consumer behaviour directly.

The theoretical proposition underpinning the figure is the Social Exchange Theory (SET), which initiates the thinking that the drivers (social, economic, and environment) can only lead to CC practices based on a higher benefit, either socially or economically or environmentally. However, for sustainable consumer behaviour to occur in CC practices for sustainable use of resources, individuals involved should be mindful and driven by ecologically sounded consumption (e.g. avoiding over-consumption, waste reduction, etc.) and engage with other consumers in CC in a manner (based on social drivers) that lead them into such behaviour (sustainable consumer behaviour). This calls for such a

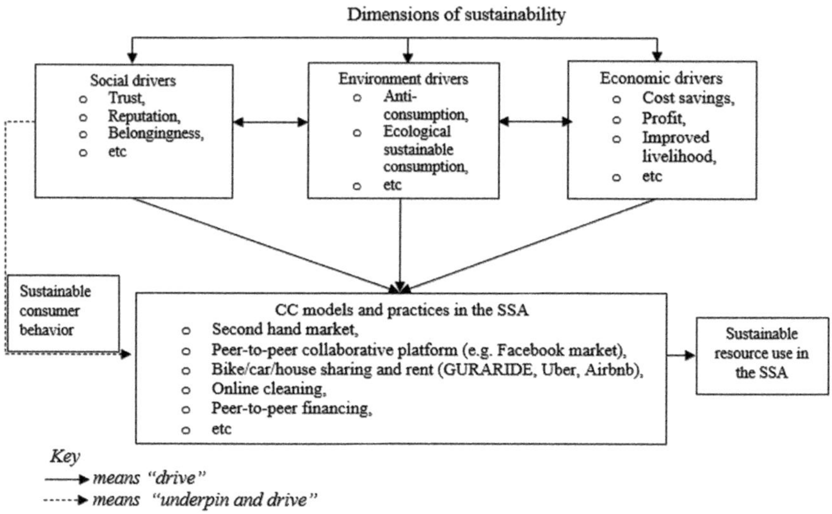

Fig. 5 Contextual framework for CC practices for sustainable resource use in the SSA (*Source* Authors' construct, 2022)

consumer to gain trust and respect from others in the continuous 'community of interactions' for which a particular CC is practised in the SSA.

8.2 Policy Discussions

Sustainable resource utilization has become a circular solution that has gained popularity in the climate change discourse (Preston et al., 2019). Moreover, the concept of consumption plays a critical role in its relational topic of sustainability that governs resource optimization. For example, with the inverse effect of resource utilization and consumption, it is relevant to identify those resources that could be well utilized. Although the practice of circular consumption (CC) is gaining momentum in the SSA, there needs to be more theoretical evidence considering the predominance of traditional consumption patterns, which is the most common marketing outlook of resource allocation in the SSA regions. This involves passive consumers who are mostly used to the capacity to consume resources or services through non-sustainable consumption behaviours.

This is very much associated with the single use of products implying a proliferation of waste and underutilization of resources rather than shared consumption and, in this sense, CC. The main realization effects of the sharing economy are manifested in changes in production and consumption patterns. From a resource efficiency perspective, the optimization effect of shared consumption reduces the need for new products and resources. And from a policy perspective, we still need to consider how the sustainability of shared use (compared to non-shared patterns) can affect CC. On this basis, it is important to see if collaborative sharing participants, especially peer network providers and consumers, achieve better social status, including opportunities for self and social development and the learning outcome that can be transferred to other economies in developing countries.

CC is often seen as a means of change and is expected to contribute to the circular economy and overcome current unsustainable consumption patterns. This concept is increasingly gaining popularity as an alternative that can be purchased and owned. Recently, rapid growth in CC participation in developed nations, for example, with the use of CC digital platforms, has been very common. These digitally enabled platforms in various sectors act as transitional mediums to reduce transaction frequency (Pouri & Hilty, 2019). Essentially, the cost between the supplier offering the goods or services and the supplier trying to buy them is truncated. CC promises resource efficiency, cost-optimization, and a sustainable e-resource utilization approach, which should be learned by other developing economies, particularly in SSA.

Policy-wise, a multi-stakeholder approach is needed in developing countries, particularly in SSA, as more than operational involvement is required to achieve circularity learning from developed economies. Instead, government commitment and involvement are essential to achieving CC adoption across economies (Mishra et al., 2019) of the SSA. Our review supports assertions of stakeholder involvement as an enabler in introducing and disseminating CC to SSA. Participation in CC through SSA can help save supplies and energy by allowing partners to reuse and recycle materials while adhering to a net zero-waste drive. In addition, CC helps energy efficiency and renewable energy to create cleaner technologies that eliminate the loss of environmental efficiency of materials and contribute to sustainability.

Beyond the SSA, the global movement towards a sustainable future is undeniable. The role of CC is essential whether it is unanimously accepted by policymakers, scholars, entrepreneurs, and other stakeholders worldwide. The current dynamism of collective consumption patterns means that the scale of the sharing economy will continue to grow and cross borders (Rahman & Tumpa, 2021). This signifies the need for peer-to-peer connectivity across developed and developing countries, driven by information, innovation, and technology for sustainable consumption.

9 Conclusion

Over the last two decades, literature has increased on various theoretical, methodological, and empirical aspects of the circular economy. Moreover, implementing the CC paradigm has become a recent strategic line of discussion. Developed economies have been interested in implementing CC on a centralized and large scale to provide long-term and sustainable solutions to the serious problems of resource depletion and environmental degradation. The study explored the concept of the CC concerning how developing countries in the SSA can create value from the role of CC as a strategic business transition with sound-cycle business models. The results show that one of the ways developing countries can move to a circular business model is through CC, which involves individuals through a chain of multiple consumptions. It concludes that CC acts as an enabler for supply chain design which facilitates the use of a more resource-efficient and sustainable way of consuming products and services. To that end, a common understanding between stakeholders and other entities in the supply chain is critical. Again, working with new suppliers, especially in close geographical proximity, creates opportunities for reduction, reuse, and recycling. This paper has provided an overview of the CC concept in literature and sustainable development strategies relevant to the SSA. This can involve governmental support for production and multiple consumption chain of resource development through promotion, regulation, monitoring, and evaluation of the success of CC implementation. The role of government was also considered important. For companies to grow cyclically, governments must create an enabling environment by helping companies easily transition from a linear to a circular model. This can be achieved by ensuring laws that promote circularity by local businesses. We propose a cooperation model that can be used as an enabler for a CC environment.

Despite the importance of this study, further studies are needed to collect and draw meaningful insight from empirical data in SSA countries that could validate the model in the context of other developed countries to test the applicability of CC across levels in the value chain with a key focus on producers, suppliers, consumers, and policy actors for better integration of CC practices in the SSA moving forward.

References

Africa Progress Panel. (2015). Power People Planet Seizing Africa's energy and climate opportunities. (2015). (ISBN 978-2-9700821-6-3). Africa Progress Panel. http://www.africaprogresspanel.org/

Agyemang, E. (2020). "Uber is here to stay": Exploring the policy implications of the Uber-Local Taxis turf war in Accra, Ghana. *Case Studies on Transport Policy, 8*(1). https://doi.org/10.1016/j.cstp.2019.12.005

Airbnb. (2017). *Africa insight report*. Retrieved from Airbnb.Com: http://www.Airbnb.Com

Anunike, O. A. O., & Onuegbu, O. (2020). Youth's exposure and utilization of internet advertisements in Awka, Anambra State. *Nnamdi Azikiwe University Journal of Communication and Media Studies, 1*(2).

Amores Salvado, J. (2013). *Environmental innovation and firm performance: A natural resource-based view*. Palgrave Macmillan.

Audu, J. (2018). *Technology adoption in Democratic Republic of Congo (DRC): An empirical study investigating factors that influence online shopping adoption* [A Master's Thesis]. University of Ottawa, Canada. https://doi.org/10.20381/ruor-21621; http://hdl.handle.net/10393/37349

Bardhi, F., & Eckhardt, G. (2012). Access-based consumption: The case of car sharing. *Journal of Consumer Research, 39*, 881–898. https://academic.oup.com/jcr/article/39/4/881/179830910.1086/666376

Barreiro-Gen, M. (2017). *Driving factors of the collaborative consumption in the European Union scene*. EDaSS Research Group on Economic Development and Social Sustainability Department of Economy, University of A Coruna http://hdl.handle.net/10419/213775

Baumeister, C., Scherer, A., & Wangenheim, F. (2015). Branding access offers: The importance of product brands, ownership status, and spill-over effects to parent brands. *Journal of the Academy of Marketing Science, 43*, 574–588 https://doi.org/10.1007/s11747-015-0440-y https://rdcu.be/c0E2k

Belk, R. W. (2014a). Ownership and collecting. In R. O. Frost & G. Steketee (Eds.), *The Oxford handbook of hoarding and acquiring*. Oxford University Press https://doi.org/10.1093/oxfordhb/9780199937783.001.0001

Belk, R. W. (2014b). You are what you can access: Sharing and collaborative consumption online. *Journal of Business Research, 67*(8), 1595–1600. https://doi.org/10.1016/j.jbusres.2013.10.01

Binninger, A. S., Ourahmoune, N., & Robert, I. (2015). Collaborative consumption and sustainability: A discursive analysis of consumer representations and collaborative website narratives. *Journal of Applied Business Research (JABR), 31*(3), 969–986. https://doi.org/10.19030/jabr.v31i3.9229

Bose, U. (2019). ICT, marketing technology and e-commerce in unstable economies (Cases of South Sudan, Syria, Democratic Republic of Congo, Yemen, Somalia). *International Journal of All Research Writings, 1*(5), 7–14. http://www.ijarw.com/PublishedPaper/IJARW1075.pdf

Botsman, R., & Rogers, R. (2010). *What's mine is yours. The rise of collaborative consumption* (1st ed.). https://tantor.com/SellSheets/1920_MineIsYours.pdf

Botsman, R., & Rogers, R. (2011). *What's mine is yours: How collaborative consumption is changing the way we live.* Collins.

Cao, M., & Zhang, Q. (2011). Supply chain collaboration: Impact on collaborative advantage and firm performance. *Journal of Operations Management, 29*(3), 163–180. https://doi.org/10.1016/j.jom.2010.12.008

Carmody, P., & Fortuin, A. (2019). "Ride-sharing", virtual capital and impacts on labor in Cape Town, South Africa. *African Geographical Review, 38*(3), 196–208. https://www.tandfonline.com/doi/abs/10.1080/19376812.2019.1607149; https://doi.org/10.1080/19376812.2019.1607149

Clark, R. (2020). Global fashion exchange launches new digital swapping systems. *Forbes*. https://www.forbes.com/sites/roddyclarke/2020/05/27/global-fashion-exchange-launches-new-digital-swapping-platform/?sh=403dfl056dfl. Accessed 27 May 2022.

Cohen, B., & Munoz, P. (2016). Sharing cities and sustainable consumption and production: Towards an integrated framework. *Journal of Cleaner Production, 134*, 87–97. https://doi.org/10.1016/j.jclepro.2015.07.133

Cropanzano, R., & Mitchell, M. S. (2005). Social exchange theory: An interdisciplinary review. *Journal of Management, 31*(6), 874–900. https://doi.org/10.1177/0149206305279602

Daily Nation. (2018, December 20). *Drivers' group launches Bebabeba taxi ap.* https://www.nation.co.ke/business/Drivers--group-launches-Bebabeba-taxiapp/996-4904510-jo44t5z/index.html

Dall Pizzol, H., Ordovás de Almeida, S., & Do Cout Soares, M. (2017). Collaborative consumption: A proposed scale for measuring the construct applied to a carsharing setting. *Sustainability, 9*(5), 703. https://doi.org/10.3390/su9050703

Degadjor, M. (2018). UBER branding saga: Why Ghana's MoT and GPRTU got it wrong. *Graphic News Paper*. https://www.graphic.com.gh/features/opinion/uber-branding-saga-why-ghana-mot-and-gprtu-got-it-wrong.html

Dillahunt, T. R., & Malone, A. R. (2015). The promise of the sharing economy among disadvantaged communities. In *Proceedings of the 33rd annual ACM conference on human factors in computing systems* (pp. 2285–2294). https://doi.org/10.1145/2702123.2702189

Dreyer, B., Lüdeke-Freund, F., Hamann, R., & Faccer, K. (2017). Upsides and downsides of the sharing economy: Collaborative consumption business models' stakeholder value impacts and their relationship to context. *Technological Forecasting and Social Change, 125*, 87–104. https://doi.org/10.1016/j.techfore.2017.03.036

Dzimiri, M. T., Thamage, K. A., Mapharing, M., Basuhi, E., & Radikoko, I. (2018). Consumer Perceptions towards Online Retailing in Botswana: A Case Study of the University of Botswana. *Transactions on Networks and Communications, 6*(5), 70. https://doi.org/10.14738/tnc.65.5292

Dzisi, E. K., Ackaah, W., Aprimah, B. A., & Adjei, E. (2020). Understanding demographics of ride-sourcing and the factors that underlie its use among young people. *Scientific African, 7*, e00288. https://doi.org/10.1016/j.sciaf.2020.e00288

Eckhardt, G., & Bardhi, F. (2015). Liquid consumption. *ACR Asia-Pacific Advances*. http://www.acrwebsite.org/volumes/1018857/volumes/ap11/AP-11

Emerson, R. M. (1962). Power-dependence relations. *American Sociological Review*, 31–41. https://doi.org/10.4324/9780429302824

Engelman, R. (2013). Beyond sustainababble. In *State of the world 2013* (pp. 3–16). Island Press. https://doi.org/10.5822/978-1-61091-458-1_1

Erdiaw-Kwasie, M. O., Abunyewah, M., Yusif, S., & Erdiaw-Kwasie, A. (2023). Does circular economy knowledge matter in sustainable service provision? A moderation analysis. *Journal of Cleaner Production, 383*, 135429.

Flaherty, K. E., & Pappas, J. M. (2009). Expanding the sales professional's role: A strategic re-orientation? *Industrial Marketing Management, 38*(7), 806–813. https://doi.org/10.1016/j.indmarman.2008.04.007

Folasade, J. O., & Ogungbe, M. A. (2020). Awareness and usability of digitalized marketing platforms of real estate properties in Nigeria. *International Journal of Sales & Marketing Management (IJSMM), 9*, 47–54. https://www.scribd.com/document/516422509/23-06-2021-1624443473-7-Ijsmm-5-Ijsmm-Foc-Dec-2020-Awareness-and-Userbility-of-Digitalized-Marketing-Platforms-of-Real-Estate-Properties-in-Nig

Folorunso, R. O., & Momoh, M. (2015). Assessment of success factors of online shopping patronages in Nigeria: A study of Jumia and Konga shoppers in selected cities in Niger state, Nigeria. *Sahel Analyst Journal of Management Sciences, 13*(2).

Foster, C., & Graham, M. (2015). *The internet and tourism in Rwanda: Value chains and networks of connectivity-based enterprises in Rwanda*. https://doi.org/10.35648/20.500.12413/11781/ii170

Frenken, K., & Schor, J. (2017). Putting the sharing economy into perspective. In *A research agenda for sustainable consumption governance*. Edward Elgar Publishing. https://www.nation.co.ke/business/Drivers--group-launches-Bebabeba-taxiapp/996-4904510-jo44t5z/index.html

Galuszka, J., Martin, E., Nkurunziza, A., Achieng'Oginga, J., Senyagwa, J., Teko, E., & Lah, O. (2021). East Africa's policy and stakeholder integration of informal operators in electric mobility transitions—Kigali, Nairobi, Kisumu and Dar es Salaam. *Sustainability, 13*(4), 1703. https://doi.org/10.3390/su13041703

Garrett, R. K., Weeks, B. E., & Neo, R. L. (2016). Driving a wedge between evidence and beliefs: How online ideological news exposure promotes political misperceptions. *Journal of Computer-Mediated Communication, 21*(5), 331–348. https://doi.org/10.1111/jcc4.12164

Gaus, G. (2012). Property and ownership. In D. Estlund (Ed.), *Oxford handbook of political philosophy* (pp. 93–112). Oxford University Press.

Giddy, J. K. (2020). Travel patterns of Uber users in South Africa. In *New directions in South African tourism geographies* (pp. 113–127). Springer, Cham. https://doi.org/10.1007/978-3-030-29377-2_7

Gladek, E. (2017). *The seven pillars of the circular economy*. http://www.metabolic.nl/the-seven-pillars-of-the-circular-economy/

Godelnik, R. (2017). Millennials and the sharing economy: Lessons from a "buy nothing new, share everything month" project. *Environmental Innovation and Societal Transitions, 23*, 40–52. https://doi.org/10.1016/j.eist.2017.02.002

Goudin, P. (2016). *The cost of non-Europe in the sharing economy: Economic, social and legal challenges and opportunities*. https://policycommons.net/artifacts/1334311/the-cost-of-non-europe-in-the-sharing-economy/1939682/

Gruszka, K. (2017). Framing the collaborative economy—Voices of contestation. *Environmental Innovation and Societal Transitions, 23*, 92–104. https://doi.org/10.1016/j.eist.2016.09.002

Guttentag, D. (2015). Airbnb: Disruptive innovation and the rise of an informal tourism accommodation sector. *Current Issues in Tourism, 18*(12), 1192–1217. https://doi.org/10.1080/13683500.2013.827159

Guttentag, D. (2016). *Why tourists choose Airbnb: A motivation-based segmentation study underpinned by innovation concepts*. http://hdl.handle.net/10012/10684

Haase, M., & Kleinaltenkamp, M. (2011). Property rights design and market process: Implications for market theory, marketing theory, and SD Logic. *Journal of Macromarketing, 31*(2), 148–159. https://doi.org/10.1177/0276146710397662

Hailemariam, A., & Erdiaw-Kwasie, M. O. (2022). Towards a circular economy: Implications for emission reduction and environmental sustainability. *Business Strategy and the Environment*.

Hall, J., Matos, S., Sheehan, L., & Silvestre, B. (2012). Entrepreneurship and innovation at the base of the pyramid: A recipe for inclusive growth or social exclusion? *Journal of Management Studies, 49*(4), 785–812. https://doi.org/10.1111/j.1467-6486.2012.01044.x

Hamari, J., Sjöklint, M., & Ukkonen, A. (2016). The sharing economy: Why people participate in collaborative consumption. *Journal of the Association for Information Science and Technology, 67*(9), 2047–2059. https://doi.org/10.1002/asi.23552

Hawlitschek, F., Teubner, T., & Gimpel, H. (2016). Understanding the sharing economy—Drivers and impediments for participation in a peer-to-peer rental. In *2016 49th Hawaii International Conference on System Sciences (HICSS)* (pp. 4782–4791). IEEE. https://doi.org/10.1109/HICSS.2016.593

Heinrichs, H. (2013). Sharing economy: A potential new pathway to sustainability. *GAIA-Ecological Perspectives for Science and Society, 22*(4), 228–231. https://doi.org/10.14512/gaia.22.4.5

Henama, U. S., & Sifolo, P. P. S. (2017). Uber: The South Africa experience. *African Journal of Hospitality, Tourism and Leisure, 6*(2), 1–10. https://www.researchgate.net/publication/317167857_Uber_The_South_Africa_Experience

Henama, U. S. (2021). eCommerce within the tourism industry in the global south: The case of the sharing economy in South Africa. In *Digital marketing & eCommerce conference* (pp. 280–290). Springer, Cham. https://doi.org/10.1007/978-3-030-76520-0_30

Heshmati, A. (2015). A review of the circular economy and its implementation (IZA Discussion Papers, No. 9611). Institute for the Study of Labor (IZA), Bonn. https://www.econstor.eu/bitstream/10419/130297/1/dp9611.pdf

Hoang, V. N. (2014). Analysis of resource efficiency: A production frontier approach. *Journal of Environmental Management, 137*, 128–136. https://doi.org/10.1016/j.jenvman.2014.01.046

Holmberg, S. (2016). *E-Commerce in Sub-Saharan Africa: Overcoming barriers to success in an online environment: A study on e-commerce SMEs in Ghana, Kenya, South Africa, and Uganda* [Master's Thesis]. Copenhagen Business School, Denmark. https://www.google.com/url?sa=t&rct=j&q=&esrc=s&source=web&cd=&cad=rja&uact=8&ved=2ahUKEwj7zZmamdT7AhX4rZUCHdcKB9cQFnoECA8QAQ&url=https%3A%2F%2Fresearch-api.cbs.dk%2Fws%2Fportalfiles%2Fportal%2F58442372%2Fsara_holmberg.pdf&usg=AOvVaw2mqj0lfRtMWJSJAIWBcgbI

Huang, M. H., & Rust, R. T. (2011). Sustainability and consumption. *Journal of the Academy of Marketing Science, 39*(1), 40–54. https://doi.org/10.1007/s11747-010-0193-6

Jaiyeoba, O. O., Chimbise, T. T., & Roberts-Lombard, M. (2018). E-service usage and satisfaction in Botswana. *African Journal of Economic and Management Studies*. https://doi.org/10.1108/AJEMS-03-2017-0061

Jenkins, R., Molesworth, M., & Scullion, R. (2014). The messy social lives of objects: Inter-personal borrowing and the ambiguity of possession and ownership. *Journal of Consumer Behaviour, 13*(2), 131–139. https://doi.org/10.1002/cb.1469

Jiang, Y., & Kim, Y. (2015). Developing multi-dimensional green value: Extending social exchange theory to explore customers' purchase intention in green hotels–evidence from Korea. *International Journal of Contemporary Hospitality Management*. ISSN: 0959–6119. https://www.emerald.com/insight/content/doi/10.1108/IJCHM-08-2013-0383/full/html?skipTracking=true

Joung, C. B., Carrell, J., Sarkar, P., & Feng, S. C. (2013). Categorization of indicators for sustainable manufacturing. *Ecological Indicators, 24*, 148–157. https://doi.org/10.1016/j.ecolind.2012.05.030

Katende-Magezi, E. (2017). *The impact of second-hand clothes and shoes in East Africa*. https://www.cuts-geneva.org/pdf/PACT2-STUDY-The_Impact_of_Second_Hand_Clothes_and_Shoes_in_East_Africa.pdf. Accessed 8 June 2022

Kirchherr, J., Reike, D., & Hekkert, M. (2017). Conceptualizing the circular economy: An analysis of 114 definitions. *Resources, Conservation, and Recycling, 127*, 221–232. https://doi.org/10.1016/j.resconrec.2017.09.005

Kufuor, K. (2018). Uber in Ghana: Markets and institutions in the emergence of ride-sharing taxis. *LUGLJ: Lancaster University of Ghana Law Journal, 4*, 45–69. https://www.google.com/url?sa=t&rct=j&q=&esrc=s&source=web&cd=&cad=rja&uact=8&ved=2ahUKEwiR4NrFm9T7AhW8HLkGHUzNBM4QFnoECBAQAQ&url=https%3A%2F%2Frepository.uel.ac.uk%2Fdownload%2Fa0e08701204d88196de3ab3b3a881927e477c444347e02737051f617cb78c827%2F505076%2F7493.pdf&usg=AOvVaw2kYipYrVMZoY3OPY0_84zV

Laizer, L. (2020). *Performance and economic effects of the online based taxi services in Tanzania; a case of Uber Tanzania limited* [Doctoral dissertation]. Mzumbe University. http://hdl.handle.net/11192/4758

Lieder, M., & Rashid, A. (2016). Towards circular economy implementation: A comprehensive review in context of manufacturing industry. *Journal of Cleaner Production, 115*, 36–51. https://doi.org/10.1016/j.jclepro.2015.12.042

Lemille, A. (2016). Circular economy 2.0. In *Ensuring that circular economy is designed for all*. [Online] https://docs.ixstatic.com/ugd/cd2144_2f29c76bed144a50a762d28ef464fcd6.pdf

Mair, J., & Reischauer, G. (2017). Capturing the dynamics of the sharing economy: Institutional research on the plural forms and practices of sharing economy organizations. *Technological Forecasting and Social Change, 125*, 11–20. https://doi.org/10.1016/j.techfore.2017.05.023

Manavhela, M. P., & Henama, M. U. S. (2019). The sharing economy according to Uber and Airbnb in South Africa: Challenges, opportunities and immediate prospects. In *9th Advances in hospitality and tourism marketing and management conference* (p. 411).

Martin, C. J. (2016). The sharing economy: A pathway to sustainability or a nightmarish form of neoliberal capitalism? *Ecological Economics, 121*, 149e159 https://doi.org/10.1016/j.ecolecon.2015.11.027

Matzler, K., Veider, V., & Kathan, W. (2015). Adapting to the sharing economy. *MIT Sloan Management Review, 56*, 71–77. https://www.google.com/url?sa=t&rct=j&q=&esrc=s&source=web&cd=&cad=rja&uact=8&ved=2ahUKEwi7r5yjntT7AhUZqZUCHZY3AjYQFnoECCcQAQ&url=https%3A%2F%2Fwww.semanticscholar.org%2Fpaper%2FAdapting-to-the-sharing-economy-Matzler-Veider%2F3d84e5445758c47ab94ab1f3d3a215a2b1725f44&usg=AOvVaw0lQ9oJsQGDNfEeUqcV9nsL

Maziriri, E., Mapuranga, M., Mushwana, J., & Madinga, N. (2020). *Antecedents that influence the intention to use the Uber mobile application: Customer perspectives in South Africa*. https://www.learntechlib.org/p/216976/

Melia, E. (2020). *African jobs in the digital era: Export options with a focus on online labour* (Discussion Paper No. 3/2020). https://www.econstor.eu/handle/10419/214203

Meltzer, J. (2014). *Supporting the internet as a platform for international trade: Opportunities for small and medium-sized enterprises and developing countries* (No. 69). Brookings Institute. https://papers.ssrn.com/sol3/papers.cfm?abstract_id=2400578

Mishra, J. L., Chiwenga, K. D., & Ali, K. (2019). Collaboration as an enabler for circular economy: A case study of a developing country. *Management Decision*. https://doi.org/10.1108/MD-10-2018-1111

Moher, D., Liberati, A., Tetzlaff, J., Altman, D. G., Altman, D., Antes, G., ... & Tugwell, P. (2009). Preferred reporting items for systematic reviews and meta-analyses: The PRISMA statement (Chinese edition). *Journal of Chinese Integrative Medicine, 7*(9), 889–896. https://doi.org/10.7326/0003-4819-151-4-200908180-00135

Möhlmann, M. (2015). Collaborative consumption: Determinants of satisfaction and the likelihood of using a sharing economy option again. *Journal of Consumer Behaviour, 14*, 193–207. https://doi.org/10.1002/cb.1512

Morozov, E. (2015, June 7). Where Uber and Amazon rule: Welcome to the world of the platform. *The Guardian*. https://www.theguardian.com/technology/2015/jun/07/facebook-uber-amazon-platform-economy. Accessed 8 June 2022

Mtaho, A. B., & Mselle, L. (2014). Securing mobile money services in Tanzania: a case of Vodacom M-Pesa. *International Journal of Computer Science and Network Solutions, 2*, 1–11. https://www.google.com/url?sa=t&rct=j&q=&esrc=s&source=web&cd=&cad=rja&uact=8&ved=2ahUKEwj78_iuodT7AhVPI7kGHUKdDbsQFnoECBMQAQ&url=https%3A%2F%2Fwww.semanticscholar.org%2Fpaper%2FSecuring-Mobile-Money-Services-in-Tanzania%253A-A-Case-Basigie-Mselle%2F58ed5ac2a7aa0fb9dcf526823a2b3da80b3e94d7&usg=AOvVaw031GFKOC7xCKcKrPwJdXLX

Mulligan, G. (2013). M-Pesa now available for online payments in Kenya. *Tanzania: PayGate*. https://www.google.com/url?sa=t&rct=j&q=&esrc=s&source=web&cd=&cad=rja&uact=8&ved=2ahUKEwid3cLVodT7AhWXKLkGHdsNDHkQFnoECA8QAQ&url=https%3A%2F%2Fventuresafrica.com%2Fm-pesa-now-available-for-online-payments-in-kenya-tanzania-paygate%2F&usg=AOvVaw2fStaizcb4Nix5FNcgAlbo

Murage, A. C. N., Mutisya, M. M., & Muthengi, S. (2020). Influence of emerging peer to peer accommodation on budget hotels competitiveness in Nairobi City County, Kenya. *Journal of Hospitality and Tourism Management, 3*(1). https://ir-library.ku.ac.ke/bitstream/handle/123456789/22695/Influence%20of%20Emerging%20Peer%20To%20Peer%20Accommodation.pdf?sequence=1

Murray, A., Skene, K., & Haynes, K. (2017). The circular economy: An interdisciplinary exploration of the concept and application in a global context. *Journal of Business Ethics, 140*(3), 369–380. https://doi.org/10.1007/s10551-015-2693-2

Mvungi, S. (2019). *Factors affecting the adoption of E-payment systems on ride sharing business in Tanzania: A case of Uber: Dar Es Salaam, Tanzania* [Doctoral Dissertation]. Mzumbe University. http://hdl.handle.net/11192/3306

Naudé, M. (2011). Sustainable development in companies: Theoretical dream or implementable reality. *Corporate Ownership & Control, 8*(4), 352–364. https://pdfs.semanticscholar.org/50d6/bde4d2c9d4eef6c23eb1fc87e4d9c2956999.pdf

Okedi, R. H., & Yano, M. R. (2020). Industry forces influencing the performance of online application-based taxi drivers in Kenya: A descriptive case study of Uber, Bolt and Little Cabs drivers. *African Journal of Emerging Issues, 2*(10), 38–56. https://ajoeijournals.org/sys/index.php/ajoei/article/view/134

Olasanmi, O. O. (2019). Online shopping and customers' satisfaction in Lagos State, Nigeria. *American Journal of Industrial and Business Management, 9*(6), 1446. https://www.scirp.org/html/10-2121503_93444.htm

Olotewo, J. (2017). Examining the antecedents of in-store and online purchasing behaviour: a case of Nigeria. *Journal of Marketing Research and Case Studies*, 1–16. https://ibimapublishing.com/wp-content/uploads/articles/JMRCS/2017/668316/668316.pdf

Oreva, O. (2016). *The state of ridesharing in Africa*. https://techcabal.com/2016/11/16/the-state-of-ridesharing-in-africa/. Accessed 31 May 2022

Ozanne, L. K., & Ballantine, P. W. (2010). Sharing as a form of anti-consumption? An examination of toy library users. *Journal of Consumer Behaviour, 9*(6), 485–498. https://doi.org/10.1002/cb.334

Papadopoulos, S., & Eck, S. V. (2021). Disruptive technologies and taxi rides in South Africa: What is the 'Uber' Uproar About? In *Global perspectives on legal challenges posed by ridesharing companies* (pp. 199–229). Springer, Singapore. https://doi.org/10.1007/978-981-15-7035-3_10

Parsons, A. (2014). The sharing economy: A short introduction to its political evolution. *Open Democracy, 5*, 1–8. https://www.opendemocracy.net/en/transformation/sharing-economy-short-introduction-to-its-political-evolution/

Perren, R., & Grauerholz, L. (2015). Collaborative consumption. *International Encyclopedia of the Social & Behavioral Sciences, 4*(2). https://www.researchgate.net/profile/Liz-Grauerholz/publication/303960052_Collaborative_Consumption/links/5aa19868aca272d448b375e6/Collaborative-Consumption.pdf

Perren, R., Massiah, C., & He, X. (2014). The next generation of peer-to-peer (P2P) marketplaces: Conceptualizing the dynamic interaction between firms and consumers in facilitating exchanges. In *American Marketing Association Winter Educator's Conference, Orlando FL* (Vol. 21).

Philip, H., Ozanne, L., & Ballantine, P. (2015). Examining temporary disposition and acquisition in peer-to-peer renting. *Journal of Marketing Management, 31*, 1310–1332. https://doi.org/10.1080/0267257X.2015.1013490

Preston, F., & Lehne, J. (2017). *A wider circle? The circular economy in developing countries*. https://policycommons.net/artifacts/1423261/a-wider-circle-the-circular-economy-in-developing-countries/2037525/

Preston, F., Lehne, J., & Wellesley, L. (2019). An inclusive circular economy. *Priorities for Developing Countries*. https://www.chathamhouse.org/sites/default/files/publications/research/2019-05-22-Circular%20Economy.pdf

Pouri, M. J., & Hilty, L. M. (2019). Conceptualizing the digital sharing economy in the context of sustainability. *Sustainability, 10*, 4453. https://doi.org/10.3390/su10124453

Rahman, M. A., & Tumpa, T. (2021). Collaborative consumption for a sustainable future. *Rethinking Governance, 1*(1), 93–100. https://www.researchgate.net/profile/Md-Azzajur-Rahman/publication/349522210_COLLABORATIVE_CONSUMPTION_FOR_SUSTAINABLE_FUTURE/links/603a4a3a92851c4ed59df5ff/COLLABORATIVE-CONSUMPTION-FOR-SUSTAINABLE-FUTURE.pdf

Schor, J. (2016). Debating the sharing economy. *Journal of Self-Governance and Management Economics, 4*(3), 7–22. https://doi.org/10.3390/su8111212

Schandl, H., & West, J. (2010). Resource use and resource efficiency in the Asia-Pacific region. *Global Environmental Change, 20*(4), 636–647. https://doi.org/10.1016/j.gloenvcha.2010.06.003

Schuitema, G., & De Groot, J. I. M. (2014). Green consumerism: The influence of product attributes and values on purchasing intentions. *Journal of Consumer Behaviour*. https://doi.org/10.1002/cb.1501

Schut, E., Crielaard, M., & Mesman, M. (2015). *Circular economy in the Dutch construction sector: A perspective for the market and government*. http://hdl.handle.net/10029/595297

Sigala, M. (2014). Customer involvement in sustainable supply chain management: A research framework and implications in tourism. *Cornell Hospitality Quarterly, 55*(1), 76–88. https://doi.org/10.1177/1938965513504030

Singh, S., & Singh, A. (2018). Interplay of organizational justice, psychological empowerment, organizational citizenship behavior, and job satisfaction in the context of circular economy. *Management Decision*. https://doi.org/10.1108/MD-09-2018-0966

Simmons, R. O. B., Effah, J., & Boateng, R. (2019). *Digital innovation and taxi services: The case of Uru in Ghana*. https://scholar.archive.org/work/kgs36tioyndybgmvoywofgtt2y/access/wayback/https://aisel.aisnet.org/cgi/viewcontent.cgi?article=1132&context=amcis2019

Taeihagh, A. (2017). Crowdsourcing, sharing economies, and development. *Journal of Developing Societies, 33*(2), 191–222. https://doi.org/10.1177/0169796X17710072

Tsai, M. T., Cheng, N. C., & Chen, K. S. (2011). Understanding online group buying intention: The roles of sense of virtual community and technology acceptance factors. *Total Quality Management & Business Excellence, 22*(10), 1091–1104. https://doi.org/10.1080/14783363.2011.614870

Tussyadiah, I. P. (2015). An exploratory study on drivers and deterrents of collaborative consumption in travel. In *Information and communication technologies in tourism* (pp. 817–830). https://doi.org/10.1007/978-3-319-14343-9_59

United Nations General Assembly (UNGA). (2015). *Resolution adopted by the General Assembly on 11 September 2015*. A/RES/69/315 15 September 2015. United Nations.

Vasirani, M., & Ossowski, S. (2012). A collaborative model for participatory load management in the smart grid. In *Workshop on AI problems and approaches for intelligent environments* (p. 21). https://www.lirmm.fr/ecai2012/images/stories/ecai_doc/pdf/workshop/W20_ProceedingsAI4IE.pdf#page=29

Vaughan, R., & Daverio, R. (2016). *Assessing the size and presence of the collaborative economy in Europe*. Publications Office of the European Union. http://grupo.us.es/iwpr/wp-content/uploads/2017/12/INFORME-PWC-COMISI%C3%93N.pdf

Visser, G., Erasmus, I., & Miller, M. (2017). Airbnb: The emergence of a new accommodation type in Cape Town, South Africa. *Tourism Review International, 21*(2), 151–168. https://doi.org/10.3727/154427217X14912408849458

Wang, Y., Xiang, D., Yang, Z., & Ma, S. S. (2019). Unraveling customer sustainable consumption behaviors in sharing economy: A socio-economic approach based on social exchange theory. *Journal of Cleaner Production, 208*, 869–879. https://doi.org/10.1016/j.jclepro.2018.10.139

Watkins, R. D., Denegri-Knott, J., & Molesworth, M. (2016). The relationship between ownership and possession: Observations from the context of digital virtual goods. *Journal of Marketing Management, 32*(1–2), 44–70. https://doi.org/10.1080/0267257X.2015.1089308

Welsum, D. (2016). *Sharing is caring? Not quite: Some observations about the "sharing economy."* (World Development Report 2016). http://hdl.handle.net/10986/23648

Wetzel, H. A., Hammerschmidt, M., & Zablah, A. R. (2014). Gratitude versus entitlement: A dual-process model of the profitability implications of customer prioritization. *Journal of Marketing, 78*(2), 1–19. https://doi.org/10.1509/jm.12.0167

World Commission for Environment and Development (WCED). (1987). *Our common future, the Brundtland Report*. Oxford, The UK. http://netzwerk-n.org/wp-content/uploads/2017/04/0_Brundtland_Report-1987-Our_Common_Future.pdf

World Economic Forum. (2015). *The Africa competitiveness report*. World Economic Forum, World Bank, OECD. https://www.google.com/url?sa=t&rct=j&q=&esrc=s&source=web&cd=&cad=rja&uact=8&ved=2ahUKEwic5Ja_rdT7AhUbFLkGHasHCjsQFnoECA0QAQ&url=https%3A%2F%2Fwww3.weforum.org%2Fdocs%2FWEF_ACR_2015%2FAfrica_Competitiveness_Report_2015.pdf&usg=AOvVaw2xT6Dd43lDme3ZdNq1WmM

World Economic Forum (WEC). (2017). *Collaboration in cities: From sharing to 'sharing economy'* (White paper). https://www3.weforum.org/docs/White_Paper_Collaboration_in_Cities_report_2017.pdf. Accessed 9 June 2022

Wright, C. Y., Godfrey, L., Armiento, G., Haywood, L. K., Inglesi-Lotz, R., Lyne, K., & Schwerdtle, P. N. (2019). Circular economy and environmental health in low-and middle-income countries. *Globalization and Health, 15*(1), 1–5. https://doi.org/10.1186/s12992-019-0501-y

Xiaoli, G. U. O., & Yang, Y. U. (2015). A residential smart power utilization strategy based on cloud computing. *Automation of Electric Power Systems, 39*(17), 114–119.

Conceptual Insights on Circular Strategies and SDGs

Circular Economy Research and Practice: Past, Present and Future

Michael Odei Erdiaw-Kwasie, Amila Kasun Sampath Udage Kankanamge, Matthew Abunyewah, G. M. Monirul Alam, Angela Erdiaw-Kwasie, Michael Afful Andoh, and Kofi Kusi Owusu-Ansah

1 Introduction

Consistent with the observed transformations in circular economy research and practice (Chizaryfard et al., 2021; De los Rios & Charnley, 2017; Hobson & Lynch, 2016; Shooshtarian et al., 2022; Zhijun & Nailing, 2007), research-to-practice gaps continue to exist, and progress in this subject has been quite slow (Barreiro-Gen & Lozano, 2020; Geissdoerfer et al., 2018; Homrich et al., 2018; Kouhizadeh et al., 2020). Furthermore, while the literature focuses on developing proactive methods and tools for circular economy implementation, corporate practitioners' awareness and use of these tools and insights are still lacking (Bocken et al., 2019; Kalmykova et al., 2018; Walker et al., 2022). Thus,

M. O. Erdiaw-Kwasie · A. K. S. Udage Kankanamge
Business and Accounting Discipline, Faculty of Arts and Society, Charles Darwin University, Darwin, NT 0810, Australia
e-mail: michael.erdiaw-kwasie@cdu.edu.au

M. O. Erdiaw-Kwasie and G. M. M. Alam (eds.), *Circular Economy Strategies and the UN Sustainable Development Goals*, Sustainable Development Goals Series,
https://doi.org/10.1007/978-981-99-3083-8_3

understanding how research shapes practice and vice versa is deemed fundamental but more important in times of crisis. Factors contributing to this problem include lapses in communication between circular economy researchers and practitioners, as exemplified by the COVID-19 pandemic unveiling ingrained gaps characterizing research-to-practice transitions (Ibn-Mohammed et al., 2021; Su & Urban, 2021; Wuyts et al., 2020). Related to this, Wuyts et al. (2020) share that COVID-19 is putting companies to the test for how well their circular economy initiatives are evidence-based and responsive to societal issues, particularly at times of both crisis and recovery.

Thus far, the extant literature on the link between the circular economy and COVID-19 has focused on either research or organizational practices (Homrich et al., 2018; Ibn-Mohammed et al., 2021; Sharma et al., 2021). Thus, studies that offer holistic perspectives on how research shapes practice and the associated gaps are critical. The following gaps in the literature are noted. First, a comprehensive study on circular economy research and practice synergy and COVID-19 is currently lacking, and relevant studies are fragmented. For instance, Ibn-Mohammed et al. (2021) limited their analysis to circular economy research. They revealed four key areas where COVID-19 had challenged circular economy research: stakeholders, societal risk, supply chain resilience, and material closed loops. Similarly, Wuyts et al. (2020) focused on organizational

M. Abunyewah
College of Human and Health Sciences, Charles Darwin University, Darwin, NT 0909, Australia

G. M. M. Alam
Bangabandhu Sheikh Mujibur Rahman Agricultural University (BSMRAU), Gazipur 1706, Bangladesh

A. Erdiaw-Kwasie (✉)
College of Arts and Sciences, American University, Washington, DC, USA
e-mail: kwasieangela@gmail.com

M. A. Andoh
School of Engineering, Bundoora East Campus, Building 251, Level 3, Melbourne, VIC, Australia

K. K. Owusu-Ansah
College of Arts and Sciences, University of Wyoming, Laramie, WY, USA

practices. They argued that the COVID-19 pandemic offers huge opportunities for businesses to shift towards more authentic circular practices and contribute to addressing urgent social and environmental challenges worldwide. Second, limited literature on circular economy research and practice synergy models exists.

Both circular economy research (design) and circular economy practice (implementation) present their own unique sets of social and technical challenges (Bressanelli et al., 2019; Ferasso et al., 2020; Galvão et al., 2018; Hossain et al., 2020; Lehner et al., 2020; Ormazabal et al., 2018; Rajput & Singh, 2019). This may explain why research-to-practice transition gaps persist despite the availability of information on the circular economy (Böckel et al., 2021; MahmoumGonbadi et al., 2021). Furthermore, the effects of synergy process models on existing gaps remain unexplored. Based on the above discussion, this paper seeks to achieve the following objectives: (i) explore emerging research-to-practice transition gaps, (ii) propose a circular economy synergy process model, and (iii) examine the dynamics and effects of the circular economy synergy process model in the post-COVID-19 era.

The rest of the paper is structured as follows. Section 2 explains the basic concept of transitions, and Sect. 3 discusses emerging circular economy research and practice issues within the context of the COVID-19 pandemic. Section 4 provides a discussion of emerging circular economy research-to-practice transition issues, while Sect. 5 presents a discussion on the new circular economy synergy process model. The paper's concluding section offers suggestions with academic and corporate implications.

2 Understanding Transition Research

The domain of transition research has experienced rapid growth in recent years. Many studies in this area have focused on investigating the characteristics of historical transitions and applying the obtained insights to developing frameworks that guide ongoing transitions (Erdiaw-Kwasie et al., 2017; Papachristos et al., 2018). The field started to broaden recently as new scholars from different disciplinary backgrounds became more engaged. Prime examples of disciplines dabbling in transition research include management (Bögel et al., 2019; Gatti et al., 2019), economics (Hafner et al., 2020), geography (Fazey et al., 2018; Wanner et al., 2018), and sociology (Papachristos et al., 2018). Walrave et al.

(2018), for example, focused on action research and process-oriented approaches to study transition problems. In this respect, one main driver behind the emergence of research-to-practice transitions has been the search for new insights into achieving a desirable evidence-based practice. Loorbach et al. (2017) demonstrate that mutual learning and cooperation of both researchers and practitioners within the transition field can facilitate a working synergy.

Transition research acknowledges that multiple actors from various institutional backgrounds (e.g. researchers, corporate managers, employees) and issues of power asymmetries are inherent in the research-to-practice transition (Avelino & Wittmayer, 2016; Hölscher et al., 2018). Other important elements in transition studies include supporting and developing shared insights and the need for systemic change in a participatory manner (Köhler et al., 2019). Scholars have argued that using transition concepts and ideas can help us explain how actors from all backgrounds understand complex societal problems and act accordingly (Hölscher et al., 2018; Loorbach et al., 2017). Research-to-practice transition ideas build on the notion that actors can produce practical solutions to address societal problems, mainly by drawing on existing knowledge and research insights. Thus, this line of thinking positions the perspective of research-to-practice transitions more broadly by looking at transitional dynamics between different actors in the context of persistent problems. To Loorbach et al. (2017), transitional dynamics may lead to learning, behavioural, and systemic change.

The field of circularity has evolved rapidly over the past five decades, with linear models shifting towards circular ones (Ghisellini & Ulgiati, 2020; Hailemariam & Erdiaw-Kwasie, 2023; McCarthy et al., 2018). As a result, the focus of circular economy research has shifted significantly beyond conventional recycling towards the reuse and dissemination CE culture to revisit its social and political dimensions. Moreover, it has become a distinct area of research into the nonlinear dynamics of societal and corporate change, which is highly relevant to policymaking and issues in the context of grand societal and corporate challenges (Ciliberto et al., 2021; Ghisellini & Ulgiati, 2020). This is reflected well in how the circular economy field has co-evolved with societal change and corporate discourse. Furthermore, firms increasingly recognize the strategic benefits that stem from the integration of resource use in their circular actions (Bocken et al., 2016; Preston, 2012; Rizos et al., 2016). Recent studies have shown that circular economy actions have been considered one of the very salient ingredients to determine a firm's green image and competitiveness (Kazancoglu et al., 2018; Moktadir et al., 2018; Ormazabal et al., 2018). Following this shift, some companies have begun to redefine their circular economy policies to prioritize the environment.

In line with the above, the goal of a research-to-practice transition framework is to guarantee the continuity of research ideas into practice. In viewing the transition in this field, a new conceptualization of circular economy implies shifting its main function as an internal management tool towards a broader understanding of the business and its triple bottom line—planet, people, and profit. These processes suggest conceptual strategies that can help develop an evidence base for transition issues and effective practices.

3 Overview of Circular Economy: Definitions, Emerging Research, and Practice Thematic Gaps

During the industrial revolution period of the seventeenth century, large amounts of linear products were manufactured by ignoring the harm to the planet through the production process (Mehrotra & Jaladi, 2022). According to Joshi and Gupta (2019), the underline reasons for encouraging the linear business models are the simplicity of raw material acquisition, production, consumption, and disposal after usage. However, the "take-make-dispose" or the linear model follows more generic value chain practices, which direct to waste in the production chain, end-of-life waste, residual energy loss, erosion of ecosystem services, and reduction of the earth's natural capital (EMF, 2015). Later, scholars and corporate sector recognized those linear models are unsustainable (Borrello et al., 2017; Erdiaw-Kwasie et al., 2023; Paul et al., 2022; Vinante et al., 2021). Then, researchers and practitioners thought about a concept that supports mitigating the drawbacks of the linear model (Ghisellini et al., 2016; Jonck et al., 2018). As a solution, they suggested the Circular Economy (CE) model by linking the concept of resource cycling to increase the longevity of the products in the economy (Murray et al., 2017).

The CE concept is not newly emerged (Anaruma et al., 2021). The concept was first introduced by two environment economists Pearce and Turner in 1989, who published the book "Economics of Natural Resources and the Environment". They discussed a closed loop of material flow rather than the open system within the economy (Pearce & Turner, 1989). Nevertheless, major schools of thought emerged in the

1970s (EMF, 2015) that are central to CE and continued the studies such as the looping economy (Stahel & Reday-Mulvey, 1981), materials balance model (Pearce & Turner, 1989), ecological industries (Lifset & Graedel, 1997), natural capitalism (Hawken et al., 2010), and blue economy (Pauli, 2010). After years, EMF (2015) made great strides to contribute to the CE by highlighting the importance of a circular model rather than a linear one due to economic losses, natural system degradation, structural waste, price risk, and supply risk of the existing system. The literature shows the progress of these historical concepts and theories, and how new methodological paradigms were advanced over the past. For instance, the organizations encourage to reuse and recycling of materials in the processes of production, distribution, and consumption to achieve sustainable development of that entity (Kirchherr et al., 2018) and three principles of CE such as reduce, reuse, and recycle (Wu & Deng, 2012) that stimulates the adaptation of circular business practices (Tura et al., 2019). The reducing, reusing, and recycling are commonly referred to as the "3R" principles which have high recognition in searching for a system with a higher level of circularity (Reike et al., 2018). Beyond the "3R" principles, Bradley et al. (2018) mentioned the relevance of the "6R" approach to CE, and the author cited the additional 3Rs as recovery, redesign, and remanufacturing. However, in the evolution process of CE, recovery, reuse, and recycle are essential in circular supply chain systems (Hussain & Malik, 2020).

Since CE is an evolving concept, no agreed definitions exist in the literature (Harris et al., 2021). CE definitions commonly focus on minimizing wastage but lack clarity on social justice and advantages for future generations (Merli et al., 2018). Therefore, the researchers started to think CE is an indicator of the implementation of sustainable development (Kirchherr & Piscicelli, 2019; Vinante et al., 2021; Whalen et al., 2018). Consequently, CE is a contested concept with diverse thematic issues prevailing in the economy, and those thematic issues are used to define CE differently. The following examples show how diverse thematic issues differently think about CE. When focused on the waste management theme, CE is defined as a recurring system that aims to eliminate waste by converting the products at the EOL (i.e. end of life) as resources for new products (Stahel, 2016). Hence, product reprocessing and "solution economy" lead to reducing and preventing waste in the economy. When attentive to the resource preservation theme, CE is defined as an economic model that limits productivity to a certain level by respecting the natural reproduction rate of resources (Korhonen et al., 2018). Therefore, the new consumption culture (e.g. sharing economy—secondhand stores) significantly increases resource efficiency.

Similarly focused on the environment protection theme, CE is defined as a mechanism to protect the environment and prevent pollution, leading to the economy's sustainable development (Ma et al., 2014). Because closing obsolete production methods and adopting cleaner technologies are primary drivers for CE in environmental protection. Further, concentrating on the business strategy theme, CE is defined as a convincing strategy that leads to reducing both inputs of raw materials and outputs of wastes by closing economic and ecological loops of resource flows (Haas et al., 2015). Since consistent use of eco-friendly design for products needs to increase the rate of EOL (i.e. end of life) recycling, when it focuses on the 3R principles theme, CE is defined as an economic business model that is based on reducing, reusing, and recycling limited resources by substituting the concept of "end of life" to gain sustainable growth (Kirchherr et al., 2017). Because an innovative business model and responsible consumers enable CE in the economy, it is clear how different thematic issues contributed to the development of CE.

The development of the concept and its associated theories and models have been divided into thematic periods. The author adopted the previous studies of Blomsma and Brennan (2017) and Tuladhar et al. (2022) to divide CE developments into three historical phases. Tuladhar et al. (2022) provided a logical interpretation for developing the timeframe. According to them, the first period began in the 1960s when they observed the acceleration of environmental movements and consideration of negative environmental impacts from business activities. After that, the second period started in the 1990s when they witnessed the explosion of research and policymaking initiatives for sustainability. Finally, the third and last period began in the 2010s when they observed an increasing number of scholarly debates and policy reports into practice (e.g. contributions from EMF). Table 1 presents the historical overview of the CE.

3.1 Circular Economy Research: A Focus on Thematic Gaps

Subsequently the recent growth of research and publications in the field of CE, the concept still requires advanced theoretical underpinning since there is a lack of an evidence-based theoretical framework to lead the

Table 1 Summary of CE research history

Period	*Thematic Areas*	*Research achievements*
Pre 1990 era	CE Initiation	1. Seminal publications (e.g. The Economics of the Coming Spaceship Earth of Boulding in 1966) highlighted the importance of a closed economy 2. Started to propose solutions for environmental issues (e.g. self-replenishing system and 3R concept—reduce, reuse, and recycle)) 3. Supported green ideology and highlighted the significance of environmental governance disclosure Materials Balance Model—Pearce and Turner (1989), Ecological Modernization Theory—Simonis (1989)
1990s–2010s	CE Conceptualization	1. Started to critically critique the linear economy and its loopholes 2. Conceptualized concepts and theories (e.g. blue economy) to search for sustainable solutions 3. Addressed the industrial ecology concerning sustainability 4. The emergence of "waste as a resource" and "strategies for regenerative design" to ensure environmental efficiency Ecological Industries—Lifset and Graedel (1997), Cradle to Cradle—McDonough and Braungart (2003), Natural Capitalism—Hawken et al. (2010), Blue Economy—Pauli (2010)
2010s–Present	CE Operationalisation	1. Increased the attention of researchers, practitioners, and governments (e.g. the EU Directives) 2. Work has been done to ensure the applicability of the theory in practice 3. Integration of EMF's five principles when transforming industries to CE 4. Circular Business Model (CBM) is considered an emerging area Circular economy (recent development)—EMF (2015), Circular Business Model—(Bocken et al., 2018)

transition of CE (Velenturf & Purnell, 2021). Therefore, progressive theoretical support is required to address the issues within the context of the post-COVID-19 pandemic period. Table 2 illustrates the emerging thematic research gaps.

Table 2 Emerging thematic research gaps

Major issue	*Contribution of research*	*Identified research gaps*
Sustainability	Walker et al. (2022) identified two perspectives on the relationship between CE and sustainability. The first perspective says sustainability is a wider and more holistic concept than CE, while the second perspective shows no conceptual and practical difference between sustainability and CE	The conceptual clarity on the relationship between sustainability and CE is not explicit in the literature which caused the constraint on the effective use of the approach (Geissdoerfer et al., 2017). Moreover, many authors focused on the environmental aspect rather than the holistic view of sustainability (Geissdoerfer et al., 2017). Hence, it should be studied how CE initiatives complement the triple bottom-line practice (i.e. actual impact) (Geissdoerfer et al., 2017)
Economic levels	Strong sustainability practices significantly impact the CE, and its triple levels (e.g. micro, meso, and macro) (Nikolaou et al., 2021) and Nikolaou et al. (2021) analysed the key relationships between CE and sustainability practices. Further, research reveals that micro-level and meso-level collaborations (e.g. supplier alliances) are essential to meet CE goals (Franco, 2017)	The literature puts more weight on the macro level (e.g. regional and national level) and engineering/natural sciences academic grounds (Nikolaou et al., 2021). Therefore, it is highlighted the need to study the interrelationship between the three levels and have an interdisciplinary approach to strengthen the research in economic/management sciences (Nikolaou et al., 2021)
Technology	The interrelationship between emerging technologies (e.g. blockchain and IoT) and CE. In addition, a study shows that Artificial Intelligence (AI)'s deep learning facility automates the CE design process (Elghaish et al., 2022; Wilson et al., 2021)	The majority of the literature provided conceptual solutions (Elghaish et al., 2022). Hence, there is an emphasis on the need for validated case studies to provide more workable solutions in the technology field despite the recent growth in the literature in the field of CE (Elghaish et al., 2022)

(continued)

Table 2 (continued)

Major issue	*Contribution of research*	*Identified research gaps*
Environment impact	Ability to extend the product life, efficiency and environmental burdens during use, and point of obsolesces are the key determinants of the environmental impact of circular products (Loon et al., 2021). For instance, the product life extension of smartphones is beneficial from the environmental (Kwak, 2016) and economic perspective	There are limited studies focused on assessing the environmental performance of circular products (Loon et al., 2021). Since that, the researchers need to study the overall environmental impact of circular products from the design stage to the remanufacturing process of the circular product (Loon et al., 2021)

3.2 Circular Economy Practice Thematic Gaps

CE practices are growing fast, specifically with digital technology support (Jabbour et al., 2018, Rosa et al., 2019). Digital technologies transform business models into intelligent systems to invent smart products and support businesses in the transition to circularity (Anaruma et al., 2021). The author adopted the previous studies of Weng et al. (2015) and Blomsma and Brennan (2017) to divide this CE practical evidence into four phases. Table 3 presents the development pathway of the CE practices.

Still, the interest of researchers and practitioners weakens due to the difficulties of translating research findings into practice, which leads to creating gaps. A holistic, practical approach is essential in the post-COVID-19 pandemic period since the transition happens in a complex process. All societal subsystems need to change rather than solely depend on the economic system. However, Ghisellini et al. (2016) emphasized transforming existing linear business models is challenging due to many businesses, including partners in the supply chain, evolving with the linear business model. Some key challenges are rigid organization structure, high start-up costs, complex supply chains, challenging business-business corporations, lack of information on product design and production, lack of technology, lack of technical skills, quality compromise, and the disassembly of products is time-consuming and expensive (Antikainen et al.,

Table 3 Development pathway to CE practice

Era	*Thematic Areas*	*Practical achievements*
Pre1980 era	Environmental protection awareness	Issue: The organizations were aware of environmental protection during this period and provided end-of-pipe solutions. As a result, the authorities enacted laws for environmental protection. Further, the organizations transformed the value creation from shareholder to stakeholder. Example: Siemens AG (i.e. a multinational conglomerate) recognized the importance of environment-friendly manufacturing and compliance with legal regulations. And they established an environmental protection department and three task forces to be responsible for waste disposal, developing eco-friendly processes, and meeting compliances (Salchow, 2021)
1980s–1990s	Sustainable development	Issue: In this period, organizations were balancing the social, environmental, and economic pressure—thought of clean production and sustainable development plans. Also, governments intervened to promote those practices. Example: 3M integrated sustainability into their innovation pipeline by rolling out the "pollution prevention pays" programme (Whelan & Fink, 2016)
1990s–2010s	Circular economy 1.0 (popularity of CE)	Issue: The organizations explored green policies for their businesses and encouraged and promoted pleasant relationships between people and nature during this period. Servitization took a centre stage in corporate activities and processes. Example: MUD jeans asks the customers to return the jeans if they no longer need that item. The organization reuses those materials to prepare new jeans or vintage pairs (Tse et al., 2016)

(continued)

Table 3 (continued)

Era	*Thematic Areas*	*Practical achievements*
2010s–present	Circular economy 2.0 (implementation of CE)	Issue: In this modern era, organizations focus on CE implementation. Further, they have experience in improving ecology, gaining financial benefits (e.g. increase in revenue and profits), building customer loyalty, attracting employees, aligning with rules and regulations, and CE innovations. Example: Dell partnered with Wistron GreenTech to recycle their plastic waste and reuses it in production (Tse et al., 2016). The following three business strategy perspectives explain key practices in the present context
Three business strategy practices in the Circular Economy 2.0 (most organizations practice a combination of these strategies at present)		
Strategy	**Themes**	**Practice**
Retain Product Ownership (RPO)	Organizations devoted to following sustainable practices and recycling raw materials Changed its business model from selling to leasing The producer was responsible for the products (e.g. maintenance and after-sale) when a customer used them	Some organizations went beyond the concept of selling and changed their business model to meet sustainable goals (e.g. a flooring company that launched a programme to shift the business model to leasing). In addition, Xerox is a good example to show the long-term lease practices of their printers and photocopiers to corporate (Atasu et al., 2021)
Product Life Extension (PLE)	Organizations focused on designing goods to last longer The products had longer lifespans to reduce re-purchases Durability was the competitive advantage of charging a premium price for the products Promote sustainable value proposition of the product	With the emergence of CE practices, many organizations intended to apply this strategy to their businesses. For instance, Patagonia is an outdoor clothing business that encourages customers to repair their clothes (to increase the product's durability and reduce the environmental footprint) free of charge rather than purchase a new one (Atasu et al., 2021)

(continued)

Table 3 (continued)

Era	*Thematic Areas*	*Practical achievements*
Design for Recycling (DFR)	Organizations maximize the recoverability of materials that can be used in new products Partnered with an organization that can recover the materials best (e.g. an expert)	Evidence shows that some organizations redesign products and manufacturing processes to increase recoverability. As an example, Adidas joined Parley for the Oceans (which produces textile threads from plastic waste) to manufacture shoes and apparel and reduce plastic waste in oceans (Atasu et al., 2021)

2018; Jaeger & Upadhyay, 2020) and high start-up cost is a key challenge for many organizations in the short run due to insufficient budget to transform the existing linear model into a CE model (Liu & Bai, 2014).

An international survey revealed more than 86% of global supply chains were impacted during the COVID-19 pandemic (Remko, 2020), and buyers and suppliers had to adjust their practices to ensure the reliability of supply (Sugg, 2022). Because the textbook supply and demand risk scenarios were not applied by the organizations in their industry practice (e.g. 64% of respondents thought the situation would come back to normal) (Remko, 2020). For instance, Nike changed the product line and product mix frequently to meet the demand risk during the COVID-19 pandemic, while the research studies focused on information distribution over the supply chain, improving the IT, and inventory buffering to improve the resilience of the supply chain by managing demand risk (Remko, 2020). Table 4 illustrates emerging thematic practice gaps.

4 Towards a Circular Research-Practice Synergy Process Model

4.1 Key Constructs and Variables

Circular economy research outcomes are far from becoming part of the organizational fabric (Mendoza et al., 2019; Pieroni et al., 2021), with the websites of more than 80 per cent of Fortune 500 companies still addressing circular economy issues (Ritala et al., 2018; Veleva et al., 2017). As the development of the circular economy phenomenon in the

Table 4 Emerging thematic practice gaps

Major CE related themes	*CE-related issues shaping practices*	*Emerging practice gaps*
Circular Business Model	A systematic analysis of life cycle stages and business processes is required to recognize the type of structural waste to develop a suitable business model for CE transition (Blomsma et al., 2019). According to them, operationalization of the CE in practice is lagging (i.e. knowledge and approaches of how to make the transition for CE is emerging), and experiments support the transition to a circular business model (Bocken et al., 2018)	Organizations require to develop an inherent circular business model for their respective organizations since most of the proposed business models are unsuitable for specific business types in the market For instance, Apeel developed a layer of plant-based edible coating to apply on fresh fruits and vegetables to reduce food waste and increase the natural defences of those products. In addition, Tesco follows a "4R strategy" to align with the CE principles in the market context
Technology (e.g. Blockchain)	A digital network that supports distributing the information about the supply chain leads to creating a more circular resource flow, reduction in waste, and data-driven decision-making, which ultimately reach CE (Böckel et al., 2021). The existing blockchain technologies are criticized due to their downsides, but the disadvantages are outperformed by the advantages of blockchain technology (Zheng et al., 2018)	Practitioners need to critically evaluate the downsides of technology (blockchain) to meet the goals of the CE since the downsides of the technologies are not tested For instance, Suez (i.e. a company that manages water and waste) uses blockchain technology to keep records from the moving sludge to agricultural soil at every stage. It helps to make data-driven decisions to manage the materials

Major CE related themes	*CE-related issues shaping practices*	*Emerging practice gaps*
Remanufacturing	The reuse and remanufacturing processes are encouraged by the market for the recovered products, and that kind of market rarely presents in developing countries (Hazen et al., 2016). According to Khan et al. (2022), the organization's culture that values remanufactured products leads to creating demand for remanufactured products	Need to promote remanufactured products in the respective markets to gain a competitive advantage since most economies do not have sufficient demand for those items For instance, Renault extends the vehicle's life of vehicles and components (e.g. gearboxes) through remanufacturing. It supports achieving environmental, societal, and business benefits
Sustainable supply chain	Organizations realize that the sustainability of the business is not limited to their organizations; they expand it to the entire demand and supply chain network (Høgevold & Svensson, 2012)	Corporate leaders and stakeholders in the market need to develop a holistic plan to achieve sustainability in the entire supply chain since individual business performances are not plentiful at present For instance, Swedish Return System supports the food industry through shared reusable pallets. It reduces the single-use packages to decrease the transport cost, waste, and emissions while increasing the ease of handling in the supply chain

(continued)

Table 4 (continued)

Major CE related themes	*CE-related issues shaping practices*	*Emerging practice gaps*
Product Service System (PSS)	Public awareness to overcome the man-made environmental and health issues crisis through waste segregation is insufficient. Therefore, organizations must keep sufficient sustainable product options available in the market to attract people towards sustainable daily uses (Mehrotra & Jaladi, 2022)	Organizations need to design a sufficient number of product service combinations in the product service system to make adequate, sustainable product solutions available since limited solutions create a barrier to the adaptation process of sustainable daily usage in society For instance, thredUP launched a modern resale experience for women's and kids' textiles on the online platform, and buyers can enjoy more than 35,000 brands from them. It helps reduce carbon emissions, water, and waste while positively impacting e-commerce

academic domain becomes more complex than ever, the observed trends in corporate practices look more consistent in terms of policy approaches towards handling social and environmental issues (Gupta et al., 2021; Lahti et al., 2018; Vence & Pereira, 2019). Nevertheless, significant deviations in practice continue to exist (Fonseca et al., 2018; Van Eygen et al., 2018).

The current paper offers a synergy process model based on the intrinsic historical link in the literature between circular economy research and corporate practices. First, drawing from our literature analysis, five cross-cutting themes are generated to illustrate the theoretical interaction section of the synergy process model: research phase, transition, practice phase, circular organizational attitude, and circular economy pandemic response. These determinants are presented in Table 5.

Focusing on the three primary sections of the process model (research, process, and practice), our literature analysis indicates that these identified themes have undergone huge transformations over the past years. In

Table 5 Determinants of circular economy synergy process model

Transition research	• Demand-driven questions • Contextual relevance
Transition process	• Knowledge synthesis and exchange • Adoption • Operationalization
Practice taxonomy	• Practice with strong evidence • Practice with moderate evidence • Practice with a potential level of evidence • No evidence-based practice
Organizational circular attitude	• Obstructive • Defensive • Accommodating • Proactive
Circular pandemic response	• Volunteerism • Transformational • Integrative

terms of operationalizing the circular research-practice synergy process model, our framework identifies two external factors (Circular organizational attitude and Circular Pandemic Response) that affect the link between circular economy research and corporate practice. We argue that understanding the transition and the external environment can guide researchers and practitioners towards achieving high levels of synergy. Figure 1 shows the post-pandemic circular economy synergy process model.

- **Transition research**

Researchers must first understand the context to achieve consistency with research and corporate functional strategies. In particular, researchers must consider the local context of the circular economy implementation system by creating an awareness of the process across key corporate stakeholders (managers, practitioners, and policymakers) at an early stage. Identifying these contextual conditions and integrating them into the transition is critical. Apart from a critical attention to the research context, the phase must be characterized by demand-driven research questions based on the needs identified by relevant corporate stakeholders.

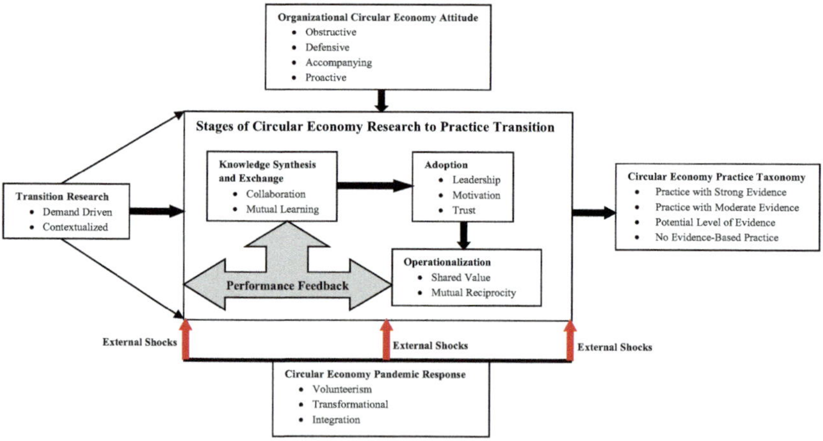

Fig. 1 Post-pandemic circular economy synergy process model

This can lead to implementation research highlighting the central stakeholder engagement paradigm in identifying key issues and formulating appropriate research questions. Furthermore, this phase must enable corporations to construct a consistent strategy with research data and outcomes over time. Therefore, this leads us to the following proposition:

Proposition 1 Transition research characterized by need-driven and contextual relevance is positively related to the performance and success rate of the research-to-practice transition.

When transition research is not demand-driven and lacks contextual relevance, its influence on the research-to-practice transition becomes minimal. Thus, this model asserts that transition research impacts the circular economy research-to-practice transition quality.

- Research-to-practice transition

The process of research-to-practice transition involves a systemic change due to the interplay of various actors at different levels and in other domains in ways that somehow interact and reinforce the system. Situating this within the circular economy context, the synergy process model acknowledges that the research-to-practice transition requires a systemic change largely impacted by institutional and behavioural factors. Applying a transition perspective, this model offers three key stages (***knowledge synthesis and exchange, adoption, and operationalization***) as the fundamental steps for promoting evidence-based practice. We argue that the transition plays a crucial role in instigating, accelerating, and triggering evidence-based practice amongst corporations. We further propose developing effective research-to-practice strategies that acknowledge interagency collaboration is critical for evidence-based practices. Finally, the model argues that weaknesses in the transition may result in practices based on moderate, potential, or no evidence. Therefore, this leads us to the following proposition:

Proposition 2 The quality of the research-to-practice transition is positively related to the level of evidence in practice.

The evidence in practice will be low if the underlying factors for each key aspect of the process—including knowledge synthesis and exchange, adoption, and operationalization—are absent. The model suggests that enabling factors, such as collaboration and mutual learning for effective knowledge synthesis and exchange, can lead to a successful transition and, ultimately, towards a strong evidence in practice.

- Practice taxonomy

To promote evidence-based practices, there has been an increasing emphasis on research focused on applying rigorous, systematic, and objective procedures. Our model indicates that the transition quality significantly impacts the following practice outcomes: *practice with a strong level of evidence, moderate level of evidence, potential level of evidence,* and *no evidence-based practice.* We argue that the quality of the transition is directly proportional to the level of evidence in practice. In other words, the identified constructs in the transition stage of the model offer more insights into evaluating the level of evidence in practice. For example, on the one hand, evidence-based practice is likely to be recorded if all the transition indicators are present in the evaluation. On the other hand, when such indicators fail to manifest, a lower level of evidence is likely to be recorded in practice. This leads us to the following proposition:

Proposition 3 Practice taxonomy outcome directly depends on the quality of the research-to-practice transition.

- Organizational circular attitude

Organizational attitude towards a circular economy can be a facilitating or limiting factor in the process of research-to-practice transition. We argue that organizational attitude towards a circular economy can be a precursor to whether or not to try a new practice and that the affective component of attitudes can impact the decision-making processes. Our model illustrates the role of circular attitudes in the research-to-practice transition. We argue that a circular organizational attitude can increase or decrease the likelihood that the transition can yield strong evidence in corporate practices. As proposed by the synergy process model, organizational attitudes towards circular adoption are influenced by the underlying perspectives, including obstructive, defensive, accommodating, or proactive. In turn, these attitudinal approaches are associated with the transition outcomes. This leads us to the following proposition:

Proposition 4 Organizational circular attitude is positively related to research-to-practice transition outcomes.

- Circular response to pandemic shocks

We propose that a major event like the COVID-19 pandemic would pose external shocks to the research-to-practice transition. Specifically, in its early phase, one of the striking aspects of the pandemic was the widespread shift in how corporations approach community engagement and environmental actions. Our model suggests that the nature of a corporation's pandemic response function influences whether or not the transition will go well. Thus, we propose that the nature of the pandemic response technique (volunteerism, transformational and integrative) has varying effects on the transition towards evidence-based practice. It seemed reasonable to expect that corporations who embarked on a voluntary response would, on average, experience greater COVID-19-related impact on the research-to-practice transition than those that take an integrative approach to handle the pandemic. As a result, corporations taking an integrative pandemic stance are more likely to experience a successful transition resulting in a *strong level of evidence in their practice* than those taking a volunteering stance to the pandemic as part of their social responsibility. This leads us to the following proposition:

Proposition 5 A corporation's pandemic response approach directly relates to circular economy research-to-practice transition outcomes.

4.2 *Application of the Circular Economy Synergy Process Model*

The framework we have developed conceptualizes synergy as the proximal outcome of circular economy research that makes corporate practices effective in tangible and measurable ways. It also identifies intrinsic factors that are likely to have a substantial impact on the ability of circular economy actions to achieve high levels of the outcome. The applications of this practical process model can help circular economy researchers, corporate managers, practitioners, and policymakers address critical research-to-practice gaps related to circular economy design and implementation.

Issue 1: Is circular economy synergy a better option for improving the capacity of researchers and corporate managers' capacity to minimise research-to-practice gaps?

The synergy model we have developed in this work can help academics and corporations involved in circular economy investments by assisting them to determine whether or not their investments in collaboration are warranted. Much of these investments are based on the reasonable but undocumented assumption that a circular economy synergy process model can offset identified research-to-practice gaps. Several reasons have been proposed to explain why it has been difficult to overcome such gaps (Jackson et al., 2014; Kristoffersen et al., 2020; Rahla et al., 2021). However, a fundamental barrier that has not been emphasized is the inability to assess the mechanism that gives circular economy synergy its unique advantage. Our framework provides a basis for measuring such a mechanism via a circular economy synergy process model.

The pilot cognitive testing of instruments based on our framework suggests that circular economy synergy and its determinants can be translated into meaningful questions for both researchers and corporate practitioners. Such instruments will make it possible to measure for the first time the extent to which circular economy synergy overcomes research-to-practice gaps and to directly test the underlying assumptions about its distinguished advantage.

Issue 2: What can be done to realize the full advantage of circular economy research and practice synergy?

The model we developed is useful in determining whether and how the circular economy synergy process model works and in helping corporations engaged in circular economy actions s to maximize their investment returns by taking full advantage of research. The circular economy literature is replete with calls for research, additional conceptualization, and new methodological tools for operationalization (Balanay & Halog, 2019; Campbell-Johnston et al., 2020; Niero & Hauschild, 2017). Of the specific needs cited in the literature, our model can contribute in three ways: (i) a better understanding of the function of circular economy research-practice synergy, (ii) the development of proximal measures that can predict the effectiveness of circular economy research-practice transition, and (iii) an improved approach to managing research-to-practice gaps. Given that the circular economy synergy is central to key themes in research and practice, applying the circular economy synergy process model is pertinent in addressing research-to-practice setbacks, particularly in pandemics.

To summarize, operationalising our circular economy synergy process model clarifies what the research-to-practice transition needs to accomplish for a circular economy initiative to reach its full potential. Such an operationalization and the determinants of circular economy synergy can stimulate constructive thinking about what steps a circular economy research translation can take to make the practice meaningful. Furthermore, going beyond either circular economy research or practice, our model can facilitate the development of diagnostic tools that can help identify the particular weaknesses and strengths of the synergy.

5 Conclusion

The circular economy research and practice synergy process model proposed in this article articulates theoretically the important elements that must be considered to ensure strong, evidence-based practice. Our aim, however, is to expand scholarly and management understanding of how research can complement practice if the transition is effectively managed. Therefore, we offer this model as a starting point for transition studies in the field of the circular economy.

In this article, we have also illustrated how circular economy research and practice have evolved into separate subsystems that operate independently of each other and pursue different agendas. Despite circular economy scholarship undergoing a transition from environmental phenomenon to social responsiveness over the past decades, the gap between circular economy research and practice remains. We thus propose a synergy to take full advantage of an effective research-to-practice transition. In sum, we argue that the circular economy synergy process model can serve as a baseline to accelerate transition studies that, in turn, would allow us to understand emerging issues in the circular economy field. We believe this could be a key resource for both researchers and managers.

References

Anaruma, J. F. P., Oliveira, J. H. C. D., Filho, F. A., Freitas, W. R. D. S., & Teixeira, A. A. (2021). The first two decades of circular economy in the 21st century: A bibliographic review. *Benchmarking: An International Journal*. https://doi.org/10.1108/bij-01-2021-0029

Antikainen, M., Uusitalo, T., & Kivikytö-Reponen, P. (2018). Digitalisation as an enabler of circular economy. *Procedia CIRP, 73*, 45–49. https://doi.org/10.1016/j.procir.2018.04.027

Atasu, A., Dumas, C., & Wassenhove, L. N. V. (2021). The circular business model. *Harvard Business Review*. Retrieved October 10, 2022, from https://hbr.org/2021/07/the-circular-business-model

Avelino, F., & Wittmayer, J. (2016). Shifting power relations in sustainability transitions: A multi-actor perspective. *Journal of Environmental Policy & Planning, 18*(5), 628–649. https://doi.org/10.1080/1523908x.2015.1112259

Balanay, R., & Halog, A. (2019). Tools for circular economy: Review and some potential applications for the Philippine textile industry. *Circular Economy in Textiles and Apparel*, 49–75.

Barreiro-Gen, M., & Lozano, R. (2020). How circular is the circular economy? Analysing the implementation of circular economy in organisations. *Business Strategy and the Environment, 29*(8), 3484–3494.

Blomsma, F., & Brennan, G. (2017). The emergence of circular economy: A new framing around prolonging resource productivity. *Journal of Industrial Ecology, 21*(3), 603–614. https://doi.org/10.1111/jiec.12603

Blomsma, F., Pieroni, M., Kravchenko, M., Pigosso, D. C., Hildenbrand, J., Kristinsdottir, A. R., Kristoffersen, E., Shahbazi, S., Nielsen, K. D., Jönbrink, A. K., Li, J., Wiik, C., & McAloone, T. C. (2019). Developing a circular strategies framework for manufacturing companies to support circular economy-oriented innovation. *Journal of Cleaner Production, 241*, 118271. https://doi.org/10.1016/j.jclepro.2019.118271

Böckel, A., Nuzum, A.-K., & Weissbrod, I. (2021). Blockchain for the circular economy: Analysis of the research-practice gap. *Sustainable Production and Consumption, 25*, 525–539.

Bocken, N., Schuit, C., & Kraaijenhagen, C. (2018). Experimenting with a circular business model: Lessons from eight cases. *Environmental Innovation and Societal Transitions, 28*, 79–95. https://doi.org/10.1016/j.eist.2018.02.001

Bocken, N., Strupeit, L., Whalen, K., & Nußholz, J. (2019). A review and evaluation of circular business model innovation tools. *Sustainability, 11*(8), 2210.

Bocken, N. M., De Pauw, I., Bakker, C., & Van Der Grinten, B. (2016). Product design and business model strategies for a circular economy. *Journal of Industrial and Production Engineering, 33*(5), 308–320.

Bögel, P. M., Pereverza, K., Upham, P., & Kordas, O. (2019). Linking socio-technical transition studies and organisational change management: Steps towards an integrative, multi-scale heuristic. *Journal of Cleaner Production, 232*, 359–368. https://doi.org/10.1016/j.jclepro.2019.05.286

Borrello, M., Caracciolo, F., Lombardi, A., Pascucci, S., & Cembalo, L. (2017). Consumers' perspective on circular economy strategy for reducing food waste. *Sustainability, 9*(1), 141. https://doi.org/10.3390/su9010141

Bradley, R., Jawahir, I., Badurdeen, F., & Rouch, K. (2018). A total life cycle cost model (TLCCM) for the circular economy and its application to post-recovery resource allocation. *Resources, Conservation and Recycling, 135*, 141–149. https://doi.org/10.1016/j.resconrec.2018.01.017

Bressanelli, G., Perona, M., & Saccani, N. (2019). Challenges in supply chain redesign for the circular economy: A literature review and a multiple case study [Article]. *International Journal of Production Research, 57*(23), 7395–7422. https://doi.org/10.1080/00207543.2018.1542176

Campbell-Johnston, K., Vermeulen, W. J., Reike, D., & Brullot, S. (2020). The circular economy and cascading: Towards a framework. *Resources, Conservation & Recycling: X, 7*, 100038.

Chizaryfard, A., Trucco, P., & Nuur, C. (2021). The transformation to a circular economy: Framing an evolutionary view. *Journal of Evolutionary Economics, 31*(2), 475–504.

Ciliberto, C., Szopik-Depczyńska, K., Tarczyńska-Łuniewska, M., Ruggieri, A., & Ioppolo, G. (2021). Enabling the circular economy transition: A sustainable lean manufacturing recipe for industry 4.0. *Business Strategy and the Environment, 30*(7), 3255–3272.

De los Rios, I. C., & Charnley, F. J. (2017). Skills and capabilities for a sustainable and circular economy: The changing role of design. *Journal of Cleaner production, 160*, 109–122.

Elghaish, F., Matarneh, S. T., Edwards, D. J., Pour Rahimian, F., El-Gohary, H., & Ejohwomu, O. (2022). Applications of industry 4.0 digital technologies towards a construction circular economy: Gap analysis and conceptual framework. *Construction Innovation, 22*(3), 647–670. https://doi.org/10.1108/ci-03-2022-0062

EMF. (2015). *Towards a circular economy: Business rationale for an accelerated transition*. Retrieved September 23, 2022, from https://ellenmacarthurfoundation.org/towards-a-circular-economy-business-rationale-for-an-accelerated-transition

Erdiaw-Kwasie, M. O., Abunyewah, M., Yusif, S., & Erdiaw-Kwasie, A. (2023). Does circular economy knowledge matter in sustainable service provision? A moderation analysis. *Journal of Cleaner Production*, 135429.

Erdiaw-Kwasie, M. O., Alam, K., & Shahiduzzaman, M. (2017). Towards understanding stakeholder salience transition and relational approach to 'better' corporate social responsibility: A case for a proposed model in practice. *Journal of Business Ethics, 144*(1), 85–101.

Fazey, I., Schäpke, N., Caniglia, G., Patterson, J. W., Hultman, J., Van Mierlo, B., Säwe, F., Wiek, A., Wittmayer, J., Aldunce, P., Waer, H. A., Battacharya, N., Bradbury, H., Carmen, E., Colvin, J., Cvitanovic, C., D'Souza, M., Gopel, M., Goldstein, B. E., . . . & Wyborn, C. (2018). Ten essentials for action-oriented and second order energy transitions, transformations and climate change research. *Energy Research & Social Science*, *40*, 54–70. https://doi.org/10.1016/j.erss.2017.11.026

Ferasso, M., Beliaeva, T., Kraus, S., Clauss, T., & Ribeiro-Soriano, D. (2020). Circular economy business models: The state of research and avenues ahead. *Business Strategy and the Environment, 29*(8), 3006–3024.

Fonseca, L. M., Domingues, J. P., Pereira, M. T., Martins, F. F., & Zimon, D. (2018). Assessment of circular economy within Portuguese organizations. *Sustainability, 10*(7), 2521.

Franco, M. A. (2017). Circular economy at the micro level: A dynamic view of incumbents' struggles and challenges in the textile industry. *Journal of Cleaner Production, 168*, 833–845. https://doi.org/10.1016/j.jclepro.2017.09.056

Galvão, G. D. A., de Nadae, J., Clemente, D. H., Chinen, G., & de Carvalho, M. M. (2018). Circular economy: Overview of barriers. *Procedia Cirp, 73*, 79–85.

Gatti, L., Seele, P., & Rademacher, L. (2019). Grey zone in—greenwash out. A review of greenwashing research and implications for the voluntary-mandatory transition of CSR. *International Journal of Corporate Social Responsibility, 4*(1). https://doi.org/10.1186/s40991-019-0044-9

Geissdoerfer, M., Morioka, S. N., de Carvalho, M. M., & Evans, S. (2018). Business models and supply chains for the circular economy. *Journal of Cleaner Production, 190*, 712–721.

Geissdoerfer, M., Savaget, P., Bocken, N. M., & Hultink, E. J. (2017). The circular economy—A new sustainability paradigm? *Journal of Cleaner Production, 143*, 757–768. https://doi.org/10.1016/j.jclepro.2016.12.048

Ghisellini, P., Cialani, C., & Ulgiati, S. (2016). A review on circular economy: The expected transition to a balanced interplay of environmental and economic systems. *Journal of Cleaner Production, 114*, 11–32. https://doi.org/10.1016/j.jclepro.2015.09.007

Ghisellini, P., & Ulgiati, S. (2020). Circular economy transition in Italy. Achievements, perspectives and constraints. *Journal of Cleaner production, 243*, 118360.

Gupta, H., Kumar, A., & Wasan, P. (2021). Industry 4.0, cleaner production and circular economy: An integrative framework for evaluating ethical and sustainable business performance of manufacturing organizations. *Journal of Cleaner production, 295*, 126253.

Haas, W., Krausmann, F., Wiedenhofer, D., & Heinz, M. (2015). How circular is the global economy?: An assessment of material flows, waste production, and recycling in the European Union and the world in 2005. *Journal of Industrial Ecology, 19*(5), 765–777. https://doi.org/10.1111/jiec.12244

Hafner, M., Yerushalmi, E., Stepanek, M., Phillips, W. D., Pollard, J., Deshpande, A., Whitmore, M., Millard, F., Subel, S., & Van Stolk, C. (2020). Estimating the global economic benefits of physically active populations over 30 years (2020–2050). *British Journal of Sports Medicine, 54*(24), 1482–1487. https://doi.org/10.1136/bjsports-2020-102590

Hailemariam, A., & Erdiaw-Kwasie, M. O. (2022). Towards a circular economy: Implications for emission reduction and environmental sustainability. *Business Strategy and the Environment*.

Harris, S., Martin, M., & Diener, D. (2021). Circularity for circularity's sake? Scoping review of assessment methods for environmental performance in the circular economy. *Sustainable Production and Consumption, 26*, 172–186. https://doi.org/10.1016/j.spc.2020.09.018

Hawken, P., Lovins, A. B., & Lovins, H. L. (2010). *Natural Capitalism* (2nd ed.). Routledge.

Hazen, B. T., Mollenkopf, D. A., & Wang, Y. (2016). Remanufacturing for the circular economy: An examination of consumer switching behavior. *Business Strategy and the Environment, 26*(4), 451–464. https://doi.org/10.1002/bse.1929

Hobson, K., & Lynch, N. (2016). Diversifying and de-growing the circular economy: Radical social transformation in a resource-scarce world. *Futures, 82*, 15–25.

Høgevold, N. M., & Svensson, G. (2012). A business sustainability model: A European case study. *Journal of Business & Industrial Marketing, 27*(2), 142–151. https://doi.org/10.1108/08858621211197001

Hölscher, K., Wittmayer, J., & Loorbach, D. (2018). Transition versus transformation: What's the difference? *Environmental Innovation and Societal Transitions, 27*, 1–3. https://doi.org/10.1016/j.eist.2017.10.007

Homrich, A. S., Galvao, G., Abadia, L. G., & Carvalho, M. M. (2018). The circular economy umbrella: Trends and gaps on integrating pathways. *Journal of Cleaner Production, 175*, 525–543.

Hossain, M. U., Ng, S. T., Antwi-Afari, P., & Amor, B. (2020). Circular economy and the construction industry: Existing trends, challenges and prospective framework for sustainable construction. *Renewable and Sustainable Energy Reviews, 130*, 109948.

Hussain, M., & Malik, M. (2020). Organizational enablers for circular economy in the context of sustainable supply chain management. *Journal of Cleaner Production, 256*, 120375. https://doi.org/10.1016/j.jclepro.2020.120375

Ibn-Mohammed, T., Mustapha, K., Godsell, J., Adamu, Z., Babatunde, K., Akintade, D., Acquaye, A., Fujii, H., Ndiaye, M., & Yamoah, F. (2021). A critical analysis of the impacts of COVID-19 on the global economy and ecosystems and opportunities for circular economy strategies. *Resources, Conservation and Recycling, 164*, 105169.

Jabbour, A. B. L. D. S., Jabbour, C. J. C., Godinho Filho, M., & Roubaud, D. (2018). Industry 4.0 and the circular economy: a proposed research agenda and original roadmap for sustainable operations. *Annals of Operations Research, 270*(1–2), 273–286. https://doi.org/10.1007/s10479-018-2772-8

Jackson, M., Lederwasch, A., & Giurco, D. (2014). Transitions in theory and practice: Managing metals in the circular economy. *Resources, 3*(3), 516–543.

Jaeger, B., & Upadhyay, A. (2020). Understanding barriers to circular economy: Cases from the manufacturing industry. *Journal of Enterprise Information Management, 33*(4), 729–745. https://doi.org/10.1108/jeim-02-2019-0047

Jonck, A. V., Ribeiro, J. M. P., Silva, S. A. D., Anhalt, T. C., & Guerra, J. B. S. O. D. A. (2018). Circular economy: A review. *Estado, Sociedade E Sustentabilidade: Debates Interdisciplinares X*, 24–38. https://doi.org/10.19177/978-85-8019-207-0.23-37

Joshi, A. D., & Gupta, S. M. (2019). Evaluation of design alternatives of end-of-life products using internet of things. *International Journal of Production Economics, 208*, 281–293. https://doi.org/10.1016/j.ijpe.2018.12.010

Kalmykova, Y., Sadagopan, M., & Rosado, L. (2018). Circular economy–From review of theories and practices to development of implementation tools. *Resources, Conservation and Recycling, 135*, 190–201.

Kazancoglu, Y., Kazancoglu, I., & Sagnak, M. (2018). A new holistic conceptual framework for green supply chain management performance assessment based on circular economy. *Journal of Cleaner Production, 195*, 1282–1299.

Khan, S., Haleem, A., & Khan, M. I. (2022). A grey-based framework for circular supply chain management: A forward step towards sustainability. *Management of Environmental Quality: An International Journal, 33*(6), 1476–1501. https://doi.org/10.1108/meq-11-2021-0265

Kirchherr, J., Reike, D., & Hekkert, M. (2017). Conceptualizing the circular economy: An analysis of 114 definitions. *Resources, Conservation and Recycling, 127*, 221–232. https://doi.org/10.1016/j.resconrec.2017.09.005

Kirchherr, J., Piscicelli, L., Bour, R., Kostense, E., Muller, J., Huibrechtse-Truijens, A., & Hekkert, M. (2018). Barriers to the circular economy: Evidence from the European Union (EU). *Ecological Economics, 150*, 264–272. https://doi.org/10.1016/j.ecolecon.2018.04.028

Kirchherr, J., & Piscicelli, L. (2019). Towards an education for the circular economy (ECE): Five teaching principles and a case study. *Resources, Conservation and Recycling, 150*, 104406. https://doi.org/10.1016/j.resconrec.2019.104406

Köhler, J., Sovacool, B. K., Kern, F., Markard, J., Wieczorek, A. J., Alkemade, F., Avelino, F., Bergek, A., Boons, F., Fünfschilling, L., Hess, D. C., Holtz, G., Hyysalo, S., Jenkins, K., Kivimaa, P., Martiskainen, M., McMeekin, A., Mühlemeier, M. S., Nykvist, B., . . . & Wells, P. E. (2019). An agenda for sustainability transitions research: State of the art and future directions. *Environmental Innovation and Societal Transitions, 31*, 1–32. https://doi.org/10.1016/j.eist.2019.01.004

Korhonen, J., Honkasalo, A., & Seppälä, J. (2018). Circular economy: The concept and its limitations. *Ecological Economics, 143*, 37–46. https://doi.org/10.1016/j.ecolecon.2017.06.041

Kouhizadeh, M., Zhu, Q., & Sarkis, J. (2020). Blockchain and the circular economy: Potential tensions and critical reflections from practice. *Production Planning & Control, 31*(11–12), 950–966.

Kristoffersen, E., Blomsma, F., Mikalef, P., & Li, J. (2020). The smart circular economy: A digital-enabled circular strategies framework for manufacturing companies. *Journal of Business Research, 120*, 241–261.

Kwak, M. (2016). Assessing the greenness of product lifetime extension. *ICIC Express Letters, 7*, 491–496. https://doi.org/10.24507/icicelb.07.02.491.

Lahti, T., Wincent, J., & Parida, V. (2018). A definition and theoretical review of the circular economy, value creation, and sustainable business models: Where are we now and where should research move in the future? *Sustainability, 10*(8), 2799.

Lehner, M., Mont, O., Mariani, G., & Mundaca, L. (2020). Circular economy in home textiles: Motivations of IKEA consumers in Sweden. *Sustainability, 12*(12), 5030.

Lifset, R., & Graedel, T. E. (1997). Industrial ecology: Goals and definitions. *A Handbook of Industrial Ecology, 10*(4337/9781843765479), 00009.

Liu, Y., & Bai, Y. (2014). An exploration of firms' awareness and behavior of developing circular economy: An empirical research in China. *Resources, Conservation and Recycling, 87*, 145–152. https://doi.org/10.1016/j.resconrec.2014.04.002

Loon, P. V., Diener, D., & Harris, S. (2021). Circular products and business models and environmental impact reductions: Current knowledge and knowledge gaps. *Journal of Cleaner Production, 288*, 125627. https://doi.org/10.1016/j.jclepro.2020.125627

Loorbach, D., Frantzeskaki, N., & Avelino, F. (2017). Sustainability transitions research: Transforming science and practice for societal change. *Annual Review of Environment and Resources, 42*(1), 599–626. https://doi.org/10.1146/annurev-environ-102014-021340

Ma, S. H., Wen, Z. G., Chen, J. N., & Wen, Z. C. (2014). Mode of circular economy in China's iron and steel industry: A case study in Wu'an city. *Journal of Cleaner Production, 64*, 505–512. https://doi.org/10.1016/j.jclepro.2013.10.008

MahmoumGonbadi, A., Genovese, A., & Sgalambro, A. (2021). Closed-loop supply chain design for the transition towards a circular economy: A systematic literature review of methods, applications and current gaps. *Journal of Cleaner Production, 323*, 129101.

McCarthy, A., Dellink, R., & Bibas, R. (2018). The macroeconomics of the circular economy transition: A critical review of modelling approaches.

McDonough, W., & Braungart, M. (2003). Towards a sustaining architecture for the 21st century: the promise of cradle-to-cradle design. *Industry and Environment*, 13–16.

Mehrotra, S., & Jaladi, S. R. (2022). How start-ups in emerging economies embrace circular business models and contribute towards a circular economy. *Journal of Entrepreneurship in Emerging Economies*. https://doi.org/10.1108/jeee-10-2021-0410

Mendoza, J. M. F., Gallego-Schmid, A., & Azapagic, A. (2019). Building a business case for implementation of a circular economy in higher education institutions. *Journal of Cleaner Production, 220*, 553–567.

Merli, R., Preziosi, M., & Acampora, A. (2018). How do scholars approach the circular economy? A systematic literature review. *Journal of Cleaner Production, 178*, 703–722. https://doi.org/10.1016/j.jclepro.2017.12.112

Moktadir, M. A., Rahman, T., Rahman, M. H., Ali, S. M., & Paul, S. K. (2018). Drivers to sustainable manufacturing practices and circular economy: A perspective of leather industries in Bangladesh. *Journal of Cleaner Production, 174*, 1366–1380.

Murray, A., Skene, K., & Haynes, K. (2017). The circular economy: An interdisciplinary exploration of the concept and application in a global context. *Journal of Business Ethics, 140*(3), 369–380. https://doi.org/10.1007/s10551-015-2693-2

Niero, M., & Hauschild, M. Z. (2017). Closing the loop for packaging: Finding a framework to operationalize circular economy strategies. *Procedia Cirp, 61*, 685–690.

Nikolaou, I. E., Jones, N., & Stefanakis, A. (2021). Circular economy and sustainability: The past, the present and the future directions. *Circular Economy and Sustainability*. https://doi.org/10.1007/s43615-021-00030-3

Ormazabal, M., Prieto-Sandoval, V., Puga, R., & Jaca, C. (2018). Circular economy in Spanish SMEs: Challenges and opportunities. *Journal of Cleaner Production, 185*, 157–167.

Papachristos, A. V., Brazil, N., & Cheng, T. (2018). Understanding the crime gap: Violence and inequality in an American city. *City & Community, 17*(4), 1051–1074. https://doi.org/10.1111/cico.12348

Paul, T., Islam, N., Mondal, S., & Rakshit, S. (2022). RFID-integrated blockchain-driven circular supply chain management: A system architecture for B2B tea industry. *Industrial Marketing Management, 101*, 238–257. https://doi.org/10.1016/j.indmarman.2021.12.003

Pauli, G. A. (2010). *The blue economy: 10 years, 100 innovations, 100 million jobs*. Paradigm Publications.

Pearce, D. W., & Turner, R. K. (1989). *Economics of natural resources and the environment*. Johns Hopkins University Press.

Pieroni, M. P., McAloone, T. C., & Pigosso, D. C. (2021). Circular economy business model innovation: Sectorial patterns within manufacturing companies. *Journal of Cleaner Production, 286*, 124921.

Preston, F. (2012). *A global redesign? Shaping the circular economy*.

Rahla, K. M., Mateus, R., & Bragança, L. (2021). Implementing circular economy strategies in buildings—From theory to practice. *Applied System Innovation, 4*(2), 26.

Rajput, S., & Singh, S. P. (2019). Industry 4.0—Challenges to implement circular economy. *Benchmarking: An International Journal*.

Reike, D., Vermeulen, W. J., & Witjes, S. (2018). The circular economy: New or Refurbished as CE 3.0?—Exploring controversies in the conceptualization of the circular economy through a focus on history and resource value retention options. *Resources, Conservation and Recycling, 135*, 246–264. https://doi.org/10.1016/j.resconrec.2017.08.027

Remko, V. H. (2020). Research opportunities for a more resilient post-COVID-19 supply chain—closing the gap between research findings and industry practice. *International Journal of Operations & Production Management, 40*(4), 341–355. https://doi.org/10.1108/ijopm-03-2020-0165

Ritala, P., Huotari, P., Bocken, N., Albareda, L., & Puumalainen, K. (2018). Sustainable business model adoption among S&P 500 firms: A longitudinal content analysis study. *Journal of Cleaner Production, 170*, 216–226.

Rizos, V., Behrens, A., Van der Gaast, W., Hofman, E., Ioannou, A., Kafyeke, T., Flamos, A., Rinaldi, R., Papadelis, S., & Hirschnitz-Garbers, M. (2016). Implementation of circular economy business models by small and medium-sized enterprises (SMEs): Barriers and enablers. *Sustainability, 8*(11), 1212.

Rosa, P., Sassanelli, C., Urbinati, A., Chiaroni, D., & Terzi, S. (2019). Assessing relations between circular economy and industry 4.0: A systematic literature review. *International Journal of Production Research*, *58*(6), 1662–1687. https://doi.org/10.1080/00207543.2019.1680896

Salchow, C. (2021). *Environmental protection at Siemens*. Siemens. Retrieved October 23, 2022, from https://www.siemens.com/global/en/company/about/history/stories/environmental-protection-at-siemens.html

Sharma, H. B., Vanapalli, K. R., Samal, B., Cheela, V. S., Dubey, B. K., & Bhattacharya, J. (2021). Circular economy approach in solid waste management system to achieve UN-SDGs: Solutions for post-COVID recovery. *Science of the Total Environment, 800*, 149605.

Shooshtarian, S., Maqsood, T., Caldera, S., & Ryley, T. (2022). Transformation towards a circular economy in the Australian construction and demolition waste management system. *Sustainable Production and Consumption, 30*, 89–106.

Simonis, U. E. (1989). Ecological modernization of industrial society—Three strategic elements. *Economy and Ecology: Towards Sustainable Development, 119–137*,. https://doi.org/10.1007/978-94-015-7831-8_7

Stahel, W. R., & Reday-Mulvey, G. (1981). *Jobs for tomorrow: The potential for substituting manpower for energy* (1st ed.). Vantage Press.

Stahel, W. R. (2016). The circular economy. *Nature, 531*(7595), 435–438. https://doi.org/10.1038/531435a

Su, C., & Urban, F. (2021). Circular economy for clean energy transitions: A new opportunity under the COVID-19 pandemic. *Applied Energy, 289*, 116666.

Sugg, B. (2022). Circular textiles innovation during COVID-19: Not the silver lining some had hoped for. *Journal of Fashion Marketing and Management: An International Journal, 1–16*,. https://doi.org/10.1108/jfmm-07-2021-0180

Tse, T., Esposito, M., & Soufani, K. (2016). How businesses can support a circular economy. *Harvard Business Review*. Retrieved October 23, 2022, from https://hbr.org/2016/02/how-businesses-can-support-a-circular-economy

Tuladhar, A., Iatridis, K., & Dimov, D. (2022). History and evolution of the circular economy and circular economy business models. *Circular Economy and Sustainability, 87–106*,. https://doi.org/10.1016/b978-0-12-819817-9.00031-4

Tura, N., Hanski, J., Ahola, T., Ståhle, M., Piiparinen, S., & Valkokari, P. (2019). Unlocking circular business: A framework of barriers and drivers. *Journal of Cleaner Production, 212*, 90–98. https://doi.org/10.1016/j.jclepro.2018.11.202

Van Eygen, E., Laner, D., & Fellner, J. (2018). Circular economy of plastic packaging: Current practice and perspectives in Austria. *Waste Management, 72*, 55–64.

Velenturf, A. P., & Purnell, P. (2021). Principles for a sustainable circular economy. *Sustainable Production and Consumption, 27*, 1437–1457. https://doi.org/10.1016/j.spc.2021.02.018

Veleva, V., Bodkin, G., & Todorova, S. (2017). The need for better measurement and employee engagement to advance a circular economy: Lessons from Biogen's "zero waste" journey. *Journal of Cleaner Production, 154*, 517–529.

Vence, X., & Pereira, Á. (2019). Eco-innovation and circular business models as drivers for a circular economy. *Contaduría y administración, 64*(SPE1), 0–0.

Vinante, C., Sacco, P., Orzes, G., & Borgianni, Y. (2021). Circular economy metrics: Literature review and company-level classification framework. *Journal of Cleaner Production, 288*, 125090. https://doi.org/10.1016/j.jclepro.2020.125090

Walker, A. M., Opferkuch, K., Roos, E., Raggi, A., Simboli, A., Vermeulen, W. J., Caeiro, S., & Salomone, R. (2022). What is the relation between circular economy and sustainability? Answers from frontrunner companies engaged with circular economy practices. *Circular Economy and Sustainability, 2*(2), 731–758.

Walrave, B., Talmar, M., Podoynitsyna, K., Romme, A., & Verbong, G. (2018). A multi-level perspective on innovation ecosystems for path-breaking innovation. *Technological Forecasting and Social Change, 136*, 103–113. https://doi.org/10.1016/j.techfore.2017.04.011

Wanner, C., Lachin, J. M., Inzucchi, S. E., Fitchett, D., Mattheus, M., George, J. T., Woerle, H., Broedl, U. C., Von Eynatten, M., & Zinman, B. (2018). Empagliflozin and clinical outcomes in patients with type 2 diabetes mellitus, Established cardiovascular disease, and chronic kidney disease. *Circulation, 137*(2), 119–129. https://doi.org/10.1161/circulationaha.117.028268

Weng, X., Dong, Z., Wu, Q., & Qin, Y. (2015). *China's path to a green economy* (No. 978-1-78431-141-4). IIED. Retrieved October 22, 2022, from https://www.iied.org/sites/default/files/pdfs/migrate/16582IIED.pdf

Whalen, K. A., Berlin, C., Ekberg, J., Barletta, I., & Hammersberg, P. (2018). 'All they do is win': Lessons learned from use of a serious game for circular economy education. *Resources, Conservation and Recycling, 135*, 335–345. https://doi.org/10.1016/j.resconrec.2017.06.021

Whelan, T., & Fink, C. (2016). The comprehensive business case for sustainability. *Harvard Business Review*. Retrieved October 23, 2022, from https://hbr.org/2016/10/the-comprehensive-business-case-for-sustainability

Wilson, M., Paschen, J., & Pitt, L. (2021). The circular economy meets artificial intelligence (AI): Understanding the opportunities of AI for reverse logistics.

Management of Environmental Quality: An International Journal, 33(1), 9–25. https://doi.org/10.1108/meq-10-2020-0222

Wu, J. L., & Deng, Y. B. (2012). Construction waste recycling technology management based on circular economic theory. *Applied Mechanics and Materials, 260–261*, 1009–1012. https://doi.org/10.4028/www.scientific.net/amm.260-261.1009

Wuyts, W., Marin, J., Brusselaers, J., & Vrancken, K. (2020). Circular economy as a COVID-19 cure? *Resources, Conservation, and Recycling, 162*, 105016.

Zheng, Z., Xie, S., Dai, H. N., Chen, X., & Wang, H. (2018). Blockchain challenges and opportunities: A survey. *International Journal of Web and Grid Services, 14*(4), 352. https://doi.org/10.1504/ijwgs.2018.095647

Zhijun, F., & Nailing, Y. (2007). Putting a circular economy into practice in China. *Sustainability Science, 2*(1), 95–101.

STEAM Education and Innovation Learning Towards Circular Strategies

Naiyer Fatema Khanom

1 Introduction

The sustainability of life around the globe is in a severe crisis due to climate change. In the last two decades, the causes of 90% of major weather-related disasters, such as heatwaves, floods, storms, and cyclones, have been attributed to changes in the climate (Centre for Research on the Epidemiology of Disasters, 2015). These extreme weather events and related challenges of climate change disrupt economies and livelihoods and increase inequality in every corner of the world. According to recent Swiss Re Institute research, the world economy could lose 10% of its total economic value due to climate change by 2050 (Swiss Re Institute, 2021). Developing countries are unduly vulnerable to increasing pollution, health risks, poverty, and unplanned urbanisation stemming

N. F. Khanom (✉)
The University of Queensland, Brisbane, Australia
e-mail: n.khanom@uq.edu.au

Founder Institute Adelaide Chapter, Adelaide, Australia

Wisecar Private Ltd., Adelaide, SA, Australia

M. O. Erdiaw-Kwasie and G. M. M. Alam (eds.), *Circular Economy Strategies and the UN Sustainable Development Goals*,
Sustainable Development Goals Series,
https://doi.org/10.1007/978-981-99-3083-8_4

from climate change. Research from Stanford University shows that the economic inequality between developed and developing countries due to climate change has increased by 25% since 1960 (Garthwaith, 2019). Additional 100 million people could push below the poverty line by 2030 because of the climate change effect estimated by the World Bank (Acheampong et al., 2021; Jafino et al., 2020). An analysis published by PARIS (AFP) in 2019 climate change could directly cost the world economy US$7.9 trillion by the mid-century as increased drought, flooding, and crop failures hamper growth and threaten infrastructure ("Climate impact", 2019).

The Economist Intelligence Unit's (EIU) Climate Change Resilience Index, which measures the preparedness of the world's 82 largest economies, found that Africa, as a region, is less resilient to climate change, with potentially (4.7%) of its GDP expected to be lost because of climate change by 2050. This percentage of predicted loss, revealed in Fig. 1, is followed by estimates for Latin America (3.8%), the Middle East (3.7%), Eastern Europe (3%), and the Asia–Pacific region (2.6%). The most resilient areas proved to be North America, with a 1.1% loss in GDP due to climate change and Western Europe, losing 1.7%. These regions are likely to see the most negligible economic impact because both are relatively affluent and better prepared to tackle climate change from an institutional standpoint (EIU, 2019).

Being less affluent, countries and emerging market economies focus on fast growth and development to generate income, create employment, and reduce poverty. Of the development stage, most countries rely on the linear (take-make-waste) economy and typically use their available natural resources for production and create waste without any systematic and innovative plan to utilise it for reproduction. This model tends to increase climate emissions, environmental degradation, and deprivation.

The recent crises associated with the global COVID-19 pandemic prompted businesses to make dramatic changes and implement survival strategies that have disrupted the global economy. Gupta (2020) has called COVID-19 one of public health history most serious threats to humanity. The World Health Organization (WHO) estimates that 14.9 million additional deaths were associated with the COVID-19 pandemic in 2020 and 2021 (WHO, 2022). The International Labour Organisation (ILO) of the United Nations estimates confirmed that, in 2020, global labour markets hit historic lows, with 8.8% of global working hours lost relative to the fourth quarter of 2019. This figure is equivalent to 255

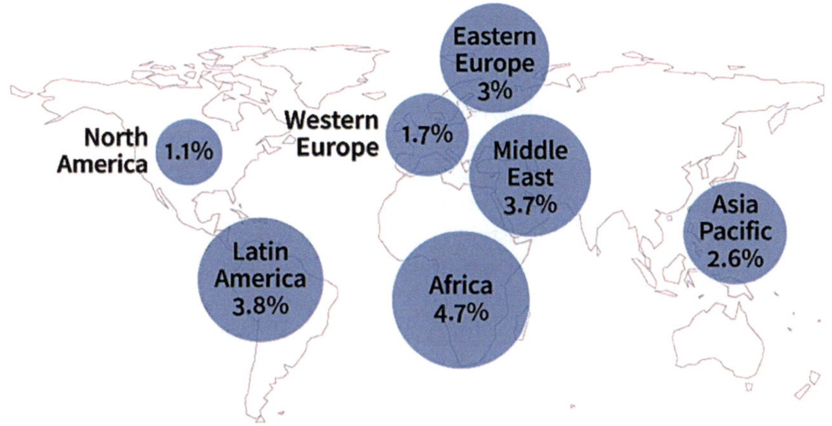

Fig. 1 World map showing the predicted average real GDP loss by 2050 by region (*Note* Adapted from Galey [2019])

million full-time jobs (assuming a 48-hour working week). Working-hour losses were around four times greater in 2020 than during the 2009 global financial crisis (ILO, 2021). These data show how seriously the pandemic affected lives and livelihoods. To protect human capital during crises, whether climate or health-related, all countries need to invest in developing sustainable economies and in making their health systems more resilient. Even though scientists have warned us about the risks of climate change and pandemics for a long time, the world has proven to be miserably unprepared for these crises. Ulrich Volz, the director of the Centre for Sustainable Finance at SOAS University of London, noted this lack of preparation concerning climate change (Edwards, 2020). Now, faced with the consequences of our failure to make the necessary provisions, we must take up the challenge to build back better to support a sustainable, resilient recovery from these climate and health shocks (OECD, 2020).

According to David Malpass, president of the World Bank Group, 'The world economy is simultaneously facing COVID-19, inflation, and policy uncertainty, with government spending and monetary policies in uncharted territory. Rising inequality and security challenges are particularly harmful to developing countries' (The World Bank, 2022, para. 3). The COVID-19 pandemic has not only had a significant impact on the lives of people and the economies of developing countries, but it has also presented them with a complex challenge: to save lives and avoid damage to livelihoods (Loayza, 2020). The economies of many developing countries rely heavily on export and remittance, which have seriously affected the prolonged economic slowdown in these places. Limited fiscal space and poor governance place the large informal sectors of developing countries in a vulnerable situation (Loayza, 2020). Countries whose economies depend on the tourism industry have also suffered severely.

COVID-19 has also had a variable effect on the education systems of countries, largely dependent on their level of income (Wajdi et al., 2020), which influences the infrastructure, health system, and policy and economic situations in a given country, all of which have a role to play to help handle a crisis. In developing countries, in particular, the disruption of children's education is likely to have an extended impact on youth skills development and, in turn, the human capital development required for sustainable economic growth. Added to which are the severe weather-related events associated with climate change—extreme floods, forest fires, hurricanes, heat waves, droughts, and storms—making it extremely challenging for emerging markets and developing economies (EMDEs) to survive and thrive in the global economy.

It is time to redesign economic approaches to growth and development to tackle the climate change emergency. Organisations need to take action with respect to the processes by which they design products and services and make decisions to follow the core principles of the circular economy: to reduce waste and pollution, to reuse raw materials at the highest possible value, and to regenerate the natural environment (Lewandowski, 2016). A circular economy is a transformational approach based on the principle that everything is connected as a resource for something else. It emphasises the development of closed-loop systems and involves taking actions that support reducing consumption, refurbishing used items, recycling waste to resources, and producing and promoting environment-friendly products to promote green industry (Bakker et al., 2010; Merli et al., 2018). Most developing countries produce a massive amount of waste, in part, because of the density of their populations. In addition, many developing countries also suffer from energy supply challenges. Applying circular economy principles enables the convergence of these two problems to create enormous potential: waste in these countries could be converted into energy (Siddique et al., 2021). According to World Bank data, the world is estimated to generate 2.01 billion tonnes of municipal solid waste annually, expected to increase to 3.40 billion tonnes by 2050 (The World Bank, n.d.). Arguably, if we do not think strategically to convert this extensive waste into a resource, the rubbish burden will be a disaster for our livelihoods and the environment.

Transforming a business to operate in a circular loop requires big ideas and execution plans that emerge from critical thinking, creativity, innovation, circular strategy, circular design, decision-making, and leadership (Ellen MacArthur Foundation, 2020). Knowledge, skills, and proper education are essential contributors to the human capital of the future workforce that will be responsible for developing circular strategies and effectively implementing them. People with essential skills and competencies are the key to transforming traditional linear economic activity into a circular economy for sustainable development (Circle Chizaryfard et al., 2021; Economy, 2008; Sustainability University, 2020). Potentially, developing countries have market opportunities in such an economy.

United Nations data indicates that young people aged between 15 and 24, numbering 1.21 billion, comprise 15.5% of the global population. Their projections suggest this youth cohort will reach 1.29 billion and 15.1% of the world's total population by 2030. Their data also shows the population growth rate is expected to remain high in the least developed 47 countries of the UN (United Nations, 2019). Developing countries can take advantage of their growing populations of young people to upskill them into competent workers capable of transforming their linear economies into circular ones. Achieving this, however, will be a considerable challenge for many of these countries because of poor policies, the lack of infrastructure, and inadequate innovation capacity and technologies (Gedam et al., 2021). These circumstances leave developing countries not to unable to adapt to the industry 4.0 technological advancement that can support the transformation.

Interdisciplinary STEAM (science, technology, engineering, arts, and mathematics) education and innovation learning promote the active use of critical thinking, creativity, collaboration, and communication to search for facts, find better ways to solve problems, create something more significant in a team, and efficiently convey ideas to fuel innovation. STEAM education develops the four Cs—critical thinking, creativity, collaboration, and communication—twenty-first-century skills essential for today's youth to gain employment in the future workplace (Stauffer, 2021). In a circular economy, human labour is viewed as different from other renewable resources, as it is creative, versatile, and adaptable, able to be educated and upskilled but destructible if unused (Stahel, 2012). The essential elements for future circular jobs are upskilling, ensuring quality work, and strengthening inclusive opportunities (Wenzel, 2020). Integrating STEAM education and innovative learning into schools can create an ecosystem in which a skilled circular workforce can be nurtured to transform businesses successfully so they can participate in circular economic activity.

The circular economy can increase employment opportunities, sustainable economic growth, and prosperity (Scheel et al., 2020). According to ILO estimates, a global net total of seven and eight million new jobs will be created by 2030 in the shift towards the circular economy (ILO, 2019). I would argue that developing countries that focus on creating new jobs in green and circular industries can generate economic growth and reduce poverty and inequality. By its design, a circular economy takes action on climate change, builds a resilient economy, and ensures social well-being by applying its circular system to utilising resources, reducing waste and pollution, and regenerating nature. Implementing circular strategies can change a business ecosystem and promote the implementation of green economic activities to make the growth of developing countries more sustainable. It is time to prepare youth for the next big market: the circular economy, which they can lead by creating innovative products and services.

This chapter seeks to build a strong theoretical and conceptual link between innovation learning and circular outcomes by reviewing and analysing available data, documents, journals, and reports. This qualitative study intends to understand the combination of STEAM education and innovation learning that is most likely to create an ecosystem to support the circular strategies that can be used to design sustainable industries for the future. This chapter also explores how STEAM education and innovation learning can raise awareness of the circular economy and inspire young people to take responsibility for tackling climate issues. The research findings reveal that STEAM education and innovation learning can help promote responsible enterprises towards circular strategies in developing countries.

2 STEAM Education Leads to Innovative Thinking

Innovation is a process of thinking differently to identify an idea and turn it into a better solution. It has the artistry to add value to humanity without destroying the ecosystem. The inventor Thomas Edison encouraged innovation by saying: 'There's a way to do it better. Find it' (New York Times, 1957). The word 'innovation' comes from the Latin *innovare* for renewing, whose root is *novus*, or new (Psichologyanswers, 2019). Innovative thinking is the skill of developing new ideas or thinking with a different approach to developing new products, services, or processes. While innovation initiatives are about creating something new, not all new things necessarily have value. Innovating for a circular economy focuses on improving or replacing products or services and on processes that create value for customers and generate business growth (Springman, 2011). Collins (2018) explained that innovation is achieved by seeking out inspiration, combining similar ideas, and solving a problem. Curiosity and inspiration drive the creativity needed to see problems in a new way and, in turn, to build new products or develop better services. We need information and knowledge to validate the ideas to understand a problem and generate ideas to fix it. Spradlin (2012) contended that innovative thinking is often engaged when one attempts to solve a problem with limited resources. The first step of this journey involves identifying the basic need that is the core of the problem. The next steps are to justify the need, contextualise the situation with available information, and establish the problem statement that can inform the creative approach to a solution.

Given that innovative thinking is a skill that involves engaging creative thought processes to develop ideas and solutions through critical thinking, problem-solving, collaboration, communication, and creativity skills, STEAM (science, technology, engineering, arts, and mathematics) education is a valuable tool to sharpen this skill. STEAM education is an approach that supports learners in developing the curiosity, creativity, and critical thinking they need to see problems holistically and solve complex problems with better ideas and strategies. STEAM education is a developing model that relates the science, technology, engineering, arts, and mathematics subjects to integrative curricula and reality (Yakman, 2008). Another popular educational approach, 'STEM', incorporates science, technology, engineering, and maths. 'STEAM' adds 'A' for arts, including humanities, languages, arts, design, and philosophy, combined to develop critical thinking skills to solve complex real-time problems. Critical thinking activates analysis of the available facts, evidence, observations, and research associated with problems and solutions to facilitate deep understanding. It can be said that thinking critically is a skill, a sum of intellectual aspects, which allows the educator to successfully carry out educational, social, or technical activities to find a better strategy to solve problems and challenges (Dumitru, 2019). STEM teaches technical skills in science, technology, engineering, and mathematics: the basic knowledge to understand the facts and cases of problems. STEAM goes a step further, developing soft skills and social skills through the study of the arts and humanities; these skills are arguably the most able to support collaborative problem-solving.

Gupta and Tiwari (2020) contend that STEAM education inspires students to solve problems creatively because of the innovative mindset nurtured by the interdisciplinary study approach. ‘Study the science of art. Study the art of science. Develop your senses—especially learn how to see. Realise that everything connects to everything else’. This famous statement from the notebooks of Leonardo da Vinci (1452–1519), as highlighted by Payes (2018), clearly explains the value of combining art and science learning to master the ability to think creatively and design innovative solutions. Leonardo da Vinci was an artist, scientist, and inventor in the Renaissance period (Museum of Science, 1996). He is an idol of creativity and innovation, admired for his mastery in STEAM fields and his design, documentation, and innovation skillset. Interdisciplinary education has positively affected students’ innovative ability and creativity to solve real-time problems (Weilin et al., 2018). Curiosity, passion, enthusiasm, and diversion are all essential artistry for innovation and designing new things. Interdisciplinary STEAM education promotes innovation skills, as described above and entrepreneurial skills, such as identifying opportunities (Indeed, 2021).

At a time when work types and professions are rapidly changing alongside technological advancement, the skills required for the future workplace are creativity, critical thinking, collaboration, information literacy, technology literacy, flexibility, initiative, productivity, and social skills (Stauffer, 2022). Innovation skills are more important than ever, and every organisation needs efficient employees with strong innovation skills (Eich, 2022). Businesses are urged to create an innovation culture that supports idea generation, analytical thinking, and collaborative action planning, where everyone can learn and contribute through knowledge and experience sharing. Arguably, the education system must also be redesigned to focus on teaching innovation to prepare today's youth as tomorrow's human capital, workers who can drive change through innovative thinking. One remarkable example of the capacity of young innovators is found in the inspiring real-life story of William Kamkwamba, who had an innovative idea to solve a problem in his community. At age 14, he built an electricity-producing windmill from spare parts and scraps to help the community in his village, Wimbe, Republic of Malawi, in Southeastern Africa. He achieved this by working from rough plans found in a local library book. Since then, he has been working to make a difference in his community and has built a solar-powered water pump that supplies drinking water in his town—a first (TED, n.d.). Knowledge-based learning, social awareness to understand the requirements, creativity, and innovation can inspire the solutions for real complex challenges, like building a windmill from scrap and limited resources. Potentially, developing countries can prepare their youth as assets with STEAM education and innovation learning to focus on various educational and training programmes in circular strategy and business transformation to create local green industry ecosystems for sustainable growth.

3 Innovation Learning Design to Embed Circular Strategy

Innovation has the incredible power to change our thinking and solve problems in the best possible ways. It may also affect the ecology and evolution of a species, as it can shape the economy, society, human behaviour, and the environment (Logan & Pepper, 2007). In addition, it has the driving force to change business strategy and policy, create employment, and facilitate adaptation to global change. To think innovatively to find out different ideas, think out of the box, and be creative (Luenendonk, 2019). Innovation learning is a process that starts with curiosity. To create an innovative learning environment, students must be encouraged to think deeply, observe, communicate, collaborate, ask questions, and think critically. Adding to the four Cs mentioned in the introduction, Fadel (2008) identified essential twenty-first-century skills for youths, including social skills, problem-solving, innovation literacy, technology and digital skills, civic and entrepreneurial literacy, global awareness, and innovation skills. Young people are arguably the most important asset to a nation as they grow up to become the human capital that builds the future of that country. Burunciuc (2018) argues that young people deserve a quality education that unlocks their potential and creativity.

Innovation learning is a continuous process that pursues the development of better products, services, strategies, and policies for sustainable growth. Learning is a cognitive process; psychologist Bandura (1977) shows this in his Social Learning Theory. It takes place in a social context and can happen through observation or instruction and by making decisions about the performance of observational learning. Cognition, environment, and behaviour mutually influence each other to drive creative thinking for innovation (Teach Thought University, 2022). Social learning theory is the core of the innovation learning process through observation, social context, and information analysis. This method is a brilliant approach to transitioning to a circular economy, which aims to find better solutions for sustainable development. To tackle the climate change challenges, in particular, the circular economy offers a holistic approach to sustainable development (Anderson, 2019), one which is clearly gaining momentum, as leaders of organisations shared their opinions on adopting and promoting circular strategies at each GreenBiz Group event throughout 2019.

At its core, the circular economy is where waste becomes material for production, minimises pollution, products and services are shared, and nature regenerates in a loop. Implementing this process requires an innovation ecosystem in which products and services can be designed to satisfy customers, save the environment, and reduce inequality. In Albert Bandura's Social Learning Theory (1977), the four main elements of attention, retention, reproduction, and motivation (Fig. 2a) may work strategically with the innovation process to achieve circular outcomes (Fig. 2b). The combined processes start with observation and then take a deep dive to get available information and acquire knowledge to understand the facts of the problem and fuel creative thinking. Amabile (1988) noted that the most crucial element of social learning theory in a business context is motivation to activate the innovation process for better solutions, such as those necessary to transform into a circular economy.

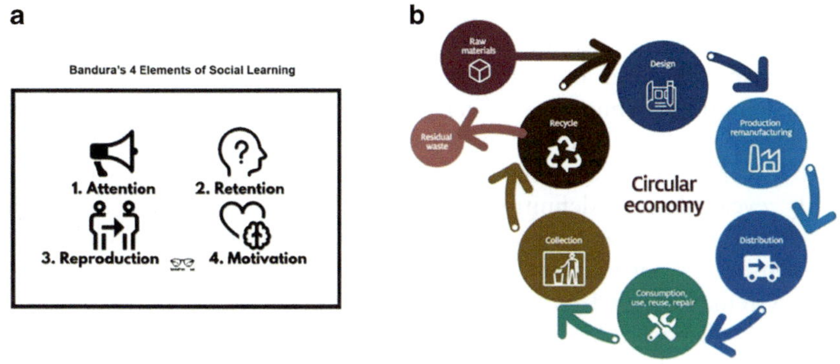

Fig. 2 a. Bandura's four elements of social learning theory. **b**. Process of a circular economy (*Note* An innovation ecosystem with the circular economy strategy to design products and services. **a**, adapted from Loveless [2022]; **b**, adapted from Australian Government [2021])

4 The Circular Economy Challenges to Change the Way of Thinking

Kenneth Ewart Boulding first mentioned the circular economy idea in his book, in which he explained in 1966 that 'we should be in a "cyclical"' system of production. In 1990, the environmental economists Pearce and Turner described the circular economy in their book *Economics of Natural Resources and the Environment* ('Circular economy', 2021; Ekins et al., 2020; Pearce & Turner, 1990). A circular economy is an innovative approach to dealing with the complex problem of waste and protecting the environment. It promotes concepts of sharing, reusing, and utilising limited resources in a system in which everything is connected, from production to consumption, in a closed loop. Transforming to a circular economy can change production and consumption strategies to save the environment and continue economic growth for sustainable development. As a result, the concept is getting attention from the public and private sectors, policymakers, and multinational companies seeking to achieve sustainable development goals effectively (EC, 2015; Lacy et al., 2014; Schroeder et al., 2019).

Since the emergence of the idea, the circular economy has meant many different things to different people. Findings of an analysis by

Kirchherr et al. (2017) of 114 definitions of the circular economy indicate that models of such an economy frequently represent a combination of reducing, reusing, and recycling activities. However, they noted that most definitions, while they reflect the linkage of this economy with the concept of sustainable development, often fail to feature that the circular economy requires a systemic change. Analysis of the findings of their study reveals that the definitions rarely include consideration of the social impact of the circular economy on equality, novel business models, and the consumers' perspectives that could result from a change in the system to a systematic closed loop. The traditional linear economy follows the take-make-dispose rules, in which raw materials are transferred into products for use, and the used products are dumped as waste. In contrast, the circular economy challenges the traditional linear economy to think in innovative ways to use waste as a resource for production and to utilise all resources circularly. The circular economy takes a systematic approach to waste, which can reshape business models, change consumer behaviour, and shift strategy and policy through innovation and creativity.

More than just a concept, the circular economy is a forward-thinking systems solution framework that aims to tackle global challenges like climate change, waste, pollution, and biodiversity loss. Systems Theory and Innovation System process are arguably the best strategic approaches to achieve the goal of transforming into a circular economy. Systems theory is an interdisciplinary theory that can be used to explain complex natural or human-made systems found in nature, society, and science and can be applied to an organisation or culture or to any high-tech industry, global policy, or informational artefact on an organism (Environment & Ecology, n.d.). This theory acknowledges that the economy, society, and environment all work interconnectedly in systems in which all components are interdependent and responsible for the consequences of the action. According to Donella Meadows (2008), an expert systems thinker, 'a system is an interconnected set of elements that is coherently organised in a way that achieves something' (p. 11). 'Systems thinking' is, essentially, the process of viewing problems holistically to understand their deeper roots and identify innovative and better solutions. System thinking also helps us understand how things influence one another as part of a whole. In the circular economy, systems thinking plays a double role: it provides the lens or frame for the conceptual understanding of a problem and is an enabling tool to help identify root causes to inform better solutions (Ellen MacArthur Foundation, 2017a, 2017b, 2017c).

Systems thinking helps support the creation of innovative products, services, processes, and policies by examining complex circular activities that influence one another within a whole and identifying patterns and consequences over time (Rogers, 2022). However, as previously noted, the transformation to a circular economy will require every organisation to develop an innovation culture and ecosystem in which people can learn, grow their mindset, and be ready to change existing linear operating models. To stimulate creative ideas and design processes to get the desired outcomes in both the short and long run—without adverse effects on the economy and climate—organisations can engage innovation systems characterised by modern technological advancement and information availability ('Innovation system', 2022). Innovation and technological developments in today's market are fundamental to action planning for the circular economy (Ellen MacArthur Foundation, 2021). Organisations can practice four innovation systems: incremental, disruptive, sustaining, and radical. These innovation processes can be used to improve existing products and services, create new business models to disrupt the current market, or transform industries and create an entirely new market (Alam et al., 2022; Erdiaw-Kwasie et al., 2022; Ottinger, 2021).

In addition to systems thinking and innovation, transforming from a linear to a circular economy requires strategic planning, which is informed by design thinking, a non-linear interactive human-centred design approach to problem-solving that can lead to innovation, which, in turn, can enable differentiation and competitive advantage (Dam, n.d.; Gibbons, 2016). Design thinking engages an interactive process that applies cognitive, strategic, and sensible design concepts to redefine problems, challenge assumptions, create innovative solutions, build prototypes, and test for execution. Design thinking uses desirability, viability, and feasibility methodology to test ideas, concepts, and hypotheses to determine whether they are worth pursuing or not (Chasanidou et al., 2015). According to Stanford University's d.school, design thinking is a creative problem-solving methodology. It can be used in teaching practice, or students can use it as a framework for real-world projects (Stanford d.school, 2022). Stanford d.school proposed a five-stage design thinking model that goes through the following steps: empathy, definition, conception, prototype, and testing. Designers focus on the human needs of a system's users and emphasise generating real-life innovative solutions that consider how real users think, feel, and behave. Organisations such as businesses are embracing design thinking to benefit from the innovative solutions it generates, bring new products and services, and create new markets. Non-profits use design thinking to develop solutions to social problems (Brown & Wyatt, 2010). Design thinking can also be applied to policymaking, processes, and distribution for sustainable growth.

Design is at the heart of any process, product, or service, and circular design has a significant role in the circular economy system. (Widmer, 2021). The principles of the circular economy are to design for utilising waste and reduce pollution, products and materials continue in use, and eventually regenerate natural systems (Velenturf & Purnell, 2021). Figures 3a and 3b show the difference between the business operation model followed by a linear economy system and the circular economy system approach.

Circular design which involves four stages—understand, define, make, and release—integrates design into the circular economy system to develop more sustainable products, services that are oriented towards extending the product lifecycle, and supply chain systems that follow the basic circular economy concept: reduce, reuse, and recycle to generate natural resources (The Circular Design Guide, 2018). Figure 4 illustrates how design thinking can maintain this circular loop by following the five-stage design steps proposed by the Stanford d.school: empathy, definition, conception, prototype, and testing. These steps contribute to redesigning more sustainable products, creative processes and policies, providing collaborative services, customer behaviour, and social change. Design thinking and human-centred design approaches are used at each stage in the system to zoom in on the user needs and zoom out to consider systemic implications, oscillating continuously. Emerging technological tools like AI (artificial intelligence), IoT (internet of things), and biomimicry will assist innovation through imagination and design thinking (The Circular Design Guide, 2018). According to the Ellen MacArthur Foundation (2017a, 2017b, 2017c), the circular economy is the next big shift for business, industry, and the global economy: now is the time for innovation to drive the transformation to the circular economy to tackle climate change and sustainable development.

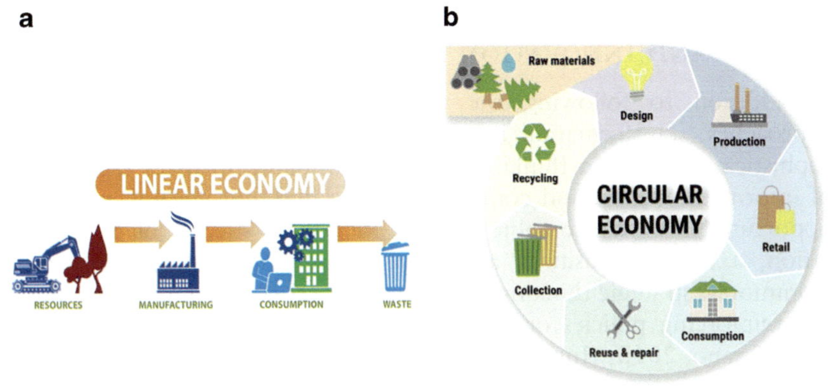

Fig. 3 **a**. Linear economy process. **b**. Business value in a circular economy (*Note* Transforming from a linear to a circular system. There is a difference between the business operation model followed by the linear economy system and the circular economy system approach. **a**. Adapted from Power [2019]; **b**. Adapted from Masterson [2022])

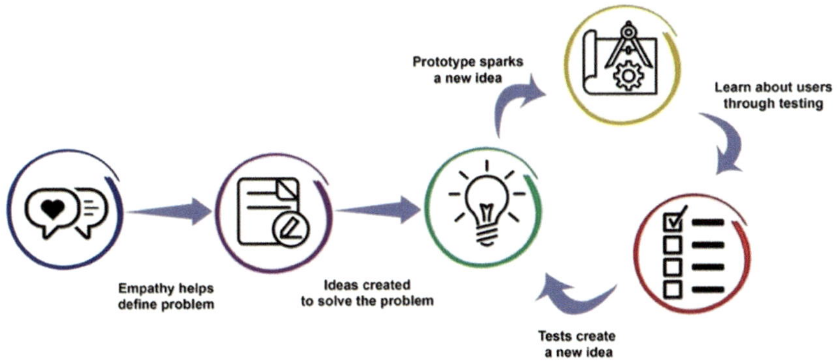

Fig. 4 Stages of design thinking (*Note* Design thinking can work with circular strategy following the Stanford d.school proposed five-stage design steps. Adapted from Interaction Design Foundation [n.d.])

5 Strategic Thinking and Innovation for a Better Future

In addition to following the aforementioned thinking and design approaches, transformation into this new economic model will require sophisticated strategic thinking. Strategic thinking involves the analysis of information, data, and complex issues to find and develop significant opportunities to create value for the long-term success of an organisation, a team, a business, a society, or an individual (Jolley, 2022). Commonly, strategic thinking leads to a clear set of coherent plans, mutually reinforcing policies or behaviours to achieve a specific competitive goal (Pisano, 2015). Innovative ideas are required for all of us to survive and thrive in a competitive, changing environment. According to Henry Mintzberg (1987), a strategy is a plan to take intentional actions to deal with a situation consciously and purposefully. Strategy is also described as the art of planning to achieve long-term goals, particularly when facing complex challenges and working with limited resources. Yuval Atsmon, Senior Partner, McKinsey & Company, shows in a diagram (Fig. 5) the three ways in which companies formulate strategy. In summary, strategy

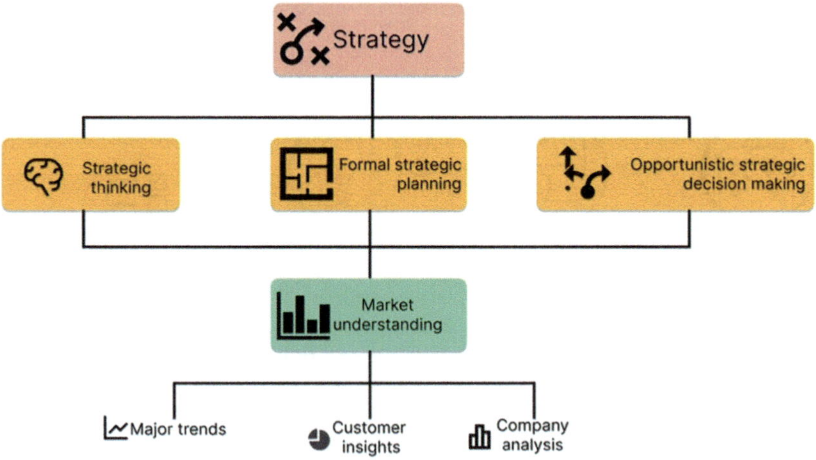

Fig. 5 Three ways in which companies formulate strategy (*Note* Three ways in which companies formulate a strategy. Adapted from Atsmon [2017])

is formulated through strategic thinking, formal strategic planning, and opportunistic strategic decision-making.

The actions and outcomes of each stage are influenced by relevant market factors, available resources, behaviour, and the environment (Atsmon, 2017).

The Greek philosopher Heraclitus (535 BC–475 BC) is quoted as saying, 'change is the only constant' (Foster, 2020). We can predict change but do not know what is coming. Strategy and innovation can shape the future, facilitate adaptation to continual changes, and support sustainable growth to minimise the risk of pull-down. During the 2021 United Nations Climate Change Conference in Glasgow, world leaders recognised that it is vital to transition to renewable energy to tackle climate change (UNECE, 2021). To do this, we need an ecosystem in which all elements of our existing take-make-waste system can be transformed to be reused as resources. As a systems solution framework, the circular economy can tackle climate change and biodiversity loss while addressing important social needs such as creating more green jobs (Ellen MacArthur Foundation, 2017a, 2017b, 2017c).

Less than 10 % of global economic activity today is circular (Houten & Ishii, 2020). This number is too low compared to the growing global linear economic (take-make-waste) activity, particularly in developing countries. Despite this relatively small percentage, the circular economy is now part of international policies, and developing countries are thinking about circular strategies and ways to create more green jobs. For example, the European Union (EU) Green Deal, with its overarching aim of making the EU climate-neutral by 2050, promotes the use of strategies to transform the economies of participating countries into circular ones for growth that is decoupled from resource use and leaves no person behind (European Commission, n.d.). Wellesley (2019) suggests that the large informal business sector in developing countries could take the opportunities presented by the circular economy model to prompt economic diversification, skills development, and value creation; circular activities like electronics waste (electric goods, phone) repairs, sustainable production, and systematic supply chain strategies could be of great value to such countries (Wellesley, 2019).

Strategic thinking and innovation are essential to transform business organisations, social innovation, the education system, and the national policy forum into a circular economy. Transitioning to renewable energy is vital to tackle climate change. Nearly two-thirds of global energy cities demand today, producing up to 80% of greenhouse gas emissions and 50% of global waste (OECD, 2019). There needs to be an immediate transition from fossil fuels to renewables like solar and wind energy for the circular flow of economic activity. The circular economy goal is sustainable production, extending product lifecycles for worthwhile consumption, and sharing the products and services. In this process, waste products become productive assets and resources, and pollution is minimised in a circular loop (UNECE, 2021).

6 Innovation on the Road to Circular Outcomes

'Innovation is the ability to see change as an opportunity—not a threat'. So said Steve Jobs (Jesson, 2020), American entrepreneur, inventor, and co-founder of Apple and other renowned companies who have sparked revolutions in technology and in customers' behavioural patterns and choices. Innovation adds value to every aspect of the economy, from producers to policymakers, services to society, supply chain processes to product distribution, and management systems to human capital resources. The industrial revolution, ushered in by innovators using materials and tools in new ways, drastically changed business models, people's lives and ways of thinking, economies, and the local and global environment. By reducing costs and increasing productivity, the industrial revolution has been described as the most intelligent revolution in human history (National Geographic, n.d.). Industrialisation has served as a key engine of development and structural transformation (Lall, 2000), and, in fact, the process of economic change from economies based on agricultural and handicraft labour and exchange to today's advanced technological era involved four industrial revolutions (Britannica, n.d.). Sadly, we are now paying the high cost of the environmental degradation associated with using the enormous opportunities presented through these revolutions in an unplanned and unorganised way.

However, the current fourth industrial revolution (4IR) era features the rise of intelligent and connected systems designed to sense, predict, and interact in real-time to make better decisions and generate innovative solutions (Sirimanne, 2022). Irwin (2016) refers to the current technological era as the golden age of innovation in which digital transformation creates enormous opportunities to implement creative ideas. Emerging technologies like educational technology, information technology, nanotechnology, biotechnology, robotics, blockchain, machine learning, IoT, and AI can contribute significantly to the design of circular products, components, and materials (McKinsey & Company, 2019). In this information-driven revolution, advanced industry 4.0 has the potential to redesign business models to be more sustainable (Khan et al., 2021), to support the creation of circular products and services (Franconi, n.d.), and to take countries into the circular economy. To shape a sustainable industry that serves green economies and creates more equitable societies, industry 4.0 needs to focus on human capital development. STEAM education, skilled-based training programmes, and climate awareness can promote the circular economy and motivate youth to think creatively to develop innovative products and services.

It is widely believed that innovation and technology can address any crisis and a new challenge, and the global COVID-19 pandemic has proven this to be the case: technology allowed us, for example, to be together even when we were isolated, physically separated from each other. At the height of the pandemic, people lost jobs, health systems broke down, and many jobs were replaced by technology. In the context of what felt like a crazy situation, many industries survived by implementing new ideas driven by systems thinking, innovation, and quick adaptation. The unique conditions of a crisis often allow innovators to think and move freely to create rapid, efficacious transformation (Clark, 2020). COVID-19 has dramatically accelerated the adoption of new technologies in organisations and by individuals for survival, and these technologies will drive change in the long run (McKinsey & Company, 2020). During COVID-19, technological trends have changed, for example, how we do business, work, produce goods, and provide medical services: we work remotely, learn and shop online, visit doctors via telehealth platforms, purchase goods with contactless payment, and even have goods delivered by drones (Xiao & Fan, 2020). Advanced technology and innovation have helped us handle difficult situations and exploit dramatic changes in all sectors of our economies. As a result, society and culture have changed to what is often called 'the new normal'.

The circular economy could and should be part of this new normal, and, like changes instigated by the pandemic, can surprise us by generating more employment opportunities, driving sustainable economic growth, and positively impacting the environment and climate change. The European Commission (EC) estimated that, in 2018, the employment of 3.9 million people was directly associated with circular economy jobs (EC, 2018, as cited in International Institute for Sustainable Development, 2020). Data from the Club of Rome found that full adoption of a circular economy would create more than 500,000 jobs in France, 400,000 in Spain, 200,000 in the Netherlands, 100,000 in Sweden, and 75,000 in Finland by 2030 (Wijkman & Skånberg, 2016, as cited in Effects of the Circular Economy on Jobs, 2020). The ILO projects the net creation of 24 million green jobs opportunity by 2030 (ILO, 2018). The circular economy is not only about taking action for environmental issues. It has a substantial positive impact on business industries, the global economy, and social well-being by creating more green jobs, reducing inequality, reducing emissions, protecting human health and biodiversity, using natural resources like forests, soil, water, air, metals, and minerals in more sustainable ways, and boosting economies by creating more and better jobs (McGinty, 2021). As previously noted, the circular economy business model concepts support reuse, repair, refurbishment, sharing, leasing, and recycling (European Parliament, 2022). In all those actions, there is massive scope for innovation to create jobs, boost economies, reduce poverty and inequality, develop human capital to protect the environment, regenerate natural resources, and tackle the climate emergency.

7 Circular Strategy Towards Sustainable Growth in Developing Countries

Globally, with rapid population growth comes rising rates of waste production. It is estimated in 2020 that 2.24 billion tonnes of solid waste are generated annually, amounting to per person per day and 0.79 kilograms (The World Bank, 2022). Due to rapid population growth and urbanisation, global waste is expected to increase to 3.40 billion tonnes by 2050, and sadly in low-income countries is expected to increase by more than three times (Kaza et al., 2018). Unfortunately, compared to developed nations, developing countries are severely threatened by waste, as over 90% of waste (Kaza et al., 2018) is typically disposed of in unregulated dumps or openly burned. Any increase in waste without a circular strategy to transform it into resources for future production poses a serious threat to climate change and can worsen health, employment, economy, safety, and subsequent natural disasters. Furthermore, the impacts of climate change have proven to be very different in countries with low and high incomes. Poor developing countries are likely to be the most adversely affected; for some countries and communities, climate change could have catastrophic effects (Dervis, 2007).

The key characteristics of developing countries include mass poverty, overpopulation growth, unemployment, poor human capital, consumption pattern, capital formation, greater dependency on agriculture, unplanned industry, poor policies, and technical backwardness (Townes, 2021). These and other significant challenges facing developing countries are interconnected in a cycle that blocks economic development. Despite that, applying innovation and current advanced technology can benefit these countries and create new products and services, well-managed supply chain systems, and policies and strategies to manage waste efficiently. It can also generate new jobs and income for sustainable development. Because developing countries commonly have large populations, which could be harnessed to supply human capital for circular workforces, and a significant amount of waste, which could serve as raw materials for subsequent development, these countries could harness a circular economy for sustainable economic growth. For example, informal recycling sectors could upskill women and educate children with vocational skills—both groups already supplying labour in an informal labour market—to reduce waste and pollution and ultimately mitigate the climate crisis.

It has been argued that, if the world only took one action in response to the climate crisis and created a circular economy in five key sectors—cement, aluminium, steel, food, and plastics—CO2 emissions could be reduced by 3.7 billion tons by 2050, which is equivalent to eliminating the current emissions from all forms of transport (Zaldívar, 2021). There is an immense opportunity to create green jobs in these key five sectors, to reduce and recycle plastics, manage food waste and regenerate the natural environment, reuse steel in the production cycle, and make better use of cement in construction and industry to reduce waste and generate energy. As previously noted, in the transformation from a linear to a circular economy, millions of new jobs will be created in the recycling, repairing, renting, and remanufacturing sectors, promoting sustainable development, and reducing climate risk. In developing countries shifting from the traditional linear economy to the circular economy, the key challenges involve access to innovative business models, products and services, policies, awareness, and implementation strategies. Knowledge and skills can build the human capital into a highly innovative, socially aware, adaptive, creative group capable of turning these challenges into opportunities to tackle the climate crisis and promote circular economic activities to protect life and livelihood.

8 Conclusion

A circular strategy is an innovative approach to business transformation that can create new employment opportunities, utilise natural resources efficiently, change consumer behaviours for sustainable economic growth, reduce social inequality, and tackle the most burning climate change issues (Brown et al., 2021). In this qualitative analysis, I have found a strong link between innovation and the circular economy model that can change global business models and economic systems for collective approaches to a better future. We live in the information age with its advanced technology, and I contend that it is easier to develop innovative products and services to support a circular economy. 'Innovation comes from necessity'; this well-known proverb came from the Socratic dialogue 'Republic', Plato wrote, 'our need will be the real creator' ('Necessity is the mother of invention', 2021). These famous statements make it clear that, in the face of our need to tackle the burning issue of the climate crisis and save biodiversity, we will think creatively to solve complex issues in innovative ways. Making the transformation from linear to circular activities will require us to take holistic approaches to arrive at systematic solutions. From the micro to the macro level, planning, design, and execution are connected and affect one another. In the entire ecosystem of a circular economy, business models, economy, and environment work interdependently as systems that function like the human body (CFI, n.d.), which must work in a balanced way so that the entire system can work adequately.

People are the centre of all of these systems, so we need to focus on human-centred change and management as we transition, taking knowledge-driven approaches and developing skills as an investment in human capital. Moreover, people are the heart of the circular economy, so teaching them about innovation for better design and holistic solutions will help countries to adopt circular business and economic practices. As Henry Mintzberg said in 1975, management is a practice where art, science, and craft meet all (Elaydi, 2015). Investing in innovation learning and STEAM education is critical to developing today's youth as assets of the circular workforce. This chapter's discussions and findings insist that circular strategies, as solutions to today's most pressing problems, come from thinking deeply; STEAM education and innovation learning can prepare young people to promote responsible enterprises towards circular strategies, especially in developing countries where overpopulation and waste can become resources rather than burdens.

The discussion identifies that developing countries face significant challenges of poor policy design, inadequate infrastructure for adapting to the modern technological era, and low-skilled workforces. There is scope for further studies on how developing countries' current education systems can enhance their human capital's innovation skills by providing additional training and advanced learning models to help people address these challenges and pursue circular strategies. In addition, further study can investigate the correlation between STEAM education and decision-making skills development to support the implementation of circular business models. Innovation learning and STEAM education may hold the key to establishing an ecosystem where the transformation into a circular economy can gain momentum.

Acknowledgements I acknowledge the contribution of my research fellow, Dr. Khandaker Jafor Ahmed, who helped to review and edit the chapter and contributed to refining the content through his valuable comments. Also, I acknowledge my technical co-founder Md Zaidul Alam, Director of Wisecar Pty Ltd, who provided his valuable thoughts on advanced technology and the innovation process of designing new products and services.

References

Acheampong, A. O., Erdiaw-Kwasie, M. O., & Abunyewah, M. (2021). Does energy accessibility improve human development? Evidence from energy-poor regions. *Energy Economics, 96*, 105165.

Alam, K., Ali, M. A., Erdiaw-Kwasie, M., Shahiduzzaman, M., Velayutham, E., Murray, P. A., & Wiesner, R. (2022). Impact of ICTs on innovation and performance of firms: Do start-ups, regional proximity and skills matter? *Sustainability, 14*(10), 5801.

Amabile, T. M. (1988). A model of creativity and innovation in organisations. *Research in Organisation Behaviour, 10*(1), 123–167.

Anderson, D. (2019, December 2). *The circular economy can be a holistic approach to addressing climate change*. GreenBiz. https://www.greenbiz.com/article/circular-economy-can-be-holistic-approach-addressing-climate-change

Atsmon, Y. (2017, May 2). *How to unleash your strategic thinking*. McKinsey & Company. https://www.mckinsey.com/business-functions/strategy-and-corporate-finance/our-insights/the-strategy-and-corporate-finance-blog/how-to-unleash-your-strategic-thinking

Australian Government. (2021). *Sustainable procurement guide: A practical guide for commonwealth entities*. Department of Agriculture, Water and the Environment. ISBN 978–1–76003–486–3.

Bakker, C. A., Wever, R., Teoh, C., & De Clercq, S. (2010). Designing cradle-to-cradle products: A reality check. *International Journal of Sustainable Engineering, 3*(1), 2–8.

Britannica. (n.d.). *Industrial revolution*. https://www.britannica.com/event/Industrial-Revolution

Brown, T., & Wyatt, J. (2010). *Design thinking for social innovation*. Stanford social innovation review. https://ssir.org/articles/entry/design_thinking_for_social_innovation

Brown, P., Von Daniels, C., Bocken, N., & Balkenende, A. (2021). A process model for collaboration in circular oriented innovation. *Journal of Cleaner Production, 286*, 125499. https://doi.org/10.1016/j.jclepro.2020.125499

Burunciuc, L. (2018, August 21). *Investing today in human capital for a brighter future tomorrow*. World Bank Blogs. https://blogs.worldbank.org/europeandcentralasia/investing-today-human-capital-brighter-future-tomorrow

Centre for Research on the Epidemiology of Disasters (CRED). (2015). *The human cost of weather related disasters (1995–2015)*.https://www.preventionweb.net/files/46796_cop21weatherdisastersreport2015.pdf

CFI. (n.d.). *Management theories*. https://corporatefinanceinstitute.com/resources/careers/soft-skills/management-theories/

Chasanidou, D., Gasparini, A. A., & Lee, E. (2015). Design thinking methods and tools for innovation. In *Design, user experience, and usability: Design discourse* (pp. 12–23). Springer. https://doi.org/10.1007/978-3-319-20886-2_2

Chizaryfard, A., Trucco, P., & Nuur, C. (2021). The transformation to a circular economy: Framing an evolutionary view. *Journal of Evolutionary Economics, 31*(2), 475–504. https://doi.org/10.1007/s00191-020-00709-0

Circle Economy. (2008). *How HR professionals can play an active role in the circular economy*. https://www.circle-economy.com/resources/how-hr-professionals-can-play-an-active-role-in-the-circular-economy

Circular economy. (2021, July). In *Wikipedia*. Retrieved June 09, 2022, from https://en.wikipedia.org/wiki/Circular_economy#History_and_aims

Clark, L. (2020, March 26). *Innovation in a time of crisis*. Harvard Business Publishing. https://www.harvardbusiness.org/innovation-in-a-time-of-crisis/#:~:text=Problem%20solving%20at%20the%20heart&text=But%20there%20is%20much%20more,to%20create%20rapid%2C%20impactful%20change

Climate impacts 'to cost world US$7.9 trillion' by 2050. (2019, November 20). *The Straits Times*. https://www.straitstimes.com/world/europe/climate-impacts-to-cost-world-us79-trillion-by-2050
Collins, B. (2018, Jun 25). 3 strategies for developing innovative thinking. *Forbes*. https://www.forbes.com/sites/bryancollinseurope/2018/06/25/innovative-thinking/?sh=225ac4b37b88
Dam, R. F. (n.d.). *The 5 stages in the design thinking process*. Interaction Design Foundation. https://www.interaction-design.org/literature/article/5-stages-in-the-design-thinking-process#:~:text=We%20focus%20on%20the%20five,%3A%20research%20your%20users'%20needs
Dervis, K. (2007). *Devastating for the world's poor: Climate change threatens the development gains already achieved*. United Nations. https://www.un.org/en/chronicle/article/devastating-worlds-poor-climate-change-threatens-development-gains-already-achieved
Dumitru, D. E. (2019). *Arts and humanities as a source of critical thinking development*. [Paper presentation]. INTCESS 2019—6th International Conference on Education and Social Sciences, Dubai, U.A.E. https://www.researchgate.net/publication/348754750_Arts_and_Humanities_as_a_Source_of_Critical_Thinking_Development
Economist Intelligence Unit (EIU). (2019, November 20). *Global economy will be 3 percent smaller by 2050 due to lack of climate resilience*. https://www.eiu.com/n/global-economy-will-be-3-percent-smaller-by-2050-due-to-lack-of-climate-resilience/
Edwards. B. (Host). (2020, November 12). *Ulrich Volz: Pandemic is but a prelude to looming climate crisis*. IMF Podcasts. https://www.imf.org/en/News/Podcasts/All-Podcasts/2020/11/12/ulrich-volz
Eich, D. J. (2022, March). *Innovation skills for the future: Insights from research reports*. Innovation Training. https://www.innovationtraining.org/innovation-skills-for-the-future/
Ekins, P., Domenech Aparisi, T., Drummond, P., Bleischwitz, R., Hughes, N., & Lotti, L. (2020). The circular economy: What, why, how and where. *Background paper for an OECD/EC Workshop on 5 July 2019 within the workshop series "Managing environmental and energy transitions for regions and cities", Paris*.
Elaydi, A. (2015, September). Management is a practice where art, science and craft meet (Mintzberg's management roles). *LinkedIn*. https://www.linkedin.com/pulse/management-practice-where-art-science-craft-meet-roles-ahmed-elaydy/
Ellen MacArthur Foundation. (2017a). *Circular design*. https://archive.ellenmacarthurfoundation.org/explore/circular-design
Ellen MacArthur Foundation. (2017b). *The circular economy in detail*. https://archive.ellenmacarthurfoundation.org/explore/the-circular-economy-in-detail

Ellen MacArthur Foundation. (2017c). *System and the circular economy*. https://archive.ellenmacarthurfoundation.org/explore/systems-and-the-circular-economy

Ellen MacArthur Foundation. (2020). *The big idea: Transitioning to a circular economy*. https://ellenmacarthurfoundation.org/videos/the-big-idea-transitioning-to-a-circular-economy

Ellen MacArthur Foundation. (2021). *Innovation deal for a circular economy.* Part of a series of case studies that exemplify elements of the Universal Circular Economy Policy Goals (2021) in practice. https://ellenmacarthurfoundation.org/innovation-deal-for-a-circular-economy

Environment and Ecology. (n.d.). *What is systems theory?* http://environment-ecology.com/general-systems-theory/137-what-is-systems-theory.html

Erdiaw-Kwasie, M. O., Abunyewah, M., Yusif, S., & Erdiaw-Kwasie, A. (2022). Does circular economy knowledge matter in sustainable service provision? A moderation analysis. *Journal of Cleaner Production*, 135429.

European Commission (EC). (2015). *Closing the loop—An EU action plan for the circular economy.* https://eur-lex.europa.eu/resource.html?uri=cellar:8a8ef5e8-99a0-11e5-b3b7-01aa75ed71a1.0012.02/DOC_1&format=PDF

European Parliament. (Updated, 2022, April 26; Created, 2015, December 2). *Circular economy: Definition, importance and benefits.* https://www.europarl.europa.eu/news/en/headlines/economy/20151201STO05603/circular-economy-definition-importance-and-benefits#:~:text=The%20circular%20economy%20is%20a,reducing%20waste%20to%20a%20minimum

European Commission. (n.d.). *Impact of shift to circular economy.* Knowledge for policy. https://knowledge4policy.ec.europa.eu/foresight/topic/changing-nature-work/impact-shift-circular-economy_en#:~:text=ILO%20estimated%20a%20global%20net,leading%20to%20positive%20environmental%20footprints

Fadel, C. (2008, May). *21st century skills: How can you prepare students for the new global economy?* https://www.oecd.org/site/educeri21st/40756908.pdf

Foster, D. T. (2020, 9 April). *Change is the only constant*. The Faculty of Medical Leadership and Management (FMLM). https://www.fmlm.ac.uk/news-opinion/change-is-the-only-constant#:~:text=Heraclitus%2C%20a%20Greek%20philosopher%2C%20is,recall%20what%20'normal'%20was

Franconi, A. (n.d.). *Circular industry 4.0: Redesigning the future.* Circular Economy Alliance. https://circulareconomyalliance.com/product/circular-industry-4-0-redesigning-the-future/

Galey, P. (2019, November 20). Climate impacts 'to cost world $7.9 trillion' by 2050. *Phys.* https://phys.org/news/2019-11-climate-impacts-world-trillion.html

Garthwaith, J. (2019, April 22). Climate change has worsened global economic inequality, Stanford study shows. *Stanford News*. https://news.stanford.edu/2019/04/22/climate-change-worsened-global-economic-inequality/

Gedam, V. V., Raut, R. D., de Sousa, A. B. L., Tanksale, A. N., & Narkhede, B. E. (2021). Circular economy practices in a developing economy: Barriers to be defeated. *Journal of Cleaner Production, 311*, 127670.

Gibbons, S. (2016). *Design thinking 101.* Nielsen Norman Group. https://www.nngroup.com/articles/design-thinking/

Gupta, S. D. (2020). Coronavirus pandemic: A serious threat to humanity. *Journal of Health Management, 22*(1), 1–2. https://doi.org/10.1177/0972063420921260

Gupta, A., & Tiwari, R. (2020). STEAM education can lead to innovation and creativity. *Hindustan Times.* https://www.hindustantimes.com/education/steam-education-can-lead-to-innovation-and-creativity/story-GcBF8wnGaqOpUNwhIj78yO.html

Houten, F. V., & Ishii, N. (2020, January 22). *The world needs a circular economy. Help us make it happen.* World Economic Forum (WEF). https://www.weforum.org/agenda/2020/01/the-world-needs-a-circular-economy-lets-make-it-happen/

International Labour Organization (ILO). (2021, January 25). *ILO monitor: COVID-19 and the world of work* (7th ed.). https://www.ilo.org/global/topics/coronavirus/impacts-and-responses/WCMS_767028/lang--en/index.htm

Indeed. (2021, May 26). *Innovation skills: Definition and examples.* https://www.indeed.com/career-advice/career-development/innovation-skills

Innovation System. (2022, June 13). In *Wikipedia*. Retrieved June 10, 2022, from https://en.wikipedia.org/wiki/Innovation_system

Interaction Design Foundation. (n. d.). Design thinking. https://www.interaction-design.org/literature/topics/design-thinking

International Labour Organization (ILO). (2018, May 14). *24 million jobs to open up in the green economy.* Newsroom. https://www.ilo.org/global/about-the-ilo/newsroom/news/WCMS_628644/lang--en/index.htm

International Labour Organization (ILO). (2019). *Skills for a greener future. Key findings.* ISBN 978-92-2-133559-7. https://www.ilo.org/wcmsp5/groups/public/---ed_emp/---ifp_skills/documents/publication/wcms_709121.pdf

International Institute for Sustainable Development. (2020). *Effects of the circular economy on jobs.* https://www.iisd.org/system/files/2020-12/circular-economy-jobs.pdf

Irwin, N. (2016, May 13). What was the greatest era for innovation? A brief guided tour. *The New York Times.* https://www.nytimes.com/2016/05/15/upshot/what-was-the-greatest-era-for-american-innovation-a-brief-guided-tour.html

Jafino, B. A., Walsh, B., Rozenberg, J., & Hallegatte, S. (2020). *Revised estimates of the impact of climate change on extreme poverty by 2030.* Policy

Research Working Paper (9417). World Bank Group. https://openknowledge.worldbank.org/bitstream/handle/10986/34555/Revised-Estimates-of-the-Impact-of-Climate-Change-on-Extreme-Poverty-by-2030.pdf

Jesson, G. (2020, October 21). *What does the future of innovation look like and who will be designing the products of the future?* PLUS X. https://plusx.space/innovation-is-the-ability-to-see-change-as-an-opportunity-not-a-threat-steve-jobs/

Jolley, C. (2022). *What is 'strategic thinking'?. Effective governance*. https://www.effectivegovernance.com.au/page/knowledge-centre/news-articles/what-is-strategic-thinking

Kaza, S., Yao, L., Bhada-Tata, B., & Woerden, F. V. (2018). *What a waste 2.0. A global snapshot of solid waste management to 2050*. The World Bank. https://openknowledge.worldbank.org/handle/10986/30317

Khan, I. S., Ahmad, M. O., & Majavaa, J. (2021). Industry 4.0 and sustainable development: A systematic mapping of triple bottom line, circular economy and sustainable business models perspectives. *Journal of Cleaner Production, 297*, 126655. https://doi.org/10.1016/j.jclepro.2021.126655

Kirchherr, J., Reike, D., & Hekkert, M. (2017). Conceptualising the circular economy: An analysis of 114 definitions. *Resources, Conservation and Recycling, 127*, 221–232.

Lacy, P., Keeble, J., McNamara, R., Rutqvist, J., Haglund, T., Cui, M., Cooper, A., Pettersson, C., Kevin, E., & Buddemeier, P. (2014). *Circular advantage: Innovative business models and technologies to create value in a world without limits to growth*. Accenture, p. 24.

Lall, S. (2000). The technological structure and performance of developing country manufactured exports, 1985–98. *Oxford Development Studies, 28*(3), 337–369. https://doi.org/10.1080/713688318

Lewandowski, M. (2016). Designing the Business models for circular economy—Towards the conceptual framework. *Sustainable, 8*(1), 43. https://doi.org/10.3390/su8010043

Loayza, N. V. (2020, May 15). *Costs and trade-offs in the fight against the COVID-19 pandemic: A developing country perspective*. World Bank Research and Policy Briefs (148535).

Logan, C. J., & Pepper, J. W. (2007). Social learning is central to innovation, in primates and beyond. *Behavioral and Brain Sciences, 30*(04), 416–417.

Loveless, B. (2022, April 12). *Bandura's social learning theory in education*. Education corner. https://www.educationcorner.com/social-learning-theory/

Luenendonk, M. (2019, September 19). *Creativity boost—How to really think out-of-the-box*. Cleverism. https://www.cleverism.com/creativity-boost-how-to-think-out-of-the-box/

Masterson, V. (2022, October, 27). *What is the circular economy and why does it matter?* World Economic Forum. https://www.weforum.org/agenda/2022/06/what-is-the-circular-economy/#:~:text=Victoria%20Masterson&text=In%20a%20circular%20economy%2C%20products,with%20the%20World%20Economic%20Forum

McGinty, D. (2021, February 3). *5 opportunities of a circular economy.* World Resources Institute. https://www.wri.org/insights/5-opportunities-circular-economy

McKinsey & Company. (2019, January 23). *Artificial intelligence and the circular economy: AI as a tool to accelerate the transition.* McKinsey Sustainability. https://www.mckinsey.com/business-functions/sustainability/our-insights/artificial-intelligence-and-the-circular-economy-ai-as-a-tool-to-accelerate-the-transition

McKinsey & Company. (2020, October 5). *How COVID-19 has pushed companies over the technology tipping point—and transformed business forever.* https://www.mckinsey.com/business-functions/strategy-and-corporate-finance/our-insights/how-covid-19-has-pushed-companies-over-the-technology-tipping-point-and-transformed-business-forever

Meadows, D. H. (2008). *Thinking in systems: A primer.* Chelsea Green Publishing. http://radicalteacher.com/uploads/MeadowsIntroOne.pdf

Merli, R., Preziosi, M., & Acampora, A. (2018). How do scholars approach the circular economy? A systematic literature review. *Journal of Cleaner Production, 178*, 703–722.

Mintzberg, H. (1987, October 1). The strategy concept I: Five Ps for strategy. *California Management Review, 30*(1), 11–24. https://doi.org/10.2307/41165263

Museum of Science. (1996). *The renaissance man.* https://www.mos.org/leonardo/

National Geographic. (n.d.). *Industrial revolution and technology.* https://education.nationalgeographic.org/resource/industrial-revolution-and-technology

Necessity is the mother of invention, 2021. (2021, September). In *Wikipedia*. Retrieved December 8, 2022, from https://en.wikipedia.org/wiki/Necessity_is_the_mother_of_invention

New York Times (Published). (1957). Quote investigator. https://quoteinvestigator.com/tag/thomas-edison/

OECD. (2019). *The circular economy in cities and regions* [Brochure]. https://www.oecd.org/cfe/regionaldevelopment/Circular-economy-brochure.pdf

OECD. (2020). *Building back better: A sustainable, resilient recovery after COVID-19.* OECD policy responses to coronavirus (COVID-19). https://www.oecd.org/coronavirus/policy-responses/building-back-better-a-sustainable-resilient-recovery-after-covid-19-52b869f5/

Ottinger, R. (2021, April 2). *Create sustainable success with the 4 types of innovation*. Fresh consulting. https://www.freshconsulting.com/insights/blog/the-4-types-of-innovation/

Payes, R. (2018, February 20). What Leonardo da Vinci can tell us about STEM vs STEAM. *Innovation & Today*. https://innotechtoday.com/leonardo-da-vinci-stem-vs-steam/

Pearce, D. W., & Turner, R. K. (1990). *Economics of natural resources and the environment*. Johns Hopkins University Press.

Pisano, G. P. (2015, June). *You need an innovation strategy*. Harvard Business Review. https://hbr.org/2015/06/you-need-an-innovation-strategy

Power. M. (2019, August 09). *Business value in a circular economy*. Locus Research. https://www.locusresearch.com/think/blog/2019/08/business-value-circular-economy

Psichologyanswers. (Original published 2019). *What is the word innovation?* https://psichologyanswers.com/library/lecture/read/69653-what-is-the-word-innovation#2

Rogers, K. (2022, August 28). *Why is systems thinking important to innovation?* bigThinking. https://bigthinking.io/why-is-systems-thinking-important-to-innovation/

Scheel, C., Aguiñaga, E., & Bello, B. (2020). Decoupling economic development from the consumption of finite resources using circular economy. A model for developing countries. *Sustainability, 12*(4), 1291. https://doi.org/10.3390/su12041291

Schroeder, P., Anggraeni, K., & Weber, U. (2019). The relevance of circular economy practices to the sustainable development goals. *Journal of Industrial Ecology, 23*(1), 77–95.

Siddique, A., Nakseemok, P., & Liu, L. (2021). Youth leadership in a circular economy: Education enabled by STEAM platform. In *An introduction to circular economy* (pp. 597–622). Springer.

Sirimanne, S. (2022). *What is 'Industry 4.0' and what will it mean for developing countries?*. World Economic Forum (WEF). https://www.weforum.org/agenda/2022/04/what-is-industry-4-0-and-could-developing-countries-get-left-behind/

Spradlin, D. (2012). *Are you solving the right problem?* Harvard Business Review. https://hbr.org/2012/09/are-you-solving-the-right-problem

Springman, J. (2011, September 15). *Drop innovation from your vocabulary*. Harvard Business Review. https://hbr.org/2011/09/drop-innovation-from-your-voca#:~:text=The%20word%20%E2%80%9Cinnovation%E2%80%9D%20comes%20from,intended%20to%20generate%20business%20growth

Stahel, W. R. (2012). *The business angle of a circular economy—Higher competitiveness, higher resource security and material efficiency. EMF volume*. The

Product-Life Institute. EMF volume. http://www.rebelalliance.eu/uploads/9/2/9/2/9292963/stahel_the_business_angle_of_a_circular_economy.pdf
Stanford d.school. (2022). *Get started with design thinking*. https://dschool.stanford.edu/resources/getting-started-with-design-thinking#:~:text=Overview,can%20run%20with%20your%20students
Stauffer, B. (2021, December 31). *What are the 4 C's of 21st century skills?* Applied Education System (AEC). https://www.aeseducation.com/blog/four-cs-21st-century-skills#:~:text=Critical%20thinking%20teaches%20students%20to,how%20to%20efficiently%20convey%20ideas.
Stauffer, B. (2022, January 10). *What are 21st century skills?* Applied Education Systems. https://www.aeseducation.com/blog/what-are-21st-century-skills
Suchek, N., Fernandes, C. I., Kraus, S., Filser, M., & Sjögrén, H. (2021). Innovation and the circular economy: A systematic literature review. *Business Strategy and the Environment, 30*(8), 3686–3702. https://doi.org/10.1002/bse.2834
Sustainability University. (2020). *8 tips for HR to contribute to circular economy*. https://sustainabilityuniversity.org/8-tips-for-hr-to-contribute-to-circular-economy/
Swiss Re Institute. (2021). *The economics of climate change: No action not an option*. Swiss Re Group. https://www.swissre.com/dam/jcr:e73ee7c3-7f83-4c17-a2b8-8ef23a8d3312/swiss-re-institute-expertise-publication-economics-of-climate-change.pdf
Teach Thought University. (2022). Learning theories: Albert Bandura's principles of social learning. https://www.teachthought.com/learning/principles-of-social-learning-theory/
TED. (n.d.). *To power his family's home, young William Kamkwamba built an electricity-producing windmill from spare parts and scrap—Starting him on a journey detailed in the book and film "The boy who harnessed the wind"*. https://www.ted.com/speakers/william_kamkwamba
The Circular Design Guide. (2018). *Now is the most exhilarating time to be an innovator*. https://www.circulardesignguide.com/
The World Bank. (n.d.). Trends in Solid Waste Management. Available at: https://datatopics.worldbank.org/what-a-waste/trends_in_solid_waste_management.htm
The World Bank. (2022, February 11). Solid waste management. https://www.worldbank.org/en/topic/urbandevelopment/brief/solid-waste-management
Townes, L. C., (2021, October 6). *Characteristics of developing countries*. Business Finance Articles. https://businessfinancearticles.org/characteristics-of-developing-countries
UNECE (The United Nations Economic Commission for Europe). (2021, November 05). *Leveraging innovation for the circular economy*. https://unece.org/circular-economy/news/leveraging-innovation-circular-economy

United Nations. (2019). *World population prospects 2019—Volume II: Demographic profiles*. Produced by the Department of Economic and Social Affairs. Population Division. Sales No. E.20.XIII.8. https://population.un.org/wpp/publications/files/wpp2019_highlights.pdf

United Nations. (n.d.). *Devastating for the world's poor: Climate change threatens the development gains already achieved*. https://www.un.org/en/chronicle/article/devastating-worlds-poor-climate-change-threatens-development-gains-already-achieved

Velenturf, A. P. M., & Purnell, P. (2021). Principles for a sustainable circular economy. *Sustainable Production and Consumption, 27*, 1437–1457.

Wajdi, M. B. N., Kuswandi, I., Al Faruq, U., Zulhijra, Z., Khairudin, K., & Khoiriyah, K. (2020). Education policy overcome coronavirus, a study of Indonesians. *Journal of Education and Technology, 3*, 96–106. https://doi.org/10.29062/edu.v3i2.42

Weilin, L., Guocheng, L., Weijian, M., & Jia, L. (2018). The influence of STEAM education on the improvement of students' creative thinking. *Advances in Social Science, Education and Humanities Research*. https://www.atlantis-press.com/proceedings/icadce-18/25900340

Wellesley, L. (2019, May 23). *How the circular economy could help developing countries grow sustainably*. Reuters Events. https://www.reutersevents.com/sustainability/how-circular-economy-could-help-developing-countries-grow-sustainably#:~:text=Developing%20countries%20are%20in%20a,value%20circular%20economy%20supply%20chains

Wenzel, E. (2020). *The 3 essential elements for the circular jobs of the future*. GreenBiz. https://www.greenbiz.com/article/3-essential-elements-circular-jobs-future

WHO. (2022, May 5). 14.9 million excess deaths associated with the COVID-19 pandemic in 2020 and 2021. *News*. https://www.who.int/news/item/05-05-2022-14.9-million-excess-deaths-were-associated-with-the-covid-19-pandemic-in-2020-and-2021

Widmer, S. (2021). *Design and the circular economy*. Ellen MacArthur Foundation. https://ellenmacarthurfoundation.org/articles/design-and-the-circular-economy

Wijkman, A., & Skånberg, K. (2016). *The circular economy and benefits for society: Jobs and climate clear winners in an economy based on renewable energy and resource efficiency*. The Club of Rome. https://clubofrome.org/wp-content/uploads/2020/03/The-Circular-Economy-and-Benefits-for-Society.pdf

Xiao, Y., & Fan, Z. (2020, April 27). *10 technology trends to watch in the COVID-19 pandemic*. World Economic Forum. https://www.weforum.org/agenda/2020/04/10-technology-trends-coronavirus-covid19-pandemic-robotics-telehealth/

Yakman, G. (2008). *STEAM education: An overview of creating a model of integrative education*. Research Gate. https://www.researchgate.net/publication/327351326_STEAM_Education_an_overview_of_creating_a_model_of_integrative_education

Zaldívar M. C. S. (2021, January 29). *3 reasons why embracing the circular economy can be powerful for middle income countries*. World Economic Forum. https://www.weforum.org/agenda/2021/01/why-middle-income-countries-should-embrace-circular-economy/

How Can Ghana Transition from a Linear to a Circular Economy of Waste Management? A Conceptual Analysis of Policy Approaches

Dennis Asare, Mohammed Abubakari, Mudasiru Mahama, Dorothea Afriyie Opoku Mensah, Ivy Agyeiwaa Amponsah, Andrews Darfour Kwakye, and King Carl Tornam Duho

1 Background

The concept of circular economy (CE) has received significant attention in academic literature (Bauwens et al., 2020; Kirchherr et al., 2017; Lowe & Genovese, 2022), policy think tanks (Ellen MacArthur Foundation, 2015), and development co-operations (European Commission, 2015) in recent time because of the sustainability risk the world faces. Over the last two decades, materials exploited to meet consumption

D. Asare (✉)
IMANI Centre for Policy and Education, Accra, Ghana
e-mail: asaredennis6@gmail.com

M. Abubakari
School of Environment and Sustainability, University of Saskatchewan, Saskatoon, Canada

M. O. Erdiaw-Kwasie and G. M. M. Alam (eds.), *Circular Economy Strategies and the UN Sustainable Development Goals*, Sustainable Development Goals Series,
https://doi.org/10.1007/978-981-99-3083-8_5

needs have increased by about 70 percent, less than 10 percent of e-waste per capita are sustainably recycled, five trillion single-use plastics are improperly disposed of each year, and enormous renewable energy resources are untapped (United Nations, 2021). These appalling characteristics of the state of the earth reflect the deteriorating effect of the open-loop systems utilised over a century and the vast threats to nature these systems pose. The risk of the current production and consumption systems is unevenly distributed, with more harm likely to be witnessed in developing economies. For instance, less than 5 percent of Ghana's 22,500 tons of waste per day are recycled (Enterprise Agency, 2019a), compared to over 70 percent of recycling by Italy and Belgium in 2020 (European Commission, 2020). Even though developing countries are the least material consumers with the lowest domestic consumption per capita, they suffer the far-reaching consequences of the linear economy model (take-make-dispose) (United Nations, 2021) (Fig. 1).

Given the apparent urgency for a substantial reduction in waste generated by the current take-produce-dispose business model (Geissdoerfer et al., 2017), the concept of circular economy has emerged as the sustainable approach to addressing the looming threat of unsustainable waste management (Bauwens et al., 2020; Erdiaw-Kwasie et al., 2023; Urbinati et al., 2017).

M. Mahama
Department of Geography and Sustainability Science, University of Energy and Natural Resources, Sunyani, Ghana

D. A. O. Mensah
Department of Agribusiness and Applied Economics, North Dakota State University, Fargo, ND, USA

I. A. Amponsah
Faculty of Economics and Social Sciences, University of Potsdam, Potsdam, Germany

A. D. Kwakye
Cranfield University, Cranfield, UK

K. C. T. Duho
Dataking Research Lab, Dataking Consulting, Accra, Ghana

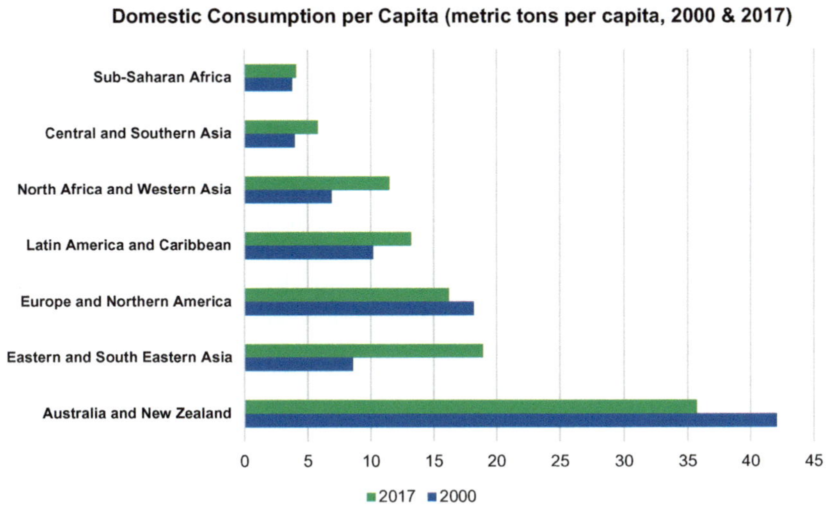

Fig. 1 Domestic consumption per capita (2000 and 2017) (*Source* United Nations [2021])

The CE is defined as an "economic system that is based on business models which replaces the end-of-life concept with reducing, alternatively reusing, recycling, and recovering materials in production or distribution and consumption processes, thus operating at the macro- to micro-level, to accomplish sustainable development, which implies creating environmental quality, economic prosperity, and social equity to benefit present and future generations" (Kirchherr et al., 2017, pp. 224–225). More importantly, the circular economy seeks to transform significantly the way resources are used by replacing linear economy systems with closed-loop approaches that ensure that products are reused, recycled, and remain in the production loop for a longer period, allowing them to create value (Erdiaw-Kwasie et al., 2023; Urbinati et al., 2017) continuously. Essentially, the CE of waste management ensures that waste does not end up at landfill sites but is reused, recycled, and transformed into secondary raw materials, thus reducing over-reliance on primary resources (Lowe & Genovese, 2022).

A transition to CE has the potential to reduce the unsustainable exploitation of finite resources and maximise the utilisation of secondary raw materials, placing CE at the centre of sustainable waste management

(Sharma et al., 2021). The Ellen MacArthur Foundation (2015) estimates that transitioning to a CE could half CO_2 emissions by 2030. Also, the European Commission estimates that full implementation of CE could cut CO_2 emissions by 25 percent (European Commission, 2015). Additionally, the CE could create about 2 million jobs and cut unemployment significantly (European Commission, 2014). In Africa, the CE of waste management could inject about US$8 billion into the economy each year (UN-HABITAT, 2021). Furthermore, sustainable treatment of organic waste, the dominant form of waste in most developing countries, can transform local agriculture, reduce greenhouse gas emissions, and create more jobs along the waste management value chain (Kaza et al., 2018).

However, CE's conceptualisation and operationalisation remain blurred and indeterminate (Geissdoerfer et al., 2017; Kirchherr et al., 2017). As a result, policymakers, analysts, and academics have come up with a myriad of conceptualisations of the concept of CE, allowing multiple interpretations and utilisation (de Jesus & Mendonça, 2018). While the CE can take several forms, the conceptual contentions concerning the "how-to" of the circular economy of waste management affect its implementation at both the national and firm level (Bauwens et al., 2020). For instance, the tendency to focus extensively on one aspect of the waste hierarchy points to the conceptual contentions about the meaning and the focus of the circular economy, as evident in the analysis of 114 definitions of a CE by Kirchherr et al. (2017), where about 79 percent of the definitions of CE focused on recycling, followed by reuse. This can significantly affect the aspect of the circular economy of waste management that receives more policy and investment focus. For example, Enterprise Agency (2019b) identified most of the circular economy of waste management efforts in Ghana are overly concentrated on recycling.

The circular economy concept is nascent in Ghana. As a result, the existing policies, regulations, and institutional frameworks still need to be harmonised and coordinated to achieve the CE of waste management (Diaz, 2017). An abundance of legal frameworks and institutions have been established; however, the focus of the institutions and laws is disconnected from a CE of waste management. Nonetheless, different aspects of the CE of waste management have developed in the formal and informal waste management industry (Enterprise Agency, 2019b). The current CE principles implemented particularly in the informal sector are undertaken under unclean conditions, disconnected from the formal sector, and do not provide decent wages (Diaz, 2017). This underlines the point that transitioning from a linear economy to a CE of waste management must be conceived as a fundamental systemic change rather than twisting some aspect of the waste management value chain to reflect the impact of CE (Kirchherr et al., 2017).

Thus, the policy, regulatory, and private sector incongruence exist because the transition to CE has yet to be understood as a fundamental systemic change of the nature of CE of waste management the country wants to achieve and the policies and business models that can deliver such objectives. Therefore, the paper sets off to understand how the transition to CE of waste management can occur and the policy approaches that can deliver a circular economy of solid waste management in Ghana and developing countries. The study addresses this question by examining how the transition can occur, and different scenarios and their potential to achieve the four main principles of the CE of waste management (Reduce, Reuse, Recycle, and Recover).

2 Methodology

This is a desk study of key literature related to the concept of CE contextualised from the perspective of solid waste management. A catalogue of existing literature on the concept of CE, theories, and conceptual approaches to transitioning to a CE was gathered mainly from Elsevier Scopus, Sage, Emerald, and Google Scholar. Reports by development corporations, policy, and consulting firms such as Ellen MacArthur Foundation, World Bank, United Nations, Africa Union, and Country Reports on Waste Management in Ghana complemented this. Based on a critical literature review, the Deep Transition Framework by Schot and Kanger (2018) was used to conceptually explain how Ghana can transition from a linear to CE of solid waste management. The scenario analysis by (Bauwens et al., 2020) was adopted to explain the practical approaches Gha and other developing countries can adopt to transition to a CE of waste management.

The Deep Transition Framework critically explains the overall fundamental change process from one system to another (Bauwens et al., 2020). The Deep Transition Framework is distinct and relevant primarily because it explains how social and technical systems interact to produce change. Its emphasis on socio-technical systems places technology and institution at the centre of the change process. The Deep Transition Framework helps understand how socio-technical systems emerge, mature, decline, and evolve into other socio-technical systems (Schot & Kanger, 2018). The current linear economy of waste management is the product of a socio-technical system, and thus the Deep Transition is useful to understand the transition to a new socio-technical system underpinned by the circular economy of waste management. Given that the concept of CE in Ghana is nascent and its possible form unknown, the scenario analysis provides insights into the possible form the transition to CE of waste management in Ghana could take and the socio-technical transformation that must occur for the CE of waste management to be attained.

3 State of Waste Management in Ghana and Developing Countries

3.1 *Waste Management Situation in Sub-Saharan Africa*

Africa's population is increasing steadily, and its cities and towns have become more urbanised. With an urban growth rate of about 3.55 percent every year, the urban population in Africa is expected to grow from the present 40 percent to about 56 percent by 2050 (United Nations Environment Programme, 2018). With such rapid population growth and urbanisation, the region's waste generation will increase and become one of the major developmental problems. Thus, a restorative and regenerative system that minimises waste creation and prolongs the life span of products in the economy is needed.

Accurate waste generation data in Africa remains a challenge. However, estimates by the World Bank indicate that the sub-Saharan Africa (SSA) region generates about 174 million tons of waste annually, with an average waste generation per capita of 0.78 kg/day (World Bank, 2018). While this is below the global average of 0.74 kg per capita/day, the spatial variations are significant, with countries like Ghana recording about 0.78 kg per capita/day while Seychelles records over 2 kg per capita/day (United Nations Environment Programme, 2018).

Organic waste forms about 57 percent of total solid waste generated in urban centres due to high fresh food preparation, mainly among low-income households. However, high-income urban centres are likely to see a significant increase in plastic and paper use linked to the growing fast food industry in urban centres (United Nations Environment Programme, 2018; World Bank, 2018). In most African countries, waste collection and transportation are emphasised more than treatment and disposal (United Nations Environment Programme, 2018). That notwithstanding, less than half of the total waste generated is collected, particularly in SSA, where only 44 percent is properly collected (Kaza et al., 2018). This indicates that most waste is generated through unapproved waste collection and treatment mechanisms. As a result, 69 percent of solid waste generated in SSA is openly dumped, 24 percent in landfill sites, and only 7 percent recycled and recovered (World Bank, 2018).

4 Waste Management in Ghana

Similarly, Ghana generates about 8.2 million tons of solid waste annually, using an annual waste generation of 22,500 tons/day of waste and 0.75 kg per capita/day (Enterprise Agency, 2019b). Organic waste constitutes about two-thirds, and about 83 percent of the total waste generated is collected. Like most developing countries, a significant proportion of the waste collected is openly dumped. The waste management industry in Ghana is complex because of the interplay between formal and informal waste management players. In high-income neighbourhoods, waste is collected by formal waste management companies that practice improved hygienic processes in collecting and treating waste. Individual waste collectors using tricycles and three-wheeled motorcycles serve low-income communities with no hygienic handling processes and play a limited role in waste treatment. Communities with poor road networks and slums continue to practice open dumping sites. A complex institutional framework with conflicting functions affects the clarity of policies towards effective waste management (See Table 1). For instance, the work of the Ministry of Environment, Science and Technology and the Ministry of Sanitation and Water Resources conflicts as they are implementing separate policies towards addressing waste management, leading to a thinly spread of resources in addressing the waste management problems. Both agencies charge separate levies (EPA Levy, ECO-Levy, and the Sanitation Levy) to address the same problem.

Formal waste management companies undertake limited waste recycling and recovery mechanisms because they can finance the investments required, leaving the informal waste management section of the market

Table 1 List of institutions in Ghana's Waste Management Sector

Institutions	*Role*
1. Ministry of Sanitation and Water Resources	Responsible for policy planning and programme implementation towards the sustainable use of water resources and waste management. The Ministry would be responsible for developing policies and programmes that provide clarity on the approach towards achieving the circular economy of waste management
2. Ministry of Finance and Economic Planning	Responsible for the passage of waste sector levies and management of Sanitation levies. Additionally, the Ministry of Finance undertakes budget approval and disbursement of funds for the Ministries, Departments, and Agencies responsible for waste management
3. Ministry of Local Government, Decentralisation and Rural Development	Performs supervisory function over all the local government institutions and influences the waste management policies and programmes of the MMDAs
4. Environmental Protection Agency	Responsible for improving and conserving environmental resources and ensuring efficient resource management
5. Ministry of Environment, Science, and Technology	Performs supervisory functions over the Environmental Protection Agency. The Ministry is also responsible for policy and programme implementation concerning the environment, science, technology, and innovation
6. Metropolitan, Municipal, and District Assemblies (MMDAs)	Responsible for waste management at the metropolitan, municipal, and district levels

Source Authors' construct, 2022

largely relying on landfill sites and open dumping. Therefore, a transition towards CE must consider creating a coherent system that integrates the different aspects of the market in the waste management value chain. Over the years, the waste management (formal or informal) industry has been dominated by the private sector, managing about 80 percent of the total waste generated across the 254 Metropolitan, Municipal, and District Assemblies (MMDAs) in Ghana (Enterprise Agency, 2019b). However, despite the poor waste handling, disposal, and treatment in Ghana, several activities that reflect the waste management hierarchy (mainly recycling) and the CE are taking place in the waste management industry. In 2019, Enterprise Agency (2019b) identified 25 established recycling companies with a combined capacity of 320 tons/day engaged in recycling flexible plastics to create dustbins, carrier bags, car mats, and buckets and paid about €1.1milion monthly to plastic waste collectors, indicating the high financial value of the CE if properly implemented.

Between 1993 and 2020, eleven separate legislations were passed in Ghana to address the waste management challenges (See Table 2). However, none of the laws targets the promotion of CE, even though some of the policies explicitly mention aspects of the 4Rs. As a result, institutions have yet to be assigned the role of promoting a circular economy. This points to the need for more policy clarity on a CE, targets, and alternative mechanisms to achieve circular futures in waste management.

Considering that the private sector plays a dominant role in the waste management sector of Ghana, a transition from the current linear waste management mechanism to a circular future would depend heavily on the private sector. However, the complexity of the industry due to the different processes of waste handling and treatment by the informal and formal sectors makes it imperative for government to define the circular targets and technology and create a regulatory environment that integrates both formal and informal sectors, allowing the players to operate synergistically.

5 Theoretical Framework

The CE of waste management is underpinned by several principles demonstrating the transition from linear to a CE. The core principles form the rules that guide the behaviour of actors in implementing policies that promote the CE, as well as measuring the progress countries or firms make towards the circular future. In the context of Ghana, where there are no existing laws guiding the transition from a linear economy, the core principles provide a basis for setting policy targets and measuring the transition. This section critically examines core principles that underlie the circular economy transition, the alternative transition processes, and the theoretical frameworks underpinning policies and business models that make the transition possible.

Table 2 Laws, policies, and regulations

Laws, policies, and regulations	*Purpose*	*Explicit recognition of circular economy*
1. Local Government Act 1993, Act 462	The Act defines local government agencies' role and decentralises waste management functions to the MMDAs	No explicit recognition of CE
2. Environmental Protection Act 1994, Act 490	The Act establishes the Environmental Protection Agency and provides guidelines on permits and levies related to environmental issues	The Act recognises recycling and recovery relevant approaches for waste management; however, it does not explicitly mention CE
3. Customs and Excise (Duties and Other Taxes) Act 2013, Act 863	The Act provides the guidelines for implementing the Environmental Excise Tax on plastics and plastic products (polythene bags and other plastic products)	There is no explicit recognition of CE; however, the law establishes a recycling fund to receive the environmental excise tax. The fund is intended to support plastic waste management firms
4. Ghana Land Fill Guidelines, 2002	The Environmental Protection Agency issues the Ghana Land Fill guidelines to control the activities of landfill sites	No explicit mention of the circular economy
5. National Environmental Sanitation Policy, 2010	The policy includes guidelines for addressing liquid, solid and hazardous waste. It also proposes increased private sector involvement in waste management through contracts, franchises, and other forms of partnerships	There is an explicit mention of the 4Rs (reduce, reuse, recycle, recover) as the alternative means for waste treatment. However, there are no linkages towards the transition to CE

Laws, policies, and regulations	*Purpose*	*Explicit recognition of circular economy*
6. National Environmental Sanitation Strategy and Action Plan, 2010	The policy was an action plan to guide the MMDAs in implementing their waste management strategies. This was also accompanied by the Strategic Environmental Investment Plan of 2010	There is an explicit mention of the 4Rs (reduce, reuse, recycle, recover) as the alternative means for waste treatment. However, there are no linkages towards the transition to CE
7. Renewable Energy Act 2011, Act 832	The Act provides the legal framework for the promotion of renewable energy. It considers issues such as feed-in tariffs for different renewable energy sources, including waste-to-energy	No explicit mention of the circular economy
8. Ghana National Climate Change Policy, 2013	This policy was designed by the Ministry of Environment, Science, Technology, and Innovation to determine effective adaptation and mitigation measures to address the climate crisis. The policy seeks to ensure sustainable landfill site management with proper methane extraction, and transition to more recycling	There is an explicit mention of the 4Rs (reduce, reuse, recycle, recover) as the alternative means for waste treatment. However, there are no linkages towards the transition to CE
9. Oxo-biodegradable Directive, 2015	The policy mandates plastic importers and producers to include an Oxo-biodegradable component to allow decomposition	No explicit mention of CE

(continued)

Table 2 (continued)

Laws, policies, and regulations	*Purpose*	*Explicit recognition of circular economy*
10. Hazardous and Electronic Waste Control Management Act 2016, Act 917	The Act establishes an Eco-Levy on new or used equipment (US$0.15–US$15) per product, depending on the time	The act envisages the establishment of an electronic waste recycling centre/plant. However, it provides no linkages to CE
11. Sanitation and Pollution Levy, 2021	The legislation imposes GHC0.10 (US$0.012) on all petroleum products in Ghana to address the sanitation and pollution challenges in Ghana	No explicit mention of CE

Source Enterprise Agency (2019a)

5.1 Core Principles of Circular Economy of Waste Management

Noting Kirchherr et al. (2017), the plenitude of definitions indicates that the transition from linear to CE can be underpinned by several principles and methods of application (Yuan et al., 2008). The vast principles and methods could also blur the identification of the fundamental principles that form the rules of the CE of waste management (Ghisellini et al., 2016). The core principles of the CE can be viewed from two main perspectives; the R-Framework (Reduce, Reuse, Recycle and Recover) and the Systems Perspective (Kirchherr et al., 2017). The R-Framework explains how the CE occurs and the observable characteristics. Zhu et al. (2010) argue that the R-Framework is the "how-to" of the circular economy and thus qualifies as the core principle of the CE of waste management. The system perspective on CE, on the other hand, argues that the transition from a linear economy to CE requires a fundamental shift from the current system to a new system. The systems perspective views the transition at three levels. The macro-level highlights adjustment in industry operations and the entire structure of the economy, the meso-level focuses on eco-industrial parks, and the micro-perspective looks at the individual firms, products, and materials to enhance the circularity (Heeres et al., 2004; Jackson et al., 2014; Li et al., 2010).

While the systems perspective shows the different levels of firm-level organisations that can deliver the transition to a CE, the system's efficiency can be measured by the extent to which waste is reduced, new resources are developed, and product lifespan is enhanced through recycling and reuse. This makes the R-Framework a good core principle to explain the CE transition. Additionally, the R-Framework is at the heart of the Circular Economy Promotion Law of the Peoples' Republic of China (Beijing Review, 2008), and the EU Waste Framework Directive (European Commission, 2008), indicating that the R-Frameworks have been used in some of the successful circular economy countries and regions. That notwithstanding, several studies have also proposed other forms of Rs to the framework from 4R (European Commission, 2008) 6R (Sihvonen & Ritola, 2015), and 9R (van Buren et al., 2016). A hierarchy beginning with Reduce as the desired outcome is inherent in all the varying forms of R-Framework. For this analysis, the 4R (Reduce, Reuse, Recycle, and Recover) is adopted as the core principle of the circular economy of waste management. Kirchherr et al. (2017) identified over 40 percent of the 114 definitions examined and concluded that the core

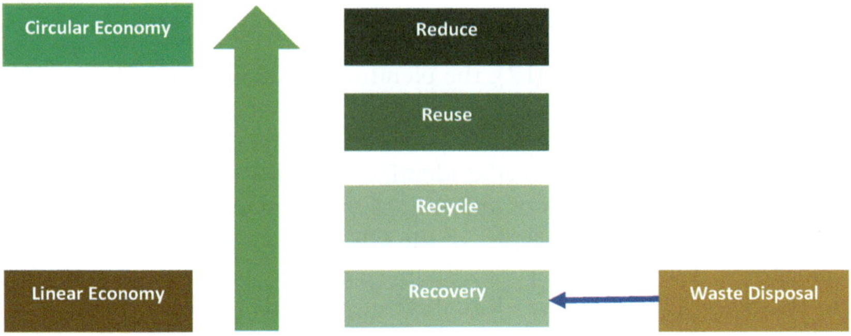

Fig. 2 The 4R-framework (*Source* Adapted from the EU Waste Framework Directive [European Commission, 2008])

principle of the circular economy was the 4R. Additionally, Kern et al. (2020), after examining the second Deep Transition towards the circular economy, concluded that the meta-rules (underlying principles) of the circular economy were the 4R framework (Fig. 2).

Based on the framework, the circular waste management economy is achieved when energy recovery from waste is maximised, products are transformed (high or low quality), product reusability is enhanced, and waste generation is reduced. This framework simplifies policy targets regarding the CE and can potentially guide businesses to identify the aspect of the waste hierarchy where value can be maximised.

5.2 *Theorising the Transition from Linear to Circular Economy of Waste Management*

This section examines the transition process from the current linear system to the circular future and the potential factors influencing the transition. The Multi-level Perspective of the Deep Transition Framework is adopted to understand how the current waste management system is transformed into a circular economy. The Deep Transition Framework developed by Schot and Kanger (2018) explains the change process across multiple systems and the mechanisms through which the systems become connected and coordinated to form a new system. The Deep Transition is a series of connected and sustained fundamental transformations of various socio-technical systems (Schot & Kanger, 2018).

The Deep Transition Framework argues that the transition of a society from one system to the other occurs within a socio-technical system. The socio-technical systems represent the interconnection between technology and institutions to create a new order. The socio-technical system comprises technology, science, regulations, culture, user practice, and production and supply networks (Bauwens et al., 2020; Erdiaw-Kwasie & Basson, 2018). The current waste management system is a configuration of the interaction between the current technology, regulation, and institutions. The socio-technical systems are underpinned by meta-rules (core principles) which drive innovation and evolution of the systems towards a particular direction (Schot & Kanger, 2018). The transition from one current system to another requires that the underlying meta-rules need to change to meet the demands of the current problems. This implies that for the current waste management system to transition to the CE; the current meta-rule (take-produce-dispose) needs to change for the core principles of the circular economy (reduce, reuse, recycle, and recover).

This places technology and institution at the heart of the transition to a CE. However, the institutions (regulations) established over time can either hamper or promote the transition to a circular economy. For instance, the current levy on flexible plastics and Eco-levy charged against the producer in Ghana do not incentivise the producers to explore new technologies that reduce waste and facilitate recycling and the reuse of secondary materials (Bauwens et al., 2020). Additionally, if the requisite technology for transitioning to the CE requires high upfront costs and research and development, it could delay the transition process. Therefore, the transition to CE must be seen in light of how technology and institutions interact effectively to drive the new waste management system.

The Multi-Level Perspective (MLP) of the Deep Transition Framework focuses on explaining the long-term and large-scale transition from one socio-technical system to another. For instance, how society transitioned from a horse-drawn carriage to an automobile and fossil fuel to renewables (Kern et al., 2020). The transition from one socio-technical system to another can be explained using three main components; ***the socio-technical regime, niches, and socio-technical landscape***. The socio-technical regime refers to the shared set of rules or routines directing the behaviour of actors on how to produce, regulate, and use technology in the socio-technical system (Schot & Kanger, 2018).

In the waste management industry, the socio-technical regime is the rules that determine the technology, regulatory framework, and policies that shape the activities linked towards incremental innovation towards a CE. As the set of rules continues aligning, coordinating, and connecting, they manifest themselves as the socio-technical system, depicting the interaction between technology and institutions. The new rules that emerge and gradually become aligned and connected are called the Niche. These can be a set of rules that propose new technology and models to address the existing system's challenges. The new set of rules tends to compete with the existing socio-technical system. In the early stages, actors in a niche or promoting the new rules are few; however, they grow over time to become the dominant rules underpinning a new socio-technical system. The core principles (reduce, reuse, recycle, and recover) and the associated technology of CE is the niche. The landscape refers to the external factors that shape the niche and regime. The external factors could be international conventions, climate crises, environmental degradation, and constraints that exert pressure for change in the socio-technical system. The external environment triggers new rules to address the looming threats the existing system poses. The increasing waste generation, resource inefficiencies, and environmental crisis constitute the social pressures that result in the new rules (reduce, reuse, recycle, and recover) to address these problems.

The MLP proposes that an interaction between the socio-technical regime, niche, and landscape leads to the transition from an existing socio-technical system to another. This implies that a combination of the new rules of CE of principles, niche technologies, and the external pressures for improved solid waste management would produce the transition from the current linear economy to the CE of waste management.

The transition can be explained in three main stages. The first stage, called the start-up phase, is where the landscape pressure or external environment raises more concerns about the internal problems caused by the existing system, creating opportunities for new rules and niche technologies. The second stage, also known as the acceleration stage, is where the niche technologies expand, attract more actors, and become the mainstream market to compete with the incumbent regime and other niches for dominance. For example, the use of recycled plastic bottles by bottling companies begins to compete with the existing system that creates non-recyclable plastic bottles. The last stage is called the stabilisation stage, where the niche technologies and socio-technical regimes have become well established, the number of actors has increased, the technologies have become more mature, and regulations are stable. This allows for the smart adoption of the technology since the institutions have become stable, and the behaviour of actors has become more predictable. New socio-technical systems also provide a blueprint for new entrants.

In the context of waste management, the transition will be driven by unsustainable resource exploitation, climate change, environmental risk, and the high cost of waste generation as the external factors that raise concerns about the sustainability risk of developing countries. At this stage, solid waste generation will not be the only concern, however, the risk excess waste poses to the environment and unsustainable production and consumption. This is similar to how the circular economy of waste management emerged as a critical policy idea in the EU during the 1980s, where rising mountains of waste generated and resource consumption became critical ecological and environmental concerns and brought waste management into the EU policy and legislation (Kern et al., 2020). Considering that over 90 percent of all waste collected in SSA will end up on open dump sites and landfill sites, the outbreak of diseases, recurrent flooding, resource depletion, and climate change have driven the call for more sustainable waste management programmes. As a result, new waste management rules underpinned by the core principles of CE and related technologies have emerged, although uncoordinated and less organised largely because of the complexity of formal and informal waste collection and management in developing countries such as Ghana.

Presently, the alternative waste management technologies that have emerged due to the CE's new rules cannot compete with the current system of open dumping and landfill sites because they are cheaper alternatives but extremely hazardous. The increased health hazards would increase the pressure for more waste reduction, reuse, waste recycling, and energy recovery from waste, paving the way for the CE rule and technologies. For instance, Enterprise Agency (2019a, 2019b) identified 25 formal recycling companies in Ghana involved in the transformation of waste into multiple semi-final and final products. The lack of clear policy direction and legislation promoting a CE contributes to the low incentive for companies to move up the waste management hierarchy towards a CE. However, with the increased interest of academics, civil society organisations, and some quasi-government institutions in the circular economy, the core principles (forming the meta-rules) are likely to be considered in future policy and legislative reviews, leading to the recognition of the CE as the new rule for waste management in Ghana and other SSA countries. As seen in the EU, the work of Potocnik on resource efficiency led to the recognition and consideration of the CE of waste management in 2012 and consequently its consideration in the 2014 Circular Economy Action Plan and the EU Action Plan for Circular Economy in 2015 (European Commission, 2015; Kern et al., 2020).

Over time, the improvement in policy and legal frameworks promoting CE and the economic case of solid waste management in Africa would potentially incentivise investment in the technologies required to climb up the waste hierarchy. For instance, United Nations Environment Programme (2018) estimates that at the base case scenario of 5 million tons of waste recovery; the financial value of recovered resources could be about US$318.6 million. Also, at about 25 percent and 50 percent recovery of the 125 million tons of waste generated in 2012, the resources recovered from moving higher the waste management hierarchy is estimated at US$1.7 billion and US$3.7 billion, respectively (United Nations Environment Programme, 2018). The economic value of recovered resources can boost investment and address the existing financing constraints of the waste management industry in Ghana and SSA. The economic value of recovered resources, combined with the improved technology and regulations and the external pressure to reduce waste, would enable the circular economy to compete with the existing waste management system, leading to the transition to the circular economy of waste management.

The Deep Transition Framework is a process where new rules emerge, compete with the existing system, and become well-coordinated and connected with the institution and technology, transitioning from one socio-technical system to another. The interactions between actors, rules (devised to regulate actors' actions), and the socio-technical system are important for the transition. Based on the MLP of the Deep Transition Framework, the transition from a linear economy to CE of waste management would occur when the existing and new laws, regulations, and policies of waste management are underpinned by the 4R framework (reduce, reuse, recycle, and recover). Ghana is at the start-up phase of the transition because the rising environmental hazards of open dumping and landfill sites coupled with increasing floods caused by indiscriminate dumping of plastics have intensified the call for more robust approaches to addressing waste issues, thus, paving the way for the circular economy.

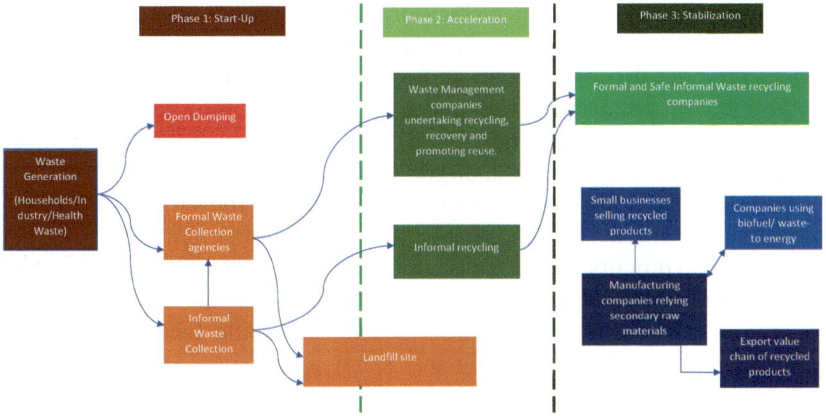

Fig. 3 Conceptual framework of the multi-level deep transition to circular economy of waste management (*Source* Authors' construct, 2022)

Additionally, several aspects of the core principles of circular economy are happening in the waste industry; however, there is no clear regulatory direction to guide the behaviour of actors (Fig. 3).

5.3 *Approaches to Accelerating the Transition to Circular Economy of Waste Management*

Policymakers can design appropriate policy responses to catalyse the transition to CE of waste management when there is a clear understanding of how the future CE would look. Geissdoerfer et al. (2017) argue that most of the existing CE literature overemphasises the impact of CE without critically analysing the form CE future may take. Bauwens et al.'s (2020) scenario analysis of the possible future of CE is adopted and contextualised in Ghana to assess how the CE transition would look like, the implications, and the policy and business solutions relevant to the paradigm shift.

Bauwens et al. (2020) argue that technology and institutions are the core drivers of change in a socio-technical system. Thus, the interaction between technology and institutions can create different environments that influence the upward progress in the waste treatment hierarchy (the core principles of the circular economy). Some aspects of the core principles of the circular waste management require low technology, while others may require high technologies to move up the waste treatment hierarchy. For instance, ***Reduce*** may require low technology and changes in the behaviour of consumers, while ***Recycling*** may require low, high technology, and behavioural changes.

Additionally, governments play critical roles in the circular economy in successful markets such as the EU and China (Bauwens et al., 2020). Governments provide policy clarity and regulations that ensure predictability in the behaviour of actors. However, excessive control of the system can distort the allocation of benefits from the CE. For instance, informal waste collectors in Ghana reach a larger proportion of the population because formal waste management companies serve high and middle income communities. Suppose the circular economy transition is centralised around only formal waste management firms. In that case, it will exclude the larger waste collectors in the system from benefiting from the economic opportunities created in the value chains. As a result, some authors advocate for a decentralised governance structure towards the transition to a CE because it is inclusive and promotes a participatory process in decision-making and policy development.

Bauwens et al. (2020) scenario analysis indicate that the interaction between institutions and technology would produce four (4) scenarios of waste management's circular future: ***Peer-to-Peer Circularity, Planned Circularity, Bottom-up Circularity, and Circular Modernism***. A Peer-to-Peer Circularity requires high-technology innovation and a decentralised governance system. This system relies greatly on collaborative platforms, smart technologies, and blockchain systems to create distributed production to enhance resource efficiency. This approach promotes waste reduction by enabling the sharing of underutilised goods between consumers. Essentially, this approach creates a market or platform for consumers to trade used goods, enhancing reuse and promoting waste management's CE. The planned circularity approach requires low technology innovation and centralised decision-making. Under this system, the circular economy principles (reduce, reuse, recycle, and recovery) are centrally implemented by the government through tight regulations and directives that determine the technology and the processes to achieve the circular economy. For example, the government can ban certain products, such as single-use plastics, and introduce legislation on production and resource extraction by manufacturing companies to reduce waste generation.

The Bottom-up Circularity requires low technology and a decentralised governance system. This approach relies greatly on the behavioural modification of consumers towards waste reduction and a shift from conventional production and consumption patterns. Consumers are incentivised to embrace behaviours that reduce waste by ensuring the reuse of waste items. The circular modernism system relies on high-technology innovation and centralised decision-making. Under this approach, the transition to circular economy decision-making rests with the government and a few large waste management companies with high—technologies targeted at higher R strategies (the core principles of circular economy). The government set the circular economy targets and create the requisite business environment for the high-tech waste generation companies to innovate. For instance, in pursuing high R strategies of the circular economy, the government invests in research and development on the key technologies, which then become the building blocks for the waste management companies to install innovative processes to achieve them (Fig. 4).

Reflecting on Ghana's current waste management situation and Bauwen et al.'s (2020) scenario analysis of the possible CE future, a planned circularity and circular modernism are plausible in Ghana. The central government has direct control and influence over the formulation

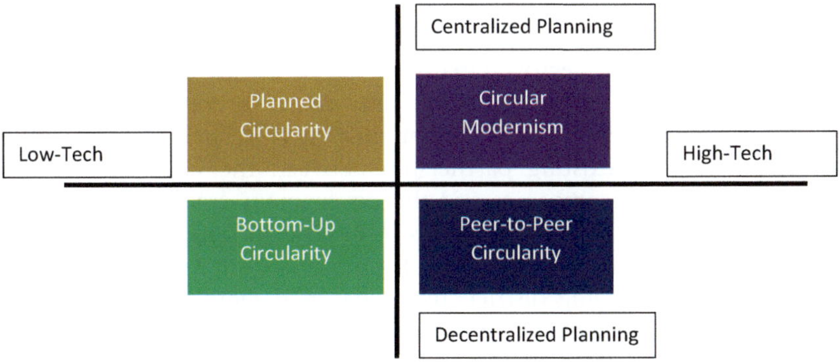

Fig. 4 Approaches/scenarios for transitioning to the circular economy of waste management (*Source* Adapted from Bauwens et al. [2020])

of laws, regulations, and guidelines, which makes a planned circularity probable in Ghana even though the policy regime is less authoritarian due to the collaborative role of decentralised agencies at the local level. Table 2 indicates the influence of the government in the waste management industry through almost a dozen laws and regulations. Even though these laws are not designed with CE in thinking and do not provide clear pathways and guidelines for waste management firms to transition to CE, they demonstrate the power of the government to implement a planned circularity in Ghana. However, the planned circularity will be achievable when the existing and new laws and regulations are underpinned by the principles of CE (such as the 4Rs) as discussed in Sect. 5.2. Additionally, successive governments have consistently attributed unsustainable waste management practices to a lack of finance to implement appropriate modern technologies. Thus, a planned circularity offers the government a low-cost approach towards transitioning to CE of waste management because the government's role under this scenario is to enforce the regulations, monitor and evaluate the implementation of the guidelines towards CE by waste management firms.

The circular modernism scenario of transition to CE of waste management is probable in Ghana because the private sector collects and manages about 80 percent of waste, and the formal sector has adopted modern technologies towards sustainable waste management in Ghana. For example, the Accra Compost and Recycling Plant (ACARP—has a waste recovery rate of 80 percent, processes 600 tons of waste per day, and produces 100 tons of compost daily) and the Kumasi Compost and Recycling Plant (KCARP—can receive about 1200 tons of MSW per day) are private sector initiatives supported by the government. They are evidence that the government creates a conducive environment for the private sector to adopt innovative technologies in waste management. The existing partnership between the large-scale formal waste management companies and the government creates the requisite partnership for the circular modernism scenario in Ghana. Additionally, the high socio-political support for formal waste management companies, including awarding district, municipal, and metropolitan contracts to large waste management companies, can help reduce the social and economical implementation cost, as indicated by Bauwens et al. (2020).

These notwithstanding, the planned circularity and circular modernism can successfully transition towards CE of waste management if it effectively integrates the informal waste management sector with the formal waste management industry. Due to the poor organisation of the informal waste sector, they are often not recognised by the government and less likely to receive support despite their enormous compliment to waste collection and recovery. The government can implement guidelines to begin regulating the activities of informal waste collectors and provide technical assistance and financial aid to organise the informal operators. This would ensure that the activities of informal waste businesses are aligned with the efforts towards CE. In addition, the government can create a regulatory framework that allows informal waste collectors to sell recyclable and recoverable materials to formal waste management companies at an economically fair price. This would prevent the unsafe recycling and recovery undertaken in the informal waste sector. A balance between planned circularity and circular modernism would ensure that government promotes CE by creating an enabling environment for the private sector to innovate and invest in high-technology equipment to achieve a sustainable waste management industry.

Given that, behavioural influences towards waste reduction and collection have not worked effectively and the weak development of collaborator platforms, the peer-to-peer circularity, and Bottom-up Circularity, which are more value-driven, are least probable and preferable. Thus, the preferable transition towards CE of waste management in Ghana would be a cocktail of planned circularity and circular modernism.

6 Policy Discussion

The adoption of the circular economy of waste management practices can have a crosscutting impact on other SDGs such as Climate Action (SDG.13), Zero Hunger (SDG.2), Decent Jobs (SDG.8), and Sustainable Cities (SDG.11). The reduction of food waste is estimated to create an economic opportunity of about US$155–405 billion by 2030 (Ellen MacArthur Foundation, 2019). The circular economy in five sectors (cement, aluminium, plastics, food, and steel) is estimated to eliminate almost half of the current emissions (Ellen MacArthur Foundation, 2019). In 2019, 25 recycling companies across the major metropolis (Accra, Tema, Kumasi, and Takoradi) were operating at a combined capacity of 360 tons of plastic waste per day and reported an average pay-out of EUR1.15million (GHC6.7million) to plastic waste collectors. This excludes the recycling and reuse initiatives within the informal sector. For instance, Agbogbloshie in Accra is well known for different forms of recycling and reuse of waste products and employed about 40,000 people in 2019 (Enterprise Agency, 2019a). D'ambrières (2019) adds that a plastic recycling plant with a capacity of 50,000 metric tons can employ an average of 30 people, indicating the economic potential of implementing circular economy principles.

To achieve these benefits of circular economy, the government must set clear CE targets and harmonise the existing legal, regulatory, and policy framework to include the core principles of the CE. This would enable businesses to pivot around technologies that are supportive of achieving the targets and ensure effective measurement of progress towards the transition. In addition, the reforms must be followed by consistent enforcement of the rules and regular monitoring and evaluation. A major challenge to achieving this reform is the need for successive governments' more political will to continue past governments' waste management programmes. Thus, non-state actors such as civil society organisations and industry players must be involved in the reform process to strengthen demand-side accountability.

Additionally, the multiplicity of institutions in the waste management industry must be addressed to enhance coherence in the performance of stateside actors. A single agency must be tasked with promoting a CE of waste management. Currently, the Ghana Innovation and Research Commercialisation Centre is a potential entry point for the private sector in waste management to develop innovative approaches to promote CE at both the formal and informal levels. However, a multi-stakeholder framework must be developed to delineate the respective functions of key stateside actors and sub-national agencies towards achieving the circular economy. For instance, the Ministry of Sanitation and the Ministry of Environment, Science, Technology, and Innovation make policies about waste management. However, the implementation rest with sub-national agencies under the Ministry of Local Government, Decentralisation, and Rural Development. Under-resourcing of sub-national institutions would lead to low prioritisation of CE by the implementing agencies, leading to weak enforcement of regulations designed to promote CE.

It is increasingly clear that the transition would depend on improved waste treatment technologies. Thus, the government must consistently invest in developing localised alternative waste treatment technologies. This will provide innovative ideas for the waste management firms to build upon, invest in local capacities, and ensure sustainable approaches. Moreover, research and development can benefit all waste industry sectors by deliberately integrating the informal and formal waste management sectors. Integrating the informal sector means their activities will be regulated to ensure that they use safe, healthy, and environmentally friendly means of collecting, transporting, and treating waste. Given the low government funding commitment to research in Ghana, this can be achieved by using the existing environmental excise tax, eco levy, and sanitation levy to provide technical and financial assistance to the informal sector and support the funding of formal waste management companies to integrate operations with the informal sector.

Government must ensure accountability in the use of waste management-related levies and taxes. For instance, stakeholders in the waste management sector must know how much Sanitation and Pollution levy, Environmental Excise Tax, and Eco-Levy has been collected and the areas of their utilisation. Moreover, these funds can be leveraged to address the financing gap and provide cheaper capital for the formal and informal waste management actors to upscale localised technologies.

7 Conclusions

This study examined theoretical and conceptual approaches that can contribute to the transition from a linear to a circular economy of solid waste management in Ghana.

It is increasingly clear that the current socio-technical system demonstrated by a multiplicity of institutions, poor policy and institutional incoherence, and relatively low technology cannot effectively deliver a transition towards CE of waste management. The existing laws, programmes, and policies need to explicitly recognise CE as the innovative approach to addressing the waste management menace in Ghana. As a result, the existing system needs to offer a clear transformative pathway to change the linear economy. Additionally, the study finds that the existing meta-rules that form the core principles of a Deep Transition are not CE-oriented. Even though there is evidence of the core principles of CE (4Rs) in the waste management sector, largely recycling, they are insufficient to deliver a transformative change towards the CE of waste management.

Furthermore, the study finds that the core principles of the circular economy must underpin all policies, regulations, and programmes on solid waste management in Ghana for the transition to occur. Now, Ghana is at the early stage of the transition, given that open dumping and unsanitary landfill sites are the cheapest waste treatment option because the circular economy technology, policies, and regulations are not well established.

Nonetheless, the high political support for formal waste management innovative programmes and the increased role of the informal waste management sector offer an opportunity to re-align institutions in the sanitation sector, reform policies, and promote CE. Furthermore, the centralisation of policies and programmes decision-making regarding waste management and the high participation of the private sector makes planned circularity and circular modernism probable and preferred scenarios of CE of waste management in Ghana.

Bibliography

Bauwens, T., Hekkert, M., & Kirchherr, J. (2020). Circular futures: What will they look like? *Ecological Economics, 175*,. https://doi.org/10.1016/j.ecolecon.2020.106703

Beijing Review. (2008). *Circular economy promotion law of the People's Republic of China*. https://www.bjreview.com/document/txt/2008-12/04/content_168428.htm

de Angelis, R. (2021a). Circular economy and paradox theory: A business model perspective. *Journal of Cleaner Production, 285*,. https://doi.org/10.1016/j.jclepro.2020.124823

de Angelis, R. (2021b). Circular economy business models: A repertoire of theoretical relationships and a research agenda. *Circular Economy and Sustainability*. https://doi.org/10.1007/s43615-021-00133-x

de Jesus, A., & Mendonça, S. (2018). Lost in transition? Drivers and barriers in the eco-innovation road to the circular economy. *Ecological Economics, 145*, 75–89. https://doi.org/10.1016/j.ecolecon.2017.08.001

Diaz, L. F. (2017). Waste management in developing countries and the circular economy. In *Waste Management and Research* (Vol. 35, Issue 1, pp. 1–2). SAGE Publications Ltd. https://doi.org/10.1177/0734242X16681406

Ellen MacArthur Foundation. (2015). *Delivering the circular economy a toolkit for policy makers*. https://ellenmacarthurfoundation.org/a-toolkit-for-policymakers

Ellen MacArthur Foundation. (2019). *How the circular economy tackles climate change*. https://emf.thirdlight.com/link/w750u7vysuy1-5a5i6n/@/preview/1?o

Enterprise Agency. (2019a). *Market survey waste and circular economy in Ghana commissioned by the Netherlands Enterprise Agency*. https://www.rvo.nl/sites/default/files/2019/08/Ghana-Market-Survey-Waste-Circular-Economy.pdf

Enterprise Agency. (2019b). *Market survey waste and circular economy in Ghana Commissioned by the Netherlands Enterprise Agency*. https://www.linkedin.com/groups/12074594/

Erdiaw-Kwasie, M. O., Abunyewah, M., Yusif, S., & Erdiaw-Kwasie, A. (2023). Does circular economy knowledge matter in sustainable service provision? A moderation analysis. *Journal of Cleaner Production*, 135429.

Erdiaw-Kwasie, M. O., & Basson, M. (2018). Reimaging socio-spatial planning: Towards a synthesis between sense of place and social sustainability approaches. *Planning Theory, 17*(4), 514–532.

European Commission. (2008). *Waste Framework Directive (2008/98/EC)*. https://www.eea.europa.eu/policy-documents/waste-framework-directive-2008-98-ec

European Commission. (2014). *Study on modelling of the economic and environmental impacts of raw material consumption*. http://www.europa.eu

European Commission. (2015). *Closing the loop – An EU action plan for the circular economy*. https://www.eea.europa.eu/policy-documents/com-2015-0614-final

European Commission. (2020). Eurostat-Waste Treatment by Type of Recovery and Disposal. File:Waste treatment by type of recovery and disposal, 2020 (% of total treatment).png—Statistics Explained. https://www.euronews.com/green/2023/06/18/italy-belgium-latvia-which-european-countries-recycle-the-most

Fogarassy, C., & Finger, D. (2020a). Theoretical and practical approaches of circular economy for business models and technological solutions. In *Resources* (Vol. 9, Issue 6, pp. 1–9). MDPI AG. https://doi.org/10.3390/resources9060076

Fogarassy, C., & Finger, D. (2020b). Theoretical and practical approaches of circular economy for business models and technological solutions. In *Resources* (Vol. 9, Issue 6, pp. 1–9). MDPI AG. https://doi.org/10.3390/resources9060076

Geissdoerfer, M., Savaget, P., Bocken, N. M. P., & Hultink, E. J. (2017). The circular economy – A new sustainability paradigm? In *Journal of Cleaner Production* (Vol. 143, pp. 757–768). Elsevier Ltd. https://doi.org/10.1016/j.jclepro.2016.12.048

Ghisellini, P., Cialani, C., & Ulgiati, S. (2016). A review on circular economy: The expected transition to a balanced interplay of environmental and economic systems. *Journal of Cleaner Production, 114*, 11–32. https://doi.org/10.1016/J.JCLEPRO.2015.09.007

Greer, R., von Wirth, T., & Loorbach, D. (2021). The waste-resource paradox: Practical dilemmas and societal implications in the transition to a circular economy. *Journal of Cleaner Production, 303*,. https://doi.org/10.1016/j.jclepro.2021.126831

Heeres, R. R., Vermeulen, W. J. V., & de Walle, F. B. (2004). Eco-industrial park initiatives in the USA and the Netherlands: First lessons. *Journal of Cleaner Production, 12*(8–10), 985–995. https://doi.org/10.1016/j.jclepro.2004.02.014

Jackson, M., Lederwasch, A., & Giurco, D. (2014). Transitions in theory and practice: Managing metals in the circular economy. *Resources, 3*(3), 516–543. https://doi.org/10.3390/resources3030516

Kalmykova, Y., Sadagopan, M., & Rosado, L. (2018). Circular economy – From review of theories and practices to development of implementation tools. *Resources, Conservation and Recycling, 135*, 190–201. https://doi.org/10.1016/j.resconrec.2017.10.034

Kaza, S., Yao, L. C., Bhada-Tata, P., & van Woerden, F. (2018). *What a waste 2.0: A global snapshot of solid waste management to 2050*. World Bank. https://doi.org/10.1596/978-1-4648-1329-0

Kern, F., Sharp, H., & Hachmann, S. (2020). Governing the second deep transition towards a circular economy: How rules emerge, align and diffuse. *Environmental Innovation and Societal Transitions, 37*, 171–186. https://doi.org/10.1016/j.eist.2020.08.008

Kirchherr, J., Reike, D., & Hekkert, M. (2017). Conceptualising the circular economy: An analysis of 114 definitions. In *Resources, Conservation and Recycling* (Vol. 127, pp. 221–232). Elsevier B.V. https://doi.org/10.1016/j.resconrec.2017.09.005

Lahti, T., Wincent, J., & Parida, V. (2018). A definition and theoretical review of the circular economy, value creation, and sustainable business models: Where are we now and where should research move in the future? In *Sustainability (Switzerland)* (Vol. 10, Issue 8). MDPI. https://doi.org/10.3390/su10082799

Li, H., Bao, W., Xiu, C., Zhang, Y., & Xu, H. (2010). Energy conservation and circular economy in China's process industries. *Energy, 35*(11), 4273–4281. https://doi.org/10.1016/j.energy.2009.04.021

Lowe, B. H., & Genovese, A. (2022). What theories of value (could) underpin our circular futures? *Ecological Economics, 195*,. https://doi.org/10.1016/j.ecolecon.2022.107382

Mathews, J. A., & Tan, H. (2011). Progress toward a circular economy in China: The drivers (and inhibitors) of eco-industrial initiative. *Journal of Industrial Ecology, 15*(3), 435–457. https://doi.org/10.1111/j.1530-9290.2011.00332.x

Merli, R., Preziosi, M., & Acampora, A. (2018). How do scholars approach the circular economy? A systematic literature review. In *Journal of Cleaner Production* (Vol. 178, pp. 703–722). Elsevier Ltd. https://doi.org/10.1016/j.jclepro.2017.12.112

Republic of Ghana. (2020). Ghana's fourth communication to the United Nations framework convention on climate change. https://unfccc.int/documents/231806

Schot, J., & Kanger, L. (2018). Deep transitions: Emergence, acceleration, stabilisation and directionality. *Research Policy, 47*(6), 1045–1059. https://doi.org/10.1016/J.RESPOL.2018.03.009

Sharma, H. B., Vanapalli, K. R., Samal, B., Cheela, V. R. S., Dubey, B. K., & Bhattacharya, J. (2021). Circular economy approach in solid waste management system to achieve UN-SDGs: Solutions for post-COVID recovery. *Science of the Total Environment, 800*, 149605. https://doi.org/10.1016/j.scitotenv.2021.149605

Sihvonen, S., & Ritola, T. (2015). Conceptualising ReX for aggregating end-of-life strategies in product development. *Procedia CIRP, 29*, 639–644. https://doi.org/10.1016/J.PROCIR.2015.01.026

UN HABITAT. (2021, March 22). *Climate change, circular economy, waste management*.

United Nations. (2021). *The sustainable development goals report*.

United Nations. (2022). Sustainable development goal 12.

United Nations Environment Programme. (2018). *Africa waste management outlook*. https://www.unep.org/ietc/resources/publication/africa-waste-management-outlook

Urbinati, A., Chiaroni, D., & Chiesa, V. (2017). Towards a new taxonomy of circular economy business models. *Journal of Cleaner Production, 168*, 487–498. https://doi.org/10.1016/j.jclepro.2017.09.047

van Buren, N., Demmers, M., van der Heijden, R., & Witlox, F. (2016). Towards a circular economy: The role of Dutch logistics industries and governments. *Sustainability, 8*(7), 647. https://doi.org/10.3390/su8070647

d'Ambrières, W. (2019). Plastics recycling worldwide: Current overview and desirable changes. *Field Actions Science Reports. The Journal of Field Actions* (Special Issue 19), 12–21. https://journals.openedition.org/factsreports/5102

World Bank. (2018). *What a waste 2.0: A global snapshot of solid waste management to 2050*. https://datatopics.worldbank.org/what-a-waste/trends_in_solid_waste_management.html#:~:text=The%20world%20generates%202.01%20billion%20tonnes%20of%20municipal,but%20ranges%20widely%2C%20from%200.11%20to%204.54%20kilograms.

World Economic Forum. (2021). *Country financing roadmap for the SDGs: Ghana*. https://www.weforum.org/reports/country-financing-roadmap-for-the-sdgs-ghana

Yuan, Z., Bi, J., & Moriguichi, Y. (2008). The circular economy: A new development strategy in China. *Journal of Industrial Ecology, 10*(1–2), 4–8. https://doi.org/10.1162/108819806775545321

Zhu, Q., Geng, Y., & Lai, K. h. (2010). Circular economy practices among Chinese manufacturers varying in environmental-oriented supply chain cooperation and the performance implications. *Journal of Environmental Management*, *91*(6), 1324–1331. https://doi.org/10.1016/J.JENVMAN.2010.02.013

Country-Level Case Studies on Circular Strategies and SDGs

Linear Economy Wolf in Circular Economy Sheep Clothing: Ride-Hailing in Ghana

Festival Godwin Boateng, Samuelson Appau, and Kingsley Tetteh Baako

1 Introduction

Africa is increasingly becoming one of the main targets for the expansionary activities of ride-hailing companies, with global giants like Uber and Bolt launching operations in the continent's cities (Acheampong, 2021; Agyemang, 2020; Dzisi et al., 2020, 2021; Kaye-Essien, 2020). This development is refreshing for the transition toward a circular economy (CE) in Africa's urban transport sector, given the foundation of ride-hailing (RH) on car sharing. The broader point here is

Parts of the data used have been published in Boateng, F. G., Appau, S., and Baako, K. T. (2022). The rise of 'smart' solutions in Africa: A review of the socio-environmental cost of the transportation and employment benefits of ride-hailing technology in Ghana. *Humanities and Social Sciences Communications, 9*(1), 1–11.

F. G. Boateng (✉)
Centre for Sustainable Urban Development, Columbia Climate School/Earth Institute, Columbia University, New York, NY, USA
e-mail: bfsvl1988@gmail.com

M. O. Erdiaw-Kwasie and G. M. M. Alam (eds.), *Circular Economy Strategies and the UN Sustainable Development Goals*, Sustainable Development Goals Series,
https://doi.org/10.1007/978-981-99-3083-8_6

that the current linear take-make-dispose economic model, which dominates global production and consumption, is widely considered socially, economically, and environmentally unsustainable. These concerns have generated calls for alternative production and consumption systems that can maintain the value of products, materials, and resources in the economy as long as possible, with a circular economy widely perceived to have that potential (Esposito et al., 2018; Kahl, 2018; MacArthur, 2013; Panwar & Niesten, 2022; Williams, 2019).

CE's definitional, conceptual, and theoretical debates are still evolving (e.g., Corvellec et al., 2021), with Kirchherr et al. (2017) counting some 114 definitions of the concept. Nevertheless, these varied definitions commonly suggest that the CE's axiology is to increase the efficiency of resources by minimizing waste and resource extraction (Kębłowski et al., 2020; Thomsen et al., 2022; Wu et al., 2018; Wuyts & Marin, 2022).

Sharing of products (as a service), especially through digital platforms are widely perceived as one of the critical pathways for realizing CE goals (e.g., Schwanholz & Leipold, 2020; Sposato et al., 2017). Advocates argue that sharing enables collaborative consumption through lending, swapping, bartering, and renting through service platforms to get more value from assets or resources (Curtis & Mont, 2020; Sposato et al., 2017). This way, sharing facilitates interesting environmental advantages, including resource-saving and waste avoidance, and, therefore, helps to overcome most of the inefficiencies of a linear economy. In this regard, on-demand ride-hailing (RH) is increasingly becoming a popular business model often presented as capable of facilitating the realization of CE goals in the transport sector (Ellen MacArthur Foundation, n.d., 2019; Schwanholz & Leipold, 2020; Sposato et al., 2017).

While private cars offer more flexible services, they are parked most of the time and carry very few people when moving. Private cars are

S. Appau
Melbourne Business School, Carlton, VIC, Australia
e-mail: S.Appau@mbs.edu

K. T. Baako
School of Property, Construction and Project Management, RMIT University Melbourne, Melbourne, VIC, Australia
e-mail: kingsley.baako@rmit.edu.au

also widely criticized for their significant negative externalities, including congestion, pollution, and road crashes (Diao et al., 2021; Habitat III, 2015; Mitchell et al., 2010; Parry et al., 2007). In this sense, private cars embody many of the environmental and social sustainability concerns of the current linear production and consumption paradigm that CE seeks to supplant, with RH often presented as a new urban transport business strategy for achieving this goal (Ellen MacArthur Foundation, n.d., 2019; Schwanholz & Leipold, 2020; Sposato et al., 2017).

RH's perceived potential to enable CE goals is linked to the claim that it creates platforms to facilitate *need-based* access to–instead of *ownership* of–private cars (Alonso-Mora et al., 2017; Cohen & Kietzmann, 2014; Fulton et al., 2017; Jin et al., 2018; Santi et al., 2014). In so doing, RH can contribute to the CE goal of promoting product use instead of ownership, which in turn, can reduce the need to invest more resources and energy in car production, with the overall result of slowing down and narrowing resource use and, to that end, facilitating the realization of another CE goal: environmental conservation.

As illustrated in Fig. 1, the perceived sustainability transition benefits of RH go beyond reducing car ownership and waste avoidance. It is perceived that by encouraging car and trip sharing; RH can ultimately help reduce the total number of cars on the roads and, hence, car-related socio-environmental troubles, including crashes, congestion, and pollution (Alonso-Mora et al., 2017; Cohen & Kietzmann, 2014). Sharing trips and vehicles also mean that existing private cars are being put to efficient or optimal use, thereby encouraging the re-use of private cars and enabling one of the key goals of CE: product re-use.

In terms of CE's social goal of employment creation, RH is said to provide opportunities for people to monetize their driving skills and private cars. In 2017, Uber co-founder Garrett Camp wrote that the company *alone* provides flexible work options to some 2 million drivers (Camp, 2017). Some recent estimates put the number of Uber drivers at 3.9 million globally (Woodcock & Graham, 2020: 51). Viewed this way, the spread of RH technology from the Global North to Africa can be expected to enable the realization of sustainability transition goals in the continent's transport sector by facilitating an overall reduction of private cars on the roads and the socio-environmental harms that they bring while creating jobs at the same time by creating opportunities for people to commercialize their private cars and driving skills.

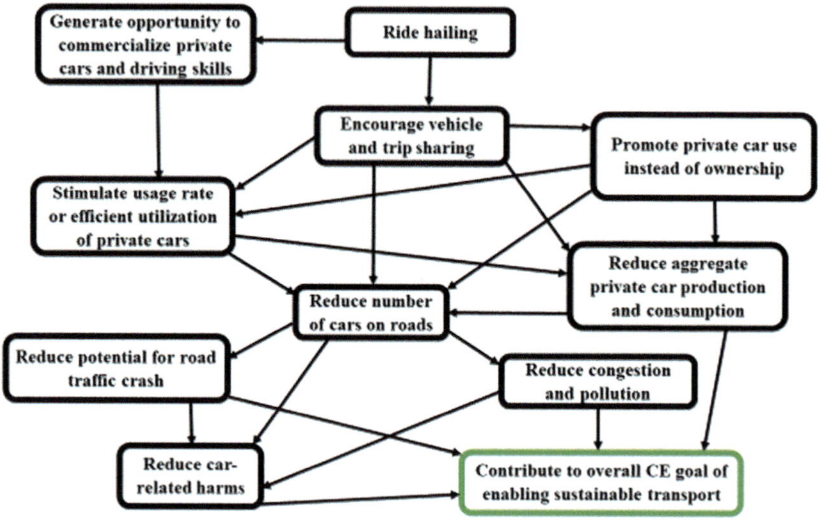

Fig. 1 Conceptual framework of the enabling influence of ride-hailing on achieving CE goals in the transport sector

The benefits of technology transfer are, however, seldom automatic. It is well-documented that when technologies free themselves from their context of discovery and spread to different contexts, they shape and are often, in turn, shaped by the socio-cultural political-economic structures of the recipient societies (Luper et al., 1991; Molnar, 1978; Samli, 1985 in Kebede et al., 2013). The resulting transformation can either enhance or undermine the viability of the technology. Therefore, the spread of RH technology into Africa's cities must be carefully studied. This will help to properly gauge whether the car-sharing practices the technology facilitates in the continent can support the realization of CE goals. Currently, there is little by way of research that examines the potential of RH to drive CE transition in Africa's transport sector. This chapter seeks to contribute to the conversation on the topic using the context of Ghana.

In addition to traditional peer-reviewed materials, the chapter relies on evidence from empirical sources. The empirical evidence was gathered by combining both purposive and snowball sampling techniques—two widely regarded qualitative sampling techniques—to engage 40 informants operating in Ghana's ride-hailing industry in various capacities as drivers, riders, car owners, and researchers. In gathering the data, the researchers first advertised the study on various social media platforms, including Facebook and WhatsApp, inviting eligible informants to participate. After volunteering information, the initial participants were asked to assist with contacts with other potential informants.

Through this, the contacts of 63 potential informants (including riders, drivers, and car owners) were received. However, upon follow-up, sixteen (16) declined participation. Of the remaining fifty-two (52), seven (7) did not participate fully—as, at the time of writing, they had not responded to some follow-up questions. Their perspectives were, therefore, not included in the final data used for the analysis.

As shown in Table 1, the majority of the rider-respondents were females, but all the driver-respondents were males. We found some additional evidence confirming male drivers' dominance in the industry. For instance, of the 15 riders who participated in the study, only 1 reported having ever met a female driver. The 11 car owners who participated in the study supply a total of 41 cars to deliver ride-hailing services; their drivers are all males. Some media reports suggest that there were only seven registered female Uber drivers in Ghana as of 2017 (B&FT Online, 2017). Similar gender dynamics apply to car owners. The over-representation of males among our driver and car owner respondents is consistent with the general gender dynamics in Ghana's RH industry (see Agyemang, 2020) and the commercial passenger transport sector generally (Boateng, 2020; Boateng et al., 2022; Okraku, 2016).

We limited our consideration of the respondents' demographic characteristics to gender because other demographic issues have been widely covered in the literature. For instance, previous studies have confirmed

Table 1 Details of participants, fieldwork, and other activities

Participants	*Number*	*Gender*		*Mode of engagement*
		Male	*Female*	
Riders	15	4	11	Face-face interviews
Drivers	11	11	0	Face-face interviews; WhatsApp messaging
Car owners	11	10	1	Face-face interviews; WhatsApp messaging
Researchers	4	4	0	Email; Zoom interview; WhatsApp messaging
Total	40			

Source Authors' fieldwork

that the younger, highly educated, and relatively high-income-earning class most patronize Ghana's RH industry (see, e.g., Acheampong et al., 2020; Agyemang, 2020; Dzisi et al., 2020). In terms of the data collection medium, as shown in Table 1, we employed a mixture of semi-structured face-to-face and Zoom interviews, email exchanges, and WhatsApp messaging. The data collection sessions varied from approximately 45 min, and close to 1 h of face-face and Zoom interviews to 3 days or more of email and WhatsApp exchanges. The ethical aspect of the study was reviewed and approved by the RMIT University College of Business Human Ethics Committee.

The evidence considered raises issues with RH potential to enable a CE transition in Ghana. Not only does RH stimulate the consumption of more private cars, but it also undermines existing shared modes of transport that have the potential to reduce private car use. Overall, RH in Ghana encourages the commercialization of private cars for the benefit of a minority while socializing the negative externalities for all in the form of increased risky driving practices, shifts from shared public transport and inundation of the roads with more highly polluting and crash-prone used private cars. The chapter contributes to the growing calls for increased scrutiny of the sustainability transition claims of so-called CE-based sharing business models (see, e.g., Curtis & Mont, 2020). This is necessary for reducing the risks of tying sustainability transition financing and other opportunities to linear economy business wolves dressed in CE sheep clothing.

The next section provides an overview of commercial passenger transport in Ghana. This is followed by 3 more sections. Section 3 considers the emergence of RH in Ghana. Section 4 attends to the study's main mandate: The extent to which RH in Ghana advances or undermines CE and sustainability transition goals. Finally, Sect. 5 discusses the findings and concludes the paper.

2 Commercial Passenger Transport in Ghana: A Brief Overview

Ghana's commercial passenger transport sector has progressively evolved into a privately run, deregulated industry with minibuses (popularly called '*tro-tro*') and shared taxis being the predominant high-volume passenger mode of transport (Boateng, 2021; Stasik & Klaeger, 2018). Nonetheless, '*Okadas*' (motorcycles) (Oteng-Ababio & Agyemang, 2012, 2015) and

'*Pragyas*' (motorized tricycles) are becoming commonplace in many cities and regions in the country (Kenu, 2021). This state of affairs has roots in the failure of successive governments, dating back to colonial times, to adequately prioritize investing in organized public transport and non-motorized modes of transport (Boateng & Klopp, 2022; Obeng-Odoom, 2013).

As with other African countries (e.g., Agbiboa, 2022; Klopp & Mitullah, 2016), millions rely on the above popular commercial transport systems to get around in Ghana. Ghana's tro-tro and taxi sector is heavily rooted in sharing–one of the main pillars of CE. While the practice of hailing or hiring a taxi (popularly called '*dropping*') to a specific location without making several stops or sharing it with other riders exists, as with the tro-tros, taxis are generally shared: multiple passengers traveling in the same direction board and share the same vehicle and driver to their respective destinations. While the popular commercial passenger transport sector provides great economic, social, and environmental sustainability benefits, concerns exist about vehicles' safety, convenience, fair pricing, and eco-friendliness of the sector.

First, the majority of tro-tros and taxis in Ghana are often old second-hand vehicles (see, e.g., Burchardt, 2015; NRSC, 2014; Obeng-Odoom, 2013). Similar dynamics exist in other African countries (see Boateng & Klopp, 2022). Some vehicles are even third-, fourth-, or even fifth-hand (see, e.g., Obeng-Odoom, 2010, 2013) and are often poorly maintained and kept on the road even as they get older, more dangerous, and more polluting. Some car owners drive their own taxi and tro-tro. Nonetheless, generally, car owners lease them to drivers to operate and, depending on the arrangement, pay an agreed-upon daily, weekly, or monthly fee (Boateng, 2020; Obeng-Odoom, 2013).

Unemployment levels in Ghana are high, while labor protections such as minimum wage, social security, health, or unemployment benefits too do not exist in the commercial passenger transport sector. These dynamics, which also exist in other African countries (Agbiboa, 2015, 2016; Rizzo, 2011), together tilt the balance of power in favor of car owners who also often seize it as an opportunity to set unrealistic end-of-day-sales or targets for the drivers (Boateng, 2020, 2021; Dotse et al., 2019; Obeng-Odoom, 2013). As a result, the drivers can only make enough revenue to cover operational costs, pay their owners, themselves, and, in the case of tro-tro drivers, their assistants by increasing the number of trips or passengers per trip. They, invariably, are forced or incentivized to drive for long hours, resort to dangerous overtaking, overload their cars, and drive at dangerously high speeds (Boateng, 2020; Obeng-Odoom, 2013). In the context of limited enforcement of road traffic regulations, this manifests in crashes, deaths, and injuries (Atuahene, 2022).

However, the challenges in Ghana's commercial passenger transport sector go beyond old polluting cars, exploitation-induced dangerous driving practices, and limited enforcement of road traffic regulations. Other problems exist, including unfair pricing. While pre-fixed fares exist, taxi and tro-tro drivers commonly take advantage of situations such as bad weather, unavailability of enough vehicles, night travels, and congestion to charge passengers in excess of set fares. The pricing problem becomes more apparent when the cars are being hired. The drivers just eyeball or guesstimate travel distance or time and, accordingly, determine transport fares. They also frequently determine fares based on their perception of the financial status of potential passengers. As one of our rider informants remarked: '*If you dress as a pauper, you might get a good discount. If you dress like somebody who is high, an elite, somebody walking around with gadgets and fancy jewelry, the price could be tripled*'. Thus, the pricing system is highly unstandardized, too erratic, and discretionary.

Further, tro-tro and taxis in Ghana are also often not GPS-enabled. Vehicles and trips are, therefore, not traceable/trackable. This increases the risk of harm (for both drivers and riders) during travel and undermines the chances of recovering lost items. Convenience, too, is a big issue. Commuters must trek to the nearest terminal or roadside to access a traditional taxi or a tro-tro. This becomes problematic in times of emergency, when the journey originates from a distant location to the nearest terminal or roadside or when one has a lot of heavy luggage. Drivers too face similar inconveniences as they have to park at designated loading stations and wait for passengers or roam the cities to search for them, costing time, car depreciation, and fuel.

The foregoing shows that while the sharing practices underlying Ghana's tro-tro and traditional taxicab sector enable some CE benefits; the sector is saddled with many challenges. These include its dominance by old polluting used cars; exploitation-induced reckless driving practices; unfair pricing; inconveniences (for both drivers and passengers); lack of opportunity to trace or track vehicles and trips, which increase the potential of being harmed or losing valuable items during travel.

3 Cometh Ride-Hailing

Against the above backdrop, the Government of Ghana signed a '*Memorandum of Understanding*' with Uber in July 2016 to enable the company to launch operations in the country. Today, other global giants, including Bolt and a number of relatively smaller RH firms, also provide RH services in the country–particularly in Accra, the capital, and Kumasi, the second biggest city (Boateng & Appau, 2022; Kufuor, 2018).

Our data show that RH is helping to address many of the long-standing problems in the commercial passenger transport sector. Consider, for instance, the sector's dominance by old and poorly maintained vehicles. Even though some riders reported that some RH vehicles are just as rickety as the traditional taxis and tro-tros, RH firms generally do not register such vehicles for RH services. RH metered-pricing system too eliminates discretion and, therefore, potential abuse or cheating. Some riders even reported that they use RH trip price estimates as benchmarks when negotiating transport fares with traditional cab drivers. Thus, as one car owner-respondent remarked, the emergence of RH has '*raised the standards [of taxi service delivery in the country]. They have upped the game in terms of pricing transparency, the quality of vehicles and the standards of service that riders receive*'.

Further, potential riders no longer have to trek or run (in cases of emergency) to bus terminals or roadsides before they can access a ride. Instead, with a smartphone, they can summon one to their doorsteps or anywhere they find themselves. '*I can sit in the comfort of my home or my office and then request a ride and go wherever I am going*', one rider-informant remarked. Drivers also enjoy similar conveniences. One of them remarked:

> *When you are a taxi driver, you drive empty, looking for passengers. But now, as an Uber or Bolt driver, you sit down when the request comes, and you go. That's a good thing. Because for a taxi, you will roam; you can't even park. If you park, you are not going to get anyone. So, you will need to roam, and waste fuel.*

Aside from convenience, the informants reported that, while price surges and gridlocks can increase cost, RH trips, compared to traditional taxicabs, are generally affordable. Acheampong et al. (2020) made similar findings. For instance, the authors found that in Accra, the capital, a typical 25–30 minute lone travel in a conventional taxi hailed on the street (known locally as 'dropping' or 'hire') can cost about GH¢80 ($13.16). But when the same trip is done via RH, it can cost around GH¢21 ($3.82). Shared tro-tro is more affordable than both of them; the same trip when done via Tro-Tro will cost an individual just GH¢4 ($0.72). Nonetheless, unlike traditional taxis and tro-tros, RH trips are *traceable/trackable*, reducing the risk of harm (for both drivers and riders) during travel and increasing the chances of recovering lost items.

Our car owner informants reported that RH has offered them lucrative investment opportunities to generate surplus income, with some of them realizing 100% returns on investment within a short period of 2 years. Thus, one of them reported:

> *I have been able to make extra income. So, monies I used to spend from my main job on certain things, I don't spend those monies on those things anymore. I use monies I get from Uber for other expenses. It's helped me a lot. I am getting extra cash for projects I am planning to work on. It's helping me to do things I want to do easily.*

Another car owner provided a more detailed insight into the opportunities RH is opening up in Ghana:

> *I like it [the RH business] because of the frequency of income. Other investments that we have in the country, for example, treasury bills, your money is stuck for the term of the investments, minimum of 91 days. You cannot use it for any other reason. You have to make sure it's there till it realizes the interest on it. But with Uber, because you keep getting the payment in weekly instalments, you are able to put it into other investments. So, you have more liquidity which you can use to unlock further investments. So, you can make much more than the percentage you typically make on treasury bills because you have more access to your money, and you can move it around to make more money for the same amount of time. So, it's definitely a better investment.*

Overall, the evidence suggests that RH is driving up the standards of road transport experience in Ghana as well as creating opportunities for private car owners to commercialize their assets profitably. Some of the long-standing problems in the commercial passenger sector, however, persist. For instance, even though women contribute immensely to the RH industry as riders (e.g., Acheampong, 2021; Acheampong et al.,

2020; Agyemang, 2020; Dzisi et al., 2020, 2021), as with the tro-tro and traditional taxi industry, men dominate RH as drivers and car owners. All of our driver-informants were males. Of the 15 rider informants, only 1 reported that they had ever met a female driver. The 11 car owner informants apply a combined total of 41 cars to the delivery of RH services; their drivers are all males. There were only 7 registered female Uber drivers in Ghana as of 2017 (see B&FT Online, 2017). Similar gender dynamics apply to car owners.

Essentially, the RH industry has patterned after the existing male-dominated gender dynamic in Ghana's commercial passenger transport sector (see e.g., Boateng, 2020; Okraku, 2016). Gender inequalities, however, are not the only long-standing problem in the existing commercial passenger transport sector that is also evident in the RH industry. There is also the issue of driver exploitation. The driver-informants reported of exploitation by not just car owners but also by RH firms. They unanimously complained that the commissions/fees charged by the RH firms are '*very high*'. In addition, the firms often seek to compensate the drivers with promotional offers, requiring them to complete a set number of trips within a certain amount of time to earn extra money. However, the problem with the offers is that they incentivize risky driving behaviors such as over-speeding as the drivers rush to complete as many trips as possible to be eligible.

Other persisting problems include limited enforcement of road traffic laws. For example, section 118 of the Road Traffic Regulations of 2012 (LI 2180) proscribes commercial drivers from driving '*for a continuous period exceeding four hours; or for a period amounting in the aggregate to more than eight hours in a period of twenty-four hours*'. However, the RH companies require drivers to take breaks only when they have worked for 12 hours, which violates the law. Nonetheless, this violation goes unchecked by regulators. Taken together, the evidence considered suggests that while RH is transforming the commercial passenger transport sector in some important positive ways; some of the long-standing problems in the sector have seeped into the RH industry. How, however, do these complicated outcomes implicate CE political and development goals and sustainability transition generally in the transport sector? Do they undermine or advance the goals and what factors mediate the outcomes? The rest of the chapter carefully considers these questions.

4 RH and CE Political and Development Goals in Ghana

We found evidence that shows that RH advances the CE goal of promoting *product/service use* instead of *ownership* (Barbu et al., 2018; Henry et al., 2021; Orasmaa et al., 2020). Thus, our data support the perceived potential of RH to facilitate *need-based* access to–instead of *ownership* of–private vehicles. '*Even though I do not have a car*', one rider-informant remarked, '*Uber makes me mobile*'. Another informant, a car owner, made similar remarks: *They [referring to RH firmsa] have allowed people to ride in a vehicle they do not own. You have almost full access to services of any typical owner. You do not have to own a car to have access to a private ride.*

By facilitating *need-based* access to private vehicles, RH advances not just the CE goal of stimulating product/service use instead of ownership, but it also contributes to the goal of preventing waste and encouraging resource re-use by reducing private car redundancy and stimulating their continued use or re-use since other people get access to the vehicles instead of the owners parking them. However, the idea behind CE goals of encouraging product sharing instead of ownership, preventing waste, and encouraging product re-use are not for their own sake. Instead, they are intended to slow down and narrow investing more resources and energy in producing and consuming even more goods. In short, the goals are intended to reduce aggregate production and consumption. It is, therefore, important to examine whether the *kind of private vehicle sharing* RH facilitates in Ghana enables or could enable achieving these goals.

Our data answer this question in the negative: the kind of private vehicle sharing RH promotes in Ghana does not seem to be adequately suited to lead to a reduction in aggregate importation and consumption of private vehicles. Most of the cars used to provide RH services in Ghana, mainly Daewoo Matiz; Hyundai i10/Eon/Getz; Kia Picanto/Morning; Toyota Yaris/Vitz/Echo/Platz and Suzuki Alto, are specifically imported for that purpose. Thus, very few people '*convert*' or commercialize existing private cars for RH services. As noted, the 11 drivers and 11 car owners who participated in the study supply a total of 41 private cars to deliver RH services. Only 4 were, hitherto, private vehicles commercialized for RH services–the remaining 37 were all imported for the express purpose of providing RH services.

Further, even though trip or ride-sharing has been a feature of Ghana's commercial passenger transport sector, where unacquainted passengers with similar origins and destinations share a ride and split the fare, the trip-sharing options recently introduced by some RH companies (e.g., *UberXShare*) are poorly patronized. Indeed, studies show that the opportunity to have *exclusive access* to vehicles instead of *sharing* them with other riders is what makes RH endearing for the younger, highly educated and relatively high-income-earning class that widely patronizes Ghana's RH industry (see, e.g., Acheampong et al., 2020; Agyemang, 2020; Dzisi et al., 2020).

The result of these dynamics is that more and more vehicles are being imported for RH business which also stays at a lower passenger occupancy rate while on the road. This worsens the traffic conditions; the cost of which riders are already paying in increased travel time and the share of income spent on transport. For example, a rider-informant reported: '*You will pick an Uber, and then you will be stuck in traffic, and then you know you are going to pay GH¢20, you get to the place, and you are paying GH¢200, GH¢500*'. The cost of the shift from shared taxis and minibuses to lower passenger occupancy RH private cars in Ghana might be more than increased congestion, travel time, and share of income spent on transport. The development can also be costly in environmental terms.

The point here is that while the quality of RH vehicles is perceived to be comparatively higher than those applied to deliver tro-tro and traditional taxi services; they all come from the same pool of *used vehicles* imported in high numbers into the country (NRSC, 2014; Obeng-Odoom, 2010, 2013; Sulemana, 2012). This does not bode well for CE and sustainability transition goals considering that used vehicles tend to be not just fault-prone but are also often highly polluting (see UNEP, 2020). By stimulating the *importation* and *usage* of more used private cars, RH can end up worsening the already troubling issue of vehicular pollution in the country (see Amegah et al., 2021; USAID, 2016).

5 Conclusion

As environmental sustainability and resource use concerns take center stage in policy and governance debates, the shift toward a CE is gathering momentum across multiple sectors. Although still nascent and developmental in its ambitions, CE initiatives' potential to arrest the unsustainable use and disposal of resources has attracted many allies. RH businesses have claimed to be such an ally and have constantly trumpeted their contribution to more efficient private car use, circulation, and redistribution while creating jobs.

In this chapter, we examined the veracity of these claims in the Ghanaian context. Our findings suggest that although RH businesses in Ghana contribute to CE goals of better resource use, they also incentivize increased resource use, potentially eliminating any gains from the intended better resource use. We conclude from the findings that RH in Ghana is not about sharing, but rather creates new forms of resource exchange that add to existing modes of commercial transportation they compete with. The result is more vehicles on the road, undermining CE goals of sustainability. Viewed this way, it can be argued that RH business models in Ghana are essentially linear economy wolves in CE sheep clothing.

The circular economy challenge with the kind of private car-sharing RH promotes in Ghana is not just about it being linear, but it is also about the distribution of the outcomes. As the evidence shows the technology has just opened an opportunity for the younger, highly educated and relatively high-income-earning class to substitute the existing inefficient public transport options (tro-tro and taxicabs) for a *new niche* high-end, affordable form of mobility founded on private cars. But it is not just the transportation benefits, the economic opportunities too disproportionately benefit a few powerful players in the industry–particularly RH firms and entrepreneurial (predominantly male) car owners. Meanwhile, the negative externalities are socialized for all in the form of increased unhealthy driving practices; shifts from shared public transport and inundation of the roads with more highly polluting and crash-prone used private cars.

A key development that bears watching is RH's potential to make the existing popular transport systems even more invisible to policy and planning actions and investments. Suppose the younger, highly educated and relatively high-income-earning class continues to shift to RH. In that case, they will have little interest in using their power–as one of the resourceful constituents of the Ghanaian society–to demand sharper policy attention to the existing popular transport systems. This can deepen the persisting neglect and disinvestment in the systems, and, in turn, incentivize an even greater shift to RH and other forms of private car use. The cost of which the Ghanaian society will pay in more human life and property losses (resulting from crashes), traffic congestion, and increased environmental damage (resulting from increased carbon emissions).

In short, if care is not taken, RH can deepen the socio-environmental vulnerabilities they are supposed to address. As a result, the likes of Curtis and Mont (2020) are calling for increased scrutiny of the sustainability transition claims of so-called CE-based sharing business models. This chapter's findings support why the call is necessary lest we risk tying sustainability transition financing and other opportunities to business models that may undermine instead of enable sustainability goals.

References

Acheampong, R. A. (2021). Societal impacts of smart, digital platform mobility services—An empirical study and policy implications of passenger safety and security in ride-hailing. *Case Studies on Transport Policy, 9*(1), 302–314.

Acheampong, R. A., Siiba, A., Okyere, D. K., & Tuffour, J. P. (2020). Mobility-on-demand: An empirical study of internet-based ride-hailing adoption factors, travel characteristics and mode substitution effects. *Transportation Research Part C: Emerging Technologies, 115*, 102638.

Agbiboa, D. E. (2015). "Policing is not work: It is stealing by force": Corrupt policing and related abuses in everyday Nigeria. *Africa Today, 62*(2), 95–126.

Agbiboa, D. E. (2016). 'No Condition IS Permanent': Informal Transport Workers and Labour Precarity in Africa's Largest City. *International Journal of Urban and Regional Research, 40*(5), 936–957.

Agbiboa, D. E. (2022). *They eat our sweat: Transport labor, corruption, and everyday survival in Urban Nigeria*. Oxford University Press.

Agyemang, E. (2020). "Uber is here to stay": Exploring the policy implications of the Uber-Local Taxis turf war in Accra, Ghana. *Case Studies on Transport Policy, 8*(1).

Alonso-Mora, J., Samaranayake, S., Wallar, A., Frazzoli, E., & Rus, D. (2017). On-demand high-capacity ride-sharing via dynamic trip-vehicle assignment. *Proceedings of the National Academy of Sciences, 114*(3), 462–467.

Amegah, A. K., Sewor, C., & Jaakkola, J. J. (2021). Cadmium exposure and risk of adverse pregnancy and birth outcomes: A systematic review and dose–response meta-analysis of cohort and cohort-based case–control studies. *Journal of Exposure Science & Environmental Epidemiology, 31*(2), 299–317.

Atuahene, K. K. (2022). *Current road safety situation: Data policy brief.* National Road Safety Authority (NRSA).

Barbu, C. M., Florea, D. L., Ogarcă, R. F., & Barbu, M. C. (2018). From ownership to access: How the sharing economy is changing the consumer behavior. *Amfiteatru Economic, 20*(48), 373–387.

Boateng, F. G. (2020). "Indiscipline" in context: A political-economic grounding for dangerous driving behaviors among Tro-Tro drivers in Ghana. *Nature: Humanities and Social Sciences Communications*, *7*(1), 1–5.

Boateng, F. G. (2021). Why Africa cannot prosecute (or even educate) its way out of road accidents: insights from Ghana. *Nature: Humanities & Social Sciences Communications, 8*, 13. https://doi.org/10.1057/s41599-020-00695-5

Boateng, F. G., & Appau, S. (2022, October 4). Blaming poor labour conditions in Ghana's transport sector on ride-hailing companies misses the deeper issues. *The Conversation Africa*.

Boateng, F. G., Appau, S., & Baako, K. T. (2022). The rise of 'smart' solutions in Africa: A review of the socio-environmental cost of the transportation and employment benefits of ride-hailing technology in Ghana. *Humanities and Social Sciences Communications, 9*(1), 1–11.

Boateng, F. G., & Klopp, J. M. (2022). Beyond bans: A political economy of used vehicle dependency in Africa. *Journal of Transport and Land Use, 15*(1), 651–670.

Burchardt, J. (2015). *Order out of chaos Self-management and public control of the paratransit sector: Case Ghana International Association for the History of Transport, Traffic and Mobility (T2M) and the Cosmobilities Network*. Santa Maria C.V. (Caserta), Italy – September 14–17, 2015. Panel theme: Shared taxi services.

Camp, G. (2017). *Uber's path forward*. https://medium.com/@gc/ubers-path-forward-b59ec9bd4ef6. Accessed 8 June 2022.

Cohen, B., & Kietzmann, J. (2014). Ride on! Mobility business models for the sharing economy. *Organization & Environment, 27*(3), 279–296.

Corvellec, H., Stowell, A. F., & Johansson, N. (2021). Critiques of the circular economy. *Journal of Industrial Ecology, 26*(2), 421–432.

Curtis, S. K., & Mont, O. (2020). Sharing economy business models for sustainability. *Journal of Cleaner Production, 266*, 121519.

Diao, M., Kong, H., & Zhao, J. (2021). Impacts of transportation network companies on urban mobility. *Nature Sustainability, 4*(6), 494–500.

Dotse, J., Nicolson, R., & Rowe, R. (2019). Behavioral influences on driver crash risks in Ghana: A qualitative study of commercial passenger drivers. *Traffic Injury Prevention, 20*(2), 134–139.

Dzisi, E. K., Obeng-Atuah, D., Ackaah, W., Tuffour, A. Y., & Aidoo, N. E. (2021). Uptake in on-demand ride-hailing for intracity long distance trip making during COVID-19. *Urban, Planning and Transport Research*, 1–12.

Dzisi, E. K., Ackaah, W., Aprimah, B. A., & Adjei, E. (2020). Understanding demographics of ride-sourcing and the factors that underlie its use among young people. *Scientific African, 7*, e00288.

Ellen MacArthur Foundation. (n.d.). *Climate change and a circular economy for transport*. https://ellenmacarthurfoundation.org/climate-change-and-a-circular-economy-for-transport. Accessed 25 Oct 2022.

Ellen MacArthur Foundation. (2019). Completing the picture: How the circular economy tackles climate change. *Ellen MacArthur Foundation Material Economics, 3*, 1–62.

Esposito, M., Tse, T., & Soufani, K. (2018). Introducing a circular economy: New thinking with new managerial and policy implications. *California Management Review, 60*(3), 5–19.

Fulton, L., Mason, J., & Meroux, D. (2017). *Three revolutions in urban transportation: How to achieve the full potential of vehicle electrification, automation, and shared mobility in urban transportation systems around the world by 2050* (No. STEPS-2050).

Habitat, I. I. I. (2015). Third United Nations conference on housing and sustainable urban development. *Journal of Education for Sustainable Development, 9*, 125.

Henry, M., Schraven, D., Bocken, N., Frenken, K., Hekkert, M., & Kirchherr, J. (2021). The battle of the buzzwords: A comparative review of the circular economy and the sharing economy concepts. *Environmental Innovation and Societal Transitions, 38*, 1–21.

Jin, S. T., Kong, H., Wu, R., & Sui, D. Z. (2018). Ridesourcing, the sharing economy, and the future of cities. *Cities, 76*, 96–104.

Kahl, C. H. (2018). *States, scarcity, and civil strife in the developing world*. Princeton University Press.

Kaye-Essien, C. W. (2020). 'Uberization' as neoliberal governmentality: A global south perspective. *Journal of Asian and African Studies, 55*(5), 716–732.

Kebede, K. Y., Ndegwah, D. J., & Kroesen, J. O. (2013). Successful contextual technology transfers and determinants of culture. In J. Kauffman & K.-M. Lee (eds.), *Handbook of Sustainable Engineering*. Springer Science.

Kębłowski, W., Lambert, D., & Bassens, D. (2020). Circular economy and the city: An urban political economy agenda. *Culture and Organization, 26*(2), 142–158.

Kenu, D. (2021). 'Pragya' rules Kumasi roads. *Graphic Online*. https://www.graphic.com.gh/news/general-news/ghana-news-pragya-rules-kumasi-roads.html. Accessed 28 Nov 2022.

Kirchherr, J., Reike, D., & Hekkert, M. (2017). Conceptualizing the circular economy: An analysis of 114 definitions. *Resources, Conservation and Recycling, 127*, 221–232.

Klopp, J. M., & Mitullah, W. (2016). "Politics, policy and paratransit: A view from Nairobi. In R. Behrens, D. McCormick, & D. Mfinanga (eds.), *Paratransit in African cities: Operations, regulations and reform*. Routledge, New York, NY.

Kufuor, K. (2018). Uber in Ghana: Markets and Institutions in the Emergence of Ride-Sharing Taxis. *LUGLJ: Lancaster University of Ghana Law Journal, 4*, 45–69.

Macarthur, E. (2013). Towards the circular economy, economic and business rationale for an accelerated transition. *Ellen macarthur Foundation: Cowes, UK*, 21–34.

Mitchell, W., Borroni-Bird, C., & Burns, L. (2010). *Reinventing the automobile: Personal Urban mobility for the 21st century*. MIT Press.

National Road Safety Commission (NRSC). (2014). *Road traffic crashes in Ghana. NRSC*. NRSC.

Obeng-Odoom, F. (2010). Drive left, look right: The political economy of urban transport in Ghana. *International Journal of Urban Sustainable Development, 1*(1–2), 33–48.

Obeng-Odoom, F. (2013). *Governance for pro-poor urban development: Lessons from Ghana*. Routledge.

Okraku, T. K. (2016). "Biribiara Wo Ne Mmerɛ" (everything has its time): Exploring changing perceptions of transportation on film from the colonial gold coast to contemporary Ghana. *Africa Today, 62*(4), 45–64.

B&FT Online. (2017). *Ghana's first female Uber driver makes it seven cars and counting*. https://thebftonline.com/20/12/2017/first-female-uber-driver-makes-it-seven-cars-and-counting/. Accessed 8 June 2022.

Orasmaa, A., Laurila, L., & Liimatainen, H. (2020). *Rethinking ownership: Producer ownership models in a circular economy*. The Finnish Innovation Fund Sitra.

Oteng-Ababio, M., & Agyemang, E. (2012). Virtue out of necessity? Urbanisation, urban growth and Okada services in Accra, Ghana. *Journal of Geography and Geology, 4*(1), 148.

Oteng-Ababio, M., & Agyemang, E. (2015). The okada war in urban Ghana: A polemic issue or policy mismatch? *African Studies Quarterly, 15*(4), 25.

Panwar, R., & Niesten, E. (2022). Jump-starting, diffusing, and sustaining the circular economy. *Business Strategy and the Environment*.

Parry, I. W., Walls, M., & Harrington, W. (2007). Automobile externalities and policies. *Journal of Economic Literature, 45*(2), 373–399.

Rizzo, M. (2011). 'Life is war': Informal transport workers and neoliberalism in Tanzania 1998–2009. *Development and Change, 42*(5), 1179–1206.

Santi, P., Resta, G., Szell, M., Sobolevsky, S., Strogatz, S. H., & Ratti, C. (2014). Quantifying the benefits of vehicle pooling with shareability networks. *Proceedings of the National Academy of Sciences, 111*(37), 13290–13294.

Schwanholz, J., & Leipold, S. (2020). Sharing for a circular economy? An analysis of digital sharing platforms' principles and business models. *Journal of Cleaner Production, 269*, 122327.

Sposato, P., Preka, R., Cappellaro, F., & Cutaia, L. (2017). Sharing economy and circular economy: How technology and collaborative consumption innovations boost closing the loop strategies. *Environmental Engineering & Management Journal (EEMJ), 16*(8).

Stasik, M., & Klaeger, G. (2018). Station waka-waka: The temporalities and temptations of (not) working in Ghanaian bus stations. *Africa Today, 65*(2), 93–110.

Sulemana, I. (2012). Assessing over-aged car legislation as an environmental policy law in Ghana. *International Journal of Business and Social Science, 3*(20).

Thomsen, T. P., Lybæk, R., & Christensen, T. B. (2022). Growth, resources and the circular economy. In *New Economies for Sustainability* (pp. 83–101). Springer, Cham.

United Nations Environment Programme (UNEP). (2020). *Used vehicles and the environment: A global overview of used light duty vehicles-flow, scale and regulation.*

USAID. (2016). *Greenhouse gas emissions in Ghana.* https://www.climatelinks.org/sites/default/files/asset/document/GHG%20Emissions%20Factsheet%20Ghana_6-17-16_edited_rev08-18-2016.pdf. Accessed 8 June 2022.

Williams, J. (2019). Circular cities. *Urban Studies, 56*(13), 2746–2762.

Woodcock, J., & Graham, M. (2020). *The gig economy. A critical introduction.* Polity.

Wu, Y., Zhu, Q., & Zhu, B. (2018). Comparisons of decoupling trends of global economic growth and energy consumption between developed and developing countries. *Energy Policy, 116*, 30–38.

Wuyts, W., & Marin, J. (2022). "Nobody" matters in circular landscapes. *Local Environment*, 1–18.

A Transition Towards a Circular Food Economy in Ghana: An Institutional and Policy Analysis

Mohammed Abubakari, Dorothea Afriyie Opoku Mensah, Ivy Agyeiwaa Amponsah, Dennis Asare, Mudasiru Mahama, and King Carl Tornam Duho

1 Introduction

The production and consumption of goods and services in most societies across the globe continue to follow the linear economy strategy, usually known as the "take-make-dispose" development model (Boon & Anuga, 2020). The linear strategy assumes that natural resources are limitless in supply and that the planet can absorb pollution and waste (Pimbert, 2015). Over the years, the application of this model has heavily relied on the over-exploitation of natural resources to stimulate global economic

M. Abubakari (✉)
School of Environment and Sustainability, University of Saskatchewan, Saskatoon, SK, Canada
e-mail: abubakarim63@gmail.com

D. A. O. Mensah
Department of Agribusiness and Applied Economics, North Dakota State University, Fargo, ND, USA

M. O. Erdiaw-Kwasie and G. M. M. Alam (eds.), *Circular Economy Strategies and the UN Sustainable Development Goals*,
Sustainable Development Goals Series,
https://doi.org/10.1007/978-981-99-3083-8_7

growth, wealth generation, utilization, and disposal of wastes which has proven to be unsustainable. However, its damaging effect on the environment, biodiversity loss, climate change, resource depletion, water scarcity, population growth, and economic issues cannot be discounted (Cobbinah et al., 2021; Hamam et al., 2021). This is the case in countries in the sub-Saharan Africa region specifically, in Ghana, where models of production (e.g. in the agricultural sector) are linear—the excessive use of natural resources to produce food leads to massive waste forming over 60% of municipal solid waste that ends up in the bin—without any safeguard measures for sustainable production (Ofori-Boateng et al., 2013). Agro-cycle agriculture, a form of agricultural practice which depends on an annual cycle of activities from land preparation, nursing, planting, and watering (crops)/feeding and raising animals, harvesting, marketing, etc., remains the principal economic activity in sub-Saharan Africa but the sector is still underdeveloped and limited by obsolete techniques, insufficient infrastructure, and funding (Boon & Anuga, 2020). As a result, the production levels of the agricultural sector in most sub-Saharan African countries like Ghana are insufficient to satisfy the increasing demand for food (Boon & Anuga, 2020).

Agriculture in Ghana is heavily dependent on rainfall and is almost exclusively represented by smallholder farmers using plots of less than 1.5 hectares. Productivity is generally low, mainly due to the use of low-input traditional farming systems and the erratic nature of rainfall in the country (Quaye et al., 2010). Whilst Ghanaian farmers are usually able to produce enough roots and tubers, they fall short in the production of cereals (rice, wheat, and maize), poultry, meat, fish, and vegetables, rendering

I. A. Amponsah
Faculty of Economics and Social Sciences, University of Potsdam, Potsdam, Germany

D. Asare
IMANI Centre for Policy and Education, Accra, Ghana

M. Mahama
Department of Geography and Sustainability Science, University of Energy and Natural Resources, Sunyani, Ghana

K. C. T. Duho
Dataking Research Lab, Dataking Consulting, Accra, Ghana

the country highly import-dependent to make up for the deficits (Agyei-Holmes et al., 2021). Specifically, the country produces about 51% of its cereal needs, 60% of its fish requirements, 50% of meat, and less than 30% of the raw materials needed for agro-based industries (Darfour & Rosentrater, 2016). This implies that the country can only produce about half of its food requirements for cereals and meat. Worsening is the fact that since most of Ghana's food production occurs in the rural areas, it is estimated that the country loses about 20–30% of cereals and legumes and about 20–50% of roots, tubers, fruits, and vegetables in storage and during transport to the urban areas (Rutten & Verma, 2014). Post-harvest cereal losses can be as high as 50–70%. It is expected that the food losses in the country can be transformed into food waste soon as a result due to urbanization and income growth (Rutten & Verma, 2014). Amid these food losses, 3.6 million Ghanaians (12% of the population) face severe and moderate food insecurities, and the food insecurities and threat of hunger are three times in rural areas compared to urban centres (Ghana Statistical Service, 2021). Ghana's agri-food sector is in dire need of transitional production and consumption patterns. The circular economy concept offers a valid pathway to enhance resource efficiency and recover value from food loss and waste (Liu et al., 2021).

A circular economy is defined as "a production and consumption model, which involves sharing, renting, reusing, repairing, renovating and recycling existing materials and products for as long as possible and reducing to the minimum of waste" (European Union, 2022, p. 1), offering a better alternative to the current model of economic development, the "take-make-dispose" model with a view to environmental, economic, and social sustainability. The advantages of circular economy systems are attributable to the reduction in the environmental impact through the minimization of waste, the redesign of products, the choice of materials, the reduction in price volatility, and the increase in economic benefits through, for example, an increase in job growth (Hamam et al., 2021). Looking at the prospects of a circular economy in advancing sustainable production and the dominance of linear models of production in Ghana's agricultural sector, this study seeks to contribute to academic and policy literature by analysing the extent of Ghana's readiness to transition its agricultural production systems from linear food production to circular food production and proposing a model for such transition with implications for the Sustainable Development Goals (SDGs), specifically goals 1 (no poverty), 2 (no hunger), and 12 (sustainable production and consumption patterns). In previous circular transition studies, Schot and Kanger (2018) identified institutions and technology as the key variables for a transition to occur, whilst Shove (2014), and Geels (2005) indicated that infrastructure and citizens' worldviews, competencies, and practices are also important variables. Considering these variables and the circular transition literature, this study uses Ghana's agricultural sector's 'institutions' and 'policies' as the analytical gateway to contribute to scholarly works on circular economy practices in sub-Saharan Africa and the developing world.

Circular economy has generally remained popular amongst multinational corporations, researchers, and policymakers in developed countries. But for developing countries, except China, the circular economy approach's potential contributions to low- and middle-income countries like Ghana and other sub-Saharan African countries have gotten comparatively little attention (Schroeder et al., 2018). Recent studies on circular economy in the agricultural sector of sub-Saharan Africa and specifically Ghana have focused on the assessment of the entire agri-food value chain, which according to the International Panel of Experts on Sustainable Food Systems (2015), entails production, processing, storage, marketing, distribution, and consumption. Such a focus in the recent studies is too broad to understand the nitty–gritty of how the transition can 'truly' be spearheaded. For example, Boon and Anuga (2020), to establish the theoretical and practical importance of circular economy for advancing sustainability, paid attention to the value chain of some selected crops in Northern Ghana, limiting the national relevance of their study to some selected regions. Also, Oppong et al. (2021) followed up with a more advanced study (at the national level) using a triple-bottom-line sustainability assessment, but the focus was still on the entire agri-food value chain. As a result, specific studies focusing on a particular component of the value chain, especially from an institution and policy perspective, are hard to identify in the circularity literature on Ghana. However, a deeper understanding of the institutional and policy framework by emphasising a particular component of the agriculture value chain can help identify specific lapses and policy direction needed for circular food transition to happen in Ghana with implications for sub-Saharan Africa. This has been the motivation for conducting this study which focuses on "production" as a component of the value chain to ascertain the degree to which it can be transformed towards circularity (and sustainability). The unique emphasis on "production" is justified by addressing the circularity constraints at the chain's production end can catalyse the entire chain. Also, Ghana is still not grounded enough in institutions, policy interventions, and priorities in the other components of the food system (e.g. marketing, distribution, consumption), not to even talk of a call for a transition to circularity in those components. It is worth noting that due to the interconnected nature of Ghana's agricultural value chain; there is a complex mix of institutions and policies that do not only cover "production" but, to some extent, the other components such as processing, storage, marketing, etc., in a holistic manner. Whilst this study focuses entirely on "production", the analysis, in some cases, partly hints at and covers the other components of the food system.

By structure, the study is divided into eight (8) interconnected sections, including this section (Sect. 1) which introduces the study and sets the agenda for it. Section 2 provides the theoretical framework. Section 3 provides the details of the study design and methodology. Section 4 analyses the institutions and policies oriented to the sustainable food production systems in Ghana. Section 5 synthesizes and draws insights from Sects. 2–4 to develop a circular food production model for Ghana. Section 6 draws the implications associated with sustainable food systems and synergizes them with the SDGs that have a direct bearing on food production, specifically goals 1 (no poverty), 2 (no hunger), and 12 (sustainable production and consumption patterns). Finally, Sect. 7 concludes the study by reflecting upon the findings to draw policy lessons that can contribute to sustainable food production in Ghana and the sub-Saharan African region at large.

2 Theoretical Framework

2.1 Circular Food System: Justifications, Principles, and Transitions

Circular food system (CFS) has gained traction in science and policy spaces following the global interest in and pursuing the sustainability agenda—the SDGs (Fassio, 2021). Goal 12—Sustainable Production and Consumption Patterns, for example, is seen to directly drive the need to transform food systems from linearity to circularity (Fassio & Tecco, 2019), which can contribute to achieving Goal 1—No Poverty, and Goal 2—No Hunger. This is necessary because the linearity of the global food system contributes significantly to climate change and environmental destruction (Shukla et al., 2019). Furthermore, it is ascertained that the food systems emit at least a quarter or more of greenhouse gases, which is likely to exacerbate, especially in developing countries (Springmann et al., 2017). The need to reduce greenhouse gas emissions in food systems at the production level, for instance, is necessary (Ripple et al., 2014), and CFS can play a key role in ensuring that.

CFS is derived from circular economy—a term that is driven by the sustainability imperatives of replacing the linear approach with the circular approach for the flow of materials and energy to cut down (or eliminate) environmental costs and enhance benefits (Barros et al., 2020; Schulte, 2013; Sillanpää & Ncibi, 2019). The circular economy remains contested in the literature due to differences in its conceptualization. For example, Kirchherr et al. (2018) identified at least 144 definitions of circular economy in the literature. Notwithstanding the diversity of its conceptualization, there is a consensus that circular economy has its root in industrial ecology—a field of study that seeks to close material loops, cut down consumption, and reduce (or eliminate) waste by replicating the natural ecosystem (Jurgilevich et al., 2016). As an attempt to produce a more generic definition of circular economy, by analysing multiple and different definitions of circular economy, Kirchherr et al. (2018, pp. 224–225) define it as " ... an economic system that is based on business models which replace the 'end-of-life' concept with **reducing**, alternatively **reusing**, **recycling** and **recovering** materials in production [...]to accomplish sustainable development, which implies creating environmental quality, economic prosperity, and social equity, to the benefit of current and future generations".

CFS is borne out of a circular economy and seeks to cut down waste in the food systems through **reusing**, using by-products, **recycling** nutrients for food production, and streamlining the consumption of food per production and processing patterns to **recover** materials and **reduce** waste (Van Zanten et al., 2019). Hence, the 4R involving Reusing, Recycling, Recovering, and Reducing are the core principles that underpin food production and are used to underlie circularity in food production in Ghana. This aligns with Kirchherr et al.'s (2018) revelation that the 4R (Reducing Reusing, Recycling, and Recovering) forms the core of the circular economy. There are a series of CFS-informed interventions that ensure the 4R principles in food production, including organic farming and manuring, sewage sludging, smart agriculture (Dong et al., 2020), and precision livestock production (Shane et al., 2016), which have been embraced in this study as they can easily be pursued in developing countries (for instance, Ghana and the SSA).

Organic farming and manuring involve the use of natural materials such as undecomposed plant materials (mulching) and animal excreta (waste) as a fertilizer (a resource) in crop production (Dong et al., 2020), feeding directly into two of the principles—reusing materials and reducing waste. Sewage sludging involves using wastewater for food production practices (Kacprzak et al., 2017), instead of outright disposal, feeding into reusing and reducing, and recovering (of water). Smart agriculture and food production practices include direct sales of farm produce between farmers and consumers, such as on-farm purchases, and the use of online farmers' stores which reduce the need for packaging, cut down the supply chains, and ensure the freshness of farm produce (Jurgilevich et al., 2016) as well as crop rotation, mixed/inter-copping, and (rain) water harvesting. Smart agriculture practices, first of all, reduce waste through direct contact with consumers as soon as possible. It also includes natural recycling and the recovering and reusing of materials through nature-based farming practices (crop rotation, water harvesting) for efficient and effective use of natural resources. Precision livestock production involves the use of input fertilizers and agrochemicals variability in soils, microclimate, and other relevant husbandry parameters such that the right amounts of resources are used at the right time and in the right place to achieve optimum performance with minimal environmental impact (Shane et al., 2016). Precision livestock methods include using animal manure and food residues as resources, and a low-cost grass-based rotational-grazing system for milk. As a CFS-informed intervention, this promotes the 4R principle in farming practices.

Integrating manuring, sewage sludging, smart agricultural practices, and precision livestock production methods in the production process implies making the 4R principles an important part of policy making and implementing CFS transition in agricultural production. This can be ensured in policy practices through offering assistance (training, finance/investment, and tax incentives) for mixed farming practices and other circular-oriented farming, promoting nature-based recycling practices as well as providing recycling facilities for converting waste during the production process, supporting the use of recovered and recycled nutrients (higher taxes on the importation of inorganic fertilizers), and encouraging sales of farm produces in local markets (Jurgilevich et al., 2016). Such policy practices for transition to occur in agricultural production rest on institutions and technology (Schot & Kanger, 2018). However, the institutional and technological challenges facing the agricultural sector in Ghana (and sub-Saharan Africa) are immense (e.g. inadequate human and material resources for institutions and the use of obsolete technology in production). This makes a transition to CFS a challenging one. De Jesus and Mendonça (2018) contend that countries must be aware of the barriers of circularity when making a transition, whilst Bauwens et al. (2020) call for the need to understand and conceptualize how the 'future' of such a transition will be for the design and implementation of mission-oriented policies. In addition, the literature also identifies infrastructure and knowledge (competencies, practices, and worldview) as important variables necessary for transition (Geels, 2005; Shove, 2014)—fundamental change, instead of a slight twist and change in a system (Kirchherr et al., 2018)—to occur.

The circularity barriers include ones associated with culture (absence of awareness and/or limited interest in the practical pursuit of circularity despite its 'fame') (Mont et al., 2017; Pheifer, 2017); regulation (limited circularity-imbibed policies, and the existence of circularity obstructing laws and regulations) (Pheifer, 2017); market (the huge cost burden in spearheading circularity vis-a-viz linearity) (Mont et al., 2017); and technology (limited technologies to support circularity and its practices) (Mont et al., 2017; Pheifer, 2017). These barriers are interconnected but vary in terms of their manifestations in geographical settings and sectors (de Jesus & Mendonça, 2018; Erdiaw-Kwasie et al., 2023; Kirchherr et al., 2018) and can be dealt with by focusing on 'institutions and 'technology' as well as paying attention to 'infrastructure' and 'education' for increased knowledge and understanding of actors involved. Considering Ghana's agricultural sector, 'institutions' tend to be the most pivotal variable to focus on in catalysing the CFS transition in agricultural production. For instance, cultural barriers in the agricultural sector manifest in the form of a lack of policy awareness and focus on the circular economy leading to the lack of policies, laws, and regulations that can promote circular-oriented technologies, infrastructure, and education and training needed for CFS transition to happen (A discussion on this is presented in Sect. 4).

Also, focusing on how the 'future' of the CFS transition will be in Ghana, four possibilities are identified in the literature based on a theoretical/conceptual framing developed by Bauwens et al. (2020): planned circularity (a transition hegemonically pushed by the government through authoritarian measures, for instance, higher taxes on the importation of inorganic fertilizers, or the ban of inorganic chemicals for farming); bottom-up sufficiency (a local-based transition usually through education and consumer engagement for sustainable behaviour and practices); circular modernism (a transition more technologically oriented involving decisions by government and few large businesses for circularity to occur, for instance, agricultural production methods based on eco-efficiency research and development); and peer-to-peer circularity (a transition more driven by technology through sharing and collaborative platforms for temporal access and use of resources and products). The conceptual model in Sect. 5 is developed through reflections on these 'plausible' futures (Bauwens et al., 2020) to inform mission-informed and innovative policies that can lead to an ideal circularity in agricultural production systems for sustainability in Ghana.

2.2 *Theoretical Basis for Circular Food Systems Transitions*

The sustainability and circularity literature is proliferated by different theoretical and conceptual ideas that can serve as a framework for circular food systems (CFS). They include industrial ecology, biomimicry, zero emissions, eco-efficiency and effectiveness, blue economy, bio-economy, cradle-to-cradle (C2C), performance economy, natural capitalism, and regenerative design (see, for instance, Andrews, 2015; Geisendorf & Pietrulla, 2018; Hailemariam & Erdiaw-Kwasie, 2022; Ogunmakinde et al., 2021) that are usually used to relate and discuss circularity. Within the context of CFS, and in relation to food production for sustainability, this study uses the C2C and natural capitalism concepts as they are useful and fitting for developing a CFS model for sustainable food production and align with the theoretical reflections on CFS (potential) barriers and 'futures' of CFS transitions in the agricultural production system of Ghana. Furthermore, the concepts of the C2C and natural capitalism are theoretically grounded by the social practice theory—materials, competence, and meaning—that can foster circularity as a form of sustainability transition in food production practices.

The first concept, C2C, is a model of production based on the 4R through a closed-loop practice of recycling, recovering, reusing, and reducing waste in practices (McDonough & Braungart, 2010). Whilst it is traditionally an industrial model of production in the design of products (McDonough et al., 2003), C2C can be contextualized to fit into food production practices not only by corporate agricultural organizations but also by smallholder farmers who are the primary food producers in Ghana (see: Chamberlin, 2008). In C2C, food production must be approached such that materials could potentially be considered nutrients (i.e. recovering) or are subjects for recycling or reusing. In this way, "… the waste of one production system is the "food" for another system, meaning "everything" can be designed, produced, used, and dissembled to be safely returned to the soil as "biological nutrients" or introduced to the production cycles as "technical nutrients" (Sillanpää & Ncibi, 2019, p. 16), thereby reducing waste substantially in farming practices. The second concept, natural capitalism, is also driven by 4R as it seeks to ensure the effective, efficient, and sustainable use of natural assets (soil, air, water, trees, minerals, etc.) for productivity in production practices and technologies (Srinivas, 2015). Its practices are closely related to C2C based on biologically oriented closed-looped production and re-use of raw materials for other production processes (Hawken et al., 2013). "Natural capitalism is inspired by nature, and it involves investing in natural capital by promoting initiatives and aiming activities at restoring and regenerating natural resources" (Sillanpää & Ncibi, 2019, p. 17). The CFS-informed interventions (manuring, sewage sludging, smart agricultural practices, and precision livestock production) are primarily related to both C2C and natural capitalism, which promote the 4R principle and can be pursued in Ghana, provided 'institutions' identify circularity barriers and deal with them based on the kind of future envisioned for a transition to occur. Also, for Ghana's food producers (farmers) to significantly uptake C2C and natural capitalism circularity practices, the institution and policy framework for food production must promote social practices and learning amongst farmers, as argued by the social practice theory.

The social practice theory suggests that practice can prevail and dominate when it is repetitively performed by people (e.g. farmers) through similar 'materials' (objects, equipment, tools, etc.), 'competence' (knowledge, skills, and understanding), and 'meaning' (benefits/value) (Shove et al., 2012). The three elements (materials, competence, and meaning) "…are integrated into the continuous enactment of the practice. The emergence, persistence, and disappearance of practice hinge on the making and breaking of the links between their defining elements" (Jauho, 2013, p. 287). On this basis, for C2C and natural capitalism practices to be effective in food production in Ghana, institutions and their policies must make farmers understand and have knowledge and skills for CFS practices in crop cultivation (i.e. competence), understand the benefits and value associated with them (i.e. meaning), and make available the resources (i.e. materials) needed for such practices to be done through social learning (i.e. farmers learning from each other in circular-driven farming practices) for successful integration of the 4Rs in farming practices. In doing so, 'institutions' (as a transition variable) can enhance farmers' competencies and practices and promote a circular worldview amongst them (a transition variable echoed by Shove, 2014 and Geels, 2005). It is also a way to deal with circularity barriers, especially cultural and technological barriers that can disrupt the transition. From the perspective of the social practice theory, without an in-depth understanding of the 4R principle through the lens of C2C and natural capitalism practices (i.e. competence for a circular transitioning in farming) and associated benefits to farmers (i.e. meaning derived from such a transition), as well as the availability of resources commonly used for such practices (i.e. materials needed for the transition) through institutional support, a circular transition in farming, might not be possible in Ghana.

Since circular 'transition' in farming/food production is promoted in this study, it becomes important to highlight the (social) effects should transition occur. This is important as social effects are usually overlooked in the literature (Köhler et al., 2019). Circular transitions are "long-term, multidimensional, and fundamental transformation processes through which established socio-technical systems shift to more sustainable modes of production ..." (Markard et al., 2012, p. 956), with multiplier (social) effects. One of the few studies on the social implications of circular transitions in developing countries identified a significant job reduction in, for instance, the textile industries (Repp et al., 2021). Another one focusing on plastic and mineral recycling argues that circular transition will produce about six million jobs, but that will occur significantly in developed countries, with Africa rather losing about one million jobs.

In the agricultural sector, little is known about the social effects of a transition to circularity, especially in developing countries such as Ghana. Whilst Boon and Anuga (2020) argue that circularity will lead to farmers' efficient use of natural resources and an increase in the quality and quantity of their outputs, they should have touched on how many farmers might move out of farming due to a circular transition. However, the fact that smallholders primarily drive food production in Ghana usually in rural communities, instead of corporate agri-investors (Food and Agricultural Organization—FAO, 2015), it becomes hard to say such farmers would lose their jobs due to circularity. But of course, farmers in contract farming with corporate investors may be affected if their services are not needed due to a circular model favouring high-level technology. Based on Boon and Anuga's (2020) study, it is likely that a significant increase in the outputs of the 'dominant' smallholder farmers will lead to higher incomes and expansion of their farmlands for which additional labour will be required (perhaps, from small-scale to medium-scale farming). Potentially, this will not lead to a decline but rather an expansion of the agricultural sector, for which many farmers will be required. Amid such outcomes, circularity-oriented policy efforts in Ghana will have to consolidate the employment gains whilst addressing potential issues that might arise, such as child labour and capitalist production. This reflection aligns with Kaza et al.'s (2018) study that identified that a circular economy in developing countries could transform local agricultural practices and create more jobs, especially in managing (agricultural) waste.

3 Materials and Methods

This study is a desk review involving a synthesized review of different but complementary sustainability literature (Basson et al., 2018; Gough et al., 2012) related to Ghana's food systems and agri-food sector policy documents. The review is based on a literature review approach that has been recommended by Cronin et al. (2008)—literature searching; literature gathering and reflection; and writing and referencing—which also feeds into the SALSA (Search, Appraisal, Synthesis, and Analysis) framework by Grant and Booth (2009) noted as a means for undertaking rigorous and guided review (Fig. 1).

The study uses academic and non-academic outlets/databases to obtain secondary documents. The academic databases used for the search were Scopus and Web of Science supplemented with Google Scholar and Google where documents not necessarily academic were obtained. The interest in not just academic publications was driven by the fact that policy documents and writings have tremendously informed circularity literature (Blomsma & Brennan, 2017). Also, the bibliographies of the documents searched for and gathered were examined, which aided in identifying other relevant documents used for the study. The following search query was used to search for documents in the academic databases (i.e. Scopus

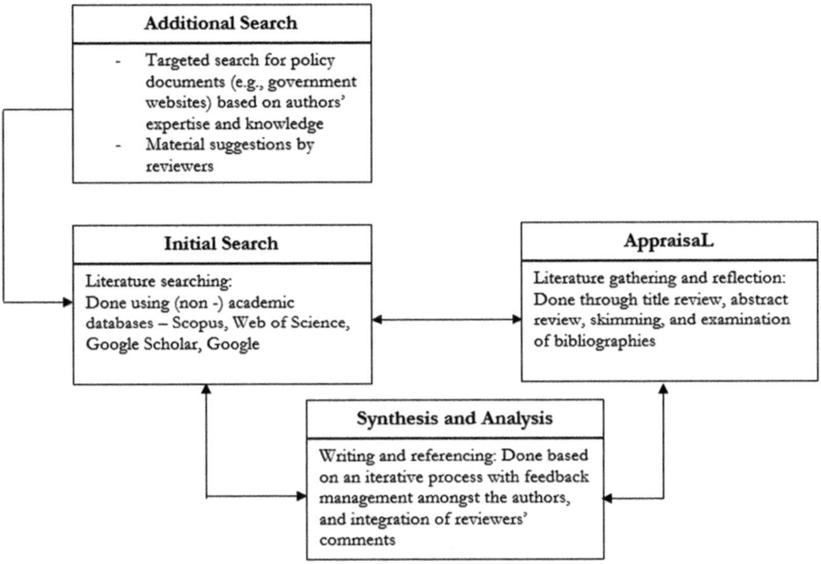

Fig. 1 The adopted literature review approach (*Source* Authors' construct [2022])

and Web of Science): "Circular Economy" AND ("agricultural production" OR "transition" OR "agri-food policy" OR "sustainability" OR "Ghana"). As a result, many publications were obtained (over 40). Still, through gathering and reflection (appraisal), a significant number (about 20 publications) were excluded because their focus was not related to the objectives of the study. For the number of papers that were included, the authors identified other relevant terms/themes related to circular economy such as circular food system, cradle-to-cradle, natural capitalism, organic farming, R-framework, R-principle, etc. As a result, the initial search provided a starting point to guide the authors in subsequent searches (i.e. additional search). In Google Scholar, different combinations of the various terms were used to gather additional documents. For example: "Circular Food System and Agricultural Production in Ghana", "Circular Food System and Agri-policies in Ghana", and "Institutions and Policies for a transition to Circular Economy in Ghana". Also, the authors obtained specific academic publications through the suggestions of reviewers, especially those that ground the theoretical and conceptual framing and depths of circularity used in the study. The initial search for academic-related documents was done between March and June 2022, whilst the additional search was done between September and October 2022.

As the authors are aware of specific government policies in the agricultural sector of Ghana due to their professional and research experiences, some specific websites were purposively used to search for policy documents. This was done between March and June 2022, the same period as the initial search for academic publications. The specific websites are the Government of Ghana's websites (the Ministry of Food and Agriculture-MOFA, The Ghana Statistical Service, and the National Development Planning Commission-NDPC) and the website of the Food and Agricultural Organization (FAO) for policy documents that summarize previous policy interventions and provide an up-to-date picture on food production and sustainability in Ghana. Through this targeted search approach, the authors obtained the following documents; the Food and Agriculture Sector Development policy-FASDEP II (2007), Ghana Shared Growth and Development Agenda-GSGDA (2010–2013), Medium-Term Agriculture Sector Investment Plan (2011–2015), the 2018 Ghana Living Standard Survey (GLSS 7), Ghana Voluntary National Review Report on the Implementation of the 2030 Agenda for Sustainable Development, the 2020 Ghana Food Security and Nutrition Monitoring System (FSNMS), Ghana's, 2020 Comprehensive Food Security and Vulnerability Analysis (CFSVA), and Investing for Food and Jobs (IFJ): An Agenda for Transforming Ghana's Agriculture (2018–2021) and Ghana's, 2020 SDGs Report.

The data obtained from the academic and policy documents were gathered, reflected upon/analysed, and organized/synthesized to inform the study's writing and referencing process, taking into account the purpose of the study. The writing and referencing process was iterative such that the authors reviewed each other's part of the work and provided inputs to solidify the study's arguments and contributions.

4 Analysis of Institutions and Policies for Sustainable Food Production in Ghana

Post-independence food production policies were premiered through price restrictions, subsidies, and strong participation of the State (Brooks et al., 2007). One prominent policy was "Operation Feed Yourself" which followed the Green Revolution in Asia and aimed at promoting self-sufficiency through export-oriented agricultural production, with the need to satisfy local industries' demands for raw materials (Vernhout, 2014). Another policy-oriented programme for food production was the Economic Recovery Programme spearheaded by the World Bank in the 1980s as part of the response to bushfires and their associated droughts in 1983 and other mounting economic pressures. The Economic Recovery Programme consisted of three phases, with the first two phases related to the agricultural sector, primarily focusing on price stabilization and food security. The idea was that lower inflation and higher prices are given to food producers would contribute to poverty reduction. The final phase focused more on agricultural exports than food production through interventions including deregulation of service and commodity markets, liberalization of export markets, and elimination of minimum prices and subsidies (Yaro et al., 2018).

The Structural Adjustment Programme (SAP) III continued the reforms in the agricultural sector during the early 1990s (Yaro et al., 2018). The forefront of agricultural priorities was to increase output, specifically for export and food security. The 2000s saw a liberal economic paradigm entrenched in the agricultural sector, implementing two Growth and Poverty Reduction Strategies with funding from the World Bank. These strategies encouraged the transition from an agrarian economy to an industrial one. Table 1 gives a snapshot of policy interventions and priorities of the agricultural sector of Ghana from the 1970s (immediately after independence) to 2010, the year in which world leaders initiated the Millennium Development Goals, which serve as the leverage point for the current global sustainability framework, the SDGs.

Since 2006, improving agricultural production and exports has been featured in most policies and strategies, allowing for mechanization and subsidizing costs for farmers as the leading approach (Food and Agricultural Organization, 2015). The Ministry of Food and Agriculture implemented programmes such as the Block Farming Programme, Agricultural Mechanization Centres, the Irrigation Development Programme, and the Fertilizer Subsidy Programme to support farmers (Ministry of Food and Agriculture, 2007). The Food and Agriculture Sector Development Policy (2007) started this journey of agricultural mechanization to increase the productivity of the average Ghanaian farmer (Ministry of Food and Agriculture, 2007). The Medium-Term Agriculture Sector

Table 1 Overview of Food Policy Paradigms in Ghana Over Time

	1970s	*1980s*	*1990s*	*2000s* *2010s*
Relevant Policies	Operation Feed Yourself	Economic Recovery Programme (ERP)	Structural Adjustment Programme (SAP)	Growth and Poverty Reduction Strategies and Millennium Development Goals (MDGs)
Broad Paradigms	Government intervention/ control to increase agricultural productivity	Reduced government intervention, privatization	Liberalization, deregulation	Industrialization & commercialization of agriculture
Food Policy Priorities	Food security, export & import substitution (Balance of Payments)	Price stabilization, removal of subsidies and controls (ERP), increase in yields, aggregate food security	Privatization, trade liberalization & deregulation (SAP), aggregate food security	Enhance private sector activity, modernization, export diversification, employment/ poverty reduction (new subsidies), market orientation and reliance on the private sector, aggregate food security

Source Thow et al. (2020)

Investment Plan (2010–2015) was the implementation plan of The Food and Agriculture Sector Development Policy (2007), and it had food security and emergency preparedness; and science and technology as applied to food and agricultural development as part of its main agricultural policy priorities (Ministry of Food & Agriculture, 2010). Also, Ghana's School Feeding Programme, a consumer-oriented agricultural policy, ensured that food needed to support the programme was locally produced and procured, providing a direct output market for smallholder farmers (Food and Agricultural Policy, 2015). The Ghana Shared Growth and Development Agenda (2010–2013) also shared in this vision propelling all other governmental strategic plans and frameworks to include agricultural extension, research, and infrastructural development (National Development Planning Commission, 2010). As indicated in Fig. 2, which is an

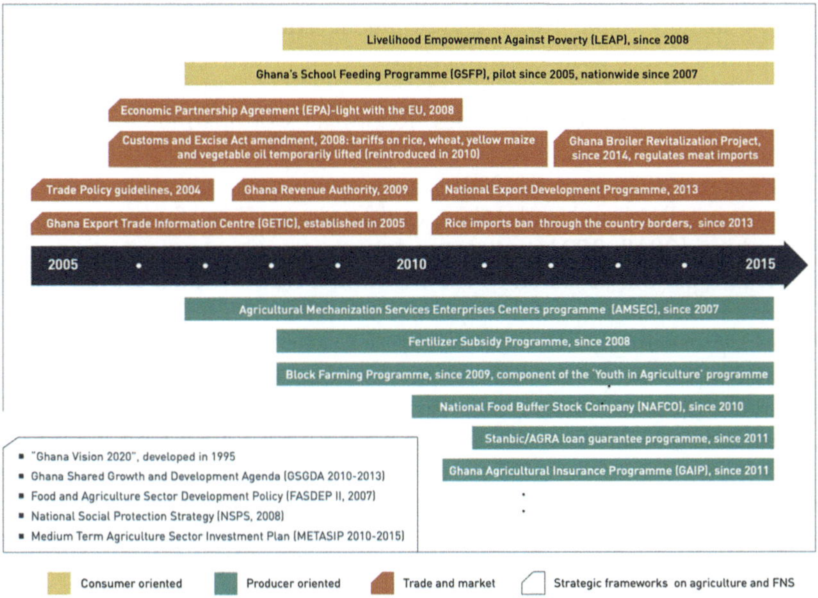

Fig. 2 Strategies and Policies related to Agriculture and Food Security in Ghana (*Source* FAO [2015])

extension of Table 1, various policy interventions with implications for the agricultural sector have taken place, including the Livelihood Empowerment Against Poverty and the Ghana School Feeding Programme, which aim to cushion food production and alleviate poverty and hunger (National Development Planning Commission, 2010).

Currently, the Ministry of Food and Agriculture—MoFA (production), the Ministry of Fisheries—MoF, (production, processing, and trade), the Ministry of Local Government, Decentralization and Rural Development—MLG&RD (agriculture development and local industries), and the Ministry of Trade and Industry—MoTI (agro-processing, retail, trade, and marketing) oversee food policy in Ghana (Thow et al., 2020). In addition, the National Development Planning Commission oversees and guides policymaking in the country and advises on general national policy goals. In recent times, agricultural policy objectives in Ghana have been aimed at economic and livelihood development. Agriculture's economic importance is evident in the Ministry of Local Government,

Decentralization and Rural Development and the Ministry of Food and Agriculture's policies and programmes, which encourage agricultural output through decentralized governance structures (Metropolitan, Municipal, and District Assemblies/Regional Coordinating Councils) and agribusiness development as part of Small and Medium-sized Enterprises Development (Thow et al., 2020).

Investing for Food and Jobs: An Agenda for Transforming Ghana's Agriculture (2018–2021) was created to put the Government of Ghana's vision into action, as stated in the Medium-Term National Development Policy Framework dubbed "*Agenda for Jobs: Creating Prosperity and Equal Opportunity for All (2018–2021)*" (Ministry of Food & Agriculture, 2018, p. 1). It also links to international development frameworks like the SDGs, the Comprehensive African Agriculture Development Programme—Malabo declaration, and the Economic Community of West Africa Agricultural Policy at the global, continental, and regional levels, respectively, to local agricultural policy implementation (Ministry of Food & Agriculture, 2018). The goal of this medium-term plan is to reform the agricultural sector by investing not less than 10% of the national budget in agriculture to achieve an annual sector growth rate of at least 6% over the plan period, as stipulated in the Malabo declaration (Ministry of Food & Agriculture, 2018). The planting for Jobs campaign is one of the government's flagship agricultural projects. Its goals include guaranteeing food security, expanding job possibilities, particularly for the youth, assuring the availability of raw materials for industry, and promoting agricultural exports.

Inferring from the reviews of policies and programmes in the agricultural sector, it appears that none of the policies and programmes has ever considered a circular transition in food production to mitigate the food losses in the sector. This implies that the circular economy concept is nascent and needs critical attention.

5 A CFS Model for Ghana: Insight from Reviews on Institutions, Policies, and Theories

Through the analysis of institutions and policies for food production in Ghana, the study gathered a lack of focus on circular food production systems. Policy efforts have generally pushed for economic well-being and prosperity, for example, livelihood improvements and poverty reduction, value addition, large-scale production and mechanization for food security, improved agricultural productivity, access to the market, and neoliberal interventions. All these efforts have general positive implications for economic (and social) sustainability. However, the consequences of food production practices on the natural environment (environmental sustainability) have not been given considerable attention. Contemporary policies by concerned institutions (e.g. Ministry of Food and Agriculture, Ministry of Local Government, Decentralization, and Rural Development, etc.) are still incentivized by economic benefits, for example, Investing for Food and Jobs policy (2018–2021), and are based on linearity in terms of food production slowing the country's progress towards the sustainability agenda.

A circular approach to policy analysis is, thus, a requisite approach for shifting from the current linear production model to a circular one in Ghana (Fig. 3) for environmentally sustainable production practices amongst farmers/producers and dealing with the losses associated with food production. The starting point is a policy thought for CFS by responsible institutions which can engineer a sense of intentionality and sensibility to circularity in policy conceptualization and design based on the 4R principle. This can lead to implementing a CFS-informed policy and associated initiatives and practices in agricultural production in Ghana that promote such circular economy principles. Such a policy and associated initiatives also call for cooperation and coordination amongst the responsible state institutions and other institutions and agencies at both the local and international levels.

The CFS policy objective must focus on a transition of food production from linearity to circularity for sustainable food production in Ghana (Barros et al., 2020; Hamam et al., 2021), with key indicators to measure transition progress taking into account (potential) circularity barriers that exist (de Jesus & Mendonça, 2018), and a future outlook concerning the kind of transition that is expected to happen—a planned, or bottom-up sufficiency, or circular modernism, or peer-to-peer, or a hybrid circularity (Bauwens et al., 2020). The indicators, taking into account the variables for transition to occur (i.e. institutions; technology; infrastructure; and competencies, practices, and worldviews of individual producers) should start with whether or not CFS policy has been formulated and implemented by the concerned state institutions, and if 'yes', what are (i) the proportion of farmers/ producers engaged in CFS policy-informed initiatives, (ii) the percentage of farmers/producers in CFS policy-informed initiatives financially supported, (iii) the proportion of farmers that have

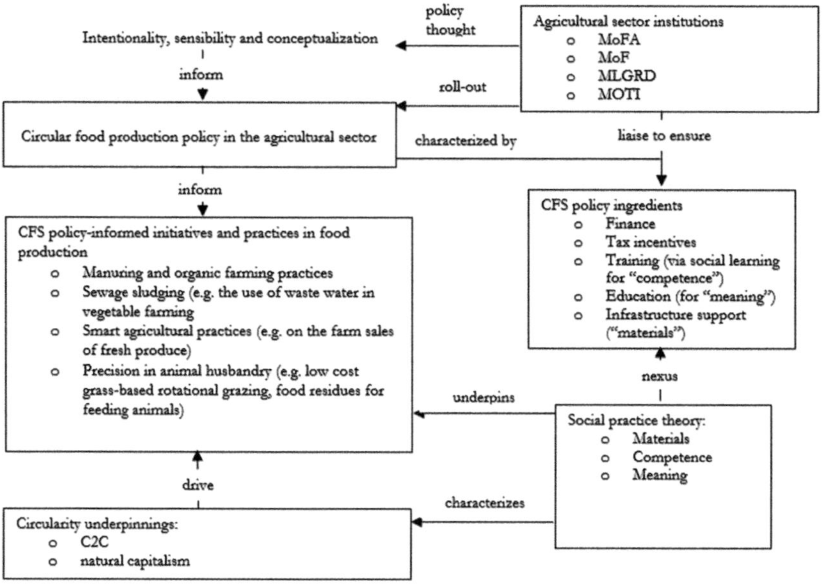

Fig. 3 A proposed CFS model for Ghana's food production (*Source* Authors' construct [2022]; insight is drawn from the literature)

been educated and trained on CFS policy-informed initiatives, and associated technologies and practices based on, for instance, C2C and natural capitalism concepts, and (iv) a percentage of farmers who practice circular food production. Based on the proposed model, these indicators can help measure the degree of shift (and transition) from linearity to circularity in food production. Ultimately, the authors perceive the transition to take a hybrid circularity approach where different circularity scenarios guide policy actions for agricultural production.

It becomes obvious that a transition can occur when farmers (producers) are financially supported. This can happen through the government's finance offerings and tax subsidies to incentivize the farmers and create an enabling environment for them to be educated and trained through social learning. Such an enabling environment will call for communications, knowledge exchanges and sharing, and interactions amongst farmers engaged in 'common' food crops and/or animal farming for learning and similar knowledge uptake and understanding of their newly adopted circular practices. Also, farmers will have to be infrastructurally supported (e.g. recycling facilities) to engage in circular food production initiatives, technologies, and practices (e.g. organic farming, manuring in farming, etc.). The CFS policy initiatives, technologies, and practices can then lead to the 4R principles where the minimization of waste through recovering, recycling, and reusing for closed-loop food production to reduce losses/waste, i.e. the concept of C2C (McDonough & Braungart, 2010). The use of 'social practice' as the crux of policy initiatives can make farmers understand the importance of the natural environment (e.g. land, water, seeds, etc.) so that they treat them as assets leading to careful utilization during their farming practices, i.e. the concept of natural capitalism (Hawken et al., 2013). With the prescribed and described CFS model for food production, Ghana can gradually assimilate circularity practices in food production at the local level for a systematic transition towards environmentally sustainable food production underpinned by the 4R principle in the production processes. In summary, the authors argue that within the policy space; the following should happen for a transition to take place in agricultural production in Ghana: (i) Carving a circular policy taking into account the (potential) barriers that disrupt circular practices, and envisioning the kind of transition expected to happen with key indicators to measure the progress towards the transition; and (ii) Creating an environment for the circular policy to 'survive' and be sustainable through cooperation and collaboration; finance for policy design, implementation, and monitoring and evaluation; and infrastructural, technological, education, and training support systems that complement and catalyse the policy impact. Agricultural sector institutions must ultimately lead all these in the country.

6 Circular Food Economy and SDGs Nexus: Implications for SDGs 1 (No Poverty), 2 (No Hunger), and 12 (Sustainable Production and Consumption Patterns) in Ghana

Circular economy activities can practically contribute (directly and indirectly) to achieving the SDGs, and this is extremely important in a country like Ghana, which ranks 110th out of 163 countries on the 2022 SDG Index with a score of 63.4. Ghana's progress for SDG 1 (no poverty) is 'moderately improving', whilst SDG 2 (no hunger) is 'stagnating'. For SDG 12 (sustainable production and consumption), Ghana is seen as 'on track or maintaining SDG achievement' (Sachs et al., 2022). Whilst Ghana is 'stagnating' in progress for SDG 2 (no hunger), a circular food economy can contribute to such a goal by reducing food losses and waste and building circular, regenerative food systems. This also has implications for progress towards SDGs 3 (good health and well-being) and 6 (clean water and sanitation). A circular approach in the agricultural supply chain has the potential to contribute to addressing SDGs 8 (decent work and economic growth), 12 (sustainable production and consumption), and 15 (life on land), with positive spill-over effects on SDG 1 (no poverty) through job creation opportunities for producers and small rural businesses as well as SDG 2 (no hunger) through strengthening food security to reduce hunger, especially in rural areas (Schröder & Raes, 2021). This shows the tripling beneficial effects of a circular food economy on sustainability, suggesting its holistic contributions to progressing towards the SDGs. Ghana can upscale its SDG performance if a transition from linearity to circularity is ensured in agricultural production. Specifically for SDG 12 (sustainable production and consumption), Ghana has made great progress in terms of the decline in post-harvest losses in agricultural production (National Development Planning Commission, 2020); however, there are still lapses (Ghana's Comprehensive Food Security and Vulnerability Analysis Report, 2020). Therefore, in this section, the authors focus on the implications of a circular food economy for SDGs 1 (No Poverty), 2 (Zero Hunger), and 12 (Sustainable Consumption and Production Patterns) in Ghana, which are the goals generally touted and seen to have more direct relationships with circular food production (Schroeder et al., 2018). The analysis starts with circular food economy as an integral component of SDG 12 (sustainable consumption and production patterns), which then sequel to making progress to first SDG 2 (no hunger) and second, SDG 1 (no poverty).

For SDG 12 (Sustainable Production and Consumption patterns), where Ghana is seen to be 'on track' with progress, the Ministry of Finance estimates that there are still significant post-harvest losses for crops including rice, sorghum, cassava, and yam, especially between the period 2010–2018[1] (Fig. 4). Also, there has been a general decline, albeit

[1] This is the current data available. Data for 2019, 2020, 2021, and 2022 are still not available.

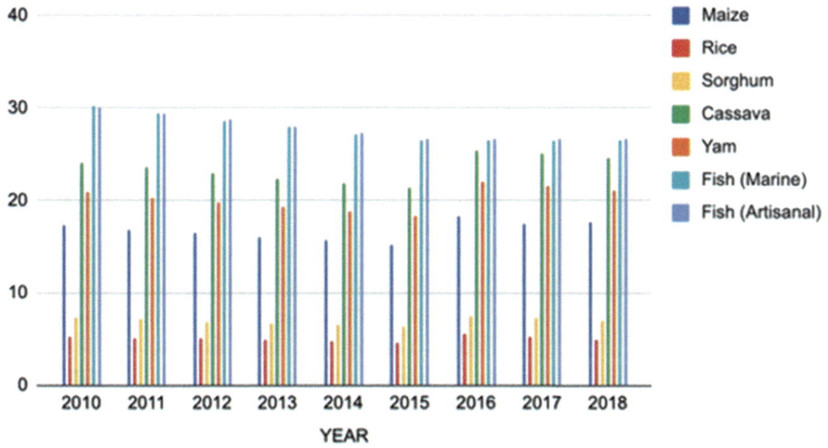

Fig. 4 Post-harvest Losses for selected food produce 2010–2018 (*Source* Ghana's Comprehensive Food Security and Vulnerability Analysis [CFSVA] Report, 2020)

at a slower pace, with artisanal fish (a loss of 26.6% in 2018). Except for maize production, which recorded a 0.2% point increase in post-harvest losses between 2017 and 2018, the produce of the other crops generally stagnated over the same period (National Development Planning Commission, 2019). Given the country's high level of food losses, the slow rate of decline is a serious cause of worry. The circular economy principles (Reuse, Recycle, Recover, and Reduce) will positively impact SDG 12, improving resource efficiency and reducing pressure on the natural environment (Schröder, 2020).

Reducing food loss and waste through responsible production and consumption will also enable Ghana to attain its SDG 2 (zero hunger). According to Ghana's, 2020 Comprehensive Food Security and Vulnerability Analysis Report, about 3.6 million people representing 11.7% of the Ghanaian population are food insecure. Out of this, 5.2% are severely food insecure, and 6.5% are moderately food insecure. Furthermore, 78% of the 3.6 million people live in rural areas compared to 22% in urban areas. This seems to suggest that the current agricultural policy interventions (e.g. Investing in Food and Jobs, Livelihood Empowerment Against Poverty, Fertilizer Subsidy Programme, etc.) cannot salvage the country from food insecurity. For instance, whilst food production of the major crops increased between 2018 and 2019 (Fig. 5), food importation likewise increased for many of these major crops (Fig. 6). For example, rice production increased by 25%, but its import also increased by 31%, implying the increase in production is still insufficient to meet the country's demand (Food Security and Nutrition Monitoring System, 2020).

Food over the years has been used as a political and economic "weapon". As Clément (2017) has argued, cereal exporters, for example, sometimes capitalize on their produce by exporting them instead of selling them in the local market leading to shortages and famine. As a result, Ghana falls short in the production of major crops (e.g. cereals) and loses about 20–30% of its cereals in storage. This problem, amongst many others, including poverty and unemployment, makes it indispensable for Ghana to strive for circular food production that can strongly contribute to sustainability. A circular food system will enable Ghana to increase food production through improved productivity per hectare and avoid food losses in storage. This will be the country to progress towards self-sufficiency in food production and significantly reduce its food imports. Consequently, food prices will stabilize, which will help achieve food security (Kwao, 2017).

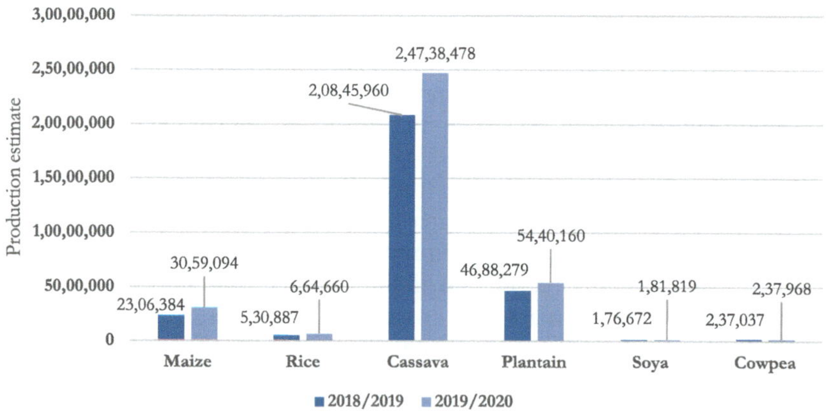

Fig. 5 Production estimates of major crops (*Source* Food Security and Nutrition Monitoring System [2020])

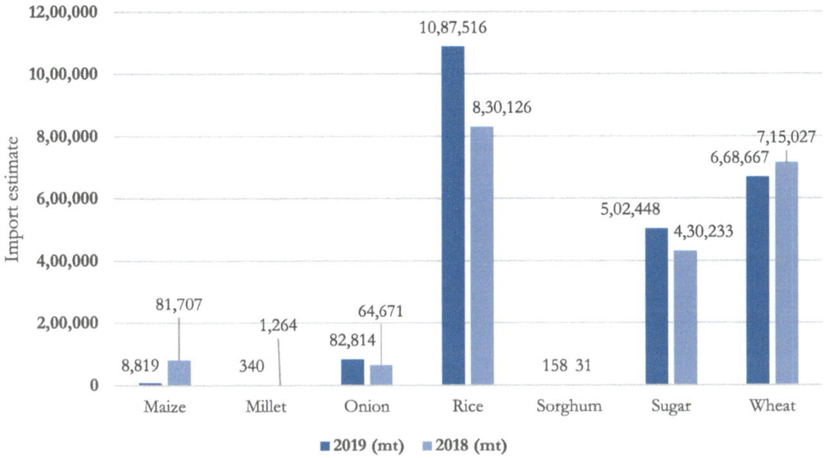

Fig. 6 Import estimates of selected crops (*Source* Food Security and Nutrition Monitoring System [2020])

Increasing food productivity per hectare whilst minimizing food losses and waste will lead Ghana to attain SDG 2 (no hunger) and have positive outcomes for farmers who are the major producers. Farmers will be able to have enough produce to sell, increase their income and improve their livelihood helping attain SDG 1 (no poverty). However, since 2012, the rate of poverty reduction in Ghana has slowed to the point of insignificance. The rate of poverty reduction in current years hovers around 0.2%, and it is estimated that 23.4% of the Ghanaian population lives in poverty (World Bank, 2020). From this estimate, the most vulnerable group to fall into this bracket are subsistence farmers in the northern regions, and poverty has deepened amongst this group since 2012. The Ghana Living Standard Survey (Ghana Statistical Service [GLSS 7], 2018) Report has indicated that the poorest households in Ghana are self-employed agricultural households—about 80% of the people from this group are poor. The main challenges faced by these individuals are low productivity, high prices of farming inputs, low educational level, limited access to agricultural extension services, poor roads, and lack of credit. Although the agricultural sector contributes significantly to the economic growth of the country, these identified challenges have reduced the sector's impact over the years from 57% in the early 1990s to 34% in 2016 (Fig. 7).

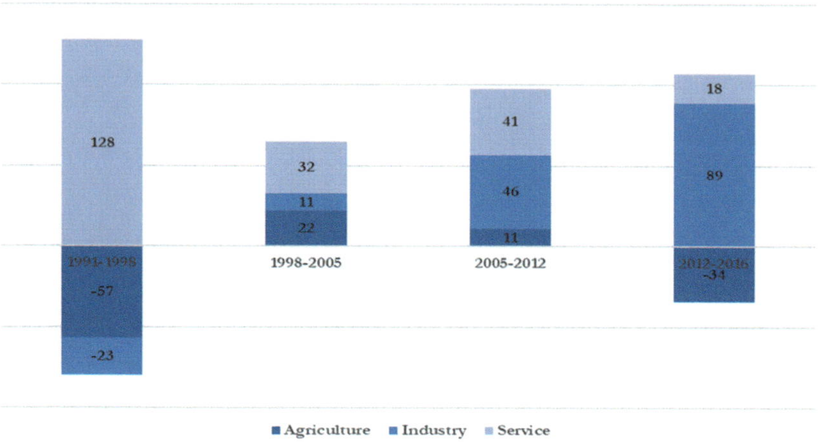

Fig. 7 Contribution to Growth in GDP per capita by sector (%) (*Source* World Bank [2020])

However, the sector remains one of the highest sources of employment for the Ghanaian labour market (40% in 2016) (Fig. 8). This is why the proposed model based on the concept of circular economy (Sect. 5, Fig. 3) stresses the need for the implementation of policies that provide maximum support for these vulnerable producers through finance and tax subsidies, education and training, and infrastructure support for acceleration to agricultural modernization through circularity-driven policy interventions. The high agricultural output that will be the outcome of the proper adoption of the model can further translate into increased income and improved livelihoods for the farmers and families, which will positively impact the efforts exerted towards poverty reduction (i.e. SDG 1) not just amongst farmers but the entire population since this group employs almost 40% of the workforce in Ghana (Ghana Statistical Service [GLSS 7], 2018).

7 Conclusion and Policy Implications

The global discourses on circular economy and specifically circular food systems within the science-policy nexus have generally attested to the sustainability benefits of circularity in food production (Barros et al., 2020; Fassio, 2021). In the current dispensation, where all countries are

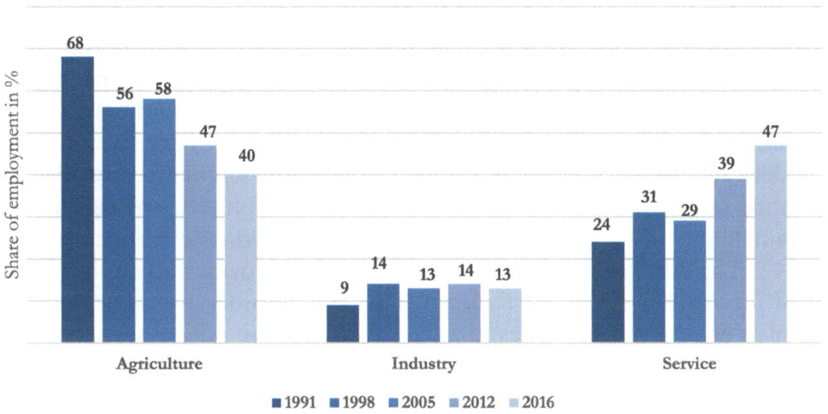

Fig. 8 Share of Employment by Sector (%) (*Source* World Bank [2020])

committed to a universal policy framework—the SDGs—for their growth and development, it becomes laudable for developing countries to intensify circular models of production for progress towards sustainable food systems. As this study indicates, Ghana as a case example of a developing country needs to transition its food production systems from linearity to circularity systematically—our circular food system model, premised on the 4R principle and based on a macro-level conceptualization of circularity can help in the change process at meso- and micro-levels with multiplier implications for achieving the SDGs, specifically SDG 1, 2, and 12. We gathered in our analysis that a circular economy model in Ghana's agricultural sector would lead to effective and efficient use of natural resources, improve the quality and yields of crops, and promote food and nutritional security, which will speed up the achievement of the SDGs. The circular economy production model can happen in Ghana when state institutions formulate and implement comprehensive circular policy guided by indicators that help assess the progress (and lapses) in transitioning from the 'dominant' linear production to the 'desired' circular production practices among farmers.

The positive outcomes associated with a circular food system for achieving sustainable agricultural production make it a necessity for a transformed institutional and policy framework of Ghana to delineate specific goals and targets of the SDGs to which the implementation of a circular food system can achieve instead of policy analysis that has no specific focus on the SDGs. While the model developed in this study is based on the contextual realities of Ghana's current institutional and policy framework, it potentially provides a "gateway" for other developing countries, especially those in sub-Saharan Africa. Already, the World Economic Forum and the Global Environment Facility have spearheaded the African Circular Economy Alliance to promote circularity in Africa to advance sustainable economic growth and development, environmental sustainability, and social welfare and progress (World Economic Forum, 2014). The African Circular Economy Alliance provides a learning platform for African countries to learn from one another by implementing circular economy modalities that are best in line with the economic strength of such countries. On this basis, our proposed model and interventions for transforming agricultural production in Ghana can be spearheaded in other countries in sub-Saharan Africa, taking into account the potential offered by the African Circular Economy Alliance.

Whilst the study advocates for circular agricultural practices, we are also mindful of the (potential) challenges that (might) serve as setbacks for promoting it in Ghana. One precondition for circular economy concepts to gain ground in Ghana and other developing countries is a sound understanding of its concepts and implementation principles, the 4R in this context. However, this is challenged in Ghana as there is limited understanding of circularity amongst key stakeholders—policy actors, agricultural value chain actors, farmers, and experts. It is, therefore, not surprising that the linear model of agricultural production still permeates policy analysis and implementation in Ghana, sub-Saharan Africa, and Africa at large (Secco et al., 2020). Again, some evidence suggests that circular food production practices are somewhat expensive, making its promotion and achievement at the initial stages in Ghana a challenge—education and training for capacity building, infrastructural and technological provisions, and policy adjustment and transformation all come with huge financial commitments (Donkor, 2019). With the agricultural budget in Ghana substantially supported by foreign actors, for example, an estimated 95% of sustainable agricultural financing in 2019 was expected to be externally financed (Assibey-Yeboah, 2019), it becomes a worrying situation in terms of how Ghana can suitably promote circularity in its agricultural food production systems. The limited circular food system practices that exist in Ghana are spearheaded through traditional farming systems as well as initiatives of some civil society organizations, albeit, they are further declining due to the lack of policy awareness and support for such practices.

Whilst a circular food system in agricultural production may be expensive, it is important to acknowledge its huge economic gains that can further strengthen Ghana's economy and catalyse its efforts in achieving sustainable development. For this reason, we agree with Boon and Anuga's (2020) argument that the challenges associated with promoting circular agricultural practices can be dealt with if Ghana "... scales up the deployment of the concept and its practices through a well-thought-out policy, strategies and concrete actions on the ground" (p. 12). This means that whilst circular food system practices might be generally expensive (e.g. treating wastewater and bioenergy use, solar mechanization in farming), not every practice is so, especially those that are based on low-technologies, and traditional knowledge systems and practices. In this study, for instance, the initiatives we suggest—manuring and organic farming practices (e.g. mulching, residue management, etc.), sewage sludging (e.g. wastewater use in vegetable farming), smart agricultural practices (e.g. on the farm sales of produce, crop rotation, mixed/inter-crop farming, (rain) water harvesting,), and precision in animal husbandry are cost-effective and efficient ways of practising circularity in farming practices through the 4R principle. On this basis, a well-thought policy framework should rather start with cost-effective and efficient circular farming practices with an emphasis on education and training for informed understanding at the initial stage of transitioning from linearity to circularity to set a strong foundation for which other expensive, high-tech circularity initiatives can gradually be mainstreamed into agricultural practices in Ghana, making it a necessity for a hybrid scenario of circular transition to be envisioned for policy actions. These prepositions/proposals are also valid for other sub-Saharan African countries in spearheading circular production practices in the agricultural sector for making the necessary progress for attaining the SDGs for their development moving forward.

We need to highlight that despite the concept of circular economy receiving global recognition and endorsement for transitioning to sustainable practices, its theoretical and conceptual understanding is still fluid and seems immature as inconsistencies and heterogeneities characterize the literature. We think this is a limitation in the circular economy literature itself. Such a limitation can make knowledge mobilization and translation from the circular economy literature a bit problematic in informing policy direction, especially in developing countries like Ghana. Perhaps, this has contributed to the snail's pace at which developing countries like Ghana adopt circularity in agricultural (and associated) practices. We hope that as circular economy and its underlying concepts gain ground in the academic literature, there will be a general-level uniformity that can meaningfully inform policy for sustainable development in developing countries. We expect this study to contribute to strengthening the growing circularity literature in the Ghanaian context that can meaningfully inform policy analysis that reflects the realities in Ghana with relevant implications for the SSA and developing countries in general. This study is limited because it relied only on secondary data. Future studies on circular food systems in Ghana focusing on 'production' through institution and policy analysis must delve deeper by considering primary data collected through engagements with concerned institutions and producers (farmers). Also, studies are needed to delve deeper into circular food transition and progress towards the SDGs by focusing on specific goals in order to concretely bring out the implications of circularity for sustainability in Ghana to guide policy direction and efforts.

References

Agyei-Holmes, A., Ankrah, D. A., & Boakye, A. A. (2021). COVID-19 and Ghana's agri-food system: an assessment of resilience. *African Geographical Review*, 1–22. https://doi.org/10.1080/19376812.2021.1971096

Andrews, D. (2015). The circular economy, design thinking, and education for sustainability. *Local Economy, 30*(3), 305–315. https://doi.org/10.1177/0269094215578226

Assibey-Yeboah, M. D. (2019). Report of the finance committee on the annual budget estimates for the Ministry of Finance for the 2020 financial year. Accra, Ghana. http://ir.parliament.gh/bitstream/handle/123456789/1562/2020_01_16_13_16_34.pdf?sequence=1&isAllowed=y

Barros, M. V., Salvador, R., de Francisco, A. C., & Piekarski, C. M. (2020). Mapping of research lines on circular economy practices in agriculture: From waste to energy. *Renewable and Sustainable Energy Reviews, 131*, 109958. https://doi.org/10.1016/j.rser.2020.109958

Basson, M., van Rensburg, H., Cuthill, M., & Erdiaw-Kwasie, M. O. (2018). Is regional government-governance nexus delivering on social sustainability promises? Empirical evidence from Moranbah in Australia. *Local Government Studies, 44*(6), 826–847.

Bauwens, T., Hekkert, M., & Kirchherr, J. (2020). Circular futures: What will they look like? *Ecological Economics, 175*, 106703. https://doi.org/10.1016/j.ecolecon.2020.106703

Blomsma, F., & Brennan, G. (2017). The emergence of circular economy: A new framing around prolonging resource productivity. *Journal of Industrial Ecology, 21*(3), 603–614. https://doi.org/10.1111/jiec.12603

Boon, E. K., & Anuga, S. W. (2020). Circular economy and its relevance for improving food and nutrition security in Sub-Saharan Africa: the case of Ghana. *Materials Circular Economy*, *2*(1), 1–14. https://doi.org/10.1007/s42824-020-00005-z

Brooks, J., Croppenstedt, A., & Aggrey-Fynn, E. (2007). Distortions to agricultural incentives in Ghana (No. 1856–2016–152683). https://ageconsearch.umn.edu/record/48523?ln=en

Chamberlin, J. (2008). *It's Small World After All: Defining Smallholder Agriculture in Ghana* (Vol. 823). Intl Food Policy Res Inst. https://books.google.com.gh/books?hl=en&lr=&id=YtLKzVJ0tN0C&oi=fnd&pg=PR5&dq=Chamberlin,+J.+(2008).+It%27s+Small+World+After+All:+Defining+Smallholder+Agriculture+in+Ghana+(Vol.+823).+Intl+Food+Policy+Res+Inst.&ots=WNpo8jtrFR&sig=yxaxyWr3Plz1X-iRmPM3qLW3v4A&redir_esc=y#v=onepage&q&f=false

Clément, A. (2017). The "food weapon": The foundations for the history of a concept (seventeenth–nineteenth centuries). *Revue de philosophie économique/Review of Economic Philosophy*, 18, 103–130. https://www.cairn-int.info/article-E_RPEC_182_0103--the-food-weapon-the-foundations.htm

Cobbinah, P. B., Asibey, M. O., Zuneidu, M. A., & Erdiaw-Kwasie, M. O. (2021). Accommodating green spaces in cities: Perceptions and attitudes in slums. *Cities, 111*, 103094.

Cronin, P., Ryan, F., & Coughlan, M. (2008). Undertaking a literature review: a step-by-step approach. *British journal of nursing*, *17*(1), 38–43. https://doi.org/10.12968/bjon.2008.17.1.28059

Darfour, B., & Rosentrater, K. A. (2016). Agriculture and food security in Ghana. In the 2016 ASABE Annual International Meeting (p. 1). *American Society of Agricultural and Biological Engineers*. https://elibrary.asabe.org/abstract.asp?aid=47207

de Jesus, A., & Mendonça, S. (2018). Lost in transition? Drivers and barriers in the eco-innovation road to the circular economy. *Ecological Economics, 145*, 75–89. https://doi.org/10.1016/j.ecolecon.2017.08.001

Dong, S., Sui, B., Shen, Y., Meng, H., Zhao, L., Ding, J., Zhou, H., Zhang, X., Cheng, H., Wang, J., Li, R., Song, L., Xu, P., & Li, J., (2020). Investigation and analysis of the linkage mechanism and whole process cost of livestock manure organic fertilizer. *International Journal of Agricultural and Biological Engineering*, *13*(2), 223–227. https://doi.org/10.25165/j.ijabe.20201302.5682

Donkor, K. (2019). Structural adjustment and mass poverty in Ghana. *Routledge, London, United Kingdom*. https://doi.org/10.4324/9780429437861

Erdiaw-Kwasie, M. O., Abunyewah, M., Yusif, S., & Erdiaw-Kwasie, A. (2023). Does circular economy knowledge matter in sustainable service provision? A moderation analysis. *Journal of Cleaner Production, 383*, 135429.

European Union (EU). (2022). Circular economy: definition, importance and benefits. Available here: https://www.europarl.europa.eu/news/en/headlines/economy/20151201STO05603/circular-economy-definition-importance-and-benefits#:~:text=The%20circular%20economy%20is%20a,products%20as%20long%20as%20possible. Accessed 16 Jun 2022.

Fassio, F. (2021). The 3 C's of the Circular Economy for Food. A conceptual framework for circular design in the food system. *DIID*. https://doi.org/10.30682/diid7321h

Fassio, F., & Tecco, N. (2019). Circular economy for food: A systemic interpretation of 40 case histories in the food system in their relationships with SDGs. *Systems, 7*(3), 43. https://doi.org/10.3390/systems7030043

Food and Agricultural Organization (FAO). (2015). FAPDA Country Fact Sheet on food and agriculture policy trends, Ghana. https://www.fao.org/3/i4490e/i4490e.pdf

Food Security and Nutrition Monitoring System (FSNMS), FSNMS. (2020). Available here: https://fscluster.org/south-sudan-rep/document/food-security-and-nutrition-monitoring-4

Ghana's Comprehensive Food Security and Vulnerability Analysis (CFSVA). (2020). Available here: https://reliefweb.int/report/ghana/ghana-2020-comprehensive-food-security-and-vulnerability-analysis-cfsva#:~:text=Per%20the%202020%20CFSVA%2C%20food,population%20of%203.6%20million%20people

Ghana Statistical Service (GSS). (2018). Ghana Living Standard Survey Round 7: Poverty Profile (2005 – 2017). Accra. https://www.statsghana.gov.gh/gssmain/fileUpload/pressrelease/GLSS7%20MAIN%20REPORT_FINAL.pdf. Accessed 28 Jun 2022.

Ghana Statistical Service (2021). Comprehensive Security and Vulnerability Analysis (CFSVA), Ghana. Available here: Ghana CFSVA Summary (statsghana.gov.gh). Accessed 28 Jun 2022.

Geels, F. W. (2005). The dynamics of transitions in socio-technical systems: A multi-level analysis of the transition pathway from horse-drawn carriages to automobiles (1860–1930). *Technology Analysis & Strategic Management, 17*(4), 445–476. https://doi.org/10.1080/09537320500357319

Geisendorf, S., & Pietrulla, F. (2018). The circular economy and circular economic concepts – A literature analysis and redefinition. *Thunderbird International Business Review, 60*(5), 771–782. https://doi.org/10.1002/tie.21924

Gough, D., Thomas, J., & Oliver, S. (2012). Clarifying differences between review designs and methods. *Systematic reviews, 1*(1), 1–9. https://doi.org/10.1186/2046-4053-1-28

Grant, M. J., & Booth, A. (2009). A typology of reviews: An analysis of 14 review types and associated methodologies. *Health Information & Libraries Journal, 26*(2), 91–108. https://doi.org/10.1111/j.1471-1842.2009.00848.x

Hailemariam, A., & Erdiaw-Kwasie, M. O. (2022). Towards a circular economy: Implications for emission reduction and environmental sustainability. *Business Strategy and the Environment*.

Hamam, M., Chinnici, G., Di Vita, G., Pappalardo, G., Pecorino, B., Maesano, G., & D'Amico, M. (2021). Circular economy models in agro-food systems: A review. *Sustainability, 13*(6), 3453.

Hawken, P., Lovins, A. B., & Lovins, L. H. (2013). *Natural capitalism: The next industrial revolution*. Routledge. https://www.smictom-alsacecentrale.fr/sites/default/files/webform/natural-capitalism-creating-the-next-industrial-revolution-paul-hawken-amory-lovins-l-hunter-lovins.pdf

International Panel of Experts on Sustainable Food Systems (IPES-Food). (2015). The new science of sustainable food systems. Report 01. https://www.ipes-food.org/_img/upload/files/NewScienceofSusFood.pdf

Jauho, M. (2013). Reviewed work(s): The dynamics of social practice. Everyday life and how it changes by Elizabeth Shove, Mika Pantzar, and Matt Watson. *Acta Sociologica, 56*(3), 287–288.

Jurgilevich, A., Birge, T., Kentala-Lehtonen, J., Korhonen-Kurki, K., Pietikäinen, J., Saikku, L., & Schösler, H. (2016). Transition towards circular economy in the food system. *Sustainability, 8*(1), 69. https://doi.org/10.3390/su8010069

Kacprzak, M., Neczaj, E., Fijałkowski, K., Grobelak, A., Grosser, A., Worwag, M., Rorat, A., Brattebo, H., Almås, Å., & Singh, B. R. (2017). Sewage sludge disposal strategies for sustainable development. *Environmental Research, 156*, 39–46. https://doi.org/10.1016/j.envres.2017.03.010

Kirchherr, J., Piscicelli, L., Bour, R., Kostense-Smit, E., Muller, J., Huibrechtse-Truijens, A., & Hekkert, M. (2018). Barriers to the circular economy: Evidence from the European Union (EU). *Ecological Economics, 150*, 264–272. https://doi.org/10.1016/j.ecolecon.2018.04.028

Köhler, J., Geels, F. W., Kern, F., Markard, J., Onsongo, E., Wieczorek, A., Alkemade, F., Avelino, F., Bergek, A., Boons, F., Fünfschilling, L., Hess, D., Holtz, G., Hyysalo, S., Jenkins, K., Kivimaa, P., Martiskainen, M., McMeekin, A., Mühlemeier, M. S., Nykvist, B., Pel, B., Raven, R., Rohracher, H., Sanden, B., Schot, J., Sovacool, B., Turnheim, B., Welch, D., & Wells, P. (2019). An agenda for sustainability transitions research: State of the art and future directions. *Environmental innovation and societal transitions, 31*, 1–32. https://doi.org/10.1016/j.eist.2019.01.004

Kwao, B. (2017). Sustainable Food Systems in northern Ghana: Assessing the influence of International Development. https://dr.library.brocku.ca/handle/10464/12884

Liu, Y., Wood, L. C., Venkatesh, V. G., Zhang, A., & Farooque, M. (2021). Barriers to sustainable food consumption and production in China: A fuzzy DEMATEL analysis from a circular economy perspective. *Sustainable Production and Consumption, 28*, 1114–1129. https://doi.org/10.1016/j.spc.2021.07.028

Markard, J., Raven, R. & Truffer, B. (2012). Sustainability transitions: An emerging field of research and its prospects. *Research Policy, 41*(6), 955–967. https://doi.org/10.1016/j.respol.2012.02.013

McDonough, W., & Braungart, M. (2010). *Cradle to cradle: Remaking the way we make things*. North point press. Available here: http://www.newtechpackaging.com/wp-content/uploads/2017/07/Cradle-to-Cradle.-McDonough-and-Braungart.pdf. Accessed 12 May 2022.

McDonough, W., Braungart, M., Anastas, P. T., & Zimmerman, J. B. (2003). Peer-reviewed: Applying the principles of green engineering to cradle-to-cradle design. *Environmental science & technology, 37*(23), 434A–441A. https://doi.org/10.1021/es0326322

Ministry of Food and Agriculture. (2007). Food and Agricultural Sector Development Policy (FASDEP II), Ghana. http://rhody.crc.uri.edu/wp-content/uploads/sites/10/2018/04/Ghana-Food-and-Agriculture-Sector-Development-Policy-2007.pdf

Ministry of Food and Agriculture. (2010). Medium-Term Agriculture Sector Investment Plan (METASIP) 2011–2015, Ghana. https://new-ndpc-static1.s3.amazonaws.com/CACHES/PUBLICATIONS/2016/04/16/Medium+Term+Agriculture+Sector+Investment+Plan+(METASIP)+2011-2015.pdf

Ministry of Food and Agriculture. (2018). Investing For Food and Jobs (IFJ): An Agenda for Transforming Ghana's Agriculture (2018–2021). https://mofa.gov.gh/site/images/pdf/National%20Agriculture%20Investment%20Plan_IFJ.pdf

Mont, O., Plepys, A., Whalen, K. & Nußholz, J. L. (2017). Business model innovation for a Circular Economy: Drivers and barriers for the Swedish industry–the voice of REES companies. https://lup.lub.lu.se/search/ws/files/33914256/MISTRA_REES_Drivers_and_Barriers_Lund.pdf

National Development Planning Commission (NDPC). (2010). Medium-Term National Development Policy Framework: Ghana Shared Growth and Development Agenda (GSGDA), 2010–2013. Ghana.

National Development Planning Commission. (2019). Ghana: Voluntary National Review Report on the Implementation of the 2030 Agenda for Sustainable Development. *NDPC: Accra, Ghana*, 60–95

National Development Planning Commission. (2020). Ghana SDGs report 2020: Ghana's 2020 Report on the Implementation of the 2030 Agenda for Sustainable Development and African Union Agenda 2063. https://ndpc.gov.gh/media/2020_SDGs_Report-Final.pdf

Ofori-Boateng, C., Lee, K. T., & Mensah, M. (2013). The prospects of electricity generation from municipal solid waste (MSW) in Ghana: A better waste management option. *Fuel Processing Technology, 110*, 94–102. https://doi.org/10.1016/j.fuproc.2012.11.008

Ogunmakinde, O. E., Sher, W., & Egbelakin, T. (2021). Circular economy pillars: a semi-systematic review. *Clean Technologies and Environmental Policy, 23*(3), 899–914. https://doi.org/10.1007/s10098-020-02012-9

Oppong, E., Oppong, M., Frimpong-Boamah, E. Obodai, M., & Koh, L. (2021). Circular economy in ghana's agri-food value chain. A triple bottom line sustainability assessment. Available here: https://www.foodresearchgh.org/images/agrifood/Technical-Report.pdf. Accessed 5 Oct 2022.

Pheifer, A. G. (2017). Barriers and enablers to circular business models. White Paper. Brielle.

Pimbert, M. P. (2015). Circular food systems. In K. Albala (Ed.), *The SAGE Encyclopedia of food issues* (pp. 232–238). SAGE Publishers.

Quaye, A. K., Hall, C. A., & Luzadis, V. A. (2010). Agricultural land-use efficiency and food crop production in Ghana. *Environment, development and sustainability, 12*(6), 967–983. https://doi.org/10.1007/s10668-010-9234-z

Repp, L., Hekkert, M., & Kirchherr, J. (2021). Circular economy-induced global employment shifts in apparel value chains: Job reduction in apparel production activities, job growth in reuse and recycling activities. *Resources, Conservation and Recycling, 171*, 105621. https://doi.org/10.1016/j.resconrec.2021.105621

Ripple, W. J., Smith, P., Haberl, H., Montzka, S. A., McAlpine, C., & Boucher, D. H. (2014). Ruminants, climate change and climate policy. *Nature Climate Change, 4*(1), 2–5. https://doi.org/10.1038/nclimate2081

Rutten, M. M., & Verma, M. (2014). The Impacts of Reducing Food Loss in Ghana: A scenario study using the global economic simulation model MAGNET (No. 2014–035). Wageningen UR. https://library.wur.nl/WebQuery/wurpubs/494067

Sachs, J., Kroll, C., Lafortune, G., Fuller, G., & Woelm, F. (2022). *Sustainable Development Report 2022.* Cambridge University Press. https://s3.amazonaws.com/sustainabledevelopment.report/2022/2022-sustainable-development-report.pdf

Schot, J., & Kanger, L. (2018). Deep transitions: Emergence, acceleration, stabilization and directionality. *Research Policy, 47*(6), 1045–1059. https://doi.org/10.1016/J.RESPOL.2018.03.009

Schröder, P. (2020). How the Circular Economy Can Help Realize the Sustainable Development Goals. https://circulareconomy.earth/publications/the-circular-economy-and-the-sdgs

Schröder, P., & Raes, J. (2021). Financing an inclusive circular economy De-risking investments. *Chatham House, London, UK*. https://www.researchgate.net/profile/Patrick-Schroeder-9/publication/353764006_Financing_an_inclusive_circular_economy_De-risking_investments_for_circular_business_models_and_the_SDGs/links/6110082d169a1a0103e9fa99/Financing-an-inclusive-circular-economy-De-risking-investments-for-circular-business-models-and-the-SDGs.pdf

Schroeder, P., Anggraeni, K., & Weber, U. (2018). The relevance of circular economy practices to the sustainable development goals. *Journal of Industrial Ecology, 23*(1), 77–95. https://doi.org/10.1111/jiec.12732

Schulte, U. G. (2013). New business models for a radical change in resource efficiency. *Environmental Innovation and Societal Transitions, 9*, 43–47. https://doi.org/10.1016/j.eist.2013.09.006

Secco, C., da Luz, L. M., Pinheiro, E., de Francisco, A. C., Puglieri, F. N., Piekarski, C. M., & Freire, F. M. C. S. (2020). Circular economy in the pig farming chain: Proposing a model for measurement. *Journal of Cleaner Production, 260*, 121003. https://doi.org/10.1016/j.jclepro.2020.121003

Shane, M. W., Holden, N. M., White, E. P., & Oldfield, T. L. (2016). *The 'circular economy' applied to the agriculture (livestock production) sector-discussion paper, presented at the Workshop on the Sustainability of the EU's Livestock Production Systems*. Belgium.

Shove, E. (2014). Putting practice into policy: Reconfiguring questions of consumption and climate change. *Contemporary Social Science, 9*(4), 415–429. https://doi.org/10.1080/21582041.2012.692484

Shove, E., Pantzar, M., & Watson, M. (2012). *The dynamics of social practice: Everyday life and how it changes*. Sage. https://books.google.com.gh/books?hl=en&lr=&id=_RjrJ-6BNyMC&oi=fnd&pg=PP2&dq=Shove,+E.,+Pantzar,+M.,+%26+Watson,+M.+(2012).+The+dynamics+of+social+practice:+Everyday+life+and+how+it+changes.+Sage.&ots=hoZuPh70sg&sig=MrsbwaV1sGsSZFdSnRFe5Nagx8U&redir_esc=y#v=onepage&q=Shove%2C%20E.%2C%20Pantzar%2C%20M.%2C%20%26%20Watson%2C%20M.%20(2012).%20The%20dynamics%20of%20social%20practice%3A%20Everyday%20life%20and%20how%20it%20changes.%20Sage.&f=false

Shukla, P. R., Skeg, J., Buendia, E. C., Masson-Delmotte, V., Pörtner, H. O., Roberts, D. C., ... & Malley, J. (2019). Climate Change and Land: an IPCC special report on climate change, desertification, land degradation, sustainable land management, food security, and greenhouse gas fluxes in terrestrial ecosystems.

Sillanpää, M., & Ncibi, C. (2019). *The circular economy: case studies about the transition from the linear economy*. Academic Press. Available here: https://books.google.ca/books?id=WRGnDwAAQBAJ&printsec=frontcover&source=gbs_ViewAPI&redir_esc=y#v=onepage&q&f=false. Accessed 12 May 2022.

Springmann, M., Mason-D'Croz, D., Robinson, S., Wiebe, K., Godfray, H. C. J., Rayner, M., & Scarborough, P. (2017). Mitigation potential and global health impacts from emissions pricing of food commodities. *Nature Climate Change, 7*(1), 69–74. https://doi.org/10.1038/nclimate3155

Srinivas H (2015) Eco-efficiency. Available here: https://www.gdrc.org/sustdev/concepts/04-e-effi.html. Accessed 12 May 2022.

Thow, A. M., Apprey, C., Winters, J., Stellmach, D., Alders, R., Aduku, L. N. E., ... & Annan, R. (2020). Understanding the impact of historical policy legacies on nutrition policy space: economic policy agendas and current food policy paradigms in Ghana. *International journal of health policy and management*, *10*(12). https://ora.ox.ac.uk/objects/uuid:cd16e62a-2f75-4797-a1c6-d4d381afd6f5/download_file?safe_filename=Thow_et_al_2021_Understanding_the_impact.pdf&file_format=pdf&type_of_work=Journal+article

Van Zanten, H. H. E., Van Ittersum, M. K., & De Boer, I. J. M. (2019). The role of farm animals in a circular food system. *Global Food Security, 21*, 18–22. https://doi.org/10.1016/j.gfs.2019.06.003

Vernhout, N. J. (2014). Development policy in Ghana. A political, economic, and social of a country in transition since independence (Bachelor's thesis). https://catalog.ihsn.org/index.php/citations/69660

WEF (2014) World Economic Forum Annual Meeting. https://www.weforum.org/events/world-economic-forum-annual-meeting-201

World Bank (2020). Ghana Poverty Assessment. Report No: AUS0001903. https://policycommons.net/artifacts/1246937/ghana-poverty-assessment/1803229/

Yaro, J. A., Teye, J. K., & Torvikey, G. D. (2018). Historical context of agricultural commercialization in Ghana: Changes in land and labor relations. *Journal of Asian and African Studies, 53*(1), 49–63. https://doi.org/10.1177/002190961665736

Macroeconomic Performance and Progress Towards a Circular Economy in Developing Countries

Abebe Hailemariam and Kris Ivanovski

1 Introduction

With the alarming issue of climate change, there is a growing interest to transition from a linear economy to a circular economy in both developing and developed countries (see, e.g., Alam et al., 2021; Blomsma et al., 2022; Chen et al., 2020; Geissdoerfer et al., 2017; Kirchherr et al., 2022; Ngan et al., 2019; Patwa et al., 2021). Furthermore, a successful transition to a circular economy and spatial planning has been identified as critical preconditions for sustainability, which is a key target in the United Nation's Sustainable Development Goals (SDGs) (Cobbinah et al., 2021; Geissdoerfer et al., 2017; Heshmati & Rashidghalam, 2021; Robaina et al., 2020).

A. Hailemariam (✉)
Bankwest Curtin Economics Centre, Curtin University, Bentley, WA, Australia
e-mail: abebe.hailemariam@curtin.edu.au

K. Ivanovski
Department of Economics, Monash University, Monash, VIC, Australia
e-mail: kris.ivanovski@monash.edu

M. O. Erdiaw-Kwasie and G. M. M. Alam (eds.), *Circular Economy Strategies and the UN Sustainable Development Goals*,
Sustainable Development Goals Series,
https://doi.org/10.1007/978-981-99-3083-8_8

Recent advances in the circular economy literature emphasise the importance of understanding circular disruption, a rapid and radical reconfiguring of the current socio-technical system towards a more sustainable model (Kirchherr et al., 2022). Along these lines, Blomsma et al. (2022) identify three important phases in the transition to a circular economy in the context of circular disruptions, including the release phase, the reorganisation phase and the eruption phase, where each phase requires unique strategies to accelerate the transition to a circular economy.

The release phase sets the initial stage of the circular economy following increased discontent with the linear economy system due to rising social and environmental costs. Accelerating this phase requires business experimentations to test ideas to identify the parts of the linear paradigm that should be deinstitutionalised, preserved and repurposed in conjunction with promoting voices of pro-CE actors with a shared vision about circular paradigm (see Blomsma et al., 2022). The reorganisation phase marks the emergence of consensus on circular solutions as a viable alternative. In this phase, the focus of business experimentation will be on the applications of viable solutions (Ansell & Bartenberger, 2016; Blomsma et al., 2022; van den Bosch, 2010). These studies suggest that strategies accelerate this phase include broadening application domains of existing and emerging technologies by scholars and researchers broadening application domains of existing and emerging technologies by scholars and researchers and effective policy design and management of circular solutions by policymakers and managers. The last phase explores all opportunities and aims to scale up circular solutions and strategies to overcome selection pressures (Ansell & Bartenberger, 2016; Blomsma et al., 2022; van den Bosch, 2010). The suggested strategies to accelerate this phase include exploring the best approaches to survive the selective pressures of the mainstream market, fostering multiple technologies uses and interactions and facilitating spillover effects of technological innovations for advancing circular solutions.

Given the importance of progressing towards a circular economy to achieve a sustainable environment, policymakers should identify the key factors determining the progress towards a circular economy. The existing studies on the relationship between economic performance and the transition to a circular economy have focused on developed countries, mainly the European Union (EU) member states. These studies provide mixed evidence on the role of economic development on recycling rates. Some studies find a positive relationship between GDP and recycling rates (e.g., Cerqueira et al., 2021, Sverko Grdic et al., 2020), while the findings of other studies show that economic development (GDP) is negatively associated with recycling rates (e.g., Neves & Marques, 2022; Önder, 2018).

Most importantly, studies on the circular economy in the case of developing countries are limited mainly due to a need for more data on circularity indicators. To fill this gap in the literature, this study assesses the effect of macroeconomic performance on the progress towards a circular economy and sustainability in developing countries. To examine how macroeconomic performance is linked to the progress towards a circular economy, we utilise advanced econometric modelling using a unique historical dataset on recycling rates and real GDP for 75 developing countries from 1995 to 2019.

The findings of this study reveal that a 1 per cent increase in real GDP is associated with an increase in recycling rates in the range of 2–3 percentage points. The analysis of the subsamples based on income group shows that real GDP has a statistically insignificant effect on low-income countries. At the same time, the estimates are statistically significant for lower-middle-income and upper-middle-income countries. The estimates are robust to various alternative model specifications.

To the best of our knowledge, this study is the first to contribute to the literature by examining the nexus between macroeconomic performance and transition to a circular economy using long historical panel data from a global panel of developing countries. In addition, the study utilises the latest brand-new historical data on recycling rates from the Environmental Performance Index database. The findings of this study will inform evidence-based policy to foster the transition to a circular economy and meet the targets of the SDGs.

2 Background of Progress Towards a Circular Economy in the Context of Developing Countries

2.1 Waste Generation

In a move towards a circular economy, effective waste management is an essential component of a country's sustainability agenda, as global solid waste generation is projected to increase by 70 per cent by 2050 (Kaza et al., 2018). Evidence shows that solid waste contributes about 5 per cent of global greenhouse gas emissions (Kaza et al., 2018) and degrades air quality through haphazard incineration (Wiedinmyer et al., 2014). Environmental and public health concerns intensified during the COVID-19 pandemic (Adyel, 2020; Silva et al., 2021). Despite the environmental and health risks, waste management policies have not kept pace with the elevated level of waste generation in all countries, regardless of income level.

The level of waste generation varies significantly across space and time. For example, as shown in Fig. 1, the estimated waste generated in developing countries in 2016 ranged from 129 million tonnes per year in the Middle East and North Africa to 334 million tonnes per year in South Asia.

Figure 2 shows that the amount of waste generated increases by income level, ranging from 93 million tonnes for low-income countries to 655

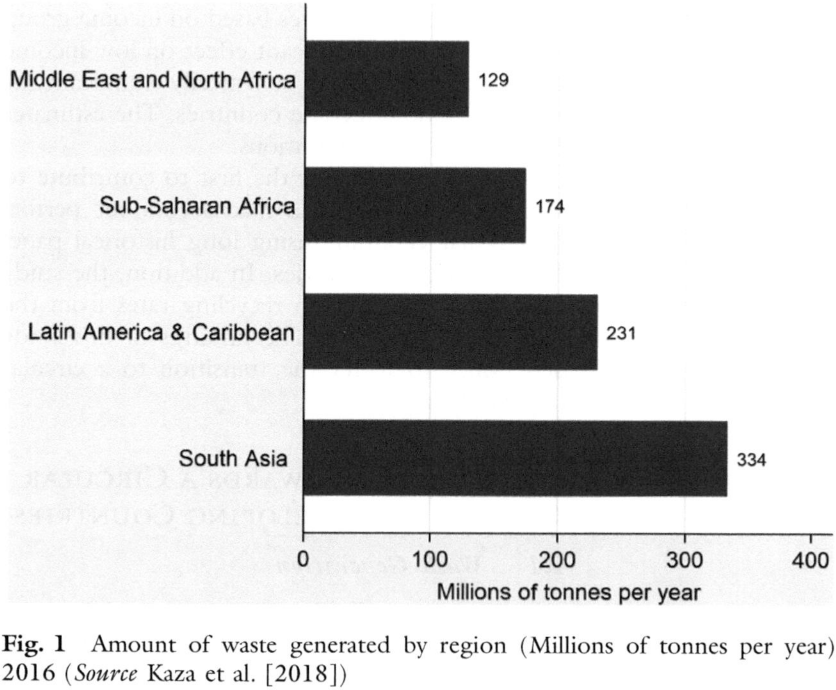

Fig. 1 Amount of waste generated by region (Millions of tonnes per year), 2016 (*Source* Kaza et al. [2018])

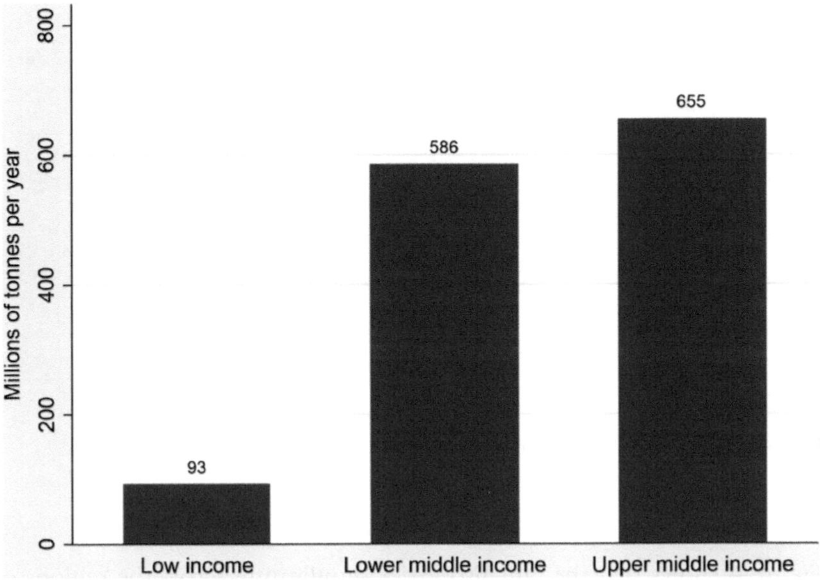

Fig. 2 Amount of waste generated by income level (Millions of tonnes per year), 2016 (*Source* Kaza et al. [2018])

million tonnes for upper-middle-income countries. In low-income countries, it is common to dump waste outside of landfills or in open areas, which leads to the leakage of waste into the surrounding environment.

Figure 3 visualises the distribution of per capita waste generated worldwide using national statistics adjusted to a common base year of 2016 for comparability across countries. As shown in the figure, low-income and middle-income countries in sub-Saharan Africa and South Asia generate the lowest quantities of per capita waste ranging from 0 to 0.99. However, in terms of the growth rate of waste generation, low- and middle-income countries are positioned for the fastest rate of growth, where waste levels are expected to triple by 2050 due to the rapid increase in population, urbanisation, and economic activities.

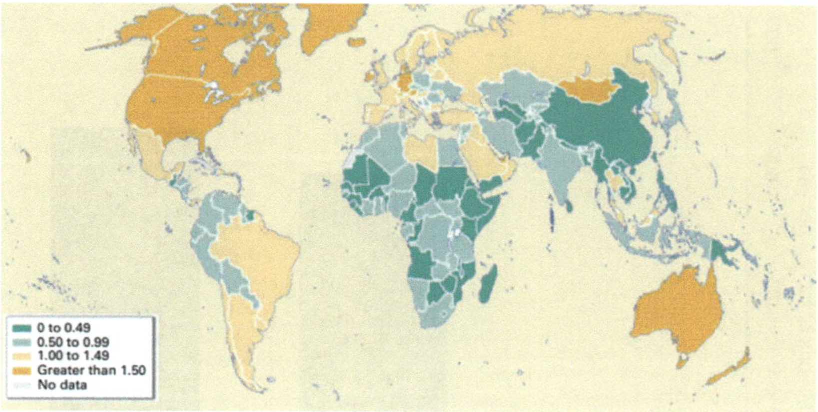

Fig. 3 Waste generation per capita (*Source* Kaza et al. [2018])

2.2 *Waste Collection*

Figure 4 shows that the rate also varies significantly across the regions of the developing world, ranging from 44 per cent in South Asia and sub-Saharan Africa to 84 per cent in Latin America and the Caribbean. The alarming issue is that the waste collection rate is the lowest in South Asia, while the quantity of waste generated is the largest in the region.

As shown in Fig. 5, waste collection rates are substantially higher for urban areas than for rural areas. This is because the waste collection is one of the most common services provided by municipalities in urban areas, which is unavailable in most rural areas. As depicted in Fig. 5, waste collection rates in lower-middle-income countries are more than twice as high in cities as in rural areas. Generally, waste collection rates increase with the level of income.

Figure 6 shows the share of countries with defined solid waste management laws or guidelines, ranging from broad environmental rules to targeted interventions. The figure shows that only about 60 per cent of low-income countries have defined solid waste management laws, suggesting that solid waste management systems are more nascent in these countries without laws and guidelines for proper waste management. The share of countries with specific solid waste management laws increases significantly in lower-middle-income countries (89 per cent) followed by upper-middle-income countries (84 per cent). According to Kaza et al.

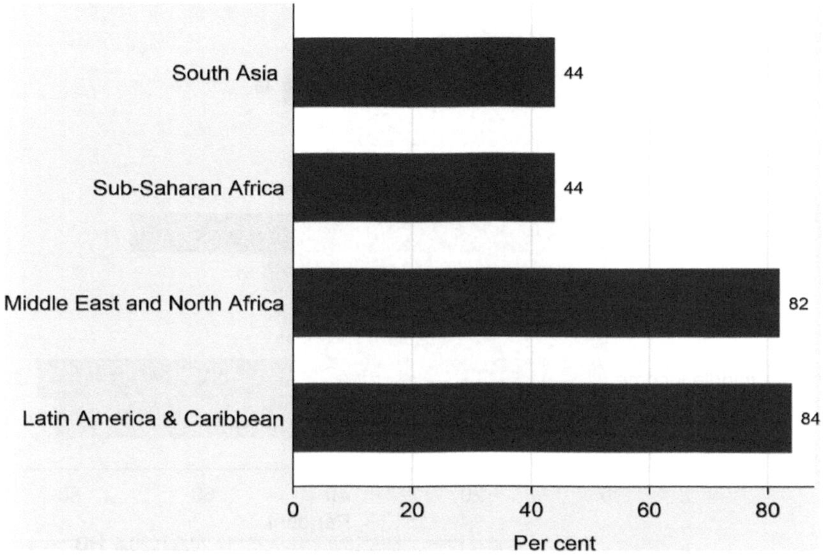

Fig. 4 Waste collection rates by region (*Source* Kaza et al. [2018])

(2018), sub-Saharan African countries constitute the vast majority of countries without waste legislation, where laws are still being developed, and enforcement is a common challenge due to inadequate and limited staffing and problems in cultural alignment with legislative goals.

According to the latest Environmental Performance Index (EPI) report by Wolf et al. (2022), sub-Saharan African and South Asian countries are placed at the bottom of the global ranking in waste management and air quality (see Fig. 7). This suggests that much has to be done in developing countries in terms of waste management and progress towards a circular economy and environmental sustainability.

2.3 Waste Recycling

Globally, only about 19 per cent of waste undergoes materials recovery through recycling and composting, while 37 per cent is disposed of in some landfill, and 33 per cent is openly dumped. The least proportion (11 per cent) is disposed of through modern incineration (Kaza et al., 2018). The EPI 2022 report developed metrics of recycling rates defined

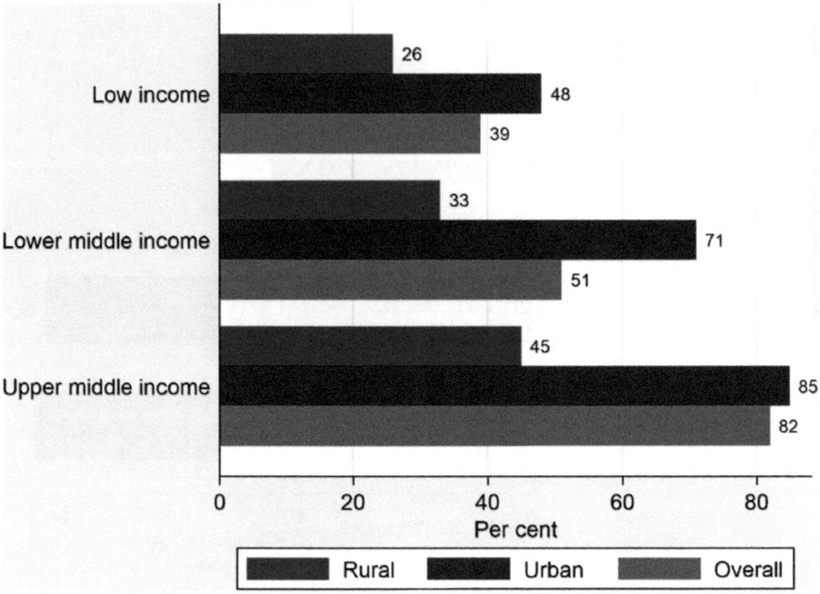

Fig. 5 Waste collection rates by income level: urban vs rural (*Source* Kaza et al. [2018])

by the proportion of post-consumer recyclable materials (such as glass, metal, plastic, and paper) each country sorts for recycling.

Figure 8 shows the historical global trends of signs of progress towards a circular economy based on the newly constructed recycling indicators from 1995 to 2020. The figure shows that the trend in the recycling rate is sloping upward in all regions despite the existence of significant disparity. Europe and Central Asia region is leading with a recycling rate estimated at more than 25 per cent in 2020, followed by East Asia and the Pacific region (about 25 per cent in 2020). On the other hand, East Asia and sub-Saharan Africa regions are identified as laggards with an estimated recycling rate of only less than 15 per cent in 2020 despite the fastest growth from the low base in 1995. In sub-Saharan Africa and South Asia, more than two-thirds of waste are dumped.

The rate of recycling varies significantly by income level (Fig. 9). As shown in Fig. 9, the recycling rate is significantly lower in low-income

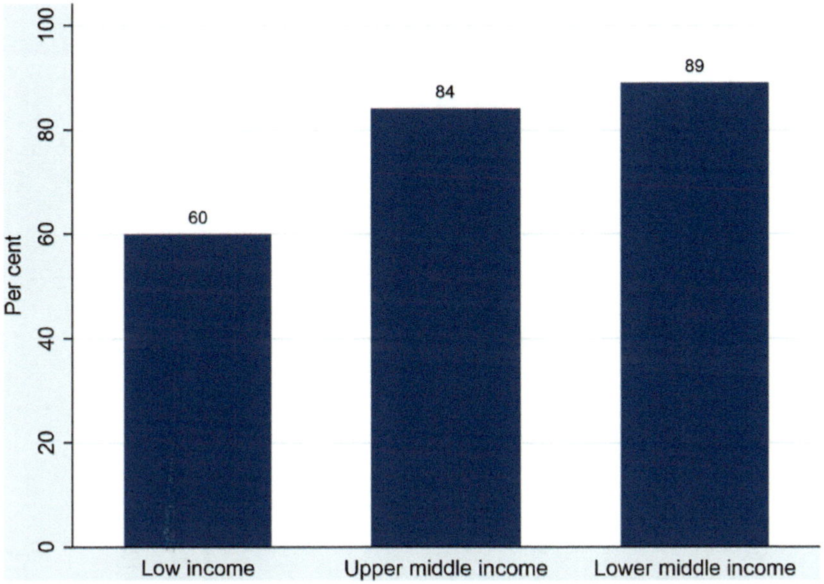

Fig. 6 Share of countries with national waste management regulation (*Source* Kaza et al. [2018])

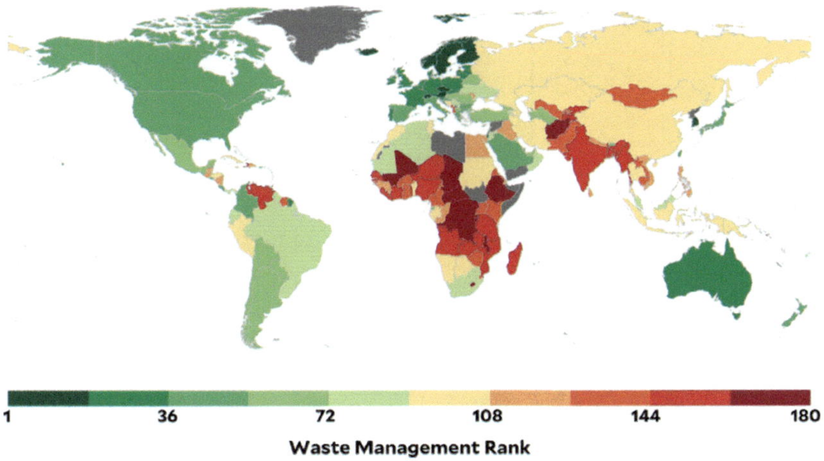

Fig. 7 Global rankings on waste management, 2022 (*Source* Adapted from Wolf et al. [2022])

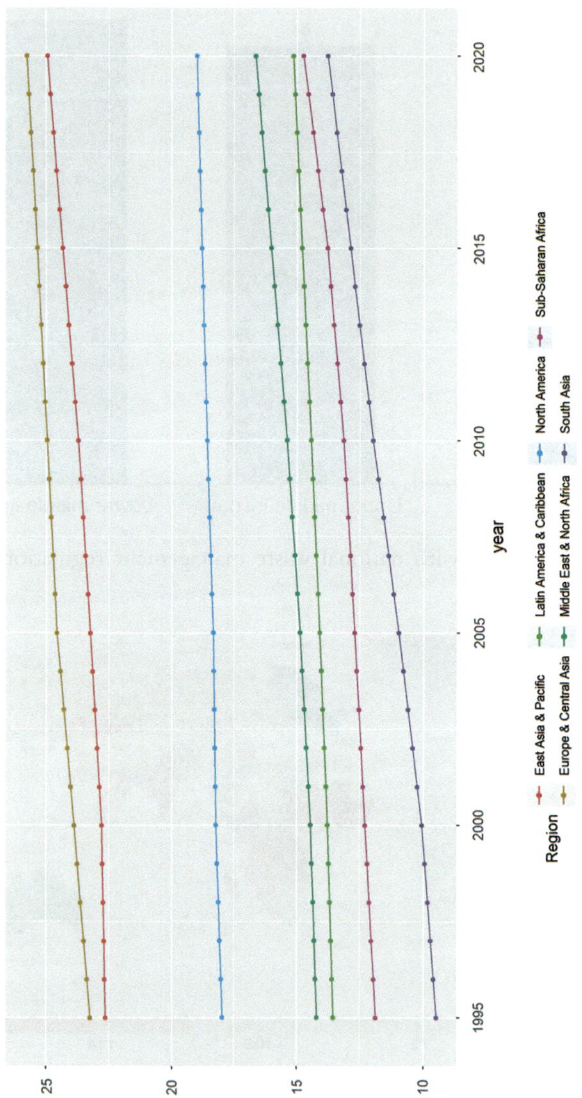

Fig. 8 Recycling rate by region, 1995–2020 (*Source* Wolf et al. [2022])

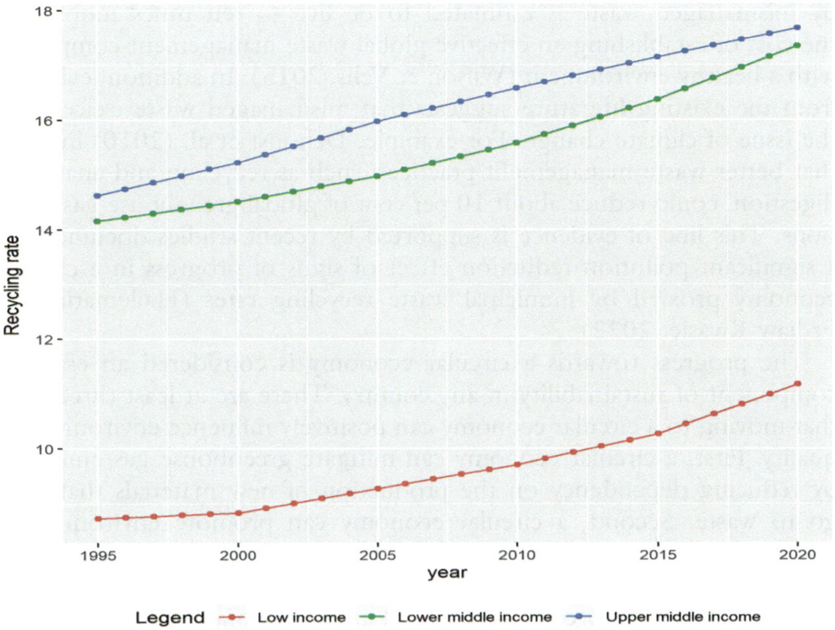

Fig. 9 Recycling rate by income level, 1995–2020 (*Source* Wolf et al. [2022])

countries despite picking up momentum in recent years. In these countries, landfills are very limited, leaving open dumping as a prevalent means of waste disposal. On the other hand, the recycling rate in lower-middle-income and upper-middle-income countries increased significantly over the recent decades, suggesting the prospect of sustainable ways of waste management as nations prosper economically.

3 Why Transition to a Circular Economy is Important for Developing Countries

Mismanaged waste has cross-cutting implications for environmental sustainability. Globally, an estimated five billion people lack access to waste disposal services or municipal solid waste collection services (Wilson & Velis, 2015). Well-founded evidence shows that mismanaged waste has severe environmental and public health costs. The cost inflicted by mismanaged waste is estimated to be five to ten times more than the cost of establishing an effective global waste management compatible with a healthy environment (Wilson & Velis, 2015). In addition, evidence from the existing literature suggests that mismanaged waste exacerbates the issue of climate change. For example, Dehoust et al. (2010) indicate that better waste management practices, such as recycling and anaerobic digestion, could reduce about 10 per cent of global greenhouse gas emissions. This line of evidence is supported by recent studies documenting a significant pollution reduction effect of signs of progress in a circular economy proxied by municipal waste recycling rates (Hailemariam & Erdiaw-Kwasie, 2022).

The progress towards a circular economy is considered an essential component of sustainability in any country. There are at least three ways that moving to a circular economy can positively influence environmental quality. First, a circular economy can mitigate greenhouse gas emissions by reducing dependency on the production of new materials that then go to waste. Second, a circular economy can promote environmental sustainability by retaining the embodied energy in products and materials through recycling, which reduces greenhouse gas emissions. Third, a circular economy improves the soil and environmental quality by storing and retaining carbon in the soil (Ellen MacArthur Foundation, 2021).

Through effective waste management and a transition to a circular economy, it is estimated that a reduction of 9.3 billion tonnes of greenhouse gases is achievable, which is equivalent to the current global emissions from all forms of transport (Ellen MacArthur Foundation, 2021). In addition, the report indicates that approximately 45 per cent of total greenhouse gas emissions are associated with making products that require less energy and resource use which can be mitigated through a circular economy. As a result, promoting a transition to a circular economy is of primary interest to policymakers as a mechanism to mitigate climate change (Agrawal et al., 2021; Antonioli et al., 2022; Barreiro-Gen & Lozano, 2020; Falk & Hagsten, 2020; Kennedy & Linnenluecke, 2022; Suchek et al., 2021).

In line with this, a wealth of evidence supports that the transition to a circular economy is linked to sustainability as the main objective is to maximise the value of materials and products by minimising and reusing waste (Bressanelli et al., 2019; Moktadir et al., 2020; Morseletto, 2020; Ünal & Shao, 2019; Urbinati et al., 2017). This is essential in the transition to net zero emissions as it reduces significant greenhouse gas emissions from using primary materials. It also creates improvements in material flow efficiency as well as extends the life of products and materials (Çimen, 2021; Munaro et al., 2020; Ness & Xing, 2017). Moreover, a circular economy can facilitate the use of environmentally friendly technologies and minimise the use of virgin materials that can reduce emissions significantly (Hailemariam et al., 2022; Joensuu et al., 2020; Mhatre et al., 2021).

4 A Brief Review of Literature on the Nexus Between Macroeconomic Performance and the Transition to a Circular Economy

This section briefly reviews related studies on the link between macroeconomic performance and progresses towards a circular economy. Most existing studies on the nexus between economic development and progress towards a circular economy are based on a sample of the EU countries. Although developing countries context could differ from developed economies, reviewing the evidence from developed countries helps contextualise our study within the literature.

The existing studies provide mixed evidence on the impact of national income on the transition to a circular economy. For example, Neves and Marques (2022) examine the drivers and barriers in the transition using annual data from a panel of 19 European Union countries for the period from 2010 to 2019. They find that higher per capita GDP is associated with a reduced recycling rate, indicating the negative effects on a circular economy. Along these lines, Önder (2018) also finds that an increase in GDP is negatively associated with recycling performance.

However, other studies' findings document evidence of the positive effect of economic development proxied by real GDP on the recycling rate of the waste. For example, Cerqueira et al. (2021) find that economic development positively affects the recycling rate. Similarly, Huang et al. (2020) find that GDP and fixed capital formation contribute to the recycling rate of solid waste in China. Sverko Grdic et al. (2020) also provide evidence for this line of argument. Their regression estimates suggest that a 1 per cent increase in GDP per capita in the EU is associated with an increase in the recycling rate of municipal waste by 0.16 per cent, packaging waste by 0.05 per cent and e-waste by 0.06 per cent.

There are also a number of drivers of a circular economy identified in the literature. For example, Neligan et al. (2022) investigate the role of digitalisation in promoting transitions to a circular economy using firm-level data from 599 German manufacturing firms and 296 industrial providers. They find evidence that digitisation is a driving force for implementing a circular economy in companies with a strong digital focus. Droege et al. (2022) highlight the importance of policy entrepreneurs in promoting a circular economy by influencing policy decision-makers and implementers to pass national policies in favour of progressing to a circular economy. A comprehensive review of drivers and barriers to the transition to a circular economy can be obtained in Govindan and Hasanagic (2018), Ranta et al. (2018) and Gusmerotti et al. (2019).

In conclusion, the evidence from the existing literature on the relationship between macroeconomic performance and recycling rates is mixed and inconclusive. Most importantly, there is a lack of evidence on the issue in the context of developing countries. This study aims to fill this gap in the literature using recently developed historical data on recycling rates for a broad range of developing countries.

5 Econometric Methodology and Data

5.1 Econometric Methodology

To examine the effects of macroeconomic performance on the progress towards a circular economy, we estimate a panel data model of the form:

$$REC_{it} = \alpha_i + \beta_1 \ln(RGDP_{it}) + \gamma X_{it} + \mu_t + \varepsilon_{it} \qquad (1)$$

where REC_{it} is recycling rate, $RGDP_{it}$ denotes real GDP for country i in year t, and *ln* indicates natural logarithm. We control for a set of covariates denoted by X_{it}, , including capital stock (K), human capital index (HC), the rate of urbanisation (URB), trade openness (TRADE) and democracy (DEMOC), as well as year fixed effect (μ_t). The last term (ε_{it}) in Eq. (1) denotes idiosyncratic error term.

5.2 Data

We use annual data for 75 developing countries for the period from 1995 to 2019. The sample countries include a wide range of nations, including low-income, lower-middle-income and upper-middle-income countries as per the World Bank's recent classification of countries by income group. The sample selection is based on the fact that we include as many countries as possible to represent a global panel of developing countries across multiple continents.

The dependent variable, REC, , as a measure of a circular economy indicator, is sourced from the Environmental Performance Indicator (EPI) database. REC is defined as recycling rates as the proportion of post-consumer recyclable materials (glass, plastic, paper, and metal) that are recycled. Data on real GDP, real capital stock and human capital index are obtained from Penn World Table version 10.0. Data on urbanisation rate and trade openness are sourced from the World Development Indicators databases of the World Bank, while the Boix-Miller-Rosato dichotomous measure of democracy is sourced from the Quality of Governance database. Table 1 presents an overview of the basic statistics for each variable.

As shown in Table 1, the mean value of the recycling rate is 13.7 per cent with a standard deviation of 8.2 per cent. This shows that there are great variations in the progress towards a circular economy in developing countries. In addition, there are also significant variations in the explanatory variables, including real GDP.

6 Empirical Results and Discussions

6.1 Basic Results

We begin our empirical analysis by discussing the observed simple correlation between real GDP and progress in a circular economy proxied by the recycling rates. Figure 10 presents the cross-country scatter plots on the relationship between average natural logs of real GDP and recycling rates. The figure shows that economic development positively correlates with countries' recycling rates.

Table 2 presents the estimation results from the benchmark OLS model. In Column (1), REC is regressed on ln (RGDP), controlling for country-fixed effects only. In the subsequent columns, we control for an additional covariate at a time to examine the sensitivity of the estimates. As shown in Table 2, the estimated coefficient on RGDP is positive and statistically significant at a 1 per cent significance level in all regressions. This shows that economic development is associated with progress towards a circular economy.

Table 1 Summary statistics: 1995–2019

Variable	*Description*	*Obs*	*Mean*	*Stdv*	*Min*	*Max*
REC	Recycling rate (per cent)	2442	13.72	8.17	0.85	56.65
RGDP	Real GDP at constant 2017 prices (in billions of 2017US$)	2442	361	1434	0.35	20,600
K	Capital stock at constant 2017 prices (in billions of 2017US$)	2417	1420	5577	1.67	102,000
HC	Human capital index	1917	2.16	0.58	1.05	3.61
URB	Urban population (per cent of total population)	2420	47.92	19.30	7.21	91.99
TRADE	Trade (per cent of GDP)	2220	77.53	37.86	0.02	348.00
DEMOC	Dichotomous democracy measure	2413	0.50	0.50	0.00	1.00

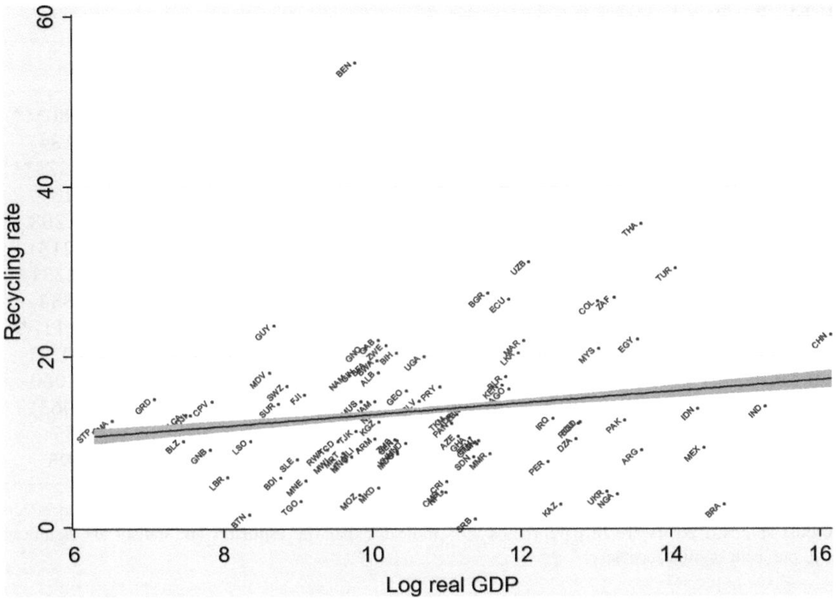

Fig. 10 Cross-country correlation between log real GDP and recycling rates, average 1995–2019

The economic interpretation of the estimates is that a 1 per cent increase in real GDP is associated with a recycling rate of about 2 percentage points after controlling for all covariates and country-fixed effects (Column 6). Regarding the covariates, the results show that real capital stock has a positive and significant effect on recycling rates. This result is consistent with the findings of Huang et al. (2020). On the other hand, trade openness has a negative and statistically significant effect on recycling rates.

A concern with the OLS approach is that the estimates could be biased and inconsistent as it cannot control for time-invariant factors. We adopt panel fixed effect approaches in our main estimations to address this issue.

6.2 *Main Results*

In using cross-country panel data, it is important to account for unobserved time-invariant factors to ensure that the estimates are not biased

Table 2 Benchmark OLS estimates

	(1)	(2)	(3)	(4)	(5)	(6)
ln RGDP	2.736***	2.146***	1.943***	1.952***	2.062***	2.080***
	(0.066)	(0.112)	(0.126)	(0.131)	(0.133)	(0.134)
ln K		0.702***	0.859***	0.812***	0.886***	0.857***
		(0.113)	(0.105)	(0.113)	(0.127)	(0.131)
HC			−0.184	−0.128	−0.229	−0.208
			(0.183)	(0.187)	(0.208)	(0.213)
ln URB				0.108	−0.292	−0.281
				(0.344)	(0.382)	(0.383)
ln TRADE					−0.125***	−0.117***
					(0.029)	(0.029)
DEMOC						−0.060
						(0.063)
Observations	2442	2417	1892	1881	1791	1785
R^2	0.993	0.993	0.995	0.995	0.995	0.995

Notes The dependent variable is the recycling rate (REC). All regressions include country-fixed effects. Robust standard errors are in parentheses. *** indicates that the estimates are statistically significant at 1 per cent significance level

and inconsistent. This is relevant in our case as we employ data from a diverse set of developing countries. Table 3 presents the estimates from panel fixed effect (Column 1) and random effect (Column 2) models that address this issue. The results show that economic development has a positive and statistically significant effect on recycling rates. Specifically, the fixed effect estimates show that a 1 per cent increase in real GDP increases recycling rates by about 2.5 percentage points. The random effect estimates in Column (2) of Table 3 show the robustness of this finding. These results support the findings of some recent studies for developed countries (e.g., Cerqueira et al., 2021; Sverko Grdic et al., 2020). The panel data estimates are quantitatively larger than the OLS estimates (with full controls), suggesting the OLS estimates' downward bias.

6.3 *Robustness Check*

While the fixed effect model is helpful in addressing the issue of estimation biases due to unobserved time-invariant factors, it is not a panacea to the issue of endogeneity bias. This is particularly important in our case as the

Table 3 Effect of economic development on circular economy: panel data estimates

	(1)	*(2)*
	Fixed Effect	Random Effect
ln RGDP	2.481***	2.413***
	(0.440)	(0.432)
Observations	1785	1785
Controls	Yes	Yes

Notes The dependent variable is the recycling rate (REC). All regressions include the full set of covariates, including real capital stock, human capital index, urbanisation rate, trade openness, democracy, and time-fixed effect. Robust standard errors are in parentheses. *** indicates that the estimates are statistically significant at 1 per cent significance level

endogeneity issue may arise because data could be measured with errors or due to the likelihood of reverse causality from REC to real GDP. Therefore, to check for the robustness of the main results to the endogeneity issue, we also estimate the effect of real GDP on REC using dynamic OLS (DOLS). The DOLS method effectively corrects endogeneity biases as it is based on group-means panel data estimators (see Kao & Chiang, 2000).

Table 4 presents the estimation results from the DOLS model with different lag *(k)* selections. The results in Table 4 confirm that the main findings are robust. Furthermore, the estimates are not sensitive to the lag selection, as seen from the results in Column (1) to Column (3) of Table 4.

6.4 *Heterogeneity*

Our empirical analysis using the full sample of developing countries suggests that an increase in real GDP is associated with progress towards a circular economy. To account for heterogeneity among countries, we also perform further analysis based on a different subsample using the World Bank's classification of countries by income group: low-income, lower middle-income, and upper-middle-income countries. The results are reported in Table 5.

The results in Table 5 show that the effect of real GDP on recycling rates is statistically indistinguishable from zero for low-income countries (Column 1). However, Column (2) and Column (3) of Table 5 show that

Table 4 Robustness check using alternative specifications: dynamic OLS (DOLS)

	(1) *k = 1*	*(2)* *k = 2*	*(3)* *k = 3*
ln RGDP	2.598***	2.573***	2.529***
	(0.284)	(0.298)	(0.313)
Controls	Yes	Yes	Yes
N	1408	1344	1280

Notes The dependent variable is the recycling rate (REC). *k* indicates a lag number. All regressions include the full set of covariates, including real capital stock, human capital index, urbanisation rate, trade openness, democracy, and time-fixed effect. Robust standard errors in parentheses. *** indicates that the estimates are statistically significant at 1 per cent significance level

Table 5 Heterogeneous effects of economic development on circular economy

	(1) *Low income*	*(2)* *Lower middle income*	*(3)* *Upper middle income*
ln RGDP	0.723 (0.774)	3.290*** (0.531)	2.493** (1.074)
Observations	271	800	714
R^2	0.890	0.757	0.778

Notes The dependent variable is the recycling rate (REC). All regressions include the full set of covariates, including real capital stock, human capital index, urbanisation rate, trade openness, democracy, and time-fixed effect. Robust standard errors in parentheses. *** and ** indicate that the estimates are statistically significant at 1 per cent and 5 per cent significance levels, respectively

the effect of real GDP on recycling rates is statistically significant. This result is in line with the Environmental Kuznets Curve (EKC) hypothesis, which posits that economic activities positively affect environmental quality after a certain threshold level of economic development.

To sum up, we find evidence that an increase in real GDP is linked with progressing towards a circular economy in lower-middle-income and upper-middle-income countries. Moreover, the results are robust to various sensitivity checks.

7 Concluding Remarks

Effective waste management is crucial for a smooth transition to a circular economy in any country with cross-cutting implications for the environment and sustainable development. This is particularly important as the concern about managed waste's environmental and public health effects has intensified in the face of the COVID-19 pandemic. Therefore, it is crucial to understand the key drivers of a circular economy in developing countries to formulate an appropriate waste management and recycling policy to transition from a linear to a circular economy.

In this chapter, we examine the effect of macroeconomic performance on a circular economy using advanced econometric modelling and utilising a recently developed historical dataset on recycling rates and real GDP for 75 developing countries over the period spanning from 1995 to 2019. The findings from the econometric analysis show that an increase in real GDP is associated with an increase in recycling rates in lower-middle-income and upper-middle-income countries. More precisely, the econometric estimates from the various specifications suggest that a 1 per cent increase in real GDP is associated with increased recycling rates in the 2–3 percentage points range.

Further, the results from the subsample analysis based on income group show that the effect of real GDP on recycling rates is statistically indistinguishable from zero in low-income countries. However, real GDP positively and statistically significant affect recycling rates in lower-middle-income and upper-middle-income countries.

The policy implication of this study is that the level of economic development in low-income countries is insufficient to induce robust recycling rates. As a result, effective waste management policies in low-income countries should be deployed to keep pace with the elevated level of waste generation in these countries, creating a bottleneck in the transition from a linear to a circular economy. Coinciding with a low level of economic development, the waste collection rate is the lowest in South Asia and sub-Saharan countries. In contrast, the growth of waste generation has been elevated in recent years. Therefore, to promote the transition to a circular economy and sustainability, policymakers in low-income countries should design effective waste collection and disposal policies using a broad range of environmental rules for targeted interventions. Particularly, it is important to build capacity in staffing and create cultural alignment with environmental legislative goals that enhance waste collection and disposal, as this is a crucial step for transitioning to a circular economy through materials recovery and recycling.

References

Adyel, T. M. (2020). Accumulation of plastic waste during COVID-19. *Science, 369*(6509), 1314–1315.

Alam, G. M., Erdiaw-Kwasie, M. O., Nagy, G. J., & Leal Filho, W. (Eds.). (2021). *Climate vulnerability and resilience in the global south: Human adaptations for sustainable futures*. Springer Nature.

Ansell, C. K., & Bartenberger, M. (2016a). Varieties of experimentalism. *Ecological Economics, 130*, 64–73.

Agrawal, R., Wankhede, V. A., Kumar, A., & Luthra, S. (2021). Analysing the roadblocks of circular economy adoption in the automobile sector: Reducing waste and environmental perspectives. *Business Strategy and the Environment, 30*(2), 1051–1066.

Antonioli, D., Ghisetti, C., Mazzanti, M., & Nicolli, F. (2022). Sustainable production: The economic returns of circular economy practices. *Business Strategy and the Environment, 26*(1), 95–115.

Barreiro-Gen, M., & Lozano, R. (2020). How circular is the circular economy? Analysing the implementation of circular economy in organisations. *Business Strategy and the Environment, 29*(8), 3484–3494.

Blomsma, F., Bauwens, T., Weissbrod, I., & Kirchherr, J. (2022). The 'need for speed': Towards circular disruption—What it is, how to make it happen and how to know it's happening. *Business Strategy and the Environment*.

Bressanelli, G., Perona, M., & Saccani, N. (2019). Challenges in supply chain redesign for the Circular Economy: A literature review and a multiple case study [Article]. *International Journal of Production Research, 57*(23), 7395–7422. https://doi.org/10.1080/00207543.2018.1542176

Cerqueira, P. A., Soukiazis, E., & Proença, S. (2021). Assessing the linkages between recycling, renewable energy and sustainable development: Evidence from the OECD countries. *Environment, Development and Sustainability, 23*(7), 9766–9791.

Chen, D. M. C., Bodirsky, B. L., Krueger, T., Mishra, A., & Popp, A. (2020). The world's growing municipal solid waste: Trends and impacts. *Environmental Research Letters, 15*(7), 074021.

Çimen, Ö. (2021). Construction and built environment in circular economy: A comprehensive literature review [Review]. *Journal of Cleaner production, 305*, Article 127180. https://doi.org/10.1016/j.jclepro.2021.127180

Cobbinah, P. B., Erdiaw-Kwasie, M. O., & Basson, M. (2021). Deconstructing Africa's urban space: Sustainable development and spatial planning challenge. In *Sustainable Urban Futures in Africa* (pp. 21–41). Routledge.

Dehoust, G., Schueler, D., Vogt, R., & Giegrich, J. (2010). Climate protection potential in the waste management sector. Examples: municipal waste and waste wood; Klimaschutzpotenziale der Abfallwirtschaft. Am Beispiel von Siedlungsabfaellen und Altholz.

Droege, H., Kirchherr, J., Raggi, A., & Ramos, T. B. (2022). Towards a circular disruption: On the pivotal role of circular economy policy entrepreneurs. *Business Strategy and the Environment.*

Ellen MacArthur Foundation (2021). Completing the picture: How the circular economy tackles climate change.

Falk, M. T., & Hagsten, E. (2020). Time for carbon neutrality and other emission reduction measures at European airports. *Business Strategy and the Environment, 29*(3), 1448–1464.

Geissdoerfer, M., Savaget, P., Bocken, N. M., & Hultink, E. J. (2017). The circular economy—A new sustainability paradigm? *Journal of Cleaner Production, 143*, 757–768.

Govindan, K., & Hasanagic, M. (2018). A systematic review on drivers, barriers, and practices towards circular economy: A supply chain perspective. *International Journal of Production Research, 56*(1–2), 278–311.

Gusmerotti, N. M., Testa, F., Corsini, F., Pretner, G., & Iraldo, F. (2019). Drivers and approaches to the circular economy in manufacturing firms. *Journal of Cleaner Production, 230*, 314–327.

Hailemariam, A., & Erdiaw-Kwasie, M. O. (2022). Towards a circular economy: Implications for emission reduction and environmental sustainability. *Business Strategy and the Environment*. https://doi.org/10.1002/bse.3229

Hailemariam, A., Ivanovski, K., & Dzhumashev, R. (2022). Does R&D investment in renewable energy technologies reduce greenhouse gas emissions? *Applied Energy, 327*, 120056.

Heshmati, A., & Rashidghalam, M. (2021). Assessment of the urban circular economy in Sweden. *Journal of Cleaner Production, 310*, 127475.

Huang, Q., Chen, G., Wang, Y., Xu, L., & Chen, W. Q. (2020). Identifying the socioeconomic drivers of solid waste recycling in China for the period 2005–2017. *Science of the Total Environment, 725*, 138137.

Joensuu, T., Edelman, H., & Saari, A. (2020). Circular economy practices in the built environment. *Journal of Cleaner Production, 276*, 124215.

Kao, C. and Chiang, M.-H. (2000). On the estimation and inference of a cointegrated regression in panel data, in Baltagi B. H. (ed.), *Advances in Econometrics: Nonstationary Panels, Panel Cointegration and Dynamic Panels*, Vol. 15, pp. 179–222.

Kaza, S., Yao, L., Bhada-Tata, P., & Van Woerden, F. (2018). *What a waste 2.0: a global snapshot of solid waste management to 2050*. World Bank Publications.

Kennedy, S., & Linnenluecke, M. K. (2022). Circular economy and resilience: A research agenda. *Business Strategy and the Environment*, 1–12. https://doi.org/10.1002/bse.3004

Kirchherr, J., Bauwens, T., & Ramos, T. B. (2022). Circular disruption: Concepts, enablers and ways ahead. *Business Strategy and the Environment*.

Mhatre, P., Panchal, R., Singh, A., & Bibyan, S. (2021). A systematic literature review on the circular economy initiatives in the European Union. *Sustainable Production and Consumption, 26*, 187–202.

Moktadir, M. A., Kumar, A., Ali, S. M., Paul, S. K., Sultana, R., & Rezaei, J. (2020). Critical success factors for a circular economy: Implications for business strategy and the environment. *Business Strategy and the Environment, 29*(8), 3611–3635.

Morseletto, P. (2020). Targets for a circular economy. *Resources, Conservation and Recycling*, *153*, Article 104553. https://doi.org/10.1016/j.resconrec.2019.104553

Munaro, M. R., Tavares, S. F., & Bragança, L. (2020). Towards circular and more sustainable buildings: A systematic literature review on the circular economy in the built environment [Review]. *Journal of Cleaner production, 260*, Article 121134. https://doi.org/10.1016/j.jclepro.2020.121134

Neligan, A., Baumgartner, R. J., Geissdoerfer, M., & Schöggl, J. P. (2022). Circular disruption: Digitalisation as a driver of circular economy business models. *Business Strategy and the Environment.*

Ness, D. A., & Xing, K. (2017). Toward a resource-efficient built environment: A literature review and conceptual model. *Journal of Industrial Ecology, 21*(3), 572–592. https://doi.org/10.1111/jiec.12586

Neves, S. A., & Marques, A. C. (2022). Drivers and barriers in the transition from a linear economy to a circular economy. *Journal of Cleaner Production, 341*, 130865.

Ngan, S. L., How, B. S., Teng, S. Y., Promentilla, M. A. B., Yatim, P., Er, A. C., & Lam, H. L. (2019). Prioritisation of sustainability indicators for promoting the circular economy: The case of developing countries. *Renewable and Sustainable Energy Reviews, 111*, 314–331.

Önder, H. (2018). The socio-economic determiners of recycling: An analysis on European countries through a macro perspective. *Amfiteatru Economic Journal, 20*(48), 405–417.

Patwa, N., Sivarajah, U., Seetharaman, A., Sarkar, S., Maiti, K., & Hingorani, K. (2021). Towards a circular economy: An emerging economies context. *Journal of Business Research, 122*, 725–735.

Ranta, V., Aarikka-Stenroos, L., Ritala, P., & Mäkinen, S. J. (2018). Exploring institutional drivers and barriers of the circular economy: A cross-regional comparison of China, the US, and Europe. *Resources, Conservation and Recycling, 135*, 70–82.

Robaina, M., Villar, J., & Pereira, E. T. (2020). The determinants for a circular economy in Europe. *Environmental Science and Pollution Research, 27*(11), 12566–12578.

Silva, A. L. P., Prata, J. C., Walker, T. R., Duarte, A. C., Ouyang, W., Barcelò, D., & Rocha-Santos, T. (2021). Increased plastic pollution due to COVID-19 pandemic: Challenges and recommendations. *Chemical Engineering Journal, 405*, 126683.

Suchek, N., Fernandes, C. I., Kraus, S., Filser, M., & Sjögrén, H. (2021). Innovation and the circular economy: A systematic literature review. *Business Strategy and the Environment, 30*(8), 3686–3702.

Sverko Grdic, Z., Krstinic Nizic, M., & Rudan, E. (2020). Circular economy concept in the context of economic development in EU countries. *Sustainability, 12*(7), 3060.

Ünal, E., & Shao, J. (2019). A taxonomy of circular economy implementation strategies for manufacturing firms: Analysis of 391 cradle-to-cradle products [Article]. *Journal of Cleaner Production, 212*, 754–765. https://doi.org/10.1016/j.jclepro.2018.11.291

Urbinati, A., Chiaroni, D., & Chiesa, V. (2017). Towards a new taxonomy of circular economy business models [Article]. *Journal of Cleaner Production, 168*, 487–498. https://doi.org/10.1016/j.jclepro.2017.09.047

Van den Bosch, S. (2010). Transition Experiments: Exploring sovan den Bosch, S.J.M. (2010, September 16). Transition Experiments: Exploring societal changes towards sustainability. Erasmus University Rotterdam. Retrieved from http://hdl.handle.net/1765/20714cietal changes towards sustainability. Erasmus University Rotterdam.

Wiedinmyer, C., Yokelson, R. J., & Gullett, B. K. (2014). Global emissions of trace gases, particulate matter, and hazardous air pollutants from open burning of domestic waste. *Environmental Science & Technology, 48*(16), 9523–9530.

Wilson, D. C., & Velis, C. A. (2015). Waste management–still a global challenge in the 21st century: An evidence-based call for action. *Waste Management & Research, 33*(12), 1049–1051.

Wolf, M. J, Emerson, J. W., Esty, D. C., de Sherbinin, A., Wendling, Z. A., et al. (2022). *2022 environmental performance index*. Yale Center for Environmental Law & Policy.

Circular Economy Practices in Mauritius: Examining the Determinants

Emmanuel Senior Tenakwah
and Emmanuel Junior Tenakwah

1 Introduction

The pressing needs of today's world require a transition to a more sustainable system to be able to address environmental, societal, and economic concerns such as resource depletion, biodiversity loss, excessive land use, high unemployment, rate of social vulnerability, supply risks, and frequent economic and financial instabilities (Banerjee & Duflo, 2011; Markard et al., 2012; Rockström et al., 2009; Sachs, 2015). The circular economy (CE) concept emerged to rescue these pressing issues. While CE is not entirely new, it has and continues to drive efforts towards sustainable development (Geissdoerfer et al., 2017). CE is referred to as an economic

---

E. S. Tenakwah (✉)
Business and Accounting Discipline, Faculty of Arts and Society, Charles Darwin University, Darwin, Australia
e-mail: emmanuel.tenakwah@cdu.edu.au

E. J. Tenakwah
Swinburne University of Technology Sydney, Parramatta, NSW, Australia

Southern Cross Institute, Parramatta, Australia

M. O. Erdiaw-Kwasie and G. M. M. Alam (eds.), *Circular Economy Strategies and the UN Sustainable Development Goals*,
Sustainable Development Goals Series,
https://doi.org/10.1007/978-981-99-3083-8_9

model that focuses on the efficient use of resources, which includes minimising waste, reduction in the use of primary resources, long-term value retention, closed loops of products, and others within the boundaries of environmental protection (Morseletto, 2020). CE is touted "as a regenerative system in which resource input and waste, emission, and energy leakage are minimised by slowing, closing, and narrowing material and energy loops thanks to the long-lasting design, maintenance, repair, reuse, remanufacturing, refurbishing, and recycling" (Geissdoerfer et al., 2017, p. 759).

CE is being promoted by several national and regional governments, such as Canada, China, Finland, France, Japan, Sweden, The Netherlands and the United Kingdom (Korhonen et al., 2018). CE has also become the backbone of industrial and environmental policies across Africa (Boon & Anuga, 2020; Desmond & Asamba, 2019; Mutezo & Mulopo, 2021), China (Fan & Fang, 2020; Zhu et al., 2019), Europe (Camilleri, 2020; Domenech & Bahn-Walkowiak, 2019; Hailemariam & Erdiaw-Kwasie, 2022) and the United States (Guerra & Leite, 2021; Syberg et al., 2021). These efforts have been backed by businesses that believe sustainable environmental and economic development hinges on circular economy models (Ghisellini et al., 2016; Murray et al., 2017). Statistics from across the world supports these assertions. For example, the European Commission estimates that the EU manufacturing sector alone stands to benefit from circular economy types of transactions, which can potentially create over € 600 billion annually (COM-European Commission, 2014; Ellen MacArthur Foundation, 2012). FICF and Mckinsey (2014) also estimate that the global economy benefits $1000 billion annually from CE models.

While the concept has attracted attention worldwide, its development and application in African regions are lagging (Mutezo & Mulopo, 2021). The concept of CE remains vague in the continent and is often not supported by legal and regulatory frameworks (Desmond & Asamba, 2019). In some cases, the mechanisms to transition into CE are not in place, raising questions about the continent's commitment to CE (Desmond & Asamba, 2019). Additionally, studies have not delved into the determinants of CE among SMEs who constitute the backbone of the African economy (Naidoo, 2020). As compared to large companies, SMEs are agile, adaptive and innovative and, as such, can contribute to the drive towards CE models (Pereira et al., 2022). It is, therefore, important for contemporary studies to examine the issue of CE in Africa. This is particularly important as many African countries battle fundamental sustainable development challenges, likely exacerbating the continent's existing challenges (Boon & Anuga, 2020). This explains why policymakers, researchers and the international community are all interested in appropriate models to overcome these challenges (Boon & Anuga, 2020).

This chapter, therefore, examines the factors influencing the adoption of CE in SMEs operating in Mauritius. We examine these drivers from three perspectives: country-level determinants, organisational-level determinants and individual-level determinants. We selected Mauritius due to the bold steps taken by the government to make SMEs innovative, sustainable and globally competitive amid the myriad of sustainable development challenges. Nevertheless, the country continues battling solid waste driven by unsustainable production and consumption patterns. While knowledge development and CE adoption seem to be at advanced stages in developed societies such as Europe, there are pieces of evidence suggesting CE adoption in regions such as Africa and Asia with countries such as Mauritius, Ghana, Egypt, Rwanda, India and Pakistan (Agyemang et al., 2019; Malik et al., 2022). Mauritius is a leading example among African nations and emerging economies to introduce a systematic regulatory framework, incentives and collaborative platform for government and other major stakeholders to dialogue and develop tools for reducing, reusing and recycling waste on the Island. These opportunities created through the process encourage individuals, NGOs and cooperatives to serve as change agents to foster a greener and sustainable Mauritius agenda (Ministry of Environment, Solid Waste Management and Climate Change, 2022).

Additionally, Mauritius is poised to transform the African continent through the African Leadership University by "selecting, developing, and placing three million ethical and entrepreneurial young leaders into highly productive roles by 2035" (Mitrofanenko et al., 2021, p. 12). ALU has already integrated CE practices into its operations through their campus design, adopting reusable plates and cutlery while transforming food waste into fertilisers (Mitrofanenko et al., 2021). This makes Mauritius an ideal context for our study.

Our study is motivated by the need to extend the theory on the determinants of CE adoption in SMEs, which is still nascent (Pereira et al., 2022; Ünal et al., 2018). In doing so, we contribute to the CE literature on SMEs (Pereira et al., 2022) by suggesting that adopting CE in competitive economies hinges on external support and pressures such as tax incentives. In addition, many scholarly works on CE have sought to explore the influence of individual factors on CE adoption, often ignoring the role of organisational characteristics. This study thus advances the literature on CE adoption by focusing on how firm-level characteristics influence CE adoption. Third, this study is devoted to deepening our understanding of factors such as governmental support individuals and firm actions simultaneously influence CE adoption, contrary to examining them in isolation (Bassi & Dias, 2020; del Rio Gonzàles, 2005).

This chapter is structured as follows: the next section reviews the literature and hypotheses. Then, the section introduces the CE concept while illustrating its importance. It also presents the arguments on the key determinants of CE and the development of hypotheses. The following section describes the research design, the sources of data, the measurement of variables and estimation techniques. The subsequent section presents the results, followed by a discussion of the key findings. The last section of the chapter offers the contributions to research and practice, the limitations and directions for future research.

2 Literature Review

2.1 Circular Economy Practices

The last decade has seen a significant rise in circular economy practices. The concept was developed due to the challenges posed by traditional practices where large quantities of cheap and easily accessible materials and energy are used in business activities (Robaina et al., 2020). These activities could have a far-reaching impact on the planet in terms of emissions to or extractions from the environment (Steffen et al., 2015). Steffen et al. (2015) argue that direct human intervention or modification may change these impacts on the environment. These interventions or modifications include a circular economy. Kalmykova et al. (2018) and Kirchherr et al. (2017) defined a circular economy as a sustainable way of producing goods and services. The circular economy concept entails preserving products, components and materials at their utmost usefulness and value to reduce waste (McKinsey Center for Business and Environment, Ellen MacArthur Foundation, 2015). Prior studies have also defined CE as a model where resourcing, purchasing, production and reprocessing operations are designed to protect the well-being of humans and the environment at large (Centobelli et al., 2021; Murray et al., 2017). From the perspective of McKinsey Center for Business and Environment, Ellen Macarthur Foundation (2015), CE practices strongly hinge on three basic principles (1) preserve and value natural capital by controlling and balancing flows of renewable resources; (2) ensure productivity is optimised through the circulation of products, components and materials relating technical and biological cycles; and (3) enhance system effectiveness by minimising the impact on human well-being, such as mobility, food, shelter, education, health and entertainment, as well as managing land use, air, water and noise pollution, the release of substances and climate change.

The adoption of CE practices has gained ground in both developed and emerging economies across the globe. Most importantly, this agenda is driven by private and public entities in various economies. Prior studies have indicated enormous benefits of adopting CE practices (Robaina et al., 2020; Steffen et al., 2015). For instance, the Circular Economy and Resource Efficiency Agency (UK) (2015) revealed that adopting CE in European Union could create about 3 million job opportunities in the labour market. Mitchell and James (2015) estimate that by 2030, the EU member countries could reduce unemployment by 520,000 and create jobs because CE includes waste management, repair, recycling activities, etc. Also, adopting CE catalyses to counter global warming triggers such as greenhouse gas emissions from the consumption of non-renewable energy and extraction and production of raw materials for manufacturing (Deloitte, 2016; Tukker et al., 2016). Additionally, adopting CE practices reduces the pressure on using natural resources, thereby limiting the environmental pressures in society (Robaina et al., 2020). For SMEs, prior studies have demonstrated that they can have an edge over others when they adopt CE practices in advanced economies (Sjoerdsma & van Weele, 2015) and emerging economies (Wu & Wu, 2015). Firms adopting CE practices enhance reuse capability (Mohammed et al., 2016) and recycling capability (Zhao et al., 2012).

2.2 *Determinants of Circular Economy Practices*

The concept of CE has attracted so much attention from various stakeholders. These stakeholders are shareholders, governments and agencies, customers, suppliers, communities and others. Social pressures emanate from governments and their agencies, communities and others. These stakeholders exert some levels of pressure on firms to adopt circular economy practices to enhance their activities. These pressures often represent the firms' awareness of the various stakeholders affected by their activities (Erdiaw-Kwasie, 2018; Rivis & Sheeran, 2003). For instance, regulatory bodies pressure firms through environmental regulations. Environmental regulations were instituted to ensure environmental awareness and improve the perceived attitudes among firms (Drake et al., 2004). Such regulations improve firms' environmental responsibility and green commitments (Roxas & Coetzer, 2012). Most importantly, these regulations focus on market demand and pressure from the community to adopt CE practices in individual firms (Johnstone & Labonne, 2009).

Previous studies have shown that CE's success largely depends on societal and economic stakeholders (Ranta et al., 2018). In these studies, it was evident that the determinant of adopting CE includes regulations, governmental subsidies, culture and firm-specific structures (Ranta et al., 2018). This study categorises determinants into individual-level, firm-level and country-level pressures.

2.3 Individual-Level Determinants

Research has shown micro-level resistance to adopting CE, with most SME managers having risk-averse behaviours towards CE adoption (Rizos et al., 2016). This suggests that adopting CE will require a change at the micro-level. The system-level perspective highlights the need for change by exercising personal agentic resources by leaders, managers and owners (Malik et al., 2022; Pereira et al., 2022). Such an approach requires implementing practices that support CE and behavioural changes (Malik et al., 2022; Rizos et al., 2016). Prior studies examining individual factors influencing CE adoption have focused on factors such as employees' characteristics, ownership and positions held in the organisation (Bassi & Dias, 2020; del Rio Gonzàles, 2005). The separation between decision-making and management and the ownership of the capital often drive these factors. Hoogendoorn et al. (2015) argue that these factors primarily influence the firm's interest in circular practices.

The results of studies examining the role of females in CE adoption have generally been positive. For example, some studies on female owners' and executives' contributions to CE practices have reported a positive impact (Ali et al., 2021; Mase et al., 2017; Ünal et al., 2018). These studies attributed the significant influence on CE to female managers' and owners' behavioural commitment, values and beliefs towards CE. Females generally ensure that their values align with organisational goals. Ali et al. (2021) and Mase et al. (2017) indicated that usually, females are more concerned about environmental issues, which motivates them to adopt business models that minimise the economic and financial consequences on the company. As such, female managers and owners support initiatives that provide more benefits to society than individual interests.

However, research shows that women are under-represented in high-value-added circular activities such as eco-design, circular product development and advanced technologies compared to low-value-added informal activities (Albaladejo et al., 2022). For example, women dominate local waste management in Africa and Asia, often working in waste banks (Asomani-Boateng, 2016; Kaza et al., 2018). Despite this underrepresentation, some studies have suggested that female-led administrators of waste banks are leveraging their self-organisational skills and values to manage solid waste (Asomani-Boateng, 2016; Kaza et al., 2018), something that seems to be crippling Mauritius. The underrepresentation of women in top positions or high-value-added CE activities will likely boost the desire of female managers and owners to spearhead the CE agenda. This suggests that female managers and owners contribute to CE adoption. In line with this, we propose that:

H1a The largest owner working as the top manager positively influences the CE practices of SMEs in Mauritius.

H1b Female ownership in the firm positively influences the CE practices of SMEs in Mauritius.

H1c Female top manager positively influences the CE practices of SMEs in Mauritius.

2.4 Organisational-Level Determinants

SMEs' major issues in adopting CE at the organisational level are the absence of close collaboration between parties and suppliers and organisational inertia (Nilakant & Ramnarayan, 2006; Rizos et al., 2016). In addition, SMEs face high costs and a lack of technical knowledge in adopting CE (Rizos et al., 2016; Shi et al., 2008). Prior studies have shown that factors that support CE adoption may be within the firm. These factors include organisational environmental commitment, internationally recognised certification and legal status (Private, public or others). Organisational environmental commitment relates to a company's values to improve environmental, financial and social performance (Chang et al., 2011). Environmental commitment integrates the willingness and engagement to implement CE. Mutz et al. (2015) indicated that from a strategic perspective, firms ensure their strategy reflects the firm's attitude, perceived financial benefits and behavioural control of their environmental commitment. Regarding international certification, firms usually experience pressure from stakeholders to ensure their products and services are of global standards. These pressures represent the firms' awareness of the various stakeholders affected by their activities (Rivis & Sheeran, 2003). These stakeholders often help the firm anticipate their expectations and resolve them. The international certification pressures significantly influence CE practices and the green behaviour of firms (Singh et al., 2016). This implies a higher level of awareness by firms regarding social recognition and financial benefits associated with the stakeholders' expectations. Most importantly, these internationally recognised quality certifications are based on market demand and the community's pressure to adopt CE practices in individual firms (Johnstone & Labonne, 2009). In line with prior studies, we argue that legal status, the corporate grouping of the firm and internationally recognised quality certification significantly influence the adoption of CE practice:

H2a The firm's legal status positively influences the CE practices of SMEs in Mauritius.

H2b Corporate grouping positively influences the CE practices of SMEs in Mauritius.

H2c Pressures from internationally recognised quality certification positively influence the CE practices of SMEs in Mauritius.

2.5 Country-Level Determinants

At a macro-level, studies on the adoption of CE by SMEs have been hindered by factors such as lack of government aid and legal compliance costs, limited support for environmental initiatives, and institutional voids such as the absence of unified platforms to drive CE (Geng & Doberstein, 2008; Rizos et al., 2016). Research has shown that policy stakeholders can influence CE adoption if they collaborate, exchange knowledge, and support SMEs, where possible (Ritzén & Sandström, 2017; Rizos et al., 2016). As a result, specific incentives have been provided by governments and other international agencies to drive CE. An example is green economic incentives. Green economic incentives are benefits offered by governments and non-governmental agencies to support firms adopting environmental initiatives for development (Agnello et al., 2015). These incentives include financial rewards to drive the design or adoption of environmental practices of firms in the country. The rewards enhance recycling techniques and sustainable transition to CE (Gunsilius, 2015). Undertaking investments in CE often comes at a higher cost to SMEs than large entities. The benefit of green economic incentives benefits SMEs that positively enhance internal capabilities and decision-making. Aside from these green economic incentives, governments in various jurisdictions have introduced tax concessions/incentives, particularly to SMEs, to support their drive towards CE (Chang et al., 2011). Such incentives reduce the tax burden of SMEs by providing them with additional fiscal space to deal with CE investments. Similarly, governments allocate contracts to SMEs with better circular practices to motivate other SMEs who have failed to adopt sustainable CE practices. In line with the above arguments, we argue that tax incentives and government contracts enhance SME investment in CE.

H3a Government contracts positively influence the CE practices of SMEs.

H3b Tax incentives positively influence the CE practices of SMEs.

3 Methodology

3.1 Research Design, Data Source and Sample Distribution

We adopt a descriptive quantitative research design per the study's objectives. This design is considered appropriate because it provides a broader perspective on a research phenomenon. As noted by previous studies (Creswell, 2016; Easterby-Smith et al., 2008), quantitative approaches benefit generalisation as the results apply to large or different groups or organisational settings. Our data was sourced from the World Bank Enterprise Survey (WBES) database. WBES is a firm-level survey conducted by the World Bank through a standard methodology to provide information about the practices of firms across countries. The data is collected

Table 1 Sample distribution

Industry	*Scale of Business*			*Total*
	Small	*Medium*	*Large*	
Manufacturing	81	39	14	134
Retail services	114	58	3	175
Non-retail services	153	53	7	263

by field experts who collect face-to-face data through a questionnaire. The WBES database provides extensive information on firms. It covers a wide array of topics on the general business environment, such as infrastructure and services, sales and supplies, technology, management practices, degree of competition, innovation, capacity, land and permits, crime, finance, business-government relations, labour, business environment, performance, and green economy. This study focuses on Mauritius battling unsustainable production and consumption patterns. Statistics show that solid waste has increased by over 100 per cent in the last 20 years. The sample for this study is 572 firms operating. The distribution of the sample is presented below (Table 1). For our dependent variables, we identified the circular practices of SMEs. For the independent variables, we identified owner/top manager, female ownership, female top manager, legal status, multi-firm, certification, government contract and tax incentives as determinants for circular economy practices.

3.2 Measurement of Variables

We present a description and measurement of variables used in the regression model. Table 2 presents these details.

3.3 Estimation Strategy

We adopt a binary logit regression as the estimation technique for the dichotomous nature of the data. Prior studies (Asteriou & Hall, 2011) argue that this technique is the most efficient in estimating binary data. To assure the robustness of our results, we adopted a binary probit regression to test the relationships using the same data.

Table 2 Measurement of variables

Variables	*Description*	*Measurement*
Dependent variable		
Circularity	Adoption of circular practices	Dummy variable, coded 1 if the firm adopted such measures and 0 otherwise
Independent variable		
Owner/top manager	The largest owner is a top manager	Dummy variable, coded 1 if the largest owner is also the top manager and 0 otherwise
Female ownership	Owned by a female	Dummy variable, coded 1 if the firm is owned by a female and 0 otherwise
Female top manager	Top manager.is a female	Dummy variable, coded 1 if the firm has a female as a top manager and 0 otherwise
Legal status	Legal status	Coded 1 if the firm is a shareholding company with shares traded on the stock market; 2 if the firm is a shareholding company with non-traded shares or shares traded privately; 3 if the firm is a sole proprietorship; 4 if the firm is a registered partnership; and 5 if the firm is a limited partnership
Multi-firm	Multi-firm corporate group	Dummy variable, coded 1 if the firm is part of a multi-firm corporate group, such as a conglomerate, holding company, or network and 0 otherwise
Certification	Internationally recognised quality certification	Dummy variable, coded 1 if the firm has an internationally recognised quality certification and 0 otherwise
Government contract	Award of government contract	Dummy variable, coded 1 if the firm has been awarded a government contract and 0 otherwise
Tax incentives	Beneficiary of tax incentives	Dummy variable, coded 1 if the firm has benefitted from tax incentives and 0 otherwise
Control variables		
Industry	The operating industry of the firm	Coded 1 if the firm is a manufacturing firm, 2 if the firm provides retail services and 3 if the firm provides non-retail services
Age	Firm Age	Log of years of operation
Size	Firm Size	Log of number of permanent employees engaged by the firm at the end of last fiscal year

3.4 *Model Specification*

The logit regression is specified to test the relationships between circularity and its determinants.:

$$Cir_{it} = \beta_0 + \beta_1 Owner_{i,t} + \beta_2 Female_{i,t} + \beta_3 Top_{i,t} + \beta_4 Legal_{i,t} + \beta_5 Multi + \beta_6 Cert + \beta_7 Govt + \beta_8 Tax + \delta_9 Controls_{i,t} + \varepsilon_{i,t} \quad (1)$$

where i represents firm, i, t represents year t, $circularity_{it}$ represents the dependent variable, β_1- β_8 represent independent variables: *Owner/top manager, Female ownership, Female top manager, legal status, multi-firm, certification, government contract and tax incentives*, δ_2 controls I and t represents a matrix of controls.

4 Results and Discussion

The main results are presented in Table 3. The first column in the table presents the baseline model (i.e. control variable), and the second column presents the control variables and the direct effects of individual factors such as owner/top manager, female ownership, and top female manager. The third column presents the control variables and the direct effects of individual factors such as legal status, multi-firm corporate group and internationally recognised certification. In contrast, the fourth column presents results on country-level factors such as government contracts and tax incentives. From the table, the individual-level variables' results do not significantly affect the adoption of circular practices in Mauritius. On the organisational-level variables, the results suggest that while multi-firm corporate group (0.831^{*}) and internationally recognised certification (1.345^{***}) are positive and significant, the effect of legal status is insignificant and therefore has no impact on the adoption of circular practices. At the country level, the award of government contracts (0.684^{*}) and tax incentives (2.576^{***}) significantly affect the adoption of circular practices. These results are also evident in the marginal effects presented in Table 4, suggesting that none of the individual-level variables significantly affect CE adoption. On the other hand, all two firm-level dummy variables (multi-firm corporate group and internationally recognised certification) had significant effects on CE adoption. For example, the results suggest that firms with internationally recognised certification (0.110^{**}) will likely lead to CE adoption. Our results also show that firms part of a multi-firm corporate group (0.012^{*}) will likely adopt CE practices. A firm

Table 3 Logit results

	1	*2*	*3*	*4*
Industry	−0.170 (0.172)	−0.146 (0.181)	−0.089 (0.160)	−0.049 (0.169)
Age	0.014 (0.012)	0.014 (0.012)	0.005 (0.012)	0.006 (0.013)
Size	0.000 (0.001)	0.000 (0.001)	0.001 (0.001)	0.002 (0.001)
Owner/top manager		−0.420 (0.380)		
Female ownership		−0.307 (0.418)		
Female top manager		0.752 (0.528)		
Legal status			0.252 (0.296)	
Multi-firm corporate group			0.201 (0.407)	
International certification			1.252*** (0.402)	
Government contract				0.684* (0.392)
Tax incentives				2.576*** (0.448)
Constant	−24.133 (23.511)	−25.896 (23.013)	−8.407 (23.820)	−12.971 (25.102)
Observations	572	572	572	572
R^2	0.008	0.020	0.047	0.118
Chi^2	1.700	5.722	18.265	38.157
Converged	1.000	1.000	1.000	1.000

Note The reported estimates are unstandardised with standard errors in parentheses. Model 1 includes controls only; Model 2 includes controls and individual-level variables; Model 3 includes controls and organisational-level variables; Model 4 includes controls and country-level variables; standard errors in parentheses $^{*}p < 0.10$, $^{**}p < 0.05$, $^{***}p < 0.01$

with a government contract or benefiting from tax incentives is likely to adopt CE. Therefore, our results support hypotheses 2b, 2c, 3a and 3b. On the other hand, we found no support for hypotheses 1a, 1b, 1c and 2a. In contrast, previous studies have generally found these variables to support adopting circular practices among SMEs in Mauritius.

The findings of our study have shown that multiple determinants drive CE practice adoption among SMEs in Mauritius. Our study confirms that corporate grouping, pressures from internationally recognised certification, government contracts and tax incentives are the significant

Table 4 Marginal effects

	Dy/dx	*Std. error*
Owner/top manager	−0.028	0.028
Female ownership	−0.018	0.024
Female top manager	0.058	0.050
Multi-firm corporate group	0.012*	0.026
International certification	0.110**	0.050
Government contract	0.043*	0.030
Tax incentives	0.394**	0.096

Note dy/dx is for discrete change of dummy variable from 0 to 1
*$p < 0.10$, **$p < 0.05$, ***$p < 0.01$

determinants of CE practices. Our study confirms that the corporate grouping and international certification support the firm's intention to adopt CE practices regarding the organisational factors. In terms of the corporate grouping, the results imply that if firms are part of a conglomerate or holding company, they are likely to emulate what other group members are doing. In most instances, where an entity in the group initiates successful initiatives, headquarters are likely to replicate among the other subsidiaries to ensure uniformity within the group. We attribute this result to the fact that irrespective of the firm size when firms undertake initiatives beneficial to stakeholders, others will mimic it. Our finding supports the study by Singh et al. (2016), who have argued that a multi-firm's commitment to CE practices successfully influences other firms in the group to adopt similar practices. In terms of internationally recognised certification, our findings suggest it ensures firms adopt globally accepted practices. Due to stakeholders' awareness of global trends, they often expect firms to follow similar practices. Therefore, getting internationally recognised certification will mean that firms adopt CE practices. Such trends have pushed many SMEs in Mauritius to adopt CE practices to remain relevant in the market and deliver sustainable products and services. Our findings confirm Rivis and Sheeran's (2003) claim that firms usually experience pressure from stakeholders to ensure their products and services are of global standards. These pressures represent the firms' awareness of the various stakeholders affected by their activities.

Regarding country-level determinants, our results suggest that initiatives put together by the government influence SMEs to adopt CE practices. For instance, government contract has a positive influence indicating that the government's decision to award contracts to SMEs who have adopted CE practices has proved successful. With this incentive, many SMEs believe the government rewards commitment to CE practices; therefore, there is a need to focus and commit resources towards these initiatives. Our results confirm the study by del Mar Alonso-Almeida et al. (2021). Their study suggests that providing financial incentives to SMEs for adopting CE practices influences others to adopt CE practices successfully. Another determinant based on our results is tax incentives provided to SMEs that adopt CE practices in Mauritius. Our results imply that a motivating factor for other SMEs to adopt CE practices is governments offering rewards to firms with tax incentives to compensate them for their commitments towards CE. These findings demonstrate the imperativeness of CE practices incentives for SMEs to pursue CE transition further. Our findings align with Kuo et al. (2015) suggesting that government CE incentives position firms on the right path towards CE. Overall, our findings regarding governmental incentives are that financial incentives are very effective tools to divert firms' attention towards adopting CE practices globally (del Mar Alonso-Almeida et al., 2021; Kuo et al., 2015).

Concerning the individual-level determinants, our results contradict other studies that have argued that individual characteristics often influence SMEs to adopt CE practices. Our results show that while these pressure factors, such as the gender of the owner, gender of top managers and others, may influence the firms to adopt sustainability practices in general, their impact on CE may not be as profound as other factors. This is not entirely surprising as previous studies on Africa and Mauritius, to be precise, have reported culturally sanctioned biases against women, which limits the efforts of women entrepreneurs and business managers in making significant business decisions such as CE (Tandrayen-Ragoobur & Kasseeah, 2017).

4.1 Robustness Test

To check the robustness of our results, probit regression was used to estimate the relationships. We present the results in Table 5. Similar to the main results, column 1 presents the baseline model (i.e. control variable), and column 2 presents the control variables and the direct effects of individual factors such as owner/top manager, female ownership, and top female manager. Column 3 presents the control variables and the direct effects of individual factors such as legal status, multi-firm corporate group, and internationally recognised certification. In contrast, column 4 presents the results on country-level factors such as government contracts and tax incentives. Consistent with the main results, the effects of the multi-firm corporate group, internationally recognised certification, the award of government contracts and tax incentives are all significant, while the remaining factors are not. Given that probit is a probability model, we further estimated the marginal effects of the models (see Table 6). The marginal effects of our dummy variables (i.e. multi-firm corporate group, international certification, government contract and tax incentives) were all significant, indicating that a change in these variables will likely improve CE adoption in firms. The results, therefore, suggest that they are robust enough to influence managerial and government decision-making.

5 Implications

5.1 Theoretical Implications

CE in the context of Africa is yet to spread to SMEs, so research on CE practices of SMEs has attracted limited research attention (Mutezo & Mulopo, 2021). While the literature review confirms and substantiates this contention, the results thus suggest that the trend is changing, and SMEs are beginning to transition from linear to circular economic models. Our study extends the literature on CE in SMEs by demonstrating the critical determinants of CE in emerging and developing countries by examining these factors across three levels (i.e. individual, organisational and country levels). Specifically, we contribute to the literature in the following ways:

First, our study contributes to stakeholder theory by considering the influence of government and competitors in adopting CE practices among SMEs in developing countries. The results have developed an initial roadmap of how stakeholders can influence the adoption of CE in businesses, particularly SMEs. While the role of stakeholders in sustainable development is well known, its influence on CE, particularly in emerging and developing countries, is rarely studied. Our results highlight this by offering insights into which category of stakeholders can drive CE adoption in SMEs, particularly in Africa.

Table 5 Probit results

	1	*2*	*3*	*4*
Industry	−0.076 (0.082)	−0.072 (0.083)	−0.043 (0.078)	−0.022 (0.084)
Age	0.006 (0.006)	0.007 (0.006)	0.001 (0.006)	0.003 (0.006)
Size	0.000 (0.000)	0.000 (0.000)	0.000 (0.001)	0.001 (0.001)
Owner/top manager		−0.204 (0.185)		
Female ownership		−0.124 (0.188)		
Female top manager		0.360 (0.249)		
Legal status			0.135 (0.124)	
Multi-firm corporate group			0.113 (0.203)	
International certification			0.661*** (0.213)	
Government contract				0.316 (0.196)
Tax incentives				1.414*** (0.260)
Constant	−10.905 (11.862)	−12.774 (11.766)	−1.881 (11.928)	−6.638 (12.283)
Observations	572	572	572	572
R^2	0.007	0.020	0.048	0.117
Chi^2	1.579	5.300	17.399	33.716
Converged	1.000	1.000	1.000	1.000

Note The reported estimates are unstandardised with standard errors in parentheses. Model 1 includes controls only; Model 2 includes controls and individual-level variables; Model 3 includes controls and organisational-level variables; Model 4 includes controls and country-level variables; standard errors in parentheses $^{*}p < 0.10$, $^{**}p < 0.05$, $^{***}p < 0.01$

Second, many scholarly works on CE have sought to explore the influence of individual factors on CE adoption, often ignoring the role of organisational characteristics. This study thus advances the literature on CE adoption by focusing on firm-level factors in competitive economies.

Third, while past studies have primarily focused on the influence of governments (Agnello et al., 2015; Gunsilius, 2015), individuals (Bassi & Dias, 2020; del Rio Gonzàles, 2005) and firm actions (Chang et al., 2011; Mutz et al., 2015), these three streams of research have developed mainly in isolation. This study, therefore, deepens our understanding of factors

Table 6 Marginal effects

	Dy/dx	*Std. error*
Owner/top manager	−0.028	0.028
Female ownership	−0.016	0.023
Female top manager	0.056	0.045
Multi-firm corporate group	0.014*	0.027
International certification	0.116**	0.050
Government contract	0.042*	0.030
Tax incentives	0.353**	0.094

Note dy/dx is for discrete change of dummy variable from 0 to 1
$^{*}p < 0.10$, $^{**}p < 0.05$, $^{***}p < 0.01$

across these three levels by examining them simultaneously. By so doing, we extend the boundaries of the determinants of CE by showing how CE adoption hinges on firm-level and governmental support.

5.2 *Practical Implications*

Our results also have some practical implications:

First, the evidence from our study suggests that firms, irrespective of their size and location, can transition from a linear economy to a CE. While our results do not provide the challenges SMEs likely encounter, the significant effect of government incentives indicates that SMEs require support from critical stakeholders to navigate the CE journey.

Our results reiterate the need for governments to increase their efforts to incentivise businesses to move towards circular economic models. This is particularly important as most SMEs face several challenges that make it difficult to transition to CE models quickly and proactively. However, as argued and confirmed through the study, CE drives the efficient use of resources which can deliver significant economic benefits to individuals, businesses, countries and the global economy.

The results also confirm that CE offers a promising contribution to the global sustainable development agenda. By doing so, it draws the attention of regulators of businesses and the chamber of SMEs to the factors that may encourage and hinder the adoption of CE. Understanding these factors will ensure that CE efforts align with the desired economic, environmental and political priorities. It also reminds practitioners to constantly monitor these drivers to ensure the achievement of CE targets.

6 Conclusion, Limitations and Future Research Directions

Our chapter is one of the few pieces of research contributing to the nascent field of CE practices in SMEs. Our results and discussions highlight the need for SMEs and all stakeholders to pay attention to CE adoption, particularly considering the world's challenges today. We show that individual-level variables do not significantly affect the adoption of CE in Mauritius. On the other hand, organisational- and country-level variables such as multi-firm corporate groups, internationally recognised certifications, the award of government contracts and tax incentives significantly affect the adoption of CE. The study suggests that stakeholders such as governments and competitors can play a significant role in the sustainable socio-economic development of countries. This can certainly be achieved through approaches that enhance resource efficiencies and may address existing cross-sectoral challenges in an integrated manner.

Our study is without limitations: first, we rely on secondary data collected by the World Bank. While this data has provided valuable insights about key drivers, future studies can draw on other databases such as Refinitiv (i.e. Thomson Reuters—Eikon) and primary sources such as interviews to delve deep into these. Second, the absence of capital and enough support, such as industrial parks and CE communities, hinders SMEs' adoption of CE models. While our results have provided evidence of how similar support can drive the adoption of CE, future research can examine when and what kind of support will be relevant for SMEs across different industries. Third, while studies have found that individual-level factors can break micro-level resistance, our results on the insignificant effects suggest that future studies can explore other individual-level factors capable of breaking down this resistance. Finally, it would be interesting for future studies to explore how SMEs adopting CE create value for their customers, given that poor value proposition hinders SMEs from adopting CE models.

References

Agnello, X., Naveen, J., Ravichandran, M., & Balamurugan, J. (2015). Clean technology and its efficacy: Strategies of environmental management. *Journal of Environmental Social Science, 2*(2), 1–10.

Agyemang, M., Kusi-Sarpong, S., Khan, S. A., Mani, V., Rehman, S. T., & Kusi-Sarpong, H. (2019). Drivers and barriers to circular economy implementation: An explorative study in Pakistan's automobile industry. *Management Decision, 57*(4), 971–994. https://doi.org/10.1108/MD-11-2018-1178

Albaladejo, M., Arribas, V., & Mirazo, P. (2022). *Why adopting a gender-inclusive approach towards Circular Economy matters*. United Nations Industrial Development Organization (UNIDO). https://iap.unido.org/articles/why-adopting-gender-inclusive-approach-towards-circular-economy-matters. Accessed 30 Oct 2022.

Ali, Q., Parveen, S., Yaacob, H., Zaini, Z., & Sarbini, N. A. (2021). COVID-19 and dynamics of environmental awareness, sustainable consumption and social responsibility in Malaysia. *Environmental Science and Pollution Research, 28*(40), 56199–56218.

Asomani-Boateng, R. (2016). Local networks: Commodity queens and the management of organic solid waste in indigenous open-air markets in Accra, Ghana. *Journal of Planning Education and Research, 36*(2), 182–194.

Asteriou, D., & Hall, S. G. (2011). *Applied econometrics* (2nd ed.). Palgrave Macmillan.

Banerjee, A., & Duflo, E. (2011). *Poor economics: A radical rethinking of the way to fight global poverty*. Public Affairs.

Bassi, F., & Dias, J. G. (2020). Sustainable development of small-and medium-sized enterprises in the European Union: A taxonomy of circular economy practices. *Business Strategy and the Environment, 29*(6), 2528–2541.

Boon, E. K., & Anuga, S. W. (2020). Circular economy and its relevance for improving food and nutrition security in Sub-Saharan Africa: The case of Ghana. *Materials Circular Economy, 2*(1), 1–14.

Camilleri, M. A. (2020). European environment policy for the circular economy: Implications for business and industry stakeholders. *Sustainable Development, 28*(6), 1804–1812.

Centobelli, P., Cerchione, R., Esposito, E., & Passaro, R. (2021). Determinants of the transition towards circular economy in SMEs: A sustainable supply chain management perspective. *International Journal of Production Economics, 242*, 108297.

Chang, Y. Y., Hughes, M., & Hotho, S. (2011). Internal and external antecedents of SMEs' innovation ambidexterity outcomes. *Management Decision, 49*(10), 1658–1676.

COM (European Commission). (2014). Towards a circular economy: A zero-waste programme for Europe, Communication from the Commission to

the European Parliament, the Council, the European Economic and Social Committee and the Committee of the Regions, Brussels.
Creswell, J. W. (2016). *Qualitative, quantitative and mixed methods approaches*.
del Mar Alonso-Almeida, M., Rodriguez-Anton, J. M., Bagur-Femenías, L., & Perramon, J. (2021). Institutional entrepreneurship enablers to promote circular economy in the European Union: Impacts on transition towards a more circular economy. *Journal of Cleaner Production, 281*, 124841.
del Río González, P. (2005). Analysing the factors influencing clean technology adoption: A study of the Spanish pulp and paper industry. *Business Strategy and the Environment, 14*(1), 20–37.
Deloitte. (2016). Deloitte sustainability-circular economy measures and potential emissions reduction in a circular economy in high-impact sectors. https://www2.deloitte.com/content/dam/Deloitte/fi/Documents/risk/Deloitte%20-%20Circular%20economy%20and%20Global%20Warming.pdf
Desmond, P., & Asamba, M. (2019). Accelerating the transition to a circular economy in Africa: Case studies from Kenya and South Africa. In *The Circular Economy and the Global South* (pp. 152–172). Routledge.
Domenech, T., & Bahn-Walkowiak, B. (2019). Transition towards a resource efficient circular economy in Europe: Policy lessons from the EU and the member states. *Ecological Economics, 155*, 7–19.
Drake, F., Purvis, M., & Hunt, J. (2004). Meeting the environmental challenge: A case of win–win or lose–win? A study of the UK baking and refrigeration industries. *Business Strategy and the Environment, 13*(3), 172–186.
Easterby-Smith, M., Golden-Biddle, K., & Locke, K. (2008). Working with pluralism: Determining quality in qualitative research. *Organisational Research Methods, 11*(3), 419–429.
Ellen MacArthur Foundation. (2012). *Towards the circular economy*. Ellen MacArthur Foundation.
Erdiaw-Kwasie, M. O. (2018). Does pressure-induced partnership really matter? Empirical modelling of stakeholder pressure and firms' CSR attitude. *Social Responsibility Journal*.
Fan, Y., & Fang, C. (2020). Circular economy development in China-current situation, evaluation and policy implications. *Environmental Impact Assessment Review, 84*, 106441.
Finland's Independence Celebration Fund (FICF, Sitra) and Mckinsey Kiertotalouden Mahdollisuudet Suomelle (2014). *The possibilities of circular economy for Finland*. SITRA, Helsinki, Finland.
Geissdoerfer, M., Savaget, P., Bocken, N. M., & Hultink, E. J. (2017). The circular economy–A new sustainability paradigm? *Journal of Cleaner Production, 143*, 757–768.
Geng, Y., & Doberstein, B. (2008). Developing the circular economy in China: Challenges and opportunities for achieving 'leapfrog development.'

The International Journal of Sustainable Development & World Ecology, 15(3), 231–239.

Ghisellini, P., Cialani, C., & Ulgiati, S. (2016). A review on circular economy: The expected transition to a balanced interplay of environmental and economic systems. *Journal of Cleaner Production, 114*, 11–32.

Guerra, B. C., & Leite, F. (2021). Circular economy in the construction industry: An overview of United States stakeholders' awareness, major challenges, and enablers. *Resources, Conservation and Recycling, 170*, 105617.

Gunsilius, E. (2015). *Economic instruments in solid waste management: Applying economic instruments for sustainable solid waste management in low-and middle-income countries*.

Hailemariam, A., & Erdiaw-Kwasie, M. O. (2022). Towards a circular economy: Implications for emission reduction and environmental sustainability. *Business Strategy and the Environment*.

Hoogendoorn, B., Guerra, D., & van der Zwan, P. (2015). What drives environmental practices of SMEs? *Small Business Economics, 44*(4), 759–781.

Johnstone, N., & Labonne, J. (2009). Why do manufacturing facilities introduce environmental management systems? Improving and/or Signaling Performance. *Ecological Economics, 68*(3), 719–730.

Kalmykova, Y., Sadagopan, M., & Rosado, L. (2018). Circular economy–From review of theories and practices to development of implementation tools. *Resources, Conservation and Recycling, 135*, 190–201. https://doi.org/10.1016/j.resconrec.2017.10.034

Kaza, S., Yao, L., Bhada-Tata, P., & Van Woerden, F. (2018). *What a waste 2.0: A global snapshot of solid waste management to 2050*. World Bank Publications.

Kirchherr, J., Reike, D., & Hekkert, M. (2017). Conceptualizing the circular economy: An analysis of 114 definitions. *Resources, Conservation and Recycling, 127*(12), 221–232. https://doi.org/10.1016/j.resconrec.2017.09.005

Korhonen, J., Honkasalo, A., & Seppälä, J. (2018). Circular economy: The concept and its limitations. *Ecological Economics, 143*, 37–46.

Kuo, L., Yu, H. C., & Chang, B. G. (2015). The signals of green governance on mitigation of climate change–evidence from Chinese firms. *International Journal of Climate Change Strategies and Management*.

Malik, A., Sharma, P., Vinu, A., Karakoti, A., Kaur, K., Gujral, H. S., & Laker, B. (2022). Circular economy adoption by SMEs in emerging markets: Towards a multilevel conceptual framework. *Journal of Business Research, 142*, 605–619.

Markard, J., Raven, R., & Truffer, B. (2012). Sustainability transitions: An emerging field of research and its prospects. *Research Policy, 41*(6), 955–967.

Mase, A. S., Gramig, B. M., & Prokopy, L. S. (2017). Climate change beliefs, risk perceptions, and adaptation behaviour among Midwestern US crop farmers. *Climate Risk Management, 15*, 8–17.

McKinsey Center for Business and Environment, Ellen MacArthur Foundation. (2015). *Growth within a circular economy vision for a competitive Europe*. McKinsey Center for Business and Environment and Ellen MacArthur Foundation. https://www.ellenmacarthurfoundation.org/assets/downloads/publications/EllenMacArthurFoundation_Growth-Within_July15.pdf

Ministry of Environment, Solid Waste Management and Climate Change. (2022). *A shared vision of shifting from the traditional linear economy to a circular economy*. https://environment.govmu.org/Pages/Circular-Economy.aspx. Accessed 30 Oct 2022.

Mitchell, P., & James, K. (2015). Economic growth potential of more circular economies. *Waste and Resources Action Programme (WRAP): Banbury, UK*.

Mitrofanenko, A., Fylakis, G., Skaalvik, J., & Lieberknecht, L. (2021). *Circular economy on the African continent: Perspectives and potential*.

Mohammed, F., Biswas, W. K., Yao, H., & Tadé, M. (2016). Identification of an environmentally friendly symbiotic process for the reuse of industrial byproduct–an LCA perspective. *Journal of Cleaner Production, 112*, 3376–3387.

Morseletto, P. (2020). Targets for a circular economy. *Resources, Conservation and Recycling, 153*, 104553.

Murray, A., Skene, K., & Haynes, K. (2017). The circular economy: An interdisciplinary exploration of the concept and application in a global context. *Journal of Business Ethics, 140*(3), 369–380.

Mutezo, G., & Mulopo, J. (2021). A review of Africa's transition from fossil fuels to renewable energy using circular economy principles. *Renewable and Sustainable Energy Reviews, 137*, 110609.

Mutz, D., Roy, M., Boiral, O., & Paillé, P. (2015). *Circular economy and informal waste management in India—A contradiction?* Institute of Development Studies, Resource Politics 2015. Retrieved September 2, 2022.

Naidoo, V. (2020). Creativity and innovation for entrepreneurs in the circular economy. In *Handbook of Research on Entrepreneurship Development and Opportunities in Circular Economy* (pp. 538–553). IGI Global.

Nilakant, V., & Ramnarayan, S. (2006). *Change management: Altering mindsets in a global context*. SAGE Publications India.

Pereira, V., Nandakumar, M. K., Sahasranamam, S., Bamel, U., Malik, A., & Temouri, Y. (2022). An exploratory study into emerging market SMEs' involvement in the circular Economy: Evidence from India's indigenous Ayurveda industry. *Journal of Business Research, 142*, 188–199.

Ranta, V., Aarikka-Stenroos, L., Ritala, P., & Mäkinen, S. J. (2018). Exploring institutional drivers and barriers of the circular economy: A cross-regional comparison of China, the US, and Europe. *Resources, Conservation and Recycling, 135*, 70–82.

Ritzén, S., & Sandström, G. Ö. (2017). Barriers to the Circular Economy–integration of perspectives and domains. *Procedia CIRP, 64*, 7–12.

Rivis, A., & Sheeran, P. (2003). Descriptive norms as an additional predictor in the theory of planned behaviour: A meta-analysis. *Current Psychology, 22*(3), 218–233.

Rizos, V., Behrens, A., Van der Gaast, W., Hofman, E., Ioannou, A., Kafyeke, T., & Topi, C. (2016). Implementation of circular economy business models by small and medium-sized enterprises (SMEs): Barriers and enablers. *Sustainability, 8*(11), 1212.

Robaina, M., Villar, J., & Pereira, E. T. (2020). The determinants for a circular economy in Europe. *Environmental Science and Pollution Research, 27*(11), 12566–12578. https://doi.org/10.1007/s11356-020-07847-9

Rockström, J., Steffen, W., Noone, K., Persson, Å., Chapin III, F. S., Lambin, E., ... & Foley, J. (2009). Planetary boundaries: Exploring the safe operating space for humanity. *Ecology and Society, 14*(2).

Roxas, B., & Coetzer, A. (2012). Institutional environment, managerial attitudes and environmental sustainability orientation of small firms. *Journal of Business Ethics, 111*(4), 461–476.

Sachs, J. D. (2015). *The age of sustainable development*. Columbia University Press.

Shi, H., Peng, S. Z., Liu, Y., & Zhong, P. (2008). Barriers to the implementation of cleaner production in Chinese SMEs: Government, industry and expert stakeholders' perspectives. *Journal of Cleaner Production, 16*(7), 842–852.

Singh, M. P., Chakraborty, A., & Roy, M. (2016). Entrepreneurial commitment, organisational sustainability and business performance of manufacturing MSMEs: Evidence from India. *International Journal of Applied Business and Economic Research, 14*(6), 4615–4631.

Sjoerdsma, M., & van Weele, A. J. (2015). Managing supplier relationships in a new product development context. *Journal of Purchasing and Supply Management, 21*(3), 192–203.

Steffen, W., Richardson, K., Rockström, J., Cornell, S. E., Fetzer, I., Bennett, E. M., ... & Sörlin, S. (2015). Planetary boundaries: Guiding human development on a changing planet. *Science, 347*(6223), 1259855. https://doi.org/10.1126/science.1259855

Syberg, K., Nielsen, M. B., Clausen, L. P. W., van Calster, G., van Wezel, A., Rochman, C., & Hansen, S. F. (2021). Regulation of plastic from a circular economy perspective. *Current Opinion in Green and Sustainable Chemistry, 29*, 100462.

Tandrayen-Ragoobur, V., & Kasseeah, H. (2017). Is gender an impediment to firm performance? Evidence from small firms in Mauritius. *International Journal of Entrepreneurial Behavior & Research*.

Tukker, A., Bulavskaya, T., Giljum, S., de Koning, A., Lutter, S., Simas, M., ... & Wood, R. (2016). Environmental and resource footprints in a global context: Europe's structural deficit in resource endowments. *Global Environmental Change*, *40*, 171–181. https://doi.org/10.1016/j.gloenvcha.2016.07.002

Ünal, E., Urbinati, A., & Chiaroni, D. (2018). Managerial practices for designing circular economy business models: The case of an Italian SME in the office supply industry. *Journal of Manufacturing Technology Management*.

Wu, J., & Wu, Z. (2015). Key supplier relationships and product introduction success: The moderating roles of self-enforcement and interdependence between buyer and supplier. *Industrial Marketing Management, 46*, 183–192.

Zhao, Y., Zang, L., Li, Z., & Qin, J. (2012). Discussion on the model of mining circular economy. *Energy Procedia, 16*, 438–443.

Zhu, J., Fan, C., Shi, H., & Shi, L. (2019). Efforts for a circular economy in China: A comprehensive review of policies. *Journal of Industrial Ecology, 23*(1), 110–118.

Circular Economy Principles and Responsible Manufacturing: Assessing Implications for Resource Conservation, Emission Reduction, Cost Performance, and Environmental Legitimacy

Charles Baah, Ebenezer Afum, Yaw Agyabeng-Mensah, and Douglas Opoku Agyeman

1 Introduction

Circular economy, defined as a way to enhance the use of natural resources, create and add value and regenerate waste while increasing corporate and consumers' responsibility to achieve sustainability (Ghisellini et al., 2016), advances closed-loop systems that use the 4R's (reduce, reuse, recycle and recover) explained as reducing resource

---

C. Baah (✉)
Business and Accounting Discipline, Faculty of Arts and Society, Charles Darwin University, Darwin, NT, Australia
e-mail: Charles.baah@cdu.edu.au

E. Afum
Business School, University of Bristol, Bristol, UK

M. O. Erdiaw-Kwasie and G. M. M. Alam (eds.), *Circular Economy Strategies and the UN Sustainable Development Goals*, Sustainable Development Goals Series,
https://doi.org/10.1007/978-981-99-3083-8_10

consumption, reusing end or by-products, recycling waste, and recovering resources to generate higher economic efficiency and decrease environmental impact (Geng & Doberstein, 2008). However, despite the economic and environmental benefits associated with the circular economy, several scholars posit that the social impacts of the circular economy, such as community, labour, and societal well-being, have been mainly ignored in the literature (Geissdoerfer et al., 2017; Kirchherr et al., 2017; Merli et al., 2018; Mies & Gold, 2021). Thus, scholars call for integrating social issues into circular economy approaches to present a holistic understanding of the concept (Geissdoerfer et al., 2017; Mies & Gold, 2021).

Similarly, Curkovic (2003) defines responsible manufacturing as an economically driven, system-wide, and combined approach to reducing and eliminating product development, use, and disposal of waste. Ellram et al. (2008) further added that responsible manufacturing facilitates and minimises adverse environmental impacts while maximising resource efficiency. However, similar to the circular economy, several scholars indicate that present responsible manufacturing approaches do not consider or integrate social issues and their corresponding implications on firm performance and society and, thus, present significant limitations in providing a comprehensive perspective on responsible manufacturing (Curkovic & Sroufe, 2016; Ellram et al., 2008; Curkovic, 2003). Niero et al. (2021) and Curkovic (2003) further highlighted that although most studies portray circular economy adoption and responsible manufacturing as vital to a firm's production and value creation processes, there are unintended side effects such as loss of jobs due to refining and removal of redundant production processes, the collapse of small firms unable to meet circular, and environmental requirements due to lack of resources resulting in diverse negative social implications for firms and society.

Moreover, contextual disparities in manufacturing practices, infrastructure, policies, legislation, and availability of resources further create

---

Y. Agyabeng-Mensah
School of Management and Marketing, Curtin University, Perth, WA, Australia

D. O. Agyeman
School of Management Operations, and Marketing, University of Wollongong, Wollongong, NSW, Australia

typical complexities and inconsistencies in guiding responsible manufacturing decisions and practices (Baah et al., 2020b; Jabbour et al., 2020; Shubham et al., 2018). For example, Jabbour et al. (2020) from the Brazilian context indicated that institutional void impacted the execution of circular economy principles. Shubham et al. (2018), in the Indian context, also found that regulatory policies, infrastructure, and bodies played significant roles in pushing manufacturing firms to engage in environmentally responsible manufacturing practices. Finally, in the Ghanaian context, Baah et al. (2021) argued that a firm's desire to achieve environmental legitimacy with stakeholders and the institutional setting contributes to adopting responsible and environmentally friendly manufacturing practices. The above assertions justify Ellram et al. (2008) and Curkovic and Sroufe (2016) claim that contextual disparities coupled with the side effects of circular and responsible manufacturing initiatives lead some firms to still engage in environmentally irresponsible manufacturing practices.

Furthermore, Santibanez Gonzalez et al. (2019) suggested that although manufacturing sectors are already consuming enormous amounts of resources, and energy, generating waste and emitting greenhouse gasses due to 'Take, Make and Dispose of' concepts grounded in linear economy production systems, these impacts are expected to worsen in 2050. This is because the sector is expected to consume about 140 billion tonnes of natural resources to meet the demands of a growing population. Thus, Lieder and Rashid (2016) proposed adopting circular economy principles as a potential approach to resolve resource dependency, waste generation, energy consumption, and emission of greenhouse gasses. In the EU manufacturing context, de Römph and Cramer (2020) indicated that linear business models promoted 'single-use' lifestyles among manufacturers and consumers. Curkovic (2003) specified that such environmentally irresponsible lifestyles could be curbed by adopting responsible manufacturing practices. Ellram et al. (2008) highlighted that responsible manufacturing identifies, quantifies, assesses, and manages the flow of environmental waste during product development, design, and production. Curkovic and Sroufe (2016) and Santibanez Gonzalez et al. (2019) indicated that circular economy and responsible manufacturing are the ultimate go-to approaches.

Deephouse et al. (2016) elucidate from the lens of the institutional theory framework that firms are driven by the desire to attain legitimacy and hence, rely on industrial isomorphism to implement responsible and environmental practices that fit within the values, norms, and beliefs of the industry. Additionally, Hart and Dowell (2011), from the natural resource-based view (NRBV) perspective, connoted that firms that preserve the environment through environmentally responsible practices earn competitive advantages that ensure cost benefits. Hart (1995) posits that, if rightly leveraged, cost advantages associated with the NRBV can boost operational performance, reputation, and sustain competitiveness. This chapter conceptualises a model that draws insights from the institutional and NRBV theories to empirically investigate (a) how circular economy principles drive responsible manufacturing (formulated using social and environmental dimensions. See Fig. 1), resource conservation, and emission reduction; (b) the influence of responsible manufacturing, resource conservation, and emission reduction on environmental legitimacy and cost performance; and (c) the mediating roles of resource conservation and emission reduction between responsible manufacturing, environmental legitimacy, and cost performance.

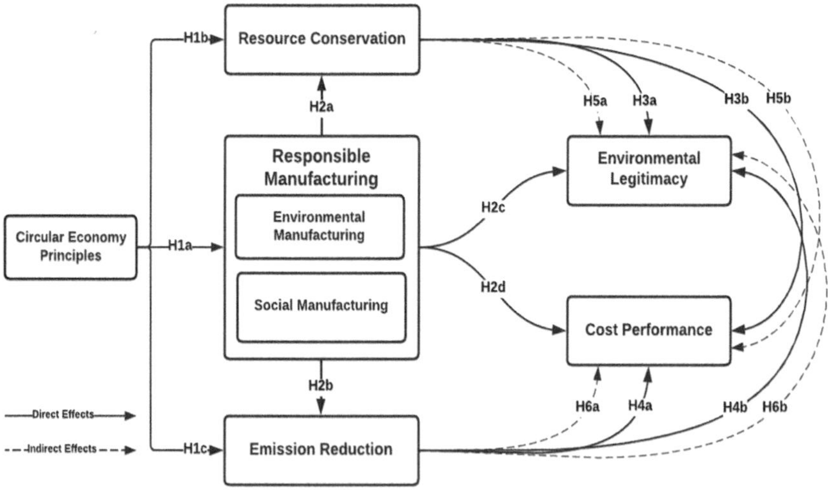

Fig. 1 Conceptual framework

Particularly, it is vital to integrate the social and environmental dimensions of responsible manufacturing due to high pressure from stakeholders for businesses to consider employee and community health and safety (Jabbour et al., 2020; Shubham et al., 2018), growing demand for companies to account for effective and efficient use of resources (Järvenpää et al., 2018) and increasing demand for businesses to establish sustainable manufacturing processes and practices that enhance economic performance without compromising environmental or social performance (Pagell & Shevchenko, 2014). While the chapter includes the social dimension of responsible manufacturing, the chapter also seeks to provide findings in the Ghanaian context, an emerging economy characterised by industrialisation (Agyabeng-Mensah et al., 2021). Besides, Jabbour et al. (2020) emphasised that circular economy studies have primarily been based in the European, American, and Asian contexts; hence, little is known about circular economy initiatives in emerging African economies economically driven by industrialisation.

Further strengthening the relevance of insights from emerging economies, this chapter focuses on SMEs, recognising their rapid rise in developing countries to constitute over 90% of firms in diverse industries (OECD, 2010). Agyabeng-Mensah et al. (2022) posited that although SMEs face crucial resource constraints, they must make critical decisions regarding adopting circular economy principles and responsible manufacturing practices, which exert pressure on resources. Shubham et al. (2018) further elaborate that implementing environmentally responsible practices and circular economy initiatives is no longer optional but mandatory. Hence, SMEs must abide by environmental legislation and requirements regardless of resource constraints. However, despite the contextual dynamics in emerging economies, empirical studies have yet to examine how SMEs engage in circular economy principles and responsible manufacturing initiatives to drive resource conservation, emission reduction, environmental legitimacy, and cost performance.

The rest of the paper is structured as follows: Sect. 2 exposes the literature review and hypotheses development. Sections 3, 4, 5, and 6 reflect the research method, result, discussion, and conclusion that highlight theoretical and managerial implications and research limitations, and future research recommendations.

2 Theoretical Framework and Hypotheses Development

2.1 Institutional and Natural Resource-Based View Theories

Jesus and Jugend (2021) explain that the circular economy balances the economy, society, and the environment through reducing, reusing, recycling, and recovering. However, the circular economy focuses on environmental and economic components neglecting the societal dimension of the concept (Erdiaw-Kwasie et al., 2023; Kirchherr et al., 2017; Mies & Gold, 2021). Ranta et al. (2018), in establishing the link between circular economy and society from the institutional theory perspective, highlight stakeholders, including the society, as key industry players that lead or coerce firms to be isomorphic, and conform to industry norms, behaviours, and practices. Deephouse (1996) explained that firms operating within an industry tend to align, adapt, and implement strategies and practices deemed acceptable in the industry resulting in what the author referred to as isomorphism. DiMaggio and Powell (1983) and Fernando and Lawrence (2014) also posit that engaging isomorphic behaviours in an industry earn a firm legitimacy with stakeholders, which according to Chen et al. (2019) is an antecedent to social and moral capital.

Baah et al. (2021), integrating the stakeholder and the institutional theories, argued that firms, regardless of the industry in which they operate, owe responsibilities to society, and as such, implementing circular economy initiatives from a social and environmentally responsible production perspective will not only reflect compliance to acceptable industrial behaviour but also aid firms accumulate stakeholder respect and legitimacy. Li et al. (2018) specified that green or environmental legitimacy signified achieving stakeholder support, loyalty, trust, and satisfaction. Tsinopoulos et al. (2018), revealing that stakeholders are crucial factors in the institutional theory, further stressed that deviations from industry-accepted practices resulted in poor stakeholder recognition and endorsements, which negatively correlate with organisational legitimacy. Given increasing stakeholder demands for circular initiatives in emerging economies, Afum et al. (2022), from a zero-waste perspective, suggested that circular economy initiatives are deemed acceptable industry norms and required behaviour. Thus, not engaging in responsible manufacturing initiatives has detrimental implications for firms.

Furthermore, Hart (1995) posits that the competitive and cost advantage benefits firms derive from environmental orientations have driven orientation towards the NRBV. The NRBV posits that organisations that positively impact the environment earn competitive advantages, enhanced goodwill, and social and moral capital (Erdiaw-Kwasie et al., 2017; Hart & Dowell, 2011; McDougall et al., 2022; Miemczyk et al., 2016). According to MacArthur (2013) and Curkovic and Sroufe (2016), adopting circular economy principles coupled with responsible manufacturing promotes waste prevention, product and service longevity, and ensures a balance between the economy and society and the environment. McDougall et al. (2022) elaborate that environmental competitiveness based on the NRBV is rooted in an organisation's ability to leverage its environmental operations. While Ellram et al. (2008) specify that environmentally responsible manufacturing, if rightly leveraged, can increase business performance by achieving competitive advantage, Gupta et al. (2020) also indicated that engaging in responsible manufacturing with the aim of sustainable development also boosted cost competitiveness and profitability.

Teece (2007, 2014) further explained that orientation towards NRBV results in developing dynamic capabilities and competencies that enable firms to integrate, build, and reconfigure green competencies to respond to rapidly evolving business environments. Helfat (2007) also argued that competencies acquired through deploying NRBV strategies improve the odds of attaining and maintaining competitive resource advantages and cost and profit performance in a volatile business environment. Singh and Del Giudice (2019) suggested that these competencies create green creativity and innovations, enabling sustainable goals to be achieved. Chen and Chang (2013) illustrated that green dynamic capabilities promote ecological balance, drive firm innovation, and align skills, motives, attitudes, and knowledge with organisational systems, processes, and practices. This chapter explains that institutional isomorphism and NRBV underpin the adoption of circular economy principles, responsible manufacturing, resource conservation, and emission reduction, which present competitive gains and capabilities leading to environmental legitimacy and increased cost performance. We describe adopted variables from the theoretical discourse and present related studies and theory-linked constructs in Tables 1, 2, and Fig. 1.

Furthermore, Roos and Agarwal (2015) suggests that adopting circular economy principles is crucial to implementing circular business model

Table 1 Description of variables

Variable	*Description*	*Source*
Circular economy principles	The adoption of reducing, reusing, recycling, and recovering techniques aimed at developing closed-loop production and consumption systems that ensure resource optimisation, economic efficiency, and value addition	MacArthur (2013), Geng and Doberstein (2008), and Ghisellini et al. (2016)
Responsible manufacturing	*Environmental manufacturing:* Integrating manufacturing approaches that incorporate environmental concerns into product design, processes, and services to ensure waste minimisation, resource efficiency, and environmental preservation *Social manufacturing:* Socially driven manufacturing approach that integrates social issues to meet the specific demands of consumers, employees, among other social community-based individuals to build up social resources, social networks, and collaborations, which lead to the efficient and flexible accomplishment of product and service lifecycle tasks	Ellram et al. (2008), Jiang et al. (2016), and Hamalainen et al. (2018)
Resource Conservation	The orientation towards enhancement, preservation and reduced dependency on natural resources mostly in the form of raw materials	MacArthur (2013), Cerchione and Bansal (2020), and Jabbour et al. (2020)
Emission reduction	An eco-friendly approach to meeting carbon tax and cap requirements while minimising greenhouse gas pollution to the environment	Saether et al. (2021) and Czerny and Letmathe (2017)
Environmental legitimacy	Favourable stakeholder perceptions or assumptions are based on industry norms and beliefs, which present an organisation's environmental behaviour as desirable, proper, or appropriate	Li et al. (2018) and Deephouse (1996)
Cost performance	Economically and environmentally driven performance targets related to material, manufacturing, and environmental costs	Burnett and Hansen (2008) and Hon (2005)

Table 2 Prior literature on circular economy principles and performance in emerging economies

Article	*Variable*	*Conclusion*	*Context*
Afum et al. (2022)	Circular principles; Cleaner production; Total quality environmental management; Zero-waste performance; Green differentiation advantage	Circular economy principles insignificantly influenced zero-waste performance and green differentiation advantage. Circular principles adoption, however, significantly correlated with cleaner production and total quality environmental management, which collectively had positive and significant impacts on zero-waste performance and green differentiation advantage	Ghana
Rodríguez-Espíndola et al. (2022)	Circular economy principles; Government support; Uncertainty; Technology; Customer; Sustainable-oriented innovation; Social, environmental and economic performances	As presented in the findings, uncertainty did not affect any of the adopted variables. However, customer pressure significantly impacted the adoption of circular economy principles. Government support is insignificantly related to the circular economy principles adoption. This finding contradicts the findings of Baah et al. (2022), who indicate regulatory stakeholders as potent influencers of circular principles adoption. Circular economy principles adoption is robustly associated with sustainable-oriented innovation, which is also significantly and robustly correlated with sustainable performance	Mexico

(continued)

Table 2 (continued)

Article	*Variable*	*Conclusion*	*Context*
Baah et al. (2022)	Organisational stakeholders influence; Regulatory stakeholder influence; Community stakeholder influence; Adoption of circular economy principles; Internal stakeholder satisfaction; External stakeholder satisfaction; Green legitimacy	All stakeholder groups positively and robustly influenced the adoption of circular economy principles, which also positively impacted both dimensions of stakeholder satisfaction and green legitimacy. In addition, internal and external stakeholder satisfaction also had positive effects on green legitimacy	Ghana
Patwa et al. (2021)	3R/Extended life cycle; Ecological balance; Big data/information flow; Circular economy adoption; Consumer behaviour; Government policies	The results illustrated that the 3R (reduce, reuse, and recycle)/extended life cycle of products influences the adoption of the circular economy. However, ecological balance weakly influenced the adoption of the circular economy. Big data/information flow, consumer behaviour, and government policies positively predicted the adoption of the circular economy. Additionally, consumer behaviour and government policies highly influence implementing the 3R/extended life cycle of products and ecological balance	India

Article	*Variable*	*Conclusion*	*Context*
Agyabeng-Mensah et al. (2021)	Circular economy performance target; Inter-organisational learning; Intra-organisational learning; Lean manufacturing; Lean product development; Organisational identity; Zero-waste practices	The study concluded that while intra-organisational learning did not significantly influence circular economy target performance, its impacts on inter-organisational learning and organisational identity were significant. Inter-organisational learning also had insignificant effects on circular economy target performance and organisational identity but significantly influenced zero-waste practices, lean product development, and manufacturing practices, which also improved circular economy target performance and organisation identity	Ghana
Jabbour et al. (2020)	Stakeholder pressure; Barriers; Motivators; Principles of Circular Economy; Economic, Environmental and Social performances	Stakeholder pressures positively influenced motivators but negatively correlated with barriers to adopting circular economy principles. Principles of circular economy positively and robustly influence economic, environmental, and social performances	Brazil

and innovation. For instance, Linder and Williander (2017) connote that the benefits of engaging circular economy principles serve as enablers of circular business models. These benefits include manufacturing cost savings (Baah et al., 2020b), reduced environmental impact (Afum et al., 2021; Baah et al., 2022; Baah et al., 2020a), competitive advantage (Afum et al., 2022; Baah et al., 2023), improved stakeholder satisfaction, and improved brand (Linder & Williander, 2017). Roos (2014) also highlighted leadership, collaboration, creativity, and customer demands as crucial enablers of circular business models. Nußholz (2017) also emphasised innovation as a crucial factor to achieve the desired outcomes of circular business models.

Contrary to these enablers, Linder and Williander (2017) and Roos (2014) agree that technological costs and expertise, regulation complexities, fear of change, risk of economic complications, inefficient end product retrieval, risk of low sales revenue, lack of supporting regulation, and partner restrictions are significant barriers to circular business models. According to several scholars, these barriers create uncertainties and require huge investments, which SMEs need (Agyabeng-Mensah et al., 2022; Baah et al., 2022). Hence, SMEs tend to resist the adoption of circular and responsible manufacturing initiatives. However, this chapter draws insights from the institutional and NRBV to posit that manufacturing SMEs may not resist but align with circular economy principles and responsible manufacturing initiatives to acquire stakeholder support and reap the potential benefits of circular and responsible manufacturing.

2.2 Adoption of Circular Economy Principles, Responsible Manufacturing, Resource Conservation, and Emission Reduction

Geissdoerfer et al. (2018) claim that most businesses and supply chains adopt principles that align with a circular economy because of the competitive gains associated with engaging mechanisms that minimise negative environmental impacts. However, two circular economy literature streams suggest inconclusive results, especially in the emerging economy context. While one stream provides empirical findings, which support the assertion that the adoption of circular economy practices significantly correlates with environmental initiatives (Baah et al., 2022; Jabbour et al., 2020; Rodríguez-Espíndola et al., 2022), the other stream suggests a weak and insignificant relationship between circular principles adoption and environmental practices (Afum et al., 2022; Petit-Boix & Leipold, 2018; Zhijun & Nailing, 2007).

For instance, Jabbour et al. (2020) found that adopting circular economy principles robustly and positively influenced the environmental performance of firms, which was measured using reduced emission, waste generation, resource efficiency, and consumption of hazardous materials. Rodríguez-Espíndola et al. (2022) also found that circular economy principles boosted sustainable innovation, improving environmental initiatives. Contrary to these findings, Afum et al. (2022) established a weak and insignificant effect in exploring the relationship between circular principles adoption and zero-waste performance. Petit-Boix and Leipold (2018), in reviewing how environmental research aligned with local practices, indicated that factors such as technology, infrastructure, and policies contributed to the misalignment between firm practices and research. The inconsistencies in prior literature suggest the need to explore the effects of circular principles adoption more, especially in emerging economies.

Moreover, Corvellec et al. (2022), in critiquing the current approach to the circular economy, explain that current approaches do not integrate social issues into CE implementation. Mies and Gold (2021) argue that circular economy principles have environmental and social manufacturing implications, although the latter is largely ignored. For example, in a systematic review, Padilla-Rivera et al. (2020) reveal a connection between adopting circular economy principles, employment, training, well-being, job creation, and community satisfaction, among others. However, these connections lack empirical evidence. Bridging this gap, the chapter investigates how adopting circular principles influences integrated environmental and socially responsible manufacturing.

MacArthur (2013) highlighted that circular economy principles strongly correlate with energy and resource conservation due to the circular economy's focus on resource efficiency and waste minimisation. In the Chinese context, Li et al. (2018) connoted that circular economy principles, specifically recycling and reusing, contributed robustly to greenhouse gas emissions. Gallego-Schmid et al. (2020), in the construction industry, indicate that a circular economy increases resource efficiency, which invariably leads to advanced construction practices and fewer emissions. From the above discourse, we propose that in an emerging economy context:

H1a Adoption of circular economy principles positively influences responsible manufacturing.

H1b Adoption of circular economy principles positively influences resource conservation.

H1c Adoption of circular economy principles positively influences emission reduction.

2.3 *Responsible Manufacturing, Resource Conservation, Emission Reduction, Environmental Legitimacy, and Cost Performance*

Responsible manufacturing has mainly been researched from an environmental perspective (Curkovic, 2003; Ellram et al., 2008). Curkovic and Sroufe (2016), in a review on responsible manufacturing, showed that responsible manufacturing centres on environmental practices. Using U.S. manufacturing firms as an example, the authors indicated that government regulations and the introduction of environmental standards such as ISO 14000 and 26,000, among other reporting indexes (Global Reporting Initiative [GRI], Life Cycle Assessment [LCA], among others), have contributed to the wide acceptance and adoption of the environmental approach to responsible manufacturing. Ball and MacBryde (2022), highlighting breweries, explained that environmental practices induce stakeholder trust, support, and satisfaction, which, within the institutional theory's pillars, translate to environmental legitimacy.

Boakye et al. (2020), from U.K.-based SME perspective, posited that engaging in environmental practices as captured by environmental manufacturing stimulates resource conservation due to stringent measures aimed at waste reduction and leads to avoidance of environmental sanction costs. Adomako et al. (2021) further postulated that implementing an environmental strategy that includes environmental practices improves resource conservation, emission reduction, and decreased production cost due to optimised production processes and less dependency on raw materials. However, in the literature, several scholars criticise responsible manufacturing's lack of integration of social concerns and variables (Curkovic, 2003; Curkovic & Sroufe, 2016; Ellram et al., 2008; Gupta et al., 2020). Thus, this chapter expands responsible manufacturing to include social manufacturing to provide a holistic and comprehensive understanding of the concept.

According to Hirscher et al. (2018), social manufacturing promotes different levels of user participation and ensures participatory design strategies that integrate social concerns into production practices. Jiang et al. (2016) indicated that the social approach to manufacturing aims to meet the social concerns and demands of customers and employees, among other community participants. Hamalainen et al. (2018) connoted that social manufacturing acquires social network collaborations and resource benefits, which according to Jiang et al. (2016) lead to strategic alliances that ensure improved product lifecycles. For example, Ding et al. (2016) posit that because products and services are hinged on customers' and community preferences, products and services integrate vital social issues that result in stakeholder satisfaction. Gaining stakeholder satisfaction from a manufacturing perspective has been associated with attaining green legitimacy and goodwill (Baah et al., 2021; Jiang et al., 2016).

Adomako (2020) highlighted that integrating stakeholders into firm strategies has driven firms towards environmental practices. From a social manufacturing perspective, stakeholders expect firms to engage in manufacturing practices that sustain and protect society (Jiang et al., 2016; Shang et al., 2022). Shang et al. (2022) explain that protecting society includes producing products and rendering services that integrate stakeholder preference, which lately includes environmental preservation. Thus, Martín-de Castro et al. (2020) explained from a corporate social responsibility (CSR) standpoint that is integrating and responding to social concerns preserve environmental resources. From the discourse, social manufacturing presents reputation benefits and supports the resource conservation and emission reduction agenda since these are stakeholder concerns in recent times. From the above, we propose that:

H2a Responsible manufacturing positively influences resource conservation.

H2b Responsible manufacturing positively influences emission reduction.

H2c Responsible manufacturing positively influences environmental legitimacy.

H2d Responsible manufacturing positively influences cost performance.

2.4 Resource Conservation, Environmental Legitimacy, and Cost Performance

Jabbour et al. (2020) emphasise the need to conserve resources in the face of climate change. With the manufacturing sector consuming enormous resources, mainly through manufacturing initiatives, Cerchione and Bansal (2020) assert that environmental pressures are gradually promoting the adoption of circular practices, which minimises resource consumption and waste. Shubham et al. (2018) highlight in the Indian context that response to stakeholder pressures on resource conservation guarantees a higher reputation with implications for enhanced stakeholder legitimacy. From Velenturf and Purnell's (2021) perspective, resource conservation, a crucial component of a circular economy, presents optimised environmental, material, and economic values.

Kalmykova et al. (2018) claim that resource conservation limits material costs and price volatilities while presenting opportunities for innovation and higher profitability. Jensen et al. (2020) indicate that Hyl Mobile using repurpose and reuse resource conservation approaches has extended the product lifecycles of over 50 million mobile devices and has earned $4 billion while preventing 6500 tonnes of landfill waste. Considering resource conservation as a critical indicator in attaining green investments, MacArthur (2013) posited that efficient utilisation of resources reduces production, material, and overall environmental costs for firms. Thus, we propose that:

H3a Resource conservation positively influences environmental legitimacy.

H3b Resource conservation positively influences cost performance.

2.5 Emission Reduction, Environmental Legitimacy, and Cost Performance

MacArthur (2013), as a pioneer of the circular economy agenda, expressed the need for emission reduction in current manufacturing practices. Mathews and Tan (2016) suggested that greenhouse gas emissions impact the quality of life of employees and society. From this viewpoint, emission reduction has attracted intense stakeholder attention. Saether et al. (2021), focusing on Norwegian maritime firms, established that emission-conscious firms developed stronger favourable stakeholder relations than less-conscious firms. In the Chinese context, Li et al. (2018) highlighted that firms' emission reduction is connected to stakeholder perception consistent with acceptable social norms, values, and beliefs.

Liu et al. (2021a) indicated that positive stakeholder perceptions lead to legitimacy and the establishment and maintenance of collaborative relationships with diverse stakeholder groups. Peng et al. (2015) illustrated that environmental pollution has been associated with severe environmental sanctions that increase a firm's costs. For example, the Volkswagen emissions scandal presented diverse costs and bad publicity to the firm, which it continues to deal with today. In the supply chain context, Liu et al. (2021b) connoted that low emissions mean resource efficiency and efficient manufacturing processes, which connote low material, production, and waste disposal costs. Emission reduction strategies also ensure the avoidance of environmental sanctions, as Czerny and Letmathe (2017) suggested. Thus, it can be inferred from the above discourse that emission reduction positively impacts a firm. As such, we propose that:

H4a Emission reduction positively influences cost performance.

H4b Emission reduction positively influences environmental legitimacy.

2.6 *Mediation Effects of Resource Conservation and Emission Reduction*

The discussions above establish direct relationships between the adoption of circular economy practices, responsible manufacturing, resource conservation, emission reduction, environmental legitimacy, and cost performance (Afum et al., 2022; Baah et al., 2022; Curkovic & Sroufe, 2016; Gupta et al., 2020; Hailemariam & Erdiaw-Kwasie, 2022; Jabbour et al., 2020; Jiang et al., 2016; Petit-Boix & Leipold, 2018; Rodríguez-Espíndola et al., 2022; Zhijun & Nailing, 2007). However, the mechanism for manufacturing SMEs in emerging economies to achieve environmental legitimacy and higher cost performance has not been widely explored. For example, Afum et al. (2021), in understanding manufacturing firms' ability to translate circular economy principles into zero-waste performance and green differentiation advantage, indicated the effect through the mediating mechanisms of cleaner production and Total Quality Environmental Management (TQEM) is significant and robust.

Li et al. (2018) also found that the mediating role of environmental legitimacy on carbon emission disclosure was significant. Thus, Afum et al. (2022) and Li et al. (2018) recommend exploring other mediating variables that impact performance.

This chapter explores resource conservation and emission reduction as mediating variables in responsible manufacturing, environmental legitimacy, and cost performance relationships. Adopting circular economy principles instigates resource conservation and emission reduction approaches (Cerchione & Bansal, 2020; MacArthur, 2013), promoting environmental legitimacy and cost performance (Jabbour et al., 2020; Li et al., 2018). Furthermore, Curkovic and Sroufe (2016), Ellram et al. (2008), and Jiang et al. (2019) illustrated that responsible manufacturing from both social and environmental dimensions necessitates orientation towards resource conservation and emission reduction.

Integrating environmental and social concerns into manufacturing makes a firm sensitive to reducing waste and environmental pollution through resource conservation and emission reduction. Although emission reduction and resource conservation create value in environmental and social manufacturing, how this value creation improves cost performance and environmental legitimacy needs further research. Curkovic and Sroufe (2016) suggest that environmental and social manufacturing at the firm level provides the needed platform to engage resource conservation and emission reduction mechanisms, leading to higher environmental legitimacy and cost performance. As such, we propose that:

H5a Resource conservation mediates the relationship between responsible manufacturing and environmental legitimacy.

H5b Resource conservation mediates the relationship between responsible manufacturing and cost performance.

H6a Emission reduction mediates the relationship between responsible manufacturing and environmental legitimacy.

H6b Emission reduction mediates the relationship between responsible manufacturing and cost performance.

In circular economy literature, the adoption of circular economy principles has primarily been viewed from an institutional theory perspective (Agyabeng-Mensah et al., 2022; Baah et al., 2021, 2022). The NRBV has also been used to elaborate on firms' benefits from environmental initiatives. The institutional theory is used to explain the adoption of circular economy principles and the acquisition of environmental legitimacy. In contrast, the relationships between responsible manufacturing, resource conservation, emission reduction, and cost performance are explained from the theoretical lens of NRBV. Figure 1 exposes the conceptual framework indicating the hypothesised direct and indirect relationships among the study variables.

3 Research Method

3.1 Sample, Data Collection, and Common Method Bias (CMB)

Reviewing the existing literature, this chapter developed questionnaires based on the theoretical and conceptual framework outlined in Sect. 2. The questionnaires were designed to collect primary data from a sample of 480 SMEs, randomly selected from a list of about 3008 SMEs obtained from the Association of Ghana Industries using the purposive sampling technique (Agyabeng-Mensah et al., 2021). In December 2019, questionnaires and a cover letter clarifying the chapter's objectives were emailed to selected companies. After two weeks, follow-up phone calls were made to express appreciation and serve as reminders for non-responders. As a result, 320 questionnaires distributed to manufacturing SMEs were retrieved, giving a response rate of about 66 per cent. Regarding the variables, a scrutiny of the data showed that 38 entries had missing values and were therefore excluded from the data analysis. Thus, a total of 282 questionnaires were included in the data analysis.

Common Method Bias (CMB) was assessed using Harman's one-factor test due to the inconsistencies of the excluded questionnaires and the deficiencies of the survey method of data collection highlighted in previous studies (Podsakoff et al., 2003). Consequently, following the recommendations of Podsakoff et al. (2003), Harman's one-factor test was applied. The Exploratory Factor Analysis (EPA) revealed that the single factor explained 36.4 per cent of the cumulative variance. The fact that this result is below the recommended threshold of 50 per cent indicates no CMB issues. In addition, we the Kaiser–Meyer–Olkin Measure of Sampling Adequacy (KMO) and Bartlett's Test of Sphericity to determine whether the data is adequate for factor analysis and to compare the observed correlation matrix to the identity matrix, respectively. The

Table 3 Characteristics of respondents (n = 282)

Firm characteristics	*Frequency (%)*
Number of employees	
<10	34 (12%)
11–50	38 (14%)
51–100	40 (14%)
101–500	60 (21%)
501–1000	47 (17%)
>1000	63 (22%)
Firm age	
<5 years	50 (18%)
5–10 years	76 (27%)
10–15 years	77 (27%)
15–20 years	45 (16%)
>20 years	34 (12%)
Industry	
Wood, lumber products and processing	57 (20%)
Rubber and other plastic products	73 (26%)
Food and beverages	50 (18%)
Leather products and processing	52 (18%)
Textile and fabric products	30 (11%)
Others	20 (7%)
Job Qualification	
Owner	85 (30%)
Manager	90 (32%)
Assistant Manager	52 (18%)
Departmental Heads/Supervisors	55 (20%)

outcomes demonstrated that the KMO measure of sampling adequacy was 0.846, which is acceptable. Bartlett's test revealed a significance value of 0.001, less than the 0.005 thresholds. Consequently, these findings demonstrate that the sample is adequate for factor analysis and has no correlation issues. The profile of responding SMEs is outlined in Table 3.

3.2 Measurement Model Assessment

To ensure the validity and reliability of the measurement items, we first invited experts to assess the measuring items and content of the questionnaire. These experts initially indicated that the content and questions measured their respective constructs and therefore deemed the questionnaire adequate for its purpose with minor changes related to rewording and rephrasing of the contents. These invited experts comprised three industry professionals (manufacturing sector) and three from academia. As a result, the questionnaire adopted a 5-point Likert scale, where 1 represented strongly disagree to 5 strongly agree, to measure the extent to which respondents agreed or disagreed with each item in their respective firms. Based on the recommendation of Sachdev and Verma (2004), we adopted a 5-point Likert scale since it reduces respondents' frustration level and increases response rate and quality.

Furthermore, measurement items were scrutinised during Confirmatory Factor Analysis (CFA) to ensure the model has validity and reliability. According to Hair et al. (2013), items that have factor loadings above 0.70 should be maintained in a model, and items below 0.70 should be deleted to strengthen the model path, Average Variance Extracted (AVE), R Squared (R^2), and predictive relevance (Q^2). Thus, measurement items loaded below 0.70 were deleted from the model except for RM2, RM5, CEP1, RC3, ER4, and ER5. These were not deleted due to their significant contribution to the model's predictive relevance. Partial Least Square Structural Equation Modelling (PLS-SEM) (SmartPLS software 3.0) in making data analysis and interpretation.

According to Henseler et al. (2015), a model's reliability and validity are assessed using Cronbach's Alpha (α), Composite Reliability (CR), AVE, and Heterotrait-Monotrait (HTMT) ratio. These criteria should have $\geq$0.70, $\geq$0.70, $\geq$0.60, >0.50, and <0.90, respectively, to be considered satisfactory. From Table 4, the model constructs have internal consistency reliability because the α and CR were above the recommended thresholds. The AVEs, which assess that indicators measured one underlying construct, had a recommended range of 0.512–0.766. Henseler et al. (2016) posited that the outer VIFs connoted no multicollinearity issues since all values were <3.

Additionally, the HTMT ratio, the current and more sophisticated examination of discriminant validity, was examined. From Table 5, the model achieves discriminant validity since all ratios were <0.90.

3.3 Structural Model Assessment

Hair et al. (2013) indicate the need to evaluate a structural model using its R square, which speaks to the variances of dependent variables explained by the independent variables, and Stone-Geisser's Q^2, which evaluates the model's predictive relevance. Henseler et al. (2016)

Table 4 Measurement items, reliability, and validity

Item description	*Factor loading*	*Outer VIF*
Circular Economy Principles (CEP) $\alpha = 0.823$; CR = 0.884; AVE = 0.656		
CEP1: Our firm is gradually prolonging the duration of products and their usage	0.699	1.474
CEP2: Our business is promoting the reduction of waste and rework	0.806	1.975
CEP3: Our firm is aiming at reuse, recycling, and remanufacturing initiatives with regard to product life cycles	0.829	2.015
CEP4: Our business is replacing current equipment and technologies with more modern and efficient ones	0.896	2.449
Responsible manufacturing (RM) $\alpha = 0.778$; CR = 0.845; AVE = 0.523		
RM1: Environmental concerns and goals are clearly communicated to all manufacturing personnel	0.751	1.797
RM2: Our business provides employees and managers adequate training on awareness of environmental and social concerns of internal and external stakeholders	0.684	1.534
RM3: The product and service preferences and concerns of customers and other community-based stakeholders are considered in manufacturing practices	0.731	1.944
RM4: Our company frequently compares environmental and social manufacturing practices with best and accepted industry practices	0.788	1.740
RM5: Our business ensures that product designs, development and manufacturing processes, and service delivery meet environmental standards and societal expectations	0.655	1.164
Resource Conservation (RC) $\alpha = 0.709$; CR = 0.815; AVE = 0.524		
RC1: Our company adopts alternative fuels to ensure energy saving	0.742	2.298
RC2: Our company is promoting reduction of raw material waste, among other waste management practices	0.769	2.537
RC3: Our company is aiming at resource reuse, recycling, and remanufacturing initiatives	0.680	2.088
RC4: Our company avoids the use and patronage of environmentally dangerous products	0.701	2.006
Emission Reduction (ER) $\alpha = 0.743$; CR = 0.828; AVE = 0.512		
ER1: Our company prioritises reduction of CO_2	0.756	1.861
ER2: Our company aims to comply with carbon tax and cap requirements	0.700	1.656

Item description	*Factor loading*	*Outer VIF*
ER3: Our company reduces particulate emissions	0.728	1.807
ER4: Our company reduces emissions of other greenhouse gasses	0.687	1.465
ER5: Our company frequently engages with stakeholders on meeting emission requirement	0.625	1.137
Environmental Legitimacy (EL) $\alpha = 0.847$; CR = 0.908; AVE = 0.766		
EL1: Our company, from stakeholder perspectives, is committed to meeting the industry standards with sustainable practices	0.882	2.112
EL2: Our business, from stakeholder perspectives, follows government regulations for green operating procedures in the industry	0.874	2.067
EL3: Most stakeholders approve and endorse our company's responsiveness to environmental concerns	0.870	1.975
EL4: Our company from stakeholder perspectives has one of the highest rates of environmental orientation in the industry	Deleted	-
Cost Performance (CP) $\alpha = 0.749$; CR = 0.838; AVE = 0.564		
CP1: Reduced manufacturing cost	0.739	1.523
CP2: Reduced material cost	0.737	1.748
CP3: Reduced environmental cost	0.789	2.119
CP4: Reduced inventory cost	0.738	1.206

Note α—Cronbach's Alpha; CR—Composite Reliability; AVE—Average Variance Extracted; VIF—Variance

Table 5 Discriminant validity using HTMT ratio

Construct	*CEP*	*CP*	*ER*	*EL*	*RC*	*RM*
CEP						
CP	0.880					
ER	0.774	0.895				
EL	0.767	0.541	0.462			
RC	0.333	0.345	0.228	0.291		
RM	0.792	0.834	0.836	0.586	0.406	

Table 6 Model's quality criteria

Construct	R^2	R^2 *Adjusted*	*Stone-Geisser's* Q^2
CP	0.638	0.632	0.316
EL	0.534	0.528	0.374
ER	0.584	0.580	0.255
RC	0.507	0.503	0.215
RM	0.465	0.463	0.203

stipulated that reporting the R^2 and Q^2 in PLS-SEM is essential, especially in predictive studies. Stone-Geisser's Q^2 explains that a model has predictive relevance when Q^2 values are > 0. As presented in Table 6, the results show the structural model achieves the recommended model quality, especially recognising that Q^2 values are significant > 0. The R^2 and the R^2 adjusted also show that the predictor variables were significant in explaining the variances of the endogenous variables.

4 Result

Presented in Table 7 and Fig. 2, the first hypothesis, which examines the associations between circular economy principles, responsible manufacturing, resource conservation, and emission reduction, was supported. The Beta coefficient (β) (which is the degree of change in the dependent variable for each unit change in the independent variable) was 0.682, 0.519, and 0.488 showing the robustly positive effects of circular economy principles on responsible manufacturing, resource conservation, and emission reduction, respectively. The T statistics being 22.232, 7.784, and 6.780 showed that the associations were significant at the 95 per cent confidence interval with an error level of 0.05 per cent. The results showed that while the second hypotheses, specifically 2a, 2b, and 2d, were supported by having positive and significant β coefficients of 0.249, 0.343, and 0.272, respectively, hypothesis 2c, although positive, had a T statistic value of 1.663 below the significant threshold of 1.96, and hence presented an insignificant finding. Thus, hypothesis 2c was not supported.

Hypotheses 2a, 2b, and 2d were supported by having significant T statistics values of 3.472, 4.485, and 3.366, respectively. The third hypothesis examines the effect of resource conservation on environmental legitimacy and cost performance. The results show that hypotheses 3a and 3b had positive effects, with β coefficients being 0.451 and 0.238, respectively. The T statistics were also significant, with 6.808 and 3.861 values supporting hypotheses 3a and 3b. The fourth hypothesis was also supported since the β coefficients of the influence of emission reduction on cost performance and environmental legitimacy were 0.402 and 0.258, with corresponding T statistics values of 3.861 and 5.081, respectively.

Regarding the mediating effects of resource conservation and emission reduction between responsible manufacturing, environmental legitimacy, and cost performance relationships, the findings of hypothesis 5 reveal that resource conservation fully mediated the relationship between responsible manufacturing and environmental legitimacy, having an indirect β coefficient and T statistic values of 0.112 and 3.094. This presents a full mediation because the direct effect between responsible manufacturing and environmental legitimacy was insignificant, yet the indirect effect is significant, thus supporting hypothesis 5a.

Hypothesis 5b was not supported because the results showed that resource conservation had an insignificant indirect effect ($\beta = 0.059$, $T = 2.395$) but a significant direct effect ($\beta = 0.272$, $T = 3.366$), indicating that resource conservation does not play any significant role between responsible manufacturing and cost performance, thereby presenting a no-mediation effect. Therefore, the sixth hypothesis was supported, establishing the mediating effect of emission reduction between responsible manufacturing, environmental legitimacy, and cost performance. Specifically, hypothesis 6a reveals that emission reduction partially mediated the relationship between responsible manufacturing and cost performance since both the indirect ($\beta = 0.138$, $T = 3.565$) and direct ($\beta = 0.272$, $T = 3.366$) effects were positive and significant. Concerning hypothesis 6b, emission reduction fully mediated the relationship between responsible manufacturing and environmental legitimacy, having an indirect effect β

Table 7 Hypothesis testing

Direct path	*Beta (β)*	*Standard deviation (STDEV)*	*T Statistics (|O/STDEV|)*	*P Values*	*Inner VIFs*
H1a(s): CEP → RM	0.682	0.031	22.232	0.000	1.000
H1b(s): CEP → RC	0.519	0.067	7.784	0.000	1.910
H1c(s): CEP → ER	0.488	0.072	6.780	0.000	1.910
H2a(s): RM → RC	0.249	0.072	3.472	0.001	1.910
H2b(s): RM → ER	0.343	0.076	4.485	0.000	1.910
H2c(ns): RM → EL	0.118	0.071	1.663	0.096	1.908
H2d(s): RM → CP	0.272	0.081	3.366	0.001	1.908
H3a(s): RC → EL	0.451	0.066	6.808	0.000	1.126
H3b(s): RC → CP	0.238	0.062	3.861	0.000	1.126
H4a(s): ER → CP	0.402	0.079	5.081	0.000	1.762
H4b(s): ER → EL	0.258	0.081	3.194	0.001	1.762
Indirect Path					
H5a(f): RM → RC → EL	0.112	0.036	3.094	0.002	–
H5b(n): RM → RC → CP	0.059	0.025	2.395	0.017	–
H6a(p): RM → ER → CP	0.138	0.039	3.565	0.000	–
H6b(f): RM → ER → EL	0.088	0.032	2.781	0.005	–

Note s—supported; ns—not supported; f—full mediation; n—no mediation; p—partial mediation

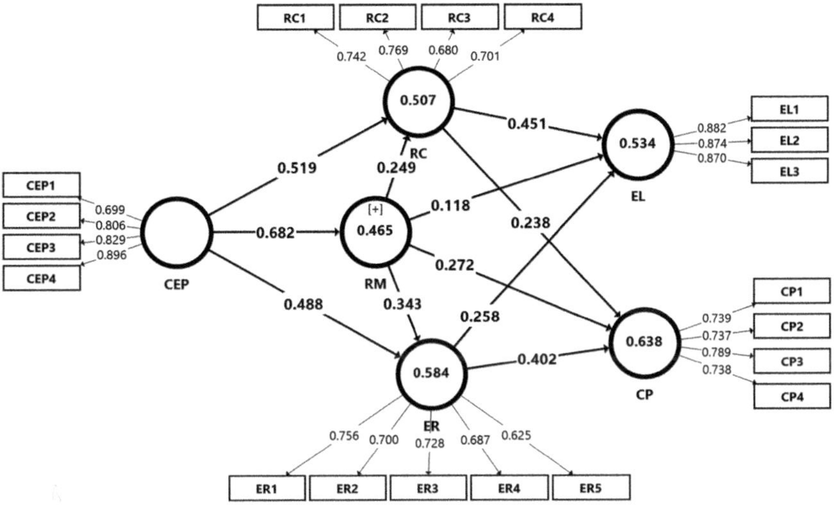

Fig. 2 Structural model

coefficient and T statistic values of 0.088 and 2.781, respectively. This result means that the relationship between responsible manufacturing, while not directly significant, is signed through the mediating mechanism of emission reduction.

5 Discussion

The results show that adopting circular economy principles had potent effects on responsible manufacturing, resource conservation, and emission reduction. Supporting hypothesis 1, the results show that circular economy principles adoption in the Ghanaian context drives orientation towards practices that conserve and preserve the environment. In line with Jabbour et al. (2019) assertion that circular principles lead to higher environmental performance, this chapter further adds empirical evidence by indicating that circular principles adoption does indeed boost responsible manufacturing, resource conservation, and emission reduction in the Ghanaian context, contrary to Afum et al.'s (2022) finding that circular principles adoption has no significant influence on zero-waste practices in the same context. The results further communicate from the institutional theory perspective that adopting circular economy principles reflects circular norms, beliefs, and values that are considered appropriate and acceptable industry behaviour. As such, firms engaging in circular practices will earn environmental legitimacy and reduce costs. Mainly, circular principles adoption had the most influence on responsible manufacturing, confirming Korhonen et al. (2018) and Geissdoerfer et al. (2017)'s assertion that circular economy orientations critically boost and improve manufacturing processes and practices.

While the findings related to hypothesis 2 corroborate the assertions of several scholars (Curkovic, 2003; Curkovic & Sroufe, 2016; Ellram et al., 2008; Jiang et al., 2016) that engaging in responsible manufacturing improves resource conservation and emission reduction. Again, the findings harmonise with Baah et al.'s (2022) explanation that engaging in environmental initiatives induced by adopting circular economy principles improves stakeholder satisfaction, translating into environmental legitimacy. MacArthur (2013) posits that circular economy initiatives have cost-reduction implications, mostly realised through sustainable manufacturing initiatives. The findings show that responsible manufacturing significantly resulted in cost savings while insignificantly leading to environmental legitimacy, as presented in Fig. 2. This result proves that responsible manufacturing has mostly been considered from the production perspective leaving out the component of meeting stakeholder social demands, which have been identified as critical predictors of environmental legitimacy in the CSR literature (Baah et al., 2021; De Roeck & Delobbe, 2012; Nielsen & Thomsen, 2018).

Hypotheses 3 and 4 highlight the significant roles of resource conservation and emission reduction in influencing environmental legitimacy and cost performance. Consistent with previous findings (Kalmykova et al., 2018; MacArthur, 2013; Velenturf & Purnell, 2021), resource conservation boosts cost savings. However, this chapter's results show that resource conservation largely influenced environmental legitimacy more than cost reduction. Furthermore, this chapter elaborates that recent stakeholder concerns about natural resource extraction have instigated manufacturing firms to ensure resource protection and efficiency. Hence, the finding that resource conservation promotes environmental legitimacy than it reduces costs is supported.

Emission reduction was more primarily associated with cost reduction than environmental legitimacy, although both associations were significant and positive. With environmental regulations and emission standards ineffective in both emerging and developed countries, the results corroborate previous findings in the sense that reducing emissions prevents sanctions and bad publicity (Peng et al., 2015; Saether et al., 2021). This reinforces Peng et al.'s (2015) claim that pollution attracts significant costs for firms, and engaging in emission reduction initiatives decreases a firm's environmental cost. Saether et al. (2021) in the maritime sector also connoted the relevance of emission reduction in promoting better stakeholder relations while improving a firm's profitability. Gaining such improved relationships with stakeholders ensures environmental legitimacy, which results in higher financial performance, as highlighted by Li et al. (2018).

Regarding the mediating effects, resource conservation fully mediated the relationship between responsible manufacturing and environmental legitimacy. This finding implies that while a firm engaging in responsible manufacturing may not necessarily earn environmental legitimacy, integrating resource conservation through the efficient sourcing and utilisation of natural resources presents dynamic mechanisms through which a firm earns environmental legitimacy with stakeholders. Establishing a no-mediation effect between responsible manufacturing and cost performance, the chapter provides results contrary to existing literature (Kalmykova et al., 2018; Velenturf & Purnell, 2021). While past literature indicates that engaging resource conservation in manufacturing promotes cost reduction, this chapter's findings show that SMEs that engage in responsible manufacturing alone increase cost performance without the mediating mechanism of resource conservation.

Emission reduction partially and fully mediated the relationship between responsible manufacturing, environmental legitimacy, and cost performance. The findings show that SMEs that engage in emission reduction practices can improve the robustness of their responsible manufacturing effects on environmental legitimacy and cost performance. Specifically, SMEs' responsible manufacturing can generate environmental legitimacy and higher cost performance when emission reduction initiatives are integrated into manufacturing strategies. The findings of this chapter provide insights into how firms can generate higher cost performance and environmental legitimacy through the direct and indirect impacts of responsible manufacturing, resource conservation, and emission reduction induced by the adoption of circular economy principles.

6 Conclusion

Several studies investigate the impact of circular principles adoption on environmental practices and performance (Afum et al., 2022; Jabbour et al., 2020; Rodríguez-Espíndola et al., 2022). However, there is a lack of empirical evidence and conclusive findings on how these principles impact responsible manufacturing. Therefore, this chapter extends responsible manufacturing to include social manufacturing to provide a holistic view of the concept, as recommended by Ellram et al. (2008) and Curkovic and Sroufe (2016). Integrating the social and environmental aspects of responsible manufacturing, the chapter contributes to the circular economy call to explore the impact of circular principles on social issues. This chapter's framework is premised on the notion that the adoption of circular economy principles impacts an integrated social and environmentally responsible manufacturing, resource conservation, and emission reduction, which cost-effectively positions a firm for higher environmental legitimacy with stakeholders (Baah et al., 2022; Li et al., 2018; Liu et al., 2021a, 2021b; Peng et al., 2015).

Overall, the study shows that in the Ghanaian context, adopting circular economy principles has significant implications for SMEs' orientations towards responsible manufacturing, emission reduction, and resource conservation, which boosts environmental legitimacy and cost performance. The robustness of the effects illustrates that circular economy principles position SMEs to operate socially and environmentally while ensuring resource preservation, reduced pollution, higher stakeholder endorsement, and reduced costs. The theoretical and practical implications of the chapter are highlighted below.

6.1 Theoretical Implications

While prior studies have explored the influence of circular economy adoptions on sustainable performance, less is known about how circular economy principles impact responsible manufacturing with an integrated social component in the context of emerging economies. This chapter contributes to the circular economy literature by examining how responsible manufacturing, resource conservation, and emission reduction translate into cost performance and environmental legitimacy. Integrating the institutional and NRBV theories, this chapter develops and explores a novel framework that extends the circular economy and responsible manufacturing literature by examining how resource conservation and emission reduction strengthen and mediate the association between responsible manufacturing, environmental legitimacy, and cost performance.

Notably, the chapter shows that resource conservation and emission reduction could be mechanisms through which SMEs attain superior performance, especially from cost reduction and stakeholder endorsement perspectives. Drawing on the NRBV theory, the chapter also confirms that adopting circular economy principles enables a firm to take advantage of environmental initiatives to develop strategic capabilities that underpin cost reduction and firm reputation. As such, this chapter indicates from institutional and NRBV theories perspectives that adopting circular economy principles while aligning with stakeholder interests, satisfaction, and endorsement aids in developing capabilities that strengthen superior performance impacts.

The chapter also contributes to the responsible manufacturing literature in that this chapter responds to the calls to integrate the social dimension of responsible manufacturing. Thus, this chapter develops a responsible manufacturing construct that integrates the social manufacturing dimension, as indicated in Fig. 1. Integrating the environmental and social manufacturing dimensions provides a holistic view of the concept. This conceptualisation and its integration into the research framework offer a pivotal approach to providing insights into how adopting circular principles impacts environmentally responsible manufacturing and an integrated construct of environmentally and socially responsible manufacturing, thereby enhancing previous knowledge and understanding of the dynamics between the two dimensions of responsible manufacturing and how these foster desired organisational outcomes, especially in the context of an emerging economy and from SMEs' perspectives.

6.2 *Practical Implications*

This chapter's findings provide business executives and managers with several suggestions for achieving cost performance and environmental legitimacy through circular economy principles adoption, which leads to responsible manufacturing, resource conservation, and emission reduction. First, managers can use this integration strategy to include social issues in manufacturing practices to earn environmental legitimacy while mitigating environmental sanction costs through stakeholder engagement and proactively addressing environmental and social concerns. Second, through resource conservation and emission reduction, businesses can achieve capabilities that boost company capabilities in achieving improved performance outcomes. Lastly, SMEs can take advantage of responsible manufacturing, emission reduction, and resource conservation practices to attract environmentally oriented investors among other stakeholders to boost collaborative and strategic alliances to achieve and sustain competitive advantage and sustainable business performance.

6.3 Limitations of the Study

Despite the contributions of the chapter, some limitations need to be addressed. First, the generalisation of chapter findings may be limited since the chapter focuses on the Ghanaian context. As such, it is recommended that similar research be conducted in other contexts to provide more insights into the interactions between circular economy principles, responsible manufacturing, resource conservation, emission reduction, environmental legitimacy, and cost performance. Again, other research can concentrate on other facets of the adopted dependent and independent variables, primarily environmental and socially responsible manufacturing dimensions, to provide a comprehensive view of their interactions and concepts. In addition, data should be collected over time to reveal the long-term interactions between variables. Lastly, while this chapter focused on SME operations in a developing economy, future research should consider other firm types and sizes.

References

Adomako, S. (2020). Environmental collaboration, sustainable innovation, and small and medium-sized enterprise growth in sub Saharan Africa: Evidence from Ghana. *Sustainable Development, 28*(6), 1609–1619.

Adomako, S., Ning, E., & Adu-Ameyaw, E. (2021). Proactive environmental strategy and firm performance at the bottom of the pyramid. *Business Strategy and the Environment, 30*(1), 422–431.

Afum, E., Agyabeng-Mensah, Y., Baah, C., Agyapong, G. K., Armas, J. A. L., & Al Farooque, O. (2022). Prioritising zero-waste performance and green differentiation advantage through the Prism of circular principles adoption: A mediated approach. *Journal of Cleaner Production*, 132182.

Afum, E., Agyabeng-Mensah, Y., Mensah, A. O., Mensah-Williams, E., Baah, C., & Dacosta, E. (2021). Internal environmental management and green human resource management: significant catalysts for improved corporate reputation and performance. *Benchmarking: An International Journal, 28*(10), 3074–3101.

Agyabeng-Mensah, Y., Afum, E., Baah, C., & Essel, D. (2022). Exploring the role of external pressure, environmental sustainability commitment, engagement, alliance and circular supply chain capability in circular economy performance. *International Journal of Physical Distribution & Logistics Management*, (ahead-of-print).

Agyabeng-Mensah, Y., Tang, L., Afum, E., Baah, C., & Dacosta, E. (2021). Organisational identity and circular economy: Are inter and intra organisational learning, lean management and zero waste practices worth pursuing? *Sustainable Production and Consumption, 28*, 648–662.

Baah, C., Afum, E., Agyabeng-Mensah, Y., & Agyeman, D. O. (2022). Stakeholder influence on adoption of circular economy principles: measuring implications for satisfaction and green legitimacy. *Circular Economy and Sustainability*, 1–21.

Baah, C., Agyabeng-Mensah, Y., Afum, E., & Kumi, C. A. (2023). Do circular economy practices accelerate CSR participation of SMEs in a stakeholder-pressured era? A network theory perspective. *Journal of Cleaner Production, 394*, 136348.

Baah, C., Agyabeng-Mensah, Y., Afum, E., & Mncwango, M. S. (2021). Do green legitimacy and regulatory stakeholder demands stimulate corporate social and environmental responsibilities, environmental and financial performance? Evidence from an emerging economy. *Management of Environmental Quality: An International Journal.*

Baah, C., Jin, Z., & Tang, L. (2020a). Organisational and regulatory stakeholder pressures friends or foes to green logistics practices and financial performance: Investigating corporate reputation as a missing link. *Journal of Cleaner Production, 247*, 119125.

Baah, C., Opoku-Agyeman, D., Acquah, I. S. K., Issau, K., & Abdoulaye, F. A. M. (2020b). Understanding the influence of environmental production practices on firm performance: a proactive versus reactive approach. *Journal of Manufacturing Technology Management.*

Ball, P., & MacBryde, J. (2022). Developing a framework for adopting environmental manufacturing practices: Learning from breweries. *Production Planning & Control, 33*(8), 758–773.

Boakye, D. J., TIngbani, I., Ahinful, G., Damoah, I., & Tauringana, V. (2020). Sustainable environmental practices and financial performance: Evidence from listed small and medium-sized enterprise in the United Kingdom. *Business Strategy and the Environment, 29*(6), 2583–2602.

Burnett, R. D., & Hansen, D. R. (2008). Ecoefficiency: Defining a role for environmental cost management. *Accounting, Organisations and Society, 33*(6), 551–581.

Cerchione, R., & Bansal, H. (2020). Measuring the impact of sustainability policy and practices in tourism and hospitality industry. *Business Strategy and the Environment, 29*(3), 1109–1126.

Chen, J., Zhang, F., Liu, L., & Zhu, L. (2019). Does environmental responsibility matter in cross-sector partnership formation? A legitimacy perspective. *Journal of Environmental Management, 231*, 612–621.

Chen, Y. S., & Chang, C. H. (2013). The determinants of green product development performance: Green dynamic capabilities, green transformational leadership, and green creativity. *Journal of Business Ethics, 116*(1), 107–119.

Corvellec, H., Stowell, A. F., & Johansson, N. (2022). Critiques of the circular economy. *Journal of Industrial Ecology, 26*(2), 421–432.

Curkovic, S. (2003). Environmentally responsible manufacturing: The development and validation of a measurement model. *European Journal of Operational Research, 146*(1), 130–155.

Curkovic, S., & Sroufe, R. (2016). A literature review and taxonomy of environmentally responsible manufacturing. *American Journal of Industrial and Business Management, 6*(03), 323.

Czerny, A., & Letmathe, P. (2017). Eco-efficiency: GHG reduction related environmental and economic performance: The case of the companies participating in the EU Emissions Trading Scheme. *Business Strategy and the Environment, 26*(6), 791–806.

De Roeck, K., & Delobbe, N. (2012). Do environmental CSR initiatives serve organisations' legitimacy in the oil industry? Exploring employees' reactions through organisational identification theory. *Journal of Business Ethics, 110*(4), 397–412.

de Römph, T. J., & Cramer, J. M. (2020). How to improve the EU legal framework in view of the circular economy. *Journal of Energy & Natural Resources Law, 38*(3), 245–260.

Deephouse, D. L. (1996). Does isomorphism legitimate? *Academy of Management Journal, 39*(4), 1024–1039.

Deephouse, D. L., Newburry, W., & Soleimani, A. (2016). The effects of institutional development and national culture on cross-national differences in corporate reputation. *Journal of World Business, 51*(3), 463–473.

DiMaggio, P. J., & Powell, W. W. (1983). The iron cage revisited: Institutional isomorphism and collective rationality in organisational fields. *American Sociological Review*, 147–160.

Ding, K., Jiang, P., Leng, J., & Cao, W. (2016). Modeling and analysing of an enterprise relationship network in the context of social manufacturing.

Proceedings of the Institution of Mechanical Engineers, Part b: Journal of Engineering Manufacture, 230(4), 752–769.

Ellram, L. M., Tate, W., & Carter, C. R. (2008). Applying 3DCE to environmentally responsible manufacturing practices. *Journal of Cleaner Production, 16*(15), 1620–1631.

Erdiaw-Kwasie, M. O., Abunyewah, M., Yusif, S., & Erdiaw-Kwasie, A. (2023). Does circular economy knowledge matter in sustainable service provision? A moderation analysis. *Journal of Cleaner Production, 383*, 135429.

Erdiaw-Kwasie, M. O., Alam, K., & Shahiduzzaman, M. (2017). Towards understanding stakeholder salience transition and relational approach to 'better' corporate social responsibility: A case for a proposed model in practice. *Journal of Business Ethics, 144*(1), 85–101.

Fernando, S., & Lawrence, S. (2014). A theoretical framework for CSR practices: Integrating legitimacy theory, stakeholder theory and institutional theory. *Journal of Theoretical Accounting Research, 10*(1), 149–178.

Gallego-Schmid, A., Chen, H. M., Sharmina, M., & Mendoza, J. M. F. (2020). Links between circular economy and climate change mitigation in the built environment. *Journal of Cleaner Production, 260*, 121115.

Geissdoerfer, M., Morioka, S. N., de Carvalho, M. M., & Evans, S. (2018). Business models and supply chains for the circular economy. *Journal of Cleaner Production, 190*, 712–721.

Geissdoerfer, M., Savaget, P., Bocken, N. M., & Hultink, E. J. (2017). The Circular Economy—A new sustainability paradigm? *Journal of Cleaner Production, 143*, 757–768.

Geng, Y., & Doberstein, B. (2008). Developing the circular economy in China: Challenges and opportunities for achieving "leapfrog development." *The International Journal of Sustainable Development & World Ecology, 15*(3), 231–239.

Ghisellini, P., Cialani, C., & Ulgiati, S. (2016). A review on circular economy: The expected transition to a balanced interplay of environmental and economic systems. *Journal of Cleaner Production, 114*, 11–32.

Gupta, H., Kusi-Sarpong, S., & Rezaei, J. (2020). Barriers and overcoming strategies to supply chain sustainability innovation. *Resources, Conservation and Recycling, 161*, 104819.

Hair, J. F., Ringle, C. M., & Sarstedt, M. (2013). Partial least squares structural equation modeling: Rigorous applications, better results and higher acceptance. *Long Range Planning, 46*(1–2), 1–12.

Hailemariam, A., & Erdiaw-Kwasie, M. O. (2022). Towards a circular economy: Implications for emission reduction and environmental sustainability. *Business Strategy and the Environment*, 1–15

Hamalainen, M., Mohajeri, B., & Nyberg, T. (2018). Removing barriers to sustainability research on personal fabrication and social manufacturing. *Journal of Cleaner Production, 180*, 666–681.

Hart, S. L. (1995). A natural-resource-based view of the firm. *Academy of Management Review, 20*(4), 986–1014.

Hart, S. L., & Dowell, G. (2011). Invited editorial: A natural-resource-based view of the firm: Fifteen years after. *Journal of Management, 37*(5), 1464–1479.

Helfat, C. E. (2007). Stylised facts, empirical research and theory development in management. *Strategic Organization, 5*(2), 185–192.

Henseler, J., Hubona, G., & Ray, P. A. (2016). Using PLS path modeling in new technology research: Updated guidelines. *Industrial Management & Data Systems, 116*(1), 2–20.

Henseler, J., Ringle, C. M., & Sarstedt, M. (2015). A new criterion for assessing discriminant validity in variance-based structural equation modeling. *Journal of the Academy of Marketing Science, 43*(1), 115–135.

Hirscher, A. L., Niinimäki, K., & Armstrong, C. M. J. (2018). Social manufacturing in the fashion sector: New value creation through alternative design strategies? *Journal of Cleaner Production, 172*, 4544–4554.

Hon, K. K. B. (2005). Performance and evaluation of manufacturing systems. *CIRP Annals, 54*(2), 139–154.

Jabbour, C. J. C., Sarkis, J., de Sousa Jabbour, A. B. L., Renwick, D. W. S., Singh, S. K., Grebinevych, O., ... & Godinho Filho, M. (2019). Who is in charge? A review and a research agenda on the 'human side' of the circular economy. *Journal of Cleaner Production, 222*, 793–801.

Jabbour, C. J. C., Seuring, S., de Sousa Jabbour, A. B. L., Jugend, D., Fiorini, P. D. C., Latan, H., & Izeppi, W. C. (2020). Stakeholders, innovative business models for the circular economy and sustainable performance of firms in an emerging economy facing institutional voids. *Journal of Environmental Management, 264*, 110416.

Järvenpää, E., Lanz, M., & Siltala, N. (2018). Formal resource and capability models supporting re-use of manufacturing resources. *Procedia Manufacturing, 19*, 87–94.

Jensen, P. D., Purnell, P., & Velenturf, A. P. (2020). Highlighting the need to embed circular economy in low carbon infrastructure decommissioning: The case of offshore wind. *Sustainable Production and Consumption, 24*, 266–280.

Jesus, G. M. K., & Jugend, D. (2021). How can open innovation contribute to circular economy adoption? Insights from a literature review. *European Journal of Innovation Management*.

Jiang, P., Leng, J., & Ding, K. (2016). Social manufacturing: A survey of the state-of-the-art and future challenges. In *2016 IEEE International Conference on Service Operations and Logistics, and Informatics (SOLI)* (pp. 12–17). IEEE.

Jiang, J., Xu, X., & Stringer, J. (2019). Optimization of process planning for reducing material waste in extrusion based additive manufacturing. *Robotics and Computer-Integrated Manufacturing, 59*, 317–325.

Kalmykova, Y., Sadagopan, M., & Rosado, L. (2018). Circular economy—From review of theories and practices to development of implementation tools. *Resources, Conservation and Recycling, 135*, 190–201.

Kirchherr, J., Reike, D., & Hekkert, M. (2017). Conceptualising the circular economy: An analysis of 114 definitions. *Resources, Conservation and Recycling, 127*, 221–232.

Korhonen, J., Nuur, C., Feldmann, A., & Birkie, S. E. (2018). Circular economy as an essentially contested concept. *Journal of Cleaner Production, 175*, 544–552.

Li, D., Huang, M., Ren, S., Chen, X., & Ning, L. (2018). Environmental legitimacy, green innovation, and corporate carbon disclosure: Evidence from CDP China 100. *Journal of Business Ethics, 150*(4), 1089–1104.

Lieder, M., & Rashid, A. (2016). Towards circular economy implementation: A comprehensive review in context of manufacturing industry. *Journal of Cleaner Production, 115*, 36–51.

Linder, M., & Williander, M. (2017). Circular business model innovation: Inherent uncertainties. *Business Strategy and the Environment, 26*(2), 182–196.

Liu, H., Zhou, J., Liu, H., & Xin, B. (2021b). Is the uncertainty of gaining legitimacy from organisational change an antecedent of employees' resistance to change? *Chinese Management Studies*.

Liu, Z., Li, W., Hao, C., & Liu, H. (2021b). Corporate environmental performance and financing constraints: An empirical study in the Chinese context. *Corporate Social Responsibility and Environmental Management, 28*(2), 616–629.

MacArthur, E. (2013). Towards the circular economy. *Journal of Industrial Ecology, 2*(1), 23–44.

Martín-de Castro, G., Amores-Salvadó, J., Navas-López, J. E., & Balarezo-Núñez, R. M. (2020). Corporate environmental reputation: Exploring its definitional landscape. *Business Ethics: A European Review, 29*(1), 130–142.

Mathews, J. A., & Tan, H. (2016). Circular economy: Lessons from China. *Nature, 531*(7595), 440–442.

McDougall, N., Wagner, B., & MacBryde, J. (2022). Competitive benefits & incentivisation at internal, supply chain & societal level circular operations in UK agri-food SMEs. *Journal of Business Research, 144*, 1149–1162.

Merli, R., Preziosi, M., & Acampora, A. (2018). How do scholars approach the circular economy? A systematic literature review. *Journal of Cleaner Production, 178*, 703–722.

Miemczyk, J., Howard, M., & Johnsen, T. E. (2016). Dynamic development and execution of closed-loop supply chains: A natural resource-based view. *Supply Chain Management: An International Journal.*

Mies, A., & Gold, S. (2021). Mapping the social dimension of the circular economy. *Journal of Cleaner Production, 321*, 128960.

Nielsen, A. E., & Thomsen, C. (2018). Reviewing corporate social responsibility communication: A legitimacy perspective. *Corporate Communications: An International Journal.*

Niero, M., Jensen, C. L., Fratini, C. F., Dorland, J., Jørgensen, M. S., & Georg, S. (2021). Is life cycle assessment enough to address unintended side effects from Circular Economy initiatives? *Journal of Industrial Ecology, 25*(5), 1111–1120.

Nußholz, J. L. (2017). Circular business models: Defining a concept and framing an emerging research field. *Sustainability, 9*(10), 1810.

OECD, & Organisation for Economic Co-operation and Development Staff. (2010). *OECD Employment Outlook 2010: Moving Beyond the Jobs Crisis*. Organisation for Economic Co-operation and Development.

Padilla-Rivera, A., Russo-Garrido, S., & Merveille, N. (2020). Addressing the social aspects of a circular economy: A systematic literature review. *Sustainability, 12*(19), 7912.

Pagell, M., & Shevchenko, A. (2014). Why research in sustainable supply chain management should have no future. *Journal of Supply Chain Management, 50*(1), 44–55.

Patwa, N., Sivarajah, U., Seetharaman, A., Sarkar, S., Maiti, K., & Hingorani, K. (2021). Towards a circular economy: An emerging economies context. *Journal of Business Research, 122*, 725–735.

Peng, B., Du, H., Ma, S., Fan, Y., & Broadstock, D. C. (2015). Urban passenger transport energy saving and emission reduction potential: A case study for Tianjin, China. *Energy Conversion and Management, 102*, 4–16.

Petit-Boix, A., & Leipold, S. (2018). Circular economy in cities: Reviewing how environmental research aligns with local practices. *Journal of Cleaner Production, 195*, 1270–1281.

Podsakoff, P. M., MacKenzie, S. B., Lee, J. Y., & Podsakoff, N. P. (2003). Common method biases in behavioral research: A critical review of the literature and recommended remedies. *Journal of Applied Psychology, 88*(5), 879.

Roos, G. (2014). Business model innovation to create and capture resource value in future circular material chains. *Resources, 3*(1), 248–274.

Roos, G., & Agarwal, R. (2015). Services innovation in a circular economy. *The handbook of service innovation* (pp. 501–520).

Ranta, V., Aarikka-Stenroos, L., Ritala, P., & Mäkinen, S. J. (2018). Exploring institutional drivers and barriers of the circular economy: A cross-regional comparison of China, the US, and Europe. *Resources, Conservation and Recycling, 135*, 70–82.

Rodríguez-Espíndola, O., Cuevas-Romo, A., Chowdhury, S., Díaz-Acevedo, N., Albores, P., Despoudi, S., Malesios, C., & Dey, P. (2022). The role of circular economy principles and sustainable-oriented innovation to enhance social, economic and environmental performance: Evidence from Mexican SMEs. *International Journal of Production Economics, 248*, 108495.

Sachdev, S. B., & Verma, H. V. (2004). Relative importance of service quality dimensions: A multisectoral study. *Journal of Services Research, 4*(1).

Saether, E. A., Eide, A. E., & Bjørgum, Ø. (2021). Sustainability among Norwegian maritime firms: Green strategy and innovation as mediators of long-term orientation and emission reduction. *Business Strategy and the Environment, 30*(5), 2382–2395.

Santibanez Gonzalez, E. D., Koh, L., & Leung, J. (2019). Towards a circular economy production system: Trends and challenges for operations management. *International Journal of Production Research, 57*(23), 7209–7218.

Shang, Y., Song, M., & Zhao, X. (2022). The development of China's Circular Economy: From the perspective of environmental regulation. *Waste Management, 149*, 186–198.

Shubham, Charan, P., & Murty, L. S. (2018). Organisational adoption of sustainable manufacturing practices in India: integrating institutional theory and corporate environmental responsibility. *International Journal of Sustainable Development & World Ecology, 25*(1), 23–34.

Singh, S. K., & Del Giudice, M. (2019). Big data analytics, dynamic capabilities and firm performance. *Management Decision.*

Teece, D. J. (2007). Explicating dynamic capabilities: The nature and microfoundations of (sustainable) enterprise performance. *Strategic Management Journal, 28*(13), 1319–1350.

Teece, D. J. (2014). The foundations of enterprise performance: Dynamic and ordinary capabilities in an (economic) theory of firms. *Academy of Management Perspectives, 28*(4), 328–352.

Tsinopoulos, C., Sousa, C. M., & Yan, J. (2018). Process innovation: Open innovation and the moderating role of the motivation to achieve legitimacy. *Journal of Product Innovation Management, 35*(1), 27–48.

Velenturf, A. P., & Purnell, P. (2021). Principles for a sustainable circular economy. *Sustainable Production and Consumption, 27*, 1437–1457.

Zhijun, F., & Nailing, Y. (2007). Putting a circular economy into practice in China. *Sustainability Science, 2*(1), 95–101.

Human Capital Transformation for Circular Economy and Sustainable Development: A Government-Linked Company Experience

Radiah Othman and *Rashid Ameer*

1 Introduction

The 2030 Agenda of the United Nations has 17 SDGs and 169 sub-targets to enable the global transition to sustainability by 2030 (Weiland et al., 2021). To that end, the principles of equity and leave-no-one-behind must be emphasized (Sachs et al., 2019). Sustainable development (SD) aims for 'development that meets the needs of the present without compromising the ability of future generations to meet their own needs' (WCED, 1987). The uptake of SD in the business community has been slow worldwide, especially in developing countries, even two decades after its introduction in the Brundtland report (Brundtland, 1987).

R. Othman (✉)
School of Accountancy, Massey University, Palmerston North, New Zealand
e-mail: r.othman@massey.ac.nz

R. Ameer
IPU New Zealand, Palmerston North, New Zealand

M. O. Erdiaw-Kwasie and G. M. M. Alam (eds.), *Circular Economy Strategies and the UN Sustainable Development Goals*,
Sustainable Development Goals Series,
https://doi.org/10.1007/978-981-99-3083-8_11

Circular Economy (CE) aims to improve the attainment of sustainability and SDGs (Geissdoerfer et al., 2017; Sauvé et al., 2016; Stahel, 2019b; Valverde & Avilés-Palacios, 2021), but what defines a CE remains contested (Galvão et al., 2018; Kirchherr et al., 2017; Panwar & Niesten, 2022). Although the literature on theoretical paradigms of the CE and SD is emerging, it lacks consensus regarding goals and performance (Geissdoerfer et al., 2017; Kirchherr et al., 2017; Suárez-Eiroa et al., 2019). CE infers fair resource utilization when human capital and natural, cultural, and manufactured stocks are used to improve the ecologic, social, and economic factors that makeup sustainability (Stahel, 2019a; Suárez-Eiroa et al., 2019).

The importance of human capital investment in developing nations within the context of SDGs has been highlighted (e.g., Debrah et al., 2018; Miyamoto, 2013; Zaric & Babic, 2020), but the link between human capital, CE, and sustainability are under-researched (e.g., Tariq et al., 2016). In the results of their 2021 survey, Marrucci et al. (2021) suggest a positive relationship between green human resource practices in business with sustainability and CE performance. They did not, however, examine the specific human capital transformation initiatives carried out by the organizations studied. Organizations must develop capabilities to accommodate CE and sustainability by mobilizing their human capital (Ma et al., 2015; Suárez-Eiroa et al., 2019). As such, it is critical to know the process, as scholars have urged the economic actors to improve their human capital management and not simply replace it with new employees (Stahel, 2019c).

The contributions of this chapter are threefold. *First*, the role of human capital is championed as being equally important as any other resource deployed for the development of the CE in a sustainability development context. Consistent with Stahel (2019b), this chapter views labour as a renewable resource and the only resource with a qualitative edge which deteriorates with time if unused. Similar to the aim of Ma et al. (2015) to guide resource-based chemical industries in SD, this chapter advocates employees' contribution in improving the sustainability performance of their employers.

Second, human capital is often described as a barrier to the CE in developing countries (e.g., Geng & Doberstein, 2010), and the employees are sometimes displaced. Therefore, an alternative perspective is essential. Vermunt et al. (2019) insisted that generalizations might be unjustified and that the barriers to a CE warrant resolution. This chapter showcases how human capital could be transformed, aiming for the development of a CE. de Jesus and Mendonça (2018) suggest that empirical understanding is important to overcome any barriers to CE. Further, the current literature is largely focused on the manufacturing sector. This chapter addresses this gap (e.g., Elia et al., 2017; Kalmykova et al., 2018; Marrucci et al., 2021; Rajput & Singh, 2019).

Third, the literature on barriers to CE is typically general and not industry based (e.g., Meys et al., 2020). Other studies in the chemical industry focus on resource efficiency and strategies for sustaining the CE (e.g., Keijer et al., 2019; Ma et al., 2015). Some studies suggest a link between technological competence and CE involvement (e.g., Arfaoui et al., 2020; de Jesus & Mendonça, 2018; Meys et al., 2020) but do not suggest how the desired level of competence could be achieved. Schot and Kanger (2016) called for more empirical evidence of actual transformational value.

In addressing these gaps in the literature, this chapter examines how human capital is transformed, measured and reported by a chemical company in transition to the CE and SDGs. The transformation has been argued as necessary for a smooth transition to a CE (de Jesus & Mendonça, 2018). The chemical sector is chosen because it is a sensitive industry, and the corporate actors in sensitive industries are perceived as the potential catalyst for sustainability. The ultimate decision-making of these employers can either increase or mitigate intended social and environmental impacts (e.g., Bebbington & Larrinaga, 2014; Kolk et al., 2018; Sinkovics et al., 2020; Soundararajan et al., 2018).

The case study company, Petronas Chemical Group (PCG), belongs to the larger Fortune 500 company Petronas Global, a Malaysian government-linked company (GLC). PCG is a unique case study for three reasons. *First*, it is one of the largest producers of chemicals in Southeast Asia. It has higher visibility to the stakeholders due to the nature of its operations and the chemical waste it produces. Second, the chemical sector sits within an industry which is the third largest industrial carbon dioxide emitter and is responsible for about 4% of total global direct emissions (Boghdadi, 2020). Third, the business case for adopting circularity in this industry has become more compelling due to pressure from stakeholders, which saw some major chemical companies adopt CE business models for sustainability (Boghdadi, 2020). PCG is one of these major chemical companies which adopts both CE and SDGs.

Second, PCG exemplifies the critical role of transforming human capital for CE and sustainability ambitions in the media and in sustainability reports (e.g., CEO Magazine, 2019). As a result, it has won various awards for human resources and sustainability practices. It has also been recognized as among Malaysia's top 100 sustainable employers (The Sun Daily, 2021; Sustainable Employer Research Centre, n.d.). This external recognition and endorsement signify exemplary achievements that many others struggle to implement, while PCG translates goals into actions and performance (Ellen MacArthur Foundation, 2015; Pauliuk, 2018).

Last, PCG provides rare insights into how a GLC in a developing country strives for both CE and SDGs. PCG is under constant public scrutiny due to the privileges it enjoys from the government and the obligations it shoulders assisting Malaysia to achieve the SDGs. Based on PCG achievements, the main research questions are, (1) How important is human capital at PCG? (2) How has human capital at PCG transformed? (3) What employee impacts have PCG reported, and how do they link to the CE and SDGs?

The next section reviews the concept of CE and SD to propose a conceptual framework for the role of human capital in these initiatives. This is followed by background information on SD goals and GLCs in Malaysia. The chapter then introduces PCG and its challenges as a GLC. The section briefly describes the research method before discussing the findings on the importance of human capital as narrated in the sustainability reports, the human capital transformation initiative undertaken, and the employee impact. The chapter also provides suggestions for PCG to monetize the employee impact as a way forward before concluding.

2 Review of Circular Economy, Sustainability Development, and Human Capital Considerations

Recent traction gained by the CE has attracted as many as 114 definitions by scholars and practitioners, primarily focusing on recycling, resource efficiency, and optimization strategies to achieve a CE (Kirchherr et al., 2017). While the social aspects have received less coverage, a CE is seen as a catalyst for SD, which allow countries, businesses, and consumers to reduce harm to the environment (Blomsma & Brennan, 2017; Geissdoerfer et al., 2017; Prieto-Sandoval et al., 2018; Sauvé et al., 2016). The CE in developed countries has focused on reducing the environmental impact of SD, while Asian counterparts and most developing countries have struggled with raising public awareness, changing industrial culture, and addressing the increasing waste crisis (Ferronato et al., 2019; Ngan et al., 2019). In these countries, government leadership is paramount to initiate the fundamental systemic shift in thinking from the traditional view of material usage and consumption. While the government in China is well ahead in terms of policies, India and Malaysia are following suit (Ngan et al., 2019).

CE and sustainability are regarded as 'two faces of the same coin' (Stahel, 2019b, p. 12). In implementing a CE within a sustainability framework, Suárez-Eiroa et al. (2019) propose a modified theoretical framework based on that of the European Commission (2018) and Ellen MacArthur Foundation (2015). While the study of Suárez-Eiroa et al. places emphasis primarily on the reproduction of materials and the extraction cycle, other scholars argue that the resources transformation process is paramount for a successful transition to a CE (Blomsma & Brennan, 2017; de Jesus & Mendonça, 2018; Kalmykova et al., 2018; Prieto-Sandoval et al., 2018). Employee training and formal and informal education could accelerate the transition to a CE (Ellen MacArthur Foundation, 2015; Ngan et al., 2019). Circularity is the principle of optimizing assets, including human capital, preserving their use values at the highest utility (Blomsma & Brennan, 2017; de Jesus & Mendonça, 2018; Stahel, 2019a), and therefore, the lack of technical support, training and misaligned incentives among employees could become barriers to a CE and sustainability (de Jesus & Mendonça, 2018).

From a human capital perspective, one could argue that employees are input resources at the point of hiring, and their future values are subject to the organizational transformation initiatives. Transformation initiatives were undertaken to protect, educate, and upskill the human capital to signal the organization's accountability to their employees (see Alam et al., 2022; Ngan et al., 2019; Rizos et al., 2015; Williams & Adams, 2013). Unlike other resources, transformed human capital can benefit the dual objectives of CE and SDGs by extending the acquired knowledge and skills beyond the workplace.

Theoretically, this approach views the potential contribution of human capital as exceeding the contribution of other renewable and non-renewable resources. This perspective expands the conflicting view of a traditional human capital theory which asserts that people's learning capacities are of comparable value to other resources involved in the production of goods and services (Lucas, 1988, 1990), but if human capital is effectively utilized, it benefits the individuals, the organization, and society at large (Schultz, 1961). Further, the latter view has yet to be extended from the SD perspective to CE. Therefore, this chapter provides a pathway for future research to examine this perspective's theoretical and practical implications.

As de Jesus and Mendonça (2018, p. 76) emphasized, a thorough understanding of the determinants is critical to address the challenges and barriers for a CE. A positive effect on employment is expected if corporations adopt a CE (Kunapatarawong & Martiniz-Ros, 2016). Further, CE, SD, and human capital are linked by a similar vision which balances and harmonizes economic, environmental, and social needs (Geissdoerfer et al., 2017; Stahel, 2019b, p. 9). A change in thinking to involve every social actor in collaboration is critical in CE (De los Rios & Charnley, 2017). Indeed, the white paper for CE solutions for chemical companies emphasizes this change in corporate thinking from the top as the enabler of the other solutions (Elsevier, 2019).[1] The business should view the re-skilling and upskilling of employees from an investment and social responsibility perspective and appreciate their role as the key drivers of a prosperous and inclusive economy (World Economic Forum, 2017). Chams and García-Blandón (2019) suggest that employee-level transformation makes the organization's transformation to attain SDGs more feasible.

Figure 1 proposes that if an organization develops its capability towards SDGs by mobilizing human capital, both CE and sustainability ambitions can benefit (Marrucci et al., 2021; Pham et al., 2019). Wade et al. (2022) stressed the importance of capabilities in research and development and stakeholder connections in circularity practices. The impact of transformation on employees extends beyond the organization to the industry and wider community, thus beyond the SDGs reporting requirement. For example, investment in periodic training and certification for health and safety enhances employees' awareness and conduct, not only potentially reducing the number of injuries and increasing savings on medical and

[1] Top five CE solutions for chemical companies include source raw materials from safe and renewable points or origins, treat internal manufacturing pipelines as sources of reusable material, refurbish asset parts and machinery rather than discarding them, form trade partnerships to exchange waste products for renewable resources, and make safety and regulatory compliance core components of company culture (Elsevier, 2019).

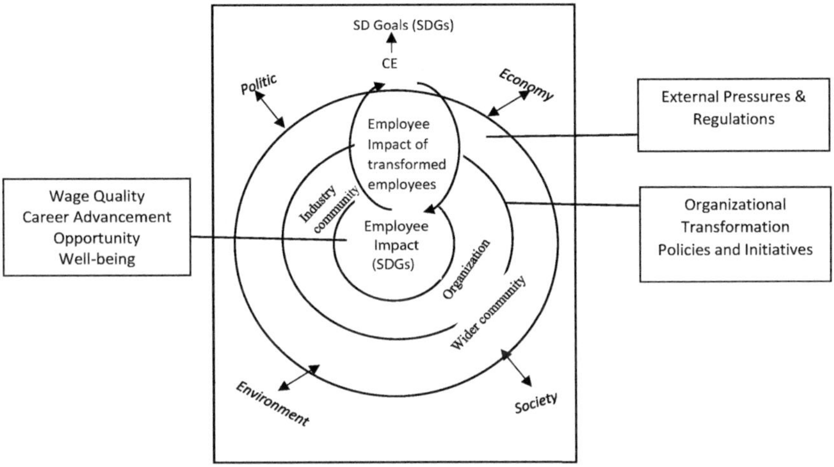

Fig. 1 Employee transformational impact and context (Adapted from Freiberg et al., 2020; Serafeim et al., 2020)

safety-related issues (as per SDGs) but may extend beyond the workplace where the employees share and practice their knowledge to their families and society, thus tripling the impact on others in wider society. Wider society may also benefit from enhanced knowledge and awareness of the conservation of resources and environmental protection as part of their daily routines in the CE. This ripple effect brought about by transformed human capital transforms the workforce, the business, and the larger community, known as a "win-win-win" strategy (World Economic Forum, 2020, p. 22).

Indeed, employees can be both the driver and the product of corporate sustainability initiatives, a notion of circularity. The value of business transforming the workforce collectively as part of the wider society is more effective than efforts by the government to attempt to transform the country's human capital for SDGs on its own. Therefore, the government is urged to work closely with businesses and civil society (Sachs et al., 2019).

Indeed, the Business and SD Commission (2016, p. 8) believe that business has a dual set of motivations when considering engagement with the SDGs. Those motivations positively drive growth by providing opportunities for innovation, market development, cost saving, efficiency, and brand building. Another set of motivations, also referred to as potential growth 'limiters', are where the SDGs outline risks to be managed and negative impacts to be reduced, mitigated, or eliminated. Thus, opportunities presented by the new solution could also be an incentive (de Jesus & Mendonça, 2018). The government can play an active role in changing business mindset through nationwide policies and providing a monitoring role for SDGs. The government needs to be involved with businesses from the very beginning of the transition stage to address the lack of synergistic governmental interventions (Barbiroli, 2011; Kirchherr et al., 2018; Zhu et al., 2015). This is especially so in the GLC context, where it controls the appointment of Board members and senior management and can exercise and influence significant decisions. These factors may accelerate or impede local industries, corporate sustainability direction, and achievements.

Employee impact is the outcome of human capital transformation initiatives. The SDG indicators infer that employment is fundamental to well-being (Bowling et al., 2010) and must be mapped at the organizational level (Nilashi et al., 2019). However, employee impact, such as wage quality, may not be a priority for all industries. For example, stakeholders in the extractive industry are more concerned with the environment and typically respond to environmental and legitimacy threats. However, with global population growth predicted to accelerate to its peak of 10.9 billion by 2100 (Chow, 2019), human-induced global warming and large-scale habitat destruction of the environment may shift the stakeholders' standpoint. Thus, today's human capital transformation may be vital to mitigate the escalating human impact on the environment. At present, considering employees as human capital, which drives the business towards SDG achievement, is implicit. Thus, the World Economic Forum (2017) urged businesses to rethink their role as consumers of 'ready-made' human capital and strive for a human-centric vision of the future of work.

The social aspect of the SDGs also requires corporates to strategize financial value creation through better employment practices (Basson & Erdiaw-kwasie, 2019; Freiberg et al., 2020; Kotsantonis & Serafeim, 2020; Nilashi et al., 2019) to provide better wage quality, opportunity, career advancement, and well-being (Serafeim et al., 2020). Sceptical management may view this as financial spending to report on how well they have managed their employees' health and well-being without realizing that investment in human capital makes economic sense. Competing priorities and the viewpoint that employees are a 'cost' to be minimized continue to impede progress. Serafeim (2020) notes this can become financially material if not managed adequately and proposes that the impact of the transformation initiatives be monetized so that management can highlight the positive impact of the spending. For instance, the monetization of savings due to reduced medical insurance paid because of better work practices should be included in corporate reporting. In this way, management sees financial spending as an investment that generates a positive impact (savings) (Serafeim, 2020). Serafeim gave the example of the estimated USD$7 billion employment impact for Novartis in 2017, which flowed from employee development, occupational safety efforts, and payment of a living wage. However, this potential milestone is far from achieved by developing countries as they struggle economically and politically to ensure that their workforce enjoys decent work (SD, 2019).

3 SDGs and GLC in Malaysia

Malaysia has formulated its own National SDG Roadmap to guide the implementation of SDGs (Karobi, 2017). The country ranks in the top three for SDG scores and is relatively better than its South-East Asian family (SD Report, 2021a) (see Table 1). Based on its scorecard, Malaysia is on track to achieve SDG1 (No Poverty), SDG8 (Decent Work and Economic Growth), and moderately improving SDG3 (Good Health and Well-being) and SDG5 (Gender Equality) (SD Report, 2021b) (see Fig. 2). Malaysia believes it has successfully transformed the economy and raised living standards as the gross national income per capita increased from US$347 in 1970 to US$10,118 in 2020 (SDG Knowledge Platform, 2021). The unemployment rate was low at 4.55% in 2020, an increase of 1.25% from pre-COVID-19 levels in 2018 (Macrotrends, 2021).

Malaysia's sustainability journey started in the 1970s when the New Economic Policy (NEP) was introduced to eradicate poverty and address societal imbalances between Bumiputra indigenes and non-Bumiputra. This positive discrimination restructuring of the social fabric was resented by non-Bumiputra, who felt deprived by the policy, but this affirmative action saw the Bumiputra Economic Community, including emergent GLCs become dominant in the economy (Jomo, 2005; Whah & Guan, 2017).

Table 1 Malaysia SDGs score and rank

	South-East Asia countries	*2021 SDG Index Score*	*2021 SDG Index Rank*
1.	Brunei Darussalam	68.3	84
2.	Burma (Myanmar)	64.9	101
3.	Cambodia	64.5	102
4.	Indonesia	66.3	97
5.	Lao PDR	63	110
6.	Malaysia	70.9	65
7.	The Philippines	64.5	103
8.	Singapore	69.9	76
9.	Thailand	74.2	43
10.	Timor-Leste	NA	NA
11.	Vietnam	72.8	51

Source SD Report (2021a, 2021b)

Fig. 2 SDG dashboards and trends—Malaysia (*Source* Sustainability Development Report [2021b])

GLCs are defined as companies with a primary commercial objective and in which with primary commercial purpose and the Malaysian government's direct controlling stake (Ting & Lean, 2011). GLCs account for 42% of the total market capitalization of Bursa Malaysia and employ around 5% of the national workforce (Khazanah, 2013; The Star, 2021). It was reported that GLCs had contributed significantly to the national economy and remain a key driver towards transforming Malaysia (Khazanah, 2012). However, not all GLCs are success stories, with some being criticized for inferior performance, weak leadership structures and placing government interventions ahead of minority shareholders (MSCI, 2017). In 2012, the high-performing 20 GLCs agreed to take the lead in undertaking transformation initiatives articulated by the government for nation-building, such as human capital investment, driving innovation, and promoting equitable distribution (Khazanah, 2013). Petronas Group reported that it was shouldering the nation-building mission of the Federal and State governments to enrich the national economies and those of the local communities concerning SDGs 3, 4, and 8 (Petronas Integrated Report, 2020, pp. 47, 54). Cooperation between the government and GLCs requires profound national and societal changes to achieve the SDGs. According to de Jesus and Mendonça (2018, p. 76), a vision articulated at the national level is an institutional driver that facilitates the processes involved by providing clear guidelines.

4 Inherent Challenges for Government-Linked Companies

The organizational approach towards its workforce will impact the nation's wider society and long-term growth (Debrah et al., 2018; Freiberg et al., 2020; World Economic Forum, 2020). As such, Wilson (2021) suggests that a political approach is required to measure and report on social and human capital. Sachs et al. (2019) insist that circularity and decoupling without lowering human well-being must underlie all SDG transformations.

The long-standing contentious issue in Malaysia is equal opportunities for ethnic minority groups, which are not in tandem with improvements made elsewhere (Adams & McPhail, 2004). Malaysia has yet to provide this information on SDG 10 (Reduced Inequality). Inequalities can result from inherent power dynamics and discrimination (IMF, 2017; Leach et al., 2018). It may also depend on whose viewpoint is being considered. From the perspective of the ruling government and the major indigenous Malay population (Bumiputera), being equal means upgrading their socio-economic status to a similar standing as the non-indigenous population. GLCs are particularly targeted to bear this aspiration by the government. However, being equal to the rest of the population means having equal access, including equity ownership in corporations and higher education across all businesses, including non-indigenous businesses, which is not in their favour.

There have been claims that GLCs have used their structural and political advantages to overcrowd other enterprises (e.g., Gomez, 2019; Lee, 2015). Petronas is accountable directly to the Malaysian Prime Minister and has implicit loyalty to the government of the day. A Member of Parliament from an ethnic minority recently proposed Petronas was answerable to the Parliament instead. Chung (2020) said the national oil company had been continuously burdened with performing national service to the government by masking Cabinet's poor fiscal management. Thus, by reporting to Parliament, transparency and accountability could be enhanced. The government also has a controlling stake, including the final say on a significant contract, strategy, financing and investment decisions, and the appointment of Petronas's Board of Directors and senior management (Said & Jaafar, 2014). The capabilities of people sitting on the GLC board as fit and proper have been questioned (e.g., Jay, 2017; Salih, 2021). The Board is a CSR (Corporate Social Responsibility) influencer (e.g., Esa & Mohd-Ghazali, 2012), but the public still raises the question of whether the national interest takes precedence over political interest (e.g., Jay, 2017; Salih, 2021). Even previous scholarly work was unable to find evidence that the GLC governing audit committees have a significant association with the quality of environmental, social, and financial reporting (e.g., Atan & Mohd-Razali, 2013; Jamil, n.d.; Mokhtar, 2012; Nelson & Jamil, 2012). This implicitly infers political power into management discourse and decision-making. We reiterate Wilson's (2021) reference to Einstein's quote, "Not everything that counts can be measured, and not everything that can be measured counts". Indeed, CE transitions need to be understood not only as a technical but also as a social and political process (Blomsma & Brennan, 2017).

5 Case Study: Petronas Chemical Groups Berhad (PCG)

Petronas is established under the Petroleum Development Act (PDA) of 1974, giving it the entire ownership of and exclusive rights to exploit Malaysia's petroleum resources. Petronas is a corporation owned by the Government of Malaysia and is directly answerable to the Prime Minister. PCG is Petronas's chemical arm specializing in manufacturing, marketing, and selling a diverse range of chemical products. It is the largest methanol producer in Asia and the fourth largest in the world. The vision of PCG is to become 'The preferred chemical company providing innovative customer solutions' (PCG Sustainability Report, 2019, 2020). It was listed on Bursa Malaysia on 26 November 2010, had a market capitalization of RM[1] 64 billion as of 20 March 2022, and is in the top five on the stock exchange. PCG also ranked in the top ten of the 200 largest companies by market capitalization in the FTSE4Good Bursa Malaysia (F4GBM) Index; an index that selects constituents with leading corporate responsibility practices following transparent and defined Environmental, Social, and Governance (ESG) criteria (FTSE Russell, 2021).

PCG has received various awards for its sustainability reporting, such as the 2019 ASEAN Corporate Governance Scorecard Awards by Minority Shareholders Watch Group, Australasian Reporting Awards, the National Annual Corporate Reports Award 2020, and the Prime Minister's Hibiscus Award 2019/2020. In addition, PCG is ranked first in Malaysia's top 100 sustainable employers (Sustainable Employer Research Centre, n.d.). In 2021, employee-related awards were received from the Royal Society for the Prevention of Accidents (RoSPA) and the 38th Malaysian Society for Occupational Safety and Health, and in 2020 PCG was voted Best Graduate Employer in Malaysia and was commended for Regional Innovation Showcase on Team Excellence (PCG Sustainability Report, 2018, 2019, 2020).

The corporate has been criticized for slow adjusting its business models to incorporate environmental consideration and circularity-oriented practices (de Jesus & Mendonça, 2018; Panwar & Niesten, 2022). However, there is evidence that PCG has reported its commitment to ensuring its business practices align with globally recognized standards for sustainable practices (PCG Sustainability Report, 2018, p. 7). It adopted the 17 SDGs in 2018 as the reference point in shaping sustainability directions and policies. KPMG, an independent third party, assured all PCG sustainability reports. As a member of the International Petroleum Industry Environmental Conservation Association (IPIECA, 2022), PCG shares similar aspirations for all 17 SDGs. Sustainability at PCG is led by the highest level of governance, the Board of Directors. The Management Committee, led by the Managing Director/CEO, oversees all sustainability matters, including direction on sustainability actions and endorses all recommended decisions. At the operational level, sustainability is the responsibility of the SD Working Committee, which comprises department heads who formulate the Group's sustainability strategy and monitor the achievement of action plans (PCG Sustainability Report, 2018, p. 13; 2019, p. 11; 2020, p. 23). In 2019, PCG identified and prioritized material matters, five of which—Innovation and Product Stewardship, Customer Relationship Management, GHG (greenhouse gas) Emissions and Climate Adaptation, and Talent Development and New Plastic Economy—as high priorities (PCG Sustainability Report, 2019, p. 13). In terms of talent development, PCG recognizes the challenge of attracting, developing, and retaining the right talents, acknowledging the lack of knowledge and competence to manage high-value assets to sustain world-class plant utilization, and the lack of capability and competency to grow in the derivatives and speciality chemicals business (PCG Sustainability Report, 2018, pp. 16–20; 2019, p. 13; 2020, p. 22).

The CE at PCG is evident in its commitment to shifting the industry towards a New Plastics Economy by investing in innovative solutions and infrastructure that will convert plastic waste into circular feedstock by 2030 (PCG Sustainability Report, 2020). Initiatives such as these are the drivers of the CE (de Jesus & Mendonça, 2018; Geissdoerfer et al., 2018; Panwar & Niesten, 2022). The aim is to minimize waste, save resources and costs, and deliver value-added products to customers. It is also in line with the Malaysian government's Towards Zero Single-Use Plastic Roadmap 2018–2030 to address plastic pollution and to promote 3R (Reduce, Reuse, and Recycle), cleanliness, and a 40% reduction in terms of emission intensity by the year 2020 against the baseline level of 2005 (PCG Sustainability Report, 2018, p. 47; 2019, p. 21). Furthermore, PCG targets to recover an equal volume of plastic waste as the production volume of polymers for the Malaysian market by 2030 and redirect this waste into the CE (PCG Sustainability Report, 2019, p. 23), a reconceptualization of waste as a resource (Wade et al., 2022).

6 Research Method

The document analysis was based on the PCG Sustainability Reports and local media narratives. The examination of PCG Sustainability Reports follows a two-step qualitative approach. First, the PCG Sustainability Reports from 2018 to 2020 were subjected to an automated content analysis text analytic tool, Leximancer 5.0, to identify these reports' most critical emerging themes. The analysis aims to reveal the stakeholders that PCG management aimed to address using the narratives in the Sustainability Reports. Second, the Sustainability Reports were examined to identify PCG human capital transformation initiatives and track down the reported SDGs related to employee impact.

Content analysis typically requires the researchers to develop a data coding scheme or derive emerging codes from the data (Crofts & Bisman, 2010). This process exposes the research method to interpretation bias. Leximancer generates a more objective and text-driven review document with reproducible and reliable concept extractions and thematic clustering and minimizes pre-conception bias (see Biesenthal & Wilden, 2014; Cheng & Edwards, 2019; Smith & Humphreys, 2006). Hence, Leximancer overcomes the subjectivity in manual data analysis and interpretation of the findings. Leximancer weights the concepts and terms of interest according to how frequently they occur in sentences containing the concept, compared to how frequently they occur elsewhere (Leximancer Guide, 2020). Previous research using Leximancer has validated that concepts of co-occurring words reflect categories that carry meaning (e.g., Thomas, 2014).

7 Findings

7.1 How Important Was Human Capital at PCG: Critical Themes in Sustainability Reports

In Leximancer, the definition of a concept is automatically learned from the text itself. Therefore, concepts frequently appear together in text attract one another strongly and are then clustered into higher-level themes (Leximancer Guide, 2021). The concept map in Leximancer is heat-mapped, so a red coloured circle indicates the most central theme in the Sustainability Reports. The concepts that appear most frequently in the text and those most connected to other concepts on the map are indicated by the size of the circles. As shown in Fig. 3, the software identified three critical themes in PCG Sustainability Reports: environmental, waste, and employee.

Coloured in red, the environment is the dominant theme, consistent with the stakeholders' expectation, of being in the chemical sector, which has implicit environmental risks. The environment circle is also the largest in size, indicating that environmental discussions dominated the narratives in the Sustainability Reports. The second largest circle is waste. PCG

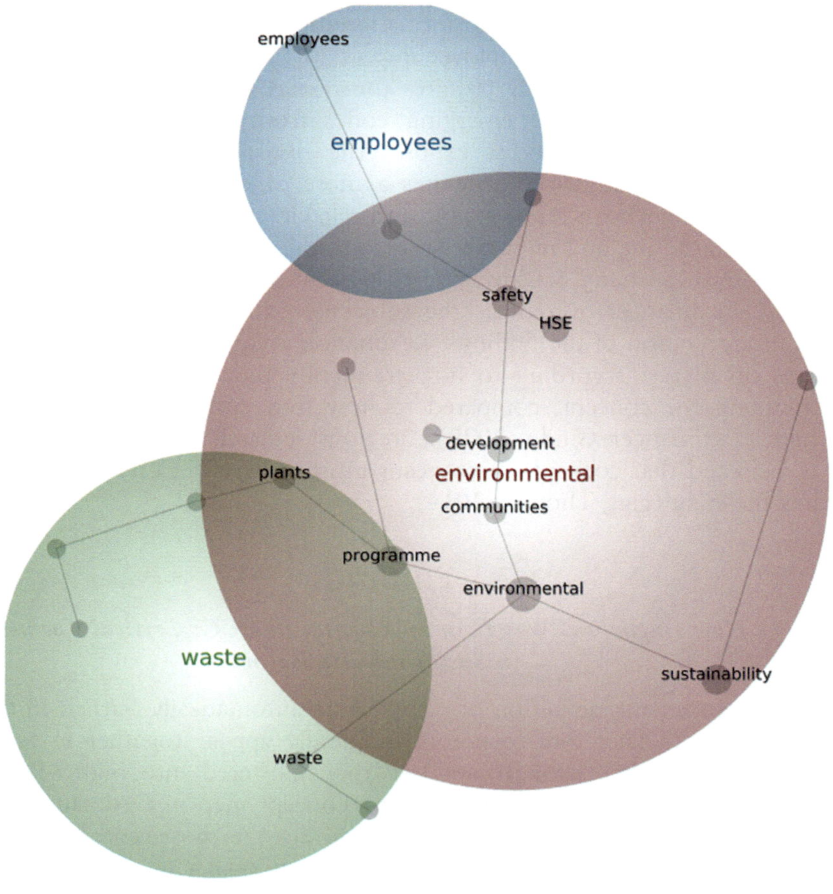

Fig. 3 PCG sustainability reports, 2018 to 2020s concept map

emphasizes its commitment towards CE by introducing its New Plastics Economy and collaborating with society and governments to undertake engagement and outreach programmes with the wider communities. As Wade et al. (2022) emphasized enhanced capabilities in innovation, experimentation, research and development, and stakeholder connections are critical. They are interconnected practices enabling waste to be reconceptualized as a resource. PCG's New Plastic Economy initiative could be seen as a proactive approach to mitigate the increasing environmental legislation, environmental standards, and waste management directives pushes for CE (de Jesus & Mendonça, 2018).

The employee circle is smaller but reflects PCG's commitment towards the well-being of its human capital and indicates how fundamental employees are to the business in achieving its goals and vision (PCG Sustainability Report, 2018, p. 9; 2019, p. 41). Nonetheless, it was evident that human capital was not the most crucial concept central to the discussion in the Sustainability Reports, consistent with CE literature (e.g., de Jesus & Mendonça, 2018; Keijer et al., 2019).

To validate further, sentiment analysis was performed to examine the tone of language in the Sustainability Reports. Leximancer automatically generates positive and negative sentiments in the text and only applies the relevant sentiment terms used consistently within the documents (Leximancer Guide, 2021, pp. 80–81). Only favourable and unfavourable sentiments were considered relevant in PCG Sustainability Reports, but only favourable ones appeared in the top ten list, as shown in Fig. 4. Regarding frequencies, Leximancer identified 302 occurrences of *favourable* sentiment in the reports, and when checked further, all were relevant to the discussions in the Sustainability Reports. This result means that PCG sets a positive tone throughout the reports. Indeed *the unfavourable theme ranked last with just nine occurrences and* 3% relevance (untabulated). It could be that PCG suppressed negative news in its Sustainability Reports, but it could also mean that they were minor as compared to the other (positive) narratives occupying the Sustainability Reports.

The sustainability theme ranked second and was 89% relevant to the discussion in the Sustainability Reports. However, when the *sustainability* theme is examined further, the *likelihood* of it being related to *stakeholders'* narratives is 31%, as shown in Fig. 5. As the transition towards a CE requires large societal changes (Vermunt et al., 2019), the role of stakeholders is more essential than ever. It is also essential that PCG's narratives

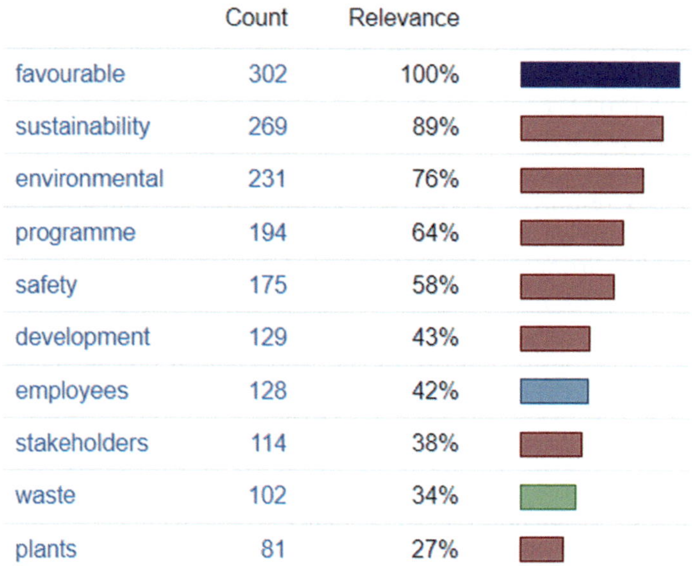

Fig. 4 Top ten concepts in PCG sustainability reports, 2018–2020

reflect its current and ongoing involvement with the stakeholders in its sustainability and CE initiatives.

The likelihood is defined as the percentage of text segments containing sustainability and the *stakeholders'* themes. The likelihood approximates the conditional probability that if the data discuss one concept; the data will also mention the other relevant concept. The total number of text context blocks across the reports in which the *sustainability* theme occurs with the *stakeholders'* theme was 35, and with the *community* theme was 230. Thus, the discussion on sustainability was addressed to the main stakeholders, for example,

> We remain firmly ***Future Focused*** and are creating value for all our stakeholders through a fundamentally resilient and sustainable business, able to capitalise on opportunities and withstand the challenges presented by a dynamically shifting global business landscape. (PCG Sustainability Report, 2018, p. 1)

	Co-Count	Likelihood
stakeholders	35	31%
communities	22	31%
development	38	29%
plants	21	26%
plastic	16	24%
environmental	53	23%
people	13	23%
waste	22	22%
programme	40	21%
safety	36	21%

Fig. 5 Top ten concepts in likelihood with sustainability concept

The discussion on *sustainability* and *employees* occurred eight times, and the likelihood of these two concepts appearing together in a block of text is only 6% (untabulated). However, employee as a theme was linked to other themes such as work (39%), culture (24%), development (19%), training (16%), and communities (15%), as shown in Fig. 6. Training courses for improving an industry's capacity in transitioning towards CE are considered rare (see Geng et al., 2010). Therefore, in-house training would benefit both the employers and employees, otherwise, they could become barriers to achieving a CE (de Jesus & Mendonça, 2018).

In sum, the PCG Sustainability Reports from 2018 to 2020 were primarily articulated for stakeholders, especially those with environmental and *waste* concerns. As stated by the CEO, "the relationship of mutual trust with the stakeholders is its social license to operate" (PCG Sustainability Report, 2018, p. 11), which clearly indicated that environmental and legitimacy threats were their main concerns. As such, radical changes

	Co-Count	Likelihood
work	28	39%
culture	13	24%
development	24	19%
training	9	16%
communities	11	15%

Fig. 6 Top five concepts related to employees' concept

are necessary for waste production and management (Vermunt et al., 2019) to move away from the traditional model of take-make-dispose (Ness, 2008; Wade et al., 2022). *Employees* came third in terms of priority of discussion in the Sustainability Reports, but overall, all reports seemed to be expressed in a positive manner. The role of employees should not be understated, as explained by Pellegrini et al. (2018). When an organization and line managers promote sustainability, the employees are more likely to internalize and make sense of sustainability to support the initiative further.

8 Human Capital Transformation Initiatives, CE, and SDGs at PCG

Transitioning to a CE in developing countries requires more than technological solutions (Corder et al., 2015; Rizos et al., 2015). The lack of government support is often cited as the main obstacle (e.g., Rizos et al., 2015). Thus, reinforcing employees' awareness of environmental sustainability and performance through human resources practices could boost an organization's resource sustainability (Sharma & Gupta, 2015). At Petronas group level, the human capital is input along with the other five capitals (social and relationship; natural; financial; manufactured; intellectual), which undergo processes to produce the output of upskilled talent (Petronas Integrated Report, 2020, p. 59). The value creation model depicts input, processes, outputs, and outcomes. As a GLC, PCG was asked to support the government's National Blue Ocean Strategy to develop human capital skills to reduce national dependency on foreign labour (PCG Sustainability Report, 2018, p. 52). Also, ongoing anti-corruption training is available to all employees to emphasize PCG's zero-tolerance policy for bribery or corruption. The impact on the ethical climate of an organization is associated with employees' perception that their company is moving towards sustainability (Guerci et al., 2015). PCG disclosed its endeavour with regard to SDG 16 for nationwide SDGs in support of the government's aspirations (PCG Sustainability Report, 2020, p. 66).

Four transformation initiatives were identified from the Sustainability Reports: recruitment, fair employment, occupational safety and health (OSH), and lifelong learning (see Fig. 7). In terms of recruitment and fair employment, the Sustainability Reports made mention of their underlying objectives to foster a productive work environment and trust among employees, suppliers, and other stakeholders (e.g., PCG Sustainability Report, 2020, p. 50). OSH was prioritized at PCG, consistent with its operation involving chemicals and other substances. Through its Health, Safety and Environment Policy, PCG reinforced the employees' responsibility to adhere to guidelines and work procedures (PCG Sustainability Report, 2019, p. 41). The company was reported to be 100% compliant in process safety management and standardization in all facilities (PCG Sustainability Report, 2020, p. 56). Indeed, improving process efficiency is fundamental to improving efficiency in production while at the same time reducing the environmental impact (Ma et al., 2015).

Education and training create the required skill base necessary for the evolution of the CE (de Jesus & Mendonça, 2018). Regarding lifelong learning, PCG relied on its internal talent and capability development programmes to provide opportunities for employees to enhance their career growth (PCG Sustainability Report, 2018, p. 50; 2019, p. 46) (see Fig. 8). The employees were also supplied with financial assistance plus up to 24 months of unpaid study leave to pursue their studies in an institution/university (PCG Sustainability Report, 2019, p. 47). If provided with monetary and non-monetary incentives, employees could be motivated to assume an active role in eco-friendly practices (Jabbar & Abid, 2014). In 2020, PCG invested a total of RM17.6 million with an average of 26 man-hours of training per employee (PCG Sustainability Report, 2020, p. 11). PCG leadership programmes aim to equip employees

Employee Transformation Initiatives

Recruitment | Fair employment | Occupational Safety and Health (OSH) | Lifelong Learning

Impact Dimensions (Freiberg et al., 2020) and PCG Reported Indicators

Wage Quality	Career Advanceme nt	Opportunity	Health & Well-being	Subjective Wellbeing	Diversity
Total number of employees	Average total hours of training per employee (man-hours)	No. of female senior managers	Total recorded case frequency (incidents per million-man hours)	% Of union member	Workforce by gender
	Training investment per employee (RM)	No. of female in Management committee	Lost time injury frequency (incidents per millions man hours)		Workforce by age group
	Training days per employee	No. of female in Board of Directors	Process safety event		
	New Hires by Gender		Total recorded cases (No. of cases)		
	Technical Professional developed (number)		Fatality (No. of cases)		
	Success rate of technical professional developed (%)		Lost time injury (No. of cases)		
	Capability assessment (%) (target and results)		Total recordable occupational illness frequency per million man-hours		

Fig. 7 Transformation initiatives, scope, and employee impact indicators (*Source* Freiberg et al., 2020 and PCG Sustainability Report, 2020)

	Employee Attrition rate (%)		% Of parental leave (female)		
	Employee Turnover rate (%)		% Of parental leave (male)		

Fig. 7 (continued)

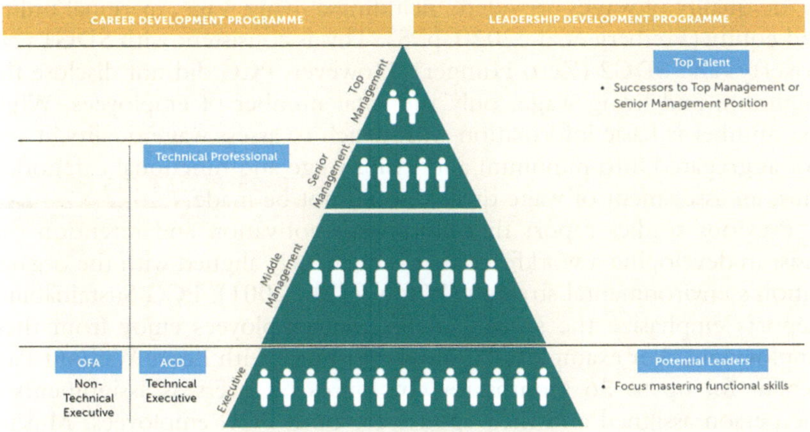

Fig. 8 Career and leadership development programme (PCG Sustainability Report, 2018, p. 53)

with foundation knowledge and make them mindful of their goals, challenges, and strengths to realize their full potential (PCG Sustainability Report, 2019, p. 47). However, Pellegrini et al. (2018) found that sustainable behaviour could be shaped if the training was mediated by affective commitment. A Human Firewall Campaign was also initiated to raise cyber employees' awareness and knowledge of social engineering scams and cybersecurity attacks alongside an anonymous whistleblowing hotline (PCG Sustainability Report, 2018, p. 57). Technical capacities are fundamental in transitioning to a CE (de Jesus & Mendonça, 2018).

8.1 Employee Impact

The six dimensions relating to employee impact: Wage Quality, Career Advancement, Opportunity, Health and Well-being, Subjective Well-being, and Diversity, following Freiberg et al. (2020), were identified from the Sustainability Reports (see Fig. 7).

8.1.1 Wage Quality

The employment conditions provided to organizational human resources could be significant in determining the success or failure of environmental improvement efforts (Daily & Huang, 2001). Wage quality is described as the quality of wages provided, including a living wage, marginal utility, and equity (Freiberg et al., 2020, p. 5). This is consistent with SDG1 (No Poverty) and SDG2 (Zero Hunger). However, PCG did not disclose the minimum and living wage, only the total number of employees. While the number is basic information with which to assess wage quality, it was not aggregated into minimum and living wage and functional categories, thus, an assessment of wage quality could not be made.

Previous studies report that employee motivation and retention can assist in developing a workforce that is effectively aligned with the organization's environmental strategy (Wright et al., 2001). PCG Sustainability Reports emphasize the social benefits that employees enjoy from their employment. For example, a 'Spousal Assistance with Leave Without Pay' benefit for up to 36 months for employees on overseas assignments if the person assigned and their spouse are both PCG employees. Muslim employees are the majority of the PCG workforce, and Muslim employees are offered leave to perform Hajj-Umrah (PCG Sustainability Report, 2018, p. 24; 2019, p. 48; 2020 p. 51). However, PCG did not disclose whether employees from other religious beliefs receive a similar privilege and how many Muslim employees had benefited from these initiatives.

8.1.2 Career Advancement

Career advancement is described by Freiberg et al. (2020, p. 5) as internal mobility resulting in increased earnings and is relevant for SDG5 (Gender Equality) and SDG10 (Reduced Inequalities). Figure 8 shows that PCG has provided a clear career advancement and succession planning pathway. Junior executives can undertake the leadership development programme, and middle management is included in the PCG succession planning programme (PCG Sustainability Report, 2018, p. 52). The accelerated capability development (ACD) and ACD Booster Programmes were tailored for the engineering competency of technical executives, a socio-technical approach as described by de Jesus and Mendonça (2018). The United Nations Department of Economic and Social Affairs (UNDESA) (2011) emphasized that technical hurdles need to be addressed to accelerate innovation, resource efficiency, and waste-reducing technologies (p. 19). PCG also provides targeted capability and skills development programmes for employees within specialized functions, for example, the digitalization of Operational Excellence Results for customer service, an analytic and intelligence tool and Sales Force Automation for marketing and sales units (PCG Sustainability Report, 2018, p. 52). As Ness (2008) stated, an integrated system opens many opportunities for integration and innovation, leading to eco-efficiency.

As can be seen in Fig. 7, the measures were primarily based on inputs such as average total hours of training per employee and the cost spent. The outcomes of these trainings, such as the success rate of technical professional and post-capability assessment, were also reported, but there was no disclosure on how many employees had been promoted based on their achievements and whether successful completion led to increased earnings (PCG Sustainability Report, 2018, p. 51; 2019, p. 47; 2020, p. 58). This additional information would be useful to link such transformation initiatives to the associated SDGs. Moreover, there was no breakdown on gender and under-represented minorities, which could be easily accessed from the human resources department. Nonetheless, Pellegrini et al. (2018) reported that rewards appeared to have no effect in heightening employees' commitment towards sustainable behaviour.

8.1.3 Opportunity

The opportunity dimension translates into employee demographics across job categories and is related to SDG10 (Reduced Inequalities) (Freiberg et al., 2020, p. 5). Guerci et al. (2015) assert that opportunity-enhancing practices improve the perceptions held by employees on corporate sustainability. Potential useful information includes how many females, minorities, and people with disabilities are in full time employment. It was mentioned that some female managers were at the Senior Management level, in a Management Committee, and on the Board of Directors, but the report did not disclose gender pay equality or people with disabilities. As such, disclosures for diversity were in lower granularity across the firm hierarchy or were missing. Grosser et al. (2008) note that detailed information on the gender pay gap remains unclear. As Adams and Harte (1998) asserted, both in their choice of disclosures and silence, corporate managers contribute to social perceptions of women and employment (p. 782). Cox (1970) famously argued that racism is a bourgeois rationalization of capital exploitation and degradation of labour.

8.1.4 Health and Well-Being

Aiming for SDG3 (Good Health and Well-being) and SDG10 (Reduced Inequality), this dimension looks at employment's impact on employees' health and well-being. This aspect includes injuries and incidents, well-being programmes, access to healthcare, paid sick leave and family-friendly workplace benefits (Freiberg et al., 2020, p. 5). Information on access to healthcare and paid sick leave was not provided in the Sustainability Reports. In addition, the reports were missing whether PCG provides health insurance or any other similar arrangement. PCG did disclose the recorded occupational illness frequency per million man-hours but most of the disclosures are related to the most basic measurement of employee health, such as the number of fatalities and total recorded incidents. The outcomes were disclosed based on lost time through injury in frequency and number. Nonetheless, Fig. 7 indicates that this category of employee impact is a priority at PCG due to the inherent nature of risk in its operation.

Kirchherr et al. (2017) found that hesitant company culture is one of the main barriers to embracing a CE. At PCG, the workplace culture placing emphasis on health and safety was represented by the 3Cs (Culture, Compliance and Competency) (PCG Sustainability Report, 2019, p. 41), the Culture of Intervention (ZeTo) Rules, "Jom Patuh dan Tegur" (Let's Comply and Intervene), and Process Safety 3Sixty to instil a proper process safety culture among the employees (PCG Sustainability Report, 2018, p. 39; 2019, p. 42). The ZeTo Rules violations tremendously improved from 25 counts in 2017 to 11 counts in 2018 (PCG Sustainability Report, 2018, pp. 21–24, 35). In 2020, the Lost Time Injury Frequency scored 0.14, better than the industry benchmark of 0.26 and zero fatalities in all operations (PCG Sustainability Report, 2020, p. 11).

In terms of well-being, PCG also provides a flexible work option which allows employees to choose their daily working hours and to complete their job at noon on the fifth day of the week upon completion of the required 39 working hours for the week (PCG Sustainability Report, 2018, p. 54; 2019, p. 54; 2020, p. 52). As a friendly employer, PCG allows pregnant women to leave early and provides maternity leave above the statutory requirement, and new fathers can also apply for parental leave (PCG Sustainability Report, 2018, 2020). PCG recent initiatives include the establishment of nursing rooms and childcare centres at the workplace (PCG Sustainability Report, 2020). Diaz-Carrion et al. (2020) suggested that higher standards of employment practices are set to cement employees' commitment to the organizational journey towards sustainability.

According to Freiberg et al. (2020), subjective well-being (or employee engagement) is intrinsically connected to job quality and achievement of SDG8 (Decent Work and Economic Growth). Therefore, it is recommended as a parallel analysis to the health and well-being category. This category includes worker's freedom of association, employee feedback systems, discrimination, and human rights grievance processes. Human rights are rights inherent to all human beings, regardless of race, sex, nationality, ethnicity, language, religion, or any other status, including the right to life and liberty, freedom from slavery and torture, freedom of opinion and expression, the right to work and education, and many more, without discrimination (UN, n.d.). PCG only reported the percentage of its union members but did not publicly publish the employee satisfaction rate. PCG asserted its commitment to human rights and zero tolerance for discrimination and provided such training to employees (PCG Sustainability Report, 2019, p. 50).

If embodied in human resources practices, sustainability principles could generate the workforce's long-term economic, physical, and social well-being. PCG periodically conducted a Social Risk Assessment to mitigate and assess the impact in each operational area (PCG Sustainability Report, 2020, p. 51). The examination of the Sustainability Reports revealed that the employees raised different concerns each year. For example, in 2018 their concerns were on occupational safety and health, capability development, employee welfare, and environmental management (PCG Sustainability Report, 2018). In 2020, the key concerns shifted to health, safety and human rights, career progression, work-life balance, and cyber security (PCG Sustainability Report, 2020, p. 13). PCG seemed to undertake an engagement approach to reach out to their employees to address these concerns on various platforms. Transformation tends to stimulate the local generation of local inputs (Bonnín-Roca et al.,

2017). The leadership team was present through informal chats with the CEO while celebrating long-serving employees. According to Krueger et al. (2021), sustainable firms can also better recruit and retain high-skilled workers. PCG management met the in-house unions regularly, especially if there were changes in human resources policies or processes (PCG Sustainability Report, 2019, p. 48). PCG was proud of its comprehensive process in handling employee grievances, and in 2018, one staff grievance was recorded and resolved (PCG Sustainability Report, 2018, p. 56). According to Zoogah (2011), if organizations focus on positive behaviours and encourage environmentally friendly practices, employees might be motivated to strengthen their environmental behaviours.

8.1.5 Diversity

Pursuant to SDG5 (Gender Equality) and SDG10 (Reduced Inequality), the diversity aspect refers to employee demographics comparative to the local population, not merely to favourable statistics on employee diversity. As Freiberg et al. (2020) explained, this indicator measures the impact of employment on socio-economic inequality and disparity in the labour force compared to the local population, but PCG has not reported this. However, it did mention a recruitment strategy which focused on hiring specialist candidates in technical and commercial areas from East Malaysia to meet and support the State Government's aspiration for employment of the local population. Thus, PCG is also championing the State government's ambition and the Federal Government directives. Furthermore, for its overseas operation, PCG reported employing local market practices and expertise to source and attract the right talent (PCG Sustainability Report, 2018, p. 56).

In its Sustainability Reports, PCG acknowledged the benefits of diversity by exchanging perspectives and experiences to push innovation. The social performance of a sustainability-inspired company is contingent upon its effectiveness in managing its human resources diversity (Jabbour & Santos, 2008). PCG reported that it envisaged a more diverse workforce encompassing a balance of gender and a good representation of different ethnicities, cultural backgrounds, and age groups (PCG Sustainability Report, 2019, p. 54; 2020, p. 50). At present, PCG has not shared complete demographic data beyond gender and age groups. Women comprised of nearly 20% of the workforce, 25.2% in Senior Management and 36.4% in Management Committees (PCG Sustainability Report, 2018, p. 54; 2019, p. 54). The female executives were strongly encouraged to participate in the Petronas Leading Women Network; a networking platform that aims to create opportunities and a safe work environment for all (PCG Sustainability Report, 2018, p. 54; 2020, p. 51). Fenwick and Bierema (2021) found that socially responsible companies focus on employee learning and promotion, employee ownership of development, and employee safety and respect.

8.2 Human Capital Transformation Initiatives and Employee Impact for CE and SDGs

Resource productivity intensification can create a competitive advantage for an organization (Ma et al., 2015). Figure 7 summarizes the employee transformation initiatives, the impact dimensions and the indicators reported by PCG in its Sustainability Reports from 2018 to 2020. As de Jesus and Mendonça (2018) emphasized de Jesus and Mendonça (2018) emphasized, a broad transformation is contingent on more than science and technology. Although the 25 indicators could be traced to the employee impact-related SDGs, they were not equally emphasized. The total number of indicators for career advancement is the same as that for the health and well-being impact dimension, indicating the current emphasis on SDG3 (Good Health & Well-being), SDG5 (Gender Equality), and SDG10 (Reduced Inequality). PCG has started improving its disclosure on the opportunity, diversity, wage quality and subjective well-being impact achievements for SDG1 (No Poverty), SDG2 (Zero Hunger), and SDG8 (Decent Work and Economic Growth).

In terms of CE, there were no specific indicators on how transformed employees were involved in the circular mission of the New Plastics Economy. According to PCG, this concept requires rethinking the future of plastics and advocating for a CE where plastics are always reused and never become waste (PCG Sustainability Report, 2018, 2019, 2020). As such, the possibility of reuse has led to technological innovation for reuse (Park & Chertow, 2014; Wade et al., 2022). PCG acknowledged its role as the producer of the main component in the production of plastics and emphasized its mission to educate the public, especially the communities around its areas of operation, to be responsible plastic users for individual well-being, the environment, and future generations (PCG Sustainability Report, 2018, p. 47; 2019, p. 53). Social awareness and environmental literacy were identified as the enabling factors for a CE (de Jesus & Mendonça, 2018). However, insufficient public education and awareness resulted in little public involvement in environmental protection in China (Geng et al., 2010).

In engaging with the communities, PCG educated the public through their employees, especially the younger generation, on the responsible use of plastics to encourage green behaviour for a sustainable future. Various community outreach programmes were organized, which focused on environmental conservation and plastics usage awareness programmes (PCG Sustainability Report, 2018, p. 8; 2019, p. 22; 2020, p. 22). As emphasized by scholars, if human capital is effectively utilized, it will benefit society as a transformation towards a sustainable society (Carrillo-Hermosilla et al., 2010; Schultz, 1961; Suárez-Eiroa et al., 2019). The initiatives also support the Government's efforts towards the Zero Single-Use Plastic Roadmap 2018–2030 and to promote the 3R (Reduce, Reuse, and Recycle) campaign and the reduction of 40% of emission intensity by the year 2025 (PCG Sustainability Report, 2018, p. 4). Geng et al. (2010) emphasized that government agencies should play a leading role. As such, PCG collaborated with government ministries, civil societies, and private entities at an estimated cost of RM150 million to move towards more sustainable solutions and the promotion of SDGs (PCG Sustainability Report, 2019, p. 45; 2020, p. 32). De los Rios and Charnley (2017) emphasize that collaborations are critical in CE.

Through their increased engagement with communities, PCG employees could become more aware of environmental issues in the nearby communities. The Sustainability Reports mentioned that employees are trained on the most up-to-date environmental best practices prior to their community involvement (PCG Sustainability Report, 2018, p. 26). Their employees have worked with the community in mangrove replanting and seedling activities to help preserve the rich biodiversity of mangrove swamp habitats in the areas of PCG plant operations. PCG also established the ecocare Environmental Education Centre in 2013, which serves as a resource centre to promote awareness, education, and research about environmental conservation among schoolchildren, teachers, and the community (PCG Sustainability Report, 2018, p. 48; 2019, p. 53). According to Kaltenborn et al. (2017), promoting social-ecological change is an important 'soft driver' (see also de Jesus & Mendonça, 2018, p. 85). Figure 7 shows that the human capital at PCG was equipped with rigorous OSH training and certification. These transformed employees were then involved with the First Aider for Every Home programme rolled out to the local communities surrounding their operations in case of emergency, including injuries, choking, or out-of-breath victims (PCG Sustainability Report, 2018, p. 46). Community outreach is often seen as a declaration of an organization's explicit commitment to sustainability (Fenwick & Bierema, 2008).

In sum, the narratives in the Sustainability Reports suggest that human capital was transformed towards SDGs and supported the CE it adopts. PCG's proactivity contradicts the literature that asserts that corporates typically respond to or in anticipation of government regulations (e.g., Andrews & deVault, 2009). The evidence suggests that the human capital undergoes a circularity process to be transformed to the values (employee impact) which benefits themselves, the organization (PCG) in supporting the New Plastics Economy, the industry by portraying good initiatives of preserving the environment, the wider communities from the education and outreach programmes, and the government through collaborative efforts to spearhead the nation's SDG efforts. The scholars call this social "circular" practice of transformative innovation (de Jesus & Mendonça, 2018, p. 85; Schot & Steinmueller, 2016). The spill-over effects of employee transformation initiatives go beyond their own achievements. Employees tend to adopt sustainable behaviour if it is valued and promoted by the organization's management (Pellegrini et al., 2018). The benefits are returned to the human capital through better employment practices and privileges they enjoy from such employment. The social implications influence "how we work but what we buy, how we think, and the way we live" (Womack et al., 1990, p. 11).

9 Way Forward: Monetizing Employee Impact

The World Economic Forum (2020) urged a better measure for human capital value at nationwide and employer levels, however, because the approaches are innovative and encapsulate a broad range of SDGs and impacts; the social impact is inherently too complex for the existing corporate accounting and reporting systems to capture (Lodhia, 2014; Scholz, 2011; Schaltegger et al., 2017, p. 118; Wiek et al., 2012). In addition, regulatory bodies have started to realize that the historical approach of disclosing costs of compensation is insufficient to understand the value and impact of human capital on current and future organizational performance (Clayton, 2019). Further, intangible assets in the form of human capital and culture are estimated to comprise more than half of a company's market value; thus, there is a growing demand to understand how companies manage and measure the impact of human capital (Klemash et al., 2019). As such, human capital metrics must be integrated with financial and operational measurements (Watson, 2020). Until recently, Harvard Business School initiated projects to measure and report on product and employee impact and encouraged businesses to practice them (Freiberg et al., 2020; Panella & Serafeim, 2021; Rouen & Serafeim, 2021; Serafeim et al., 2020). This is vital to change the mindset of management to view human capital as an investment which generates savings rather than cost. Serafeim (2020) gave the example of Novartis, which estimated its employment impact for 2017 at USD$7 billion, derived from the benefits of employee development, occupational safety efforts, and payment of a living wage.

PCG has published its achievement in various categories of employee impact, stating the way forward is to report the monetary value of the impact of the employee transformation initiatives. Employee impact does not simply mean increasing salaries (see Serafeim & Trinh, 2020, p. 5). Management needs to be convinced of the value of reporting positive employment impact (Serafeim, 2020). In fact, de Jesus and Mendonça (2018) argued that there is a need to decouple revenues from input cost to transform resource performance. The impact must be systematically measured in monetary terms using impact-weighted financial accounts (see Freiberg et al., 2020; Serafeim & Trinh, 2020). This innovation provides inclusive insights into the commitment and accountability of business for effective deployment of financial resources (de Jesus & Mendonça, 2018; Freiberg et al., 2020; Panella & Serafeim, 2021; Rouen & Serafeim, 2021; Serafeim et al., 2020) in place of conventional accounting and 'tick-box' methods of sustainability reporting (Schaltegger et al., 2017; Sinkovics et al., 2020).

The first impact is on wage quality. The wage quality impact measures the total wages paid by an organization, adjusted for a living wage benchmark, the marginal impact of income function, and wage equity within each occupation category (Serafeim et al., 2020). For example, Panella and Serafeim (2021) found that though Bank of America had a lower average salary than Facebook; its wage quality as a percentage of revenue was twice that of Facebook. In PCG's case, the absence of wage quality information indicates the need for other necessary information, such as the number of employees in each salary band and occupation category and employees' race and ethnic group disaggregated by gender. The real wage of low, medium, and high-skilled Malaysian workers and the minimum and living wage information is available from WageIndicato r.org. A living wage is based on the idea that employment should provide an adequate income to cover necessary living costs (WageIndicator, n.d.). As of 2019, the minimum wage in Malaysia was RM1050, and the living wage for a typical family (2 + 2 children) was between RM1590 to RM2300.

For the career advancement category, PCG's training information was particularly detailed on investment, hours and days per employee, and the training outcomes, such as the number and success rate of technical professionals developed. However, other information, such as the total number of employees by occupational category, average salary by occupational category, new hires by occupational category, and positions filled by internal candidates, might be useful. For example, Intel's career advancement impact was—$49 million indicating missed human capital development opportunities within the firm when internal candidates filled only 15% of positions in 2018 (Freiberg et al., 2020, p. 20; Serafeim et al., 2020).

Effective workplace health and well-being programmes affect employees' physical and mental health (OECD, 2019). Indeed, PCG provides specific information on incidents, injuries, fatality, and a number of safety events held to mitigate incidents. This information is consistent with occupational safety and health guidelines following the International Labour Organization's guidelines (e.g., ILO, 2020). However, PCG has yet to report the positive impact of its family-friendly initiatives. For instance, Bank of America was reported to have created positive health and well-being impacts, driven by its family-friendly workplace practices monetized at USD$109 million (Panella & Serafeim, 2021). Likewise, PepsiCo had a family-friendly workplace with backup childcare support and onsite childcare, a positive impact of over USD$40 million in 2018 but lacked parental leave (Panella & Serafeim, 2021). In addition, PCG is in an advantageous position by establishing onsite childcare, nursing rooms, and flexibility for pregnant women to go home earlier.

In terms of opportunity impact, globally the occupational segregation by gender and racial categories has improved but not been resolved, especially in Asian countries. According to Carter and Silva (2010), opportunity bias may begin at the point of recruitment. The opportunity impact dimension monetizes the impact of unequal access to employment within an organization using the high salary group to represent firm hierarchy (Freiberg et al., 2020). In the future, PCG might want to report the total number of employees, the number of employees in the minority group, and the average salary by each occupational category. In their analysis, Freiberg et al. (2020) and Serafeim et al. (2020) found that Asian females and Asian males drove the most considerable individual negative impacts, while Panella and Serafeim (2021) reported that PepsiCo's negative opportunity impact was due to the misrepresentation of Black male workers, who comprised 16% of the total workforce but only 6% were in the top salary tiers.

In 2020, more Malaysian males (9.13 million) were employed compared to 5.83 million females (Department of Statistics, Malaysia, 2022). In terms of diversity, PCG has taken the first step by providing information on the workforce by gender and by age group but not by minority groups and gender. At Intel, for example, Asian women were overrepresented (Freiberg et al., 2020; Serafeim et al., 2020). Similarly, no information has been provided on the locality of the employees by states in Malaysia or by local and overseas operations. For instance, both Cisco and Facebook underperformed in the diversity dimension due to significant misalignment between their workforce composition and local demographics (Panella & Serafeim, 2021). Intel, however, operates in locations with low unemployment rates; thus, it does not contribute significantly to local unemployment (Freiberg et al., 2020; Serafeim et al., 2020).

In sum, PCG has the accounting and reporting system to report the current employee impact, but this may need some adjustment if it aspires to report on the monetary impact of its employee transformation initiatives, particularly on wage quality, health and well-being, opportunity, and diversity dimensions. In addition, PCG might want to improve its reporting system by making sure issues around low data availability and comparability are resolved (Meys et al., 2020).

10 Discussion and Conclusion

As a developing country, Malaysia's rapid economic development and urbanization conflict with its environmental protection and SD goal. A CE is seen as a solution to spearhead its progress in achieving the SDGs by better management of its waste and related impacts on the economy. Though the corporate is expected to be in the driver's seat for CE, Malaysia's top-down political objective is different from that of developed countries such as those in Europe, Japan, Canada, Australia, and the US with regard to their CE implementation approach (Ghisellini et al., 2016). Focusing on human capital as one of the resources, the chapter examines the human resource transformation initiatives at Petronas Chemicals Group, a GLC. Being endorsed as the top sustainability employer, we specifically scrutinized how important human capital at PCG was as narrated by the Sustainability Reports. We also examined how human capital was transformed, employee impact was reported, and linked to CE and SDGs.

At the current pace, the analysis indicated that the environment and waste were PCG's primary concerns relative to human capital, consistent with the environmental and legitimacy threats posed by the primary stakeholders. Nonetheless, PCG had systematic and rigorous transformation initiatives and certification of its human capital. A circular business model at the firm-level is expected to galvanize CE efforts, including human capital transformation initiatives. At PCG, human capital was optimized and mobilized to benefit the organization and the wider communities, including the government's efforts towards CE and SDGs. The transformed employees' learning capacities could potentially exceed benefits from other renewable resources, thus, its role within a CE deserves similar attention, if not more, within the sustainability context. In fact, the case study suggests that human capital transformation could make achieving both CE and sustainability ambitions feasible.

While some studies predict displacement or a diminished role for employees if corporations adopt the circular approach (e.g., Repp et al., 2021), this study's findings provide evidence that they could be transformed to better support PCG in transiting to CE. Thus, transformed human capital are value-added asset, especially if they are (re)trained with the skills required for CE and SDGs. Blomsma and Brennan (2017) asserted that CE enables the capacity of the productive life of resources to be extended to create value. However, the circularity-oriented practices require PCG to radically transform its business model, including the workforce model to go beyond the conventional, technology-focused approaches (Elf et al., 2022). A CE requires a systemic change, hence, simply adding to the current linear business model might hinder its true potential (see Geissdoerfer et al., 2020; Kirchherr et al., 2017). PCG's circularity workforce model which emphasizes (new) working culture and training empowers the workforce to spread their learned capabilities on sustainability and CE beyond the workplace. The workforce development is aligned with the sustainable strategy by encouraging and rewarding environmentally friendly practices to strengthen their environmental behaviours inside and outside the workplace. This internalization of CE-focused business thinking emphasizes that the transformed workforce could heighten PCG's dynamic capabilities by creating a wider social impact. The interactions with local communities through education, awareness, and 3R projects could break through the social barriers to a CE and sustainability depicted by the lack of an informational cooperative community commitment to SDGs which are the common barriers in developing countries (e.g., Ngan et al., 2019). Ellen MacArthur Foundation (2015) suggested five ideas for a successful transition to a CE. These are tightening circles within the supply chain; entering the market early; activating the community; forming cooperation networks; and making profits from arbitrage. It seems that PCG has embedded similar strategies in its operation.

PCG transformation initiatives succeeded in creating value for the employees, and the impacts were reported in line with SDGs. The next step could be monetizing these impacts to reflect the good work and the shift in corporate thinking that human capital is an *expenses*-driven asset rather than an investment. Since the measurement of the employee impact due to transformation is still in its infancy, future research may examine the measurement and the reporting of these impacts. I management considers the feasibility of reporting these positive impacts; it might incentivize businesses to do more for their employees, indirectly spearheading SDGs.

The current literature has provided evidence of firm and nationwide level challenges transitioning towards a CE for sustainability and offers predominantly production-based and customer-focused strategies. Previous work suggests that (natural) resource transformation is a precondition to optimization of assets (Blomsma & Brennan, 2017; de Jesus & Mendonça, 2018; Prieto-Sandoval et al., 2018; Stahel, 2019a). The findings of this study bridge the gap by focusing on human resources and the CE with specific novel contributions to the dynamic nature of human capital transformation in advancing CE and SD. The alternative perspective of human resources as a CE resource acknowledges its potential to be transformed as part of circularity-oriented activities. It is proposed in this chapter that radical transformation of the workforce business models is possible beyond the traditional linear business practices, thus advancing knowledge of how circularity value can be created through the transformation of human resources and collaboration with communities and the government, which have been noted as crucial in the CE context (e.g., Geissdoerfer et al., 2018).

"The strength of the desire to create change" is paramount either at the firm or national level (Wade et al., 2022, p. 2658), and so an understanding of the barriers and resistance to CE is critical to boosting the acceptance and propagation of CE. However, the literature is predominantly western and culturally distinct from Asian countries. While Asian countries suffer from similar CE barriers (Kirchherr et al., 2018; Ngan et al, 2019; Zhu et al., 2015), the economic impact is yet to be evidenced in those countries (Panwar & Niesten, 2022). In developing countries such as Malaysia, a top-down approach is more effective if it is well-articulated by the ruling government as the institutional driver. Ngan et al. (2019) urged decision-makers and policymakers to customize action plans and incentives to convince industry players in Malaysia of the value of CE. As experienced in other developing countries, the motivation for CE and SD is often intensified if followed by monetary and non-monetary incentives (Jabbar & Abid, 2014). As in the PCG case, governmental guiding mechanisms and directives are useful in advancing CE initiatives. This chapter provides rare insights on how a GLC transitioning to a CE shoulders the government's ambitions for SDGs. Given the literature is silent about how industry players manage government expectations and directions in relation to CE and SDGs, perhaps future studies could contribute more empirical evidence on how management can navigate the need to radically transform their business thinking to simultaneously align with the government expectation and directives for SDGs at the national level. Nonetheless, effective regulation and monitoring mechanisms of CE performance are expected in the future as industries might not pursue CE if the penalty for pollutant emission is lower than the capital investment required for CE. Therefore, it is anticipated that government incentives or aggressive environmental tax may influence the starting point for industry action towards sustainability, at least in the context of developing countries.

Despite its potential, this study has several caveats and limitations. *First*, the case study findings are limited to only one GLC, and thus might not be representative on how human capital is viewed and the strategy for human capital transformation initiatives. Although GLCs benefit from a patronage-based structure and political access, the influence of the government of the day may extend to how it operates through the Board and senior management's appointment and subsequent decision-making. For example, PCG has supported the state and the federal government on their human capital and sustainability ambitions for CE and SDGs. Thus, being a GLC, it cannot isolate itself from local societal and political demands that impact how it operates, accounts, and reports information to the external stakeholders, thus bureaucratic issues still dictate its corporate direction. Future studies might want to explore how GLCs balance the demands of their main stakeholders and the government's intervention in the CE context.

Second, while Malaysia is on track to achieve SDG1 (No Poverty) and SDG8 (Decent Work and Economic Growth), PCG did not report vital information on wage quality and specific demographics of its workforce in its Sustainability Reports. Sensitive information such as wage, gender, and ethnic disparity might require the intervention, in the context of PCG, of the government of the day. PCG might have reported other measures in other mediums and reports; thus, the findings are limited only to Sustainability Reports from 2018 to 2020 as PCG started its journey in 2018.

Third, organizations differ along multiple dimensions even if they belong to the same industry, so the scope, approaches, and metrics might differ between GLCs aspiring to CE and SDGs. *Lastly*, the proposed monetization of employee impact has yet to achieve internationally agreed targets and measures, but its awareness and future undertaking may encourage business to view their employees differently. As, Ngan et al. (2019) highlighted, economic benefits and public acceptance are prominent factors affecting industry players towards CE in Malaysia.

In conclusion, it has been shown in this chapter that distinct skills, processes, and transformational activities are necessary to heighten organizational capacity and dynamic capabilities for CE by reconfiguring human resources practices. Employees' active support helps businesses adjust their business models to pursue visions of circularity and sustainability holistically and sustainably (Elf et al., 2022). Thus, lessons learnt from PCG could be extended to other developing countries in different regions of the globe so that it is possible to initiate change, either starting with or simultaneously with human resources. This chapter offers insights and guidance to other similar industries, even in developed countries, to identify promising human resource transformation activities to be adapted and adopted. While context-dependent, the case study findings are transferable to contexts where the CE transition is early. The risk of employees changing from one employer to another for sustainability reasons is less of a threat in developing countries, however, ignoring this would risk CE and sustainability ambitions which could eventually devastate financial sustainability. Therefore, the shift in business thinking for a new radical workforce model that supports business-circularity practices is essential.

Note

1. Ringgit Malaysia.

References

Primary Sources

PCG Sustainability Report. (2018). Available at https://www.petronas.com/pcg/sites/pcg/files/sustainability-reports/petronas-sustainability-report-2018.pdf. Accessed 30 Jan 2022.

PCG Sustainability Report. (2019). Available at https://www.petronas.com/pcg/sites/pcg/files/2020-04/3.%20PCG_SR_2019_02042020.pdf. Accessed 30 Jan 2022.

PCG Sustainability Report. (2020). Available at https://www.petronas.com/pcg/sites/pcg/files/sustainability-reports/PCG-2020-Sustainability-Report.pdf. Accessed 30 Jan 2022.

Petronas Integrated Report. (2020). Available at https://www.petronas.com/integrated-report-2021/. Accessed 30 Jan 2022.

Secondary Sources

Adams, C. A., & Harte, G. F. (1998). The changing portrayal of the employment of women in British banks' and retail companies' corporate annual reports. *Accounting, Organizations and Society, 23*(8), 781–812. https://doi.org/10.1016/S0361-3682(98)00028-2

Adams, C. A., & McPhail, K. J. (2004). Reporting and the Politics of Difference: (Non)Disclosure on ethnic minorities. *Abacus, 40*(3), 405–435. https://doi.org/10.1111/j.1467-6281.2004.00164.x

Alam, K., Ali, M. A., Erdiaw-Kwasie, M. O., Murray, P. A., & Wiesner, R. (2022). Digital transformation among SMEs: Does gender matter? *Sustainability, 14*(1), 535.

Andrews, C., & deVault, D. (2009). Green niche market development. *Journal of Industrial Ecology, 13*, 326–345. https://doi.org/10.1111/j.1530-9290.2009.00112.x

Arfaoui, N., Le Bas, C., Vernier, M., & Vo, L. (2020). CE practices and innovative behaviors: Lessons from empirical research in the chemical industry. *Electronic Working Paper Series, ESDES-Research Center*. Available at https://www.researchgate.net/profile/Marie-France-Vernier/publication/348565376_Circular_Economy_Practices_and_Innovative_Behaviors_Lessons_froM_Empirical_Research_in_the_Chemical_Industry_Electronic_Working_Paper_Series/links/6005570392851c13fe1c5449/Circular-Economy-Practices-and-Innovative-Behaviors-Lessons-froM-Empirical-Research-in-the-Chemical-Industry-Electronic-Working-Paper-Series.pdf. Accessed 6 Feb 2022.

Atan, R., & Mohd-Razali, N. (2013). CSR reporting by government linked companies and their corporate attributes. Available at http://ajbasweb.com/old/ajbas/2013/March/163-171.pdf. Accessed 6 Feb 2022.

Barbiroli, G. (2011). Economic consequences of the transition process toward green and sustainable economies: Costs and advantages. *International Journal of Sustainable Development & World Ecology, 18*, 17–27. https://doi.org/10.1080/13504509.2011.541592

Basson, M., & Erdiaw-Kwasie, M. O. (2019). Entrepreneurship under siege in regional communities: Evidence from Moranbah in Queensland, Australia. *Journal of Rural Studies, 66*, 77–86.

Bebbington, J., & Larrinaga, C. (2014). Accounting and SD: An exploration. *Accounting, Organizations and Society, 39*(6), 395–413.

Biesenthal, C., & Wilden, R. (2014). Multi-level project governance: Trends and opportunities. *International Journal of Project Management, 32*(8), 1291–1308.

Blomsma, F., & Brennan, G. (2017). The emergence of CE: A new framing around prolonging resource productivity. *Journal of Industrial Ecology, 21*(3), 603–614. https://doi.org/10.1111/jiec.12603

Boghdadi, I. (2020). Circularity in the chemicals industry: Drivers, challenges and solutions. Available at https://www.cleantech.com/circularity-in-the-chemicals-industry-drivers-challenges-and-solutions/. Accessed 16 Mar 2022.

Bonnín-Roca, J., Vaishnav, P., Morgan, M. G., Mendonça, J., & Fuchs, E. (2017). When risks cannot be seem: Regulating uncertainty in emerging technologies. *Research Policy, 46*, 1215–1233. https://doi.org/10.1016/j.respol.2017.05.010

Bowling, N., Eschleman, K., & Wang, Q. (2010). A meta-analytic examination of the relationship between job satisfaction and subjective well-being. *Journal of Occupational and Organizational Psychology, 83*(4), 915–934.

Brundtland, G. H. (1987). Our common future—Call for action. *Environmental Conservation, 14*(4), 291–294.

Business and SD Commission. (2016). SDGs & sectors: A review of the business opportunities. Available at SDGs & Sectors: A review of the business opportunities. Accessed 20 Apr 2022.

Carrillo-Hermosilla, J., del Río, P., & Könnölä, T. (2010). Diversity of eco-innovations: Reflections from selected case studies. *Journal of Cleaner Production, 18*, 1073–1083. https://doi.org/10.1016/j.jclepro.2010.02.014

Carter, N. M., & Silva, C. (2010). Pipeline's broken promise. *Catalyst*. https://www.catalyst.org/wp-content/uploads/2019/01/Pipelines_Broken_Promise_Final_021710.pdf. Accessed 6 Feb 2022.

CEO Magazine. (2019). In his element. *CEO Magazine*. Available at https://www.theceomagazine.com/executive-interviews/manufacturing/mahadzir-rani/. Accessed 16 Mar 2022.

Chams, N., & García-Blandón, J. (2019). On the importance of sustainable human resource management for the adoption of SD goals. *Resources, Conservation and Recycling, 141*, 109–122. https://doi.org/10.1016/j.resconrec.2018.10.006

Cheng, M., & Edwards, D. (2019). A comparative automated content analysis approach on the review of the sharing economy discourse in tourism and hospitality. *Current Issues in Tourism, 22*(1), 35–49.

Chow, D. (2019, June 18). *World's population could swell to 10.9 billion by 2100*. U.N. report finds. https://www.nbcnews.com/mach/science/world-s-population-could-swell-10-9-billion-2100-u-ncna1017791. Accessed 20 Apr 2022.

Chung, N. (2020, August 6). *Make Petronas answer to Parliament, not just PM, says MP*. https://www.freemalaysiatoday.com/category/nation/2020/08/06/make-petronas-answer-to-parliament-not-just-pm-says-mp/. Accessed 20 Apr 2022.

Clayton, J. (2019, March 28). SEC chairman Jay Clayton's remarks to the SEC Investor Advisory Committee. Available at https://www.sec.gov/news/public-statement/clayton-remarks-investor-advisory-committee-032819. Accessed 4 Feb 2022.

Corder, G. D., Golev, A., & Giurco, D. (2015). "Wealth from metal waste": Translating global knowledge on industrial ecology to metal recycling in Australia. *Minerals Engineering, 76*, 2–9. https://doi.org/10.1016/j.mineng.2014.11.004

Cox, O. C. (1970). *Caste, class and race: A study in social dynamics*. Monthly Review Modern Reader Paperbacks.

Crofts, K., & Bisman, J. (2010). Interrogating accountability: An illustration of the use of Leximancer software for qualitative data analysis. *Qualitative Research in Accounting & Management, 7*(2), 180–207. https://doi.org/10.1108/11766091011050859

Daily, B. F., & Huang, S. (2001). Achieving sustainability through attention to human resource factors in environmental management. *International Journal*

of Operations & Production Management, 21(12), 1539–1552. https://doi.org/10.1108/01443570110410892
de Jesus, A., & Mendonça, S. (2018). Lost in transition? Drivers and barriers in the eco-innovation road to the CE. *Ecological Economics, 145*, 75–89. http://www.sciencedirect.com/science/article/pii/S0921800916316597
De los Rios, C., & Charnley, F. J. S. (2017). Skills and capabilities for a sustainable and CE: The changing role of design. *Journal of Cleaner Production, 160*, 109–122. https://doi.org/10.1016/j.jclepro.2016.10.130
Debrah, Y. A., Oseghale, R. O., & Adams, K. (2018). Human capital, innovation and international competitiveness in Sub-Saharan Africa. In I. Adeleye, M. Esposito (Eds.), *Africa's competitiveness in the global economy*. AIB Sub-Saharan Africa (SSA) Series. Palgrave Macmillan. https://doi.org/10.1007/978-3-319-67014-0_9
Department of Statistics Malaysia. (2022). *Labour Market Information*. Available at https://www.dosm.gov.my/v1/index.php?r=column/ctwoByCat&parent_id=436&menu_id=Tm8zcnRjdVRNWWlpWjRlbmtlaDk1UT09. Accessed 4 Feb 2022.
Diaz-Carrion, R., López-Fernández, M., & Romero-Fernandez, P. M. (2020). Sustainable human resource management and employee engagement: A holistic assessment instrument. *Corporate Social Responsibility and Environmental Management, 27*(4), 1749–1760.
Elf, P., Werner, A., & Black, S. (2022). Advancing the CE through dynamic capabilities and extended customer engagement: Insight from small sustainable fashion enterprises in the UK. *Business Strategy and Environment, 31*(6), 2682–2699.
Elia, V., Gnoni, M. G., & Tornese, F. (2017). Measuring CE strategies through index methods: A critical analysis. *Journal of Cleaner Production, 142*, 2741–2751. https://doi.org/10.1016/j.jclepro.2016.10.196
Ellen MacArthur Foundation. (2015). Circularity indicators: An approach to measuring circularity (methodology). *Government Information Quarterly, 23*(1). https://doi.org/10.1016/j.giq.2006.04.004
Elsevier. (2019). *Whitepaper*. Available at Top-5-solutions-circular-economy_whitepaper_CHEM-RD_WEB.pdf (elsevier.com). Accessed 14 Mar 2022.
Esa, E., & Mohd-Ghazali, N. A. (2012). Corporate social responsibility and corporate governance in Malaysian government-linked companies. *Corporate Governance, 12*(3), 292–305. https://doi.org/10.1108/14720701211234564
Fenwick, T., & Bierema, L. (2008). Corporate social responsibility: Issues for human resource development professionals. *International Journal of Training and Development, 12*(1), 24–35.
Fenwick, T., & Bierema, L. (2021). Corporate social responsibility: Issues for human resource development professionals. *International Journal of*

Training and Development, 12(1), 24–35. https://doi.org/10.1111/j.1468-2419.2007.00293.x

Ferronato, N., Rada, E. C., Portillo, M. A. G., Cioca, L. I., Ragazzi, M., & Torretta, V. (2019). Introduction of the circular economy within developing regions: A comparative analysis of advantages and opportunities for waste valorization. *Journal of Environmental Management, 230*, 366–378. https://doi.org/10.1016/j.jenvman.2018.09.095

Freiberg, D., Panella, K., Serafeim, G., & Zochowski, R. (2020). *Accounting for organizational employment impact* (Harvard Business School Accounting & Management Unit Working Paper, No. 21-050, pp. 1–86).

FTSE Russell. (2021, December 31). FTSE$Good Bursa Malaysia Index. *FTSE Russell Factsheet*. https://research.ftserussell.com/Analytics/FactSheets/temp/258fd884-fc6b-4247-8d7c-4be627d561ec.pdf. Accessed 14 Mar 2022.

Galvão, G. D. A., Nadae, J., Clemente, D. H., Chinen, G., & Carvalho, M. M. (2018). CE: Overview of barriers. *Procedia CIRP, 73*, 79–85. https://doi.org/10.1016/j.procir.2018.04.011

Geissdoerfer, M., Savaget, P., Bocken, N. M. P., & Hultink, E. J. (2017). The CE—A new sustainability paradigm? *Journal of Cleaner Production, 143*, 757–768. https://doi.org/10.1016/j.jclepro.2016.12.048

Geissdoerfer, M., Morioka, S. N., de Carvalho, M. M., & Evans, S. (2018). Business models and supply chains for the circular economy. *Journal of Cleaner Production, 190*, 712–721. https://doi.org/10.1016/j.jclepro.2018.04.159

Geissdoerfer, M., Pieroni, M. P. P., Pigosso, D. C. A., & Soufani, K. (2020). Circular business models: A review. *Journal of Cleaner Production, 277*, 123741. https://doi.org/10.1016/j.jclepro.2020.123741

Ghisellini, P., Cialani, C., & Ulgiati, S. (2016). A review on CE: The expected transition to a balanced interplay of environmental and economic systems. *Journal of Cleaner Production, 114*, 11–32. https://doi.org/10.1016/j.jclepro.2015.09.007

Geng, Y., & Doberstein, B. (2010). Developing the CE in China: Challenges and opportunities for achieving 'leapfrog development.' *International Journal of Sustainable Development & World Ecology, 15*(3), 231–239.

Geng, Y., Xinbei, W., Qinghua, Z., & Hengxin, Z. (2010). Regional initiatives on promoting cleaner production in China: A case of Liaoning. *Journal of Cleaner Production, 18*, 1502–1508. https://doi.org/10.1016/j.jclepro.2010.06.028

Gomez, E. T. (2019). *Patronage-based business as usual in 'New Malaysia'*. https://www.eastasiaforum.org/2019/05/08/patronage-based-business-as-usual-in-new-malaysia/. Accessed 5 Feb 2022.

Grosser, K., Adams, C., & Moon, J. (2008). *Equal opportunity for women in the workplace: A study of corporate disclosure*. Certified Accountants Educational Trust.

Guerci, M., Radaelli, G., Siletti, E., Cirella, S., & Shani, A. B. R. (2015). The impact of human resource management practices and corporate sustainability on organizational ethical climates: An employee perspective. *Journal of Business Ethics, 126*, 325–342. https://doi.org/10.1007/s10551-013-1946-1

International Labour Organization (ILO). (2020). *The enormous burden of poor working conditions*. Available at https://www.ilo.org/moscow/areas-of-work/occupational-safety-and-health/WCMS_249278/lang--en/index.htm. Accessed 4 Feb 2022.

International Monetary Fund (IMF). (2017). *Tackling inequality*. Available at https://www.imf.org/en/Publications/FM/Issues/2017/10/05/fis cal-monitor-october-2017. Accessed 5 Feb.

International Petroleum Industry Environmental Conservation Association (IPICEA). (2022). *About us*. Available at https://www.ipieca.org/about-us/ . Accessed 6 Feb 2022.

Jabbar, M. H., & Abid, M. (2014). GHRM: Motivating employees towards organizational environmental performance. *Management Research Report, 2*(4), 267–278.

Jabbour, C. J. C., & Santos, F. C. A. (2008). The central role of human resource management in the search for sustainable organizations. *The International Journal of Human Resource Management, 19*(12), 2133–2154. https://doi.org/10.1080/09585190802479389

Jamil, N. N. (n.d.). *GLCs transformation program, audit committee and financial reporting quality: Evidence from Malaysia*. Available at https://www.virtusint erpress.org/IMG/pdf/Nurul_Nazlia_Jamil_paper.pdf. Accessed 4 Feb 2022.

Jay, B. N. (2017, October 31). *GLCs and GLICs are not doing enough to have women on their respective boards-Irwan Serigar*, shorturl.at/vKLQ1.

Jomo, K. S. (2005). Malaysia's new economic policy and 'National Unity'. In Y. Bangura & R. Stavenhagen (Eds.), *Racism and public policy*. Palgrave Macmillan. https://doi.org/10.1057/9780230554986_8

Kalmykova, Y., Sadagopan, M., & Rosado, L. (2018). Measuring CE strategies through index methods: A critical analysis. *Journal of Cleaner Production, 142*, 2741–2751. https://doi.org/10.1016/j.jclepro.2016.10.196

Kaltenborn, B. P., Linnell, J. D. C., Gómez-Baggethun, E., Lindhjem, H., Thomassen, J., & Chan, K. M. (2017). Ecosystem services and cultural values as building blocks for 'the good life': A case study in the community of Røst, Lofoten Islands, Norway. *Ecological Economics, 140*, 166–176. https://doi.org/10.1016/j.ecolecon.2017.05.003

Karobi, E. M. (2017). *Rising to the challenge: Malaysia's contribution to the SDGs*. https://www.unsdsn.org/news/2017/08/30/rising-to-the-challenge-malaysias-contribution-to-the-sdgs. Accessed 30 Jan 2022.

Keijer, T., Bakker, V., & Slootweg, J. C. (2019). Circular chemistry to enable a CE. *Nature Chemistry, 11*, 190–195. https://doi.org/10.1038/s41557-019-0226-9

Khazanah. (2013). *Khazanah Nasional FAQ*. http://www.khazanah.com.my/faq.htm. Accessed 30 Jan 2022.

Khazanah. (2012). *GLCs make further strides, continue to play key role towards transforming Malaysia*. https://www.khazanah.com.my/news_press_releases/glcs-make-further-strides-continue-to-play-key-role-towards-transforming-malaysia/. Accessed 22 Feb 2022

Kirchherr, J., Reike, D., & Hekkert, M. (2017). Resources, conservation & recycling conceptualizing the CE: An analysis of 114 definitions. *Resources, 127*, 221–232. https://doi.org/10.1016/j.resconrec.2017.09.005

Kirchherr, J., Piscicelli, L., Bour, R., Kostense-Smit, E., Muller, J., Huibrechtse-Truijens, A., & Hekkert, M. (2018). Barriers to the CE: Evidence from the European Union (EU). *Ecological Economics, 150*, 264–272.

Klemash, S., Neil, B. M., & Smith, J. C. (2019). *How and why human capital disclosures are evolving?* Harvard Law School Forum on Corporate Governance. https://corpgov.law.harvard.edu/2019/11/15/how-and-why-human-capital-disclosures-are-evolving/. Accessed 22 Feb 2022.

Kolk, A., Rivera-Santos, M., & Rufín, C. (2018). Multinationals, international business, and poverty: A cross-disciplinary research overview and conceptual framework. *Journal of International Business Policy, 1*(1/2), 92–115. https://doi.org/10.1057/s42214-018-0004-1

Kotsantonis, S., & Serafeim, G. (2020). Human capital and the future of work: Implications for investors and ESG integration. *Journal of Financial Transformation*. Available at https://www.hks.harvard.edu/centers/mrcbg/programs/growthpolicy/human-capital-and-future-work-implications-investors-and-esg. Accessed 6 Feb 2022.

Krueger, P., Metzger, D., & Wu, J. (2021). *The sustainability wage gap* (Swedish House of Finance Research Paper, No. 20-14), European Corporate Governance Institute—Finance Working Paper 718/2020, Swiss Finance Institute Research Paper No. 21-17. Available at SSRN: https://ssrn.com/abstract=3672492 or http://dx.doi.org/https://doi.org/10.2139/ssrn.3672492

Kunapatarawong, R., & Martiniz-Ros, E. (2016). Towards green growth: How does green innovation affect employment? *Research Policy, 45*, 1218–1232. https://doi.org/10.1016/j.respol.2016.03.013

Leach, M., Reyers, B., Bai, X., Brondizio, E. S., Cook, C., Díaz, S., Espindola, G., Scobie, M., Stafford-Smith, M., & Subramanian, S. M. (2018). Equity and sustainability in the Anthropocene: A social-ecological systems perspective on their intertwined futures. *Global Sustainability, 1*, 1–13. https://doi.org/10.1017/sus.2018.12

Lee, H. (2015). *Political preference crowding out enterprise in Malaysia*. Available at https://www.eastasiaforum.org/2015/01/14/political-preference-crowding-out-enterprise/. Accessed 30 Jan 2022.

Leximancer Guide. (2021). *Leximancer user guide, release 5.0*. Leximancer Pvt. Ltd.

Lodhia, S. (2014). Towards a pragmatic social accounting. *Accounting Forum, 38*(4), 288–290.

Lucas, R. (1988). On the mechanics of economic development. *Journal of Monetary Economics, 22*, 3–42.

Lucas, R. E. (1990). Why doesn't capital flow from rich to poor countries? *American Economic Review: Papers and Proceedings*, 92–96.

Ma, S., Hu, S., Chen, D., & Zhu, B. (2015). A case study of a phosphorus chemical firm's application of resource efficiency and eco-efficiency in industrial metabolism under CE. *Journal of Cleaner Production, 87*, 839–849. https://doi.org/10.1016/j.jclepro.2014.10.059

Macrotrends. (2021). Malaysia unemployment rate 1991–2022. *MacroTrends*. Accessed 30 Jan 2022.

Marrucci, L., Daddi, T., & Iraldo, F. (2021). The contribution of green human resource management to the CE and performance of environmental certified organisations. *Journal of Cleaner Production, 319*, 128859. https://doi.org/10.1016/j.jclepro.2021.128859

Mathieux, F., Ardente, F., Bobba, S., Nuss, P., Andrea Blengini, G., Alves Dias, P., Blagoeva, D., et al. (2018). *Critical raw materials and the CE*. European Commission. Available at https://www.aclima.eus/wp-content/uploads/2018/01/Report.pdf. Accessed 15 Mar 2022.

Meys, R., Frick, F., Westhues, S., Sternberg, A., Klankermayer, J., & Bardow, A. (2020). Towards a CE for plastic packaging wastes—The environmental potential of chemical recycling. *Resources, Conservation and Recycling, 162*, 105010. https://doi.org/10.1016/j.resconrec.2020.105010

Miyamoto, K. (2013). *Human capital formation and foreign direct investment in developing countries, OECD* (Working Paper No. 211). Available at http://dx.doi.org/https://doi.org/10.2139/ssrn.668505. Accessed 6 Feb 2022.

MSCI. (2017). *Corporate governance in Malaysia*. https://www.arx.cfa/~/media/59E0F0BF61F64775B7BF74FE596A74C9.ashx. Accessed 30 Jan 2022.

Mokhtar, N. (2012). Environmental reporting practices of Malaysian government linked companies. *International Journal of Economics and Management, 6*(2), 241–277.

Nelson, S. P., & Jamil, N. N. (2012). *An investigation on the audit committee's effectiveness: The case for GLCs in Malaysia*. https://papers.ssrn.com/sol3/papers.cfm?abstract_id=2020184. Accessed 4 Feb 2022.

Ness, D. (2008). Sustainable urban infrastructure in China: Towards a factor 10 improvement in resource productivity through integrated infrastructure systems. *International Journal of Sustainable Development & World Ecology, 15*, 288–301. https://doi.org/10.3843/SusDev.15.4:2

Ngan, S. L., How, B. S., Teng, S. Y., Promentilla, M. A. B., Yatim, P., Er, A. C., & Lam, H. L. (2019). Prioritization of sustainability indicators for promoting the circular economy: The case of developing countries. *Renewable and Sustainable Energy Reviews, 11*, 314–331. https://doi.org/10.1016/j.rser.2019.05.001

Nilashi, M., Rupani, P. F., Rupani, M. M., Kamyab, H., Shao, W., Ahmadi, H., Rashid, T. A., & Aljojo, N. (2019). Measuring sustainability through ecological sustainability and human sustainability: A machine learning approach. *Journal of Cleaner Production, 240*, 118162. https://doi.org/10.1016/j.jclepro.2019.118162

OECD. (2019). *The economy of wellbeing*. Available at https://www.oecd.org/officialdocuments/publicdisplaydocumentpdf/?cote=SDD/DOC(2019)2&docLanguage=En. Accessed 30 Jan 2022.

Panella, K., & Serafeim, G. (2021). *Measuring employment impact: Applications and cases* (Harvard Business School Accounting & Management Unit Working Paper, No. 21-082, pp. 1–11).

Park, J. Y., & Chertow, M. R. (2014). Establishing and testing the "reuse potential" indicator for managing wastes as resources. *Journal of Environmental Management, 137*, 45–53. https://doi.org/10.1016/j.jenvman.2013.11.053

Panwar, R., & Niesten, E. (2022). Jump-starting, diffusing, and sustaining the CE. *Business Strategy and Environment, 31*(6), 2637–2640.

Pauliuk, S. (2018). Resources, conservation & recycling critical appraisal of the CE standard BS 8001: 2017 and a dashboard of quantitative system indicators for its implementation in organizations. *Resources, Conservation & Recycling, 129*, 81–92. https://doi.org/10.1016/j.resconrec.2017.10.019

Pellegrini, C., Rizzi, F., & Frey, M. (2018). The role of sustainable human resource practices in influencing employee behavior for corporate sustainability. *Business, Strategy and the Environment, 27*(8), 1221–1232. https://doi.org/10.1002/bse.2064

Pham, T. N., Hoang, H. T., & Phan, Q. (2019). Green human resource management: A comprehensive review and future research agenda. *International Journal of Manpower*. https://doi.org/10.1108/IJM-07-2019-0350

Prieto-Sandoval, V., Jaca, C., & Ormazabal, M. (2018). Towards a consensus on the CE. *Journal of Cleaner Production, 179*, 605–615.

Rajput, S., & Singh, S. P. (2019). Connecting CE and industry 4.0. *International Journal of Information Management, 49*, 98–113.

Repp, L., Hekkert, M., & Kirchherr, J. (2021). Circular economy-induced global employment shifts in apparel value chains: Job reduction in apparel production activities, job growth in reuse and recycling activities. *Resources, Conservation and Recycling, 171*, 105621. https://doi.org/10.1016/j.resconrec.2021.105621

Rizos, V., Behrens, A., Kafyeke, T., Hirschnitz-Garbers, M., & Ioannou, A. (2015). *The CE: Barriers and opportunities for SMEs*, No. 412, CEPS Working Document, Centre for European Policy Studies (CEPS).

Rouen, E., & Serafeim, G. (2021). Impact-weighted financial accounts: A paradigm shift. Available at https://www.hks.harvard.edu/centers/mrcbg/programs/growthpolicy/impact-weighted-financial-accounts-paradigm-shift-ethan-rosen. Accessed 6 Feb 2022.

Sachs, J. D., Schmidt-Traub, G., Mazzucato, M., Messner, D., Nakicenovic, N., & Rockström, J. (2019). Six transformations to achieve the SD goals. *Nature Sustainability, 2*, 805–814. https://doi.org/10.1038/s41893-019-0352-9

Said, J., & Jaafar, N. H. B. (2014). Accountability in government linked companies: An empirical finding. *Procedia Social and Behavioral Science, 145*, 294–299.

Salih, K. (2021, January 3). *The 'politicalisation' of 'GLICs and GLCs'*. Available at https://www.nst.com.my/opinion/columnists/2021/01/654093/politicalisation-glics-and-glcs. Accessed 3 Feb 2022.

Sauvé, S., Bernard, S., & Sloan, P. (2016). Environmental sciences, SD and CE: Alternative concepts for trans-disciplinary research. *Environmental Development, 17*, 48–56. https://doi.org/10.1016/j.envdev.2015.09.002

Schaltegger, S., Etxeberria, I. Á., & Ortas, E. (2017). Innovating corporate accounting and reporting for sustainability—Attributes and challenges. *Sustainable Development, 25*(2), 113–122.

Scholz, R. (2011). *Environmental literacy in science and society: From Knowledge to Decisions*. Cambridge University Press.

Schot, J., & Steinmueller, E. (2016). *Framing innovation policy for transformative change: Innovation policy 3.0* (Working paper series). Science Policy Research Unit (SPRU), University of Sussex.

Schot, J., & Kanger, L. (2016). *Deep transitions: Emergence, acceleration, stabilization and directionality* (SPRU Working paper series No. 2016-15). Science Policy Research Unit (SPRU), University of Sussex.

Schultz, T. W. (1961). Education and economic growth. In N. B. Henry (Ed.), *Social forces influencing American education*. University of Chicago Press.

SD. (2019). *High-level political forum on SD*. Available at https://sustainabledevelopment.un.org/content/documents/23844BN_SDG_8_Decent_work.pdf. Accessed 16 Mar 2022.

SDG Knowledge Platform. (2021). *Voluntary National Review 2021—Malaysia*. https://sustainabledevelopment.un.org/memberstates/malaysia. Accessed 30 Jan 2022.

SD Report. (2021a). *Malaysia*. https://dashboards.sdgindex.org/profiles/malaysia. Accessed 30 Jan 2022.

SD Report. (2021b). *SDG dashboards and trends—Malaysia*. https://dashboards.sdgindex.org/profiles/malaysia. Accessed 30 Jan 2022.

Serafeim, G. (2020, September–October). Social-impact efforts that create real value. *Harvard Business Review*, 2–12.

Serafeim, G., & Trinh, K. (2020). *A framework for product impact-weighted accounts* (Impact-Weighted Accounts Research Report, Harvard Business School Accounting & Management Unit Working Paper 20-076). Available at https://www.hbs.edu/impact-weighted-accounts/Documents/Preliminary-Framework-for-Product-Impact-Weighted-Accounts.pdf. Accessed 6 Feb 2022.

Serafeim, G., Rischbieth, A. M., & Koh, H. K. (2020). Sustainability, business, and health. *JAMA, 24*(2), 147–148. https://jamanetwork.com/journals/jama/fullarticle/2767585.

Sharma, R., & Gupta, N. (2015). Green HRM: An innovative approach to environmental sustainability. *Proceeding of the Twelfth AIMS International Conference on Management*, 2–5.

Sinkovics, N., Sinkovics, R. R., & Archie-Acheampong, J. (2020). The business responsibility matrix: A diagnostic tool to aid the design of better interventions for achieving the SDGs. *Multinational Business Review, 29*(1), 1–20.

Smith, A. E., & Humphreys, M. S. (2006). Evaluation of unsupervised semantic mapping of natural language with Leximancer concept mapping. *Behavior Research Methods, 38*, 262–279. https://doi.org/10.3758/BF03192778

Stahel, W. R. (2019a). The CE, roots and context. . In W. R. Stahel & E. MacArthur (Eds.), *The CE—A user's guide*. Routledge. https://doi.org/10.4324/9780429259203

Stahel, W. R. (2019b). Circularity, sustainability and labour in the circular industrial economy. In W. R. Stahel & E. MacArthur (Eds.), *The CE—A user's guide*. Routledge. https://doi.org/10.4324/9780429259203

Stahel, W. R. (2019c). The invisible liability loop, labour and the role of policy. In W. R. Stahel & E. MacArthur (Eds.), *The CE—A user's guide*. Routledge. https://doi.org/10.4324/9780429259203

Soundararajan, V., Jamali, D., & Spence, L. J. (2018). Small business social responsibility: A critical multilevel review, synthesis and research agenda. *International Journal of Management Reviews, 20*(4), 934–956. https://doi.org/10.1111/ijmr.12171

Suárez-Eiroa, B., Fernández, E., Méndez-Martínez, G., & Soto-Oñate, D. (2019). Operational principles of CE for SD: Linking theory and practice. *Journal of Cleaner Production, 214*, 952–961. https://doi.org/10.1016/j.jclepro.2018.12.271

Sustainable Employer Research Centre. (n.d.). *Malaysia top 100 sustainable employer.* https://www.100sdgcorp.com/. Accessed 30 Jan 2022.

Tariq, S., Jan, F. A., & Ahmad, M. S. (2016). Green employee empowerment: A systematic literature review on state-of-art in green human resource management. *Quality & Quantity, 50*, 237–269. https://doi.org/10.1007/s11135-014-0146-0

The Star. (2021, October 30). *Greater GLC participation in nation building.* https://www.thestar.com.my/business/business-news/2021/10/30/greater-glc-participation-in-nation-building. Accessed 30 Jan 2022.

The Sun Daily. (2021, August 11). *Petronas chemicals named Malaysia's top sustainable employer*. https://www.thesundaily.my/business/petronas-chemicals-named-malaysia-s-top-sustainable-employer-DN8539647. Accessed 30 Jan 2022.

Thomas, D. A. (2014). Searching for significance in unstructured data: Text mining with Leximancer. *European Educational Research Journal, 13*(2), 235–256. https://doi.org/10.2304/eerj.2014.13.2.235

Ting, I. W. K., & Lean, H. H. (2011). Capital structure of government-linked companies in Malaysia. *Asian Academy of Management Journal of Accounting and Finance, 7*(2), 137–156.

United Nation (UN). (n.d.). *Human Rights*. Available at https://www.un.org/en/global-issues/human-rights. Accessed 30 Jan 2022.

United Nations Department of Economic and Social Affairs (UNDESA). (2011). *World economic and social survey 2011—The great green technological transformation.*

Valverde, J. M., & Avilés-Palacios, C. (2021). CE as a Catalyst for progress towards the SD goals: A positive relationship between two self-sufficient variables. *Sustainability, 13*. https://doi.org/10.3390/su132212652

Vermunt, D. A., Negro, S. O., Verweij, P. A., Kuppens, D. V., & Hekkert, M. P. (2019). Exploring barriers to implementing different circular business models. *Journal of Cleaner Production, 222*, 891–902. https://doi.org/10.1016/j.jclepro.2019.03.052

Wade, B., Meath, C., & Griffiths, A. (2022). Capabilities for circularity: Overcoming challenges to turn waste into a resource. *Business Strategy and Environment, 31*(6), 2658–2681.

WageIndicator. (n.d.). *Salary checks—World wide wage comparison*. https://wageindicator.org/. Accessed 30 Jan 2022.

Watson, W. T. (2020). Human capital as an asset: An accounting framework to reset the value of talent in the new world of work. *World Economic Forum*. Available at https://www3.weforum.org/docs/WEF_NES_HR4.0_Accounting_2020.pdf. Accessed 30 Jan 2022.

Weiland, S., Hickmann, T., Lederer, M., Marquardt, J., & Schwindenhammer, S. (2021). The 2030 agenda for SD: Transformative change through the SD goals? *Politics and Governance, 9*(1), 90–95.

Whah, C. Y., & Guan, C. T. C. (2017). *Malaysia's protracted affirmative action policy and the evolution of the Bumiputera commercial and industrial community* (Economics working paper 2015–3, Yusuf Ishak Institute, pp. 336–373). Available at ISEAS Economics Working Paper 2015–03.pdf. Accessed 6 Feb 2022.

Wiek, A., Farioli, F., Fukushi, K., & Yarime, M. (2012). Sustainability science: Bridging the gap between science and society. *Sustainability Science, 7*(1), 1–4. https://doi.org/10.1007/s11625-011-0154-0

Williams, S. J., & Adams, C. A. (2013). Moral accounting? Employee Disclosures from a Stakeholder Accountability. *26*(3), 449–495. https://doi.org/10.1108/09513571311311892

Wilson, J. P. (2021). Making information measurement meaningful: The United Nations, SD goals and the social and human capital protocol. *Information, 12*(8). https://doi.org/10.3390/info12080338

Womack, J. P., Roos, D., & Jones, D. (1990). *The machine that changed the world*. Scribner.

World Commission on Environment and Development (WCED). (1987). *Our common future*.

World Economic Forum. (2017). *5 things to know about jobs and skills in 2017*. Available at https://www.weforum.org/agenda/2017/09/what-to-know%20about-work-jobs-human-capital-2017/. Accessed 22 Feb 2022.

World Economic Forum. (2020). *We need to get netter at measuring the value of human capital-here's how to do it*. Available at https://www.weforum.org/agenda/2020/10/why-we-need-a-new-human-capital-accounting-framework-and-how-to-get-there/. Accessed 30 Jan 2022.

Wright, P. M., Dunford, B. B., & Snell, S. A. (2001). Human resources and the resource based view of the firm. *Journal of Management, 27*(6), 701–721.

Zaric, S., & Babic, V. (2020). Creating human capital responsive to requests of SD goals: Some evidence from Serbia'. *KNOWLEDGE-International Journal, 41*(1), 17–21.

Zhu, Q., Geng, Y., Sarkis, J., & Lai, K. H. (2015). Barriers to promoting eco-industrial parks development in China. *Journal of Industrial Ecology, 19*, 457–467. https://doi.org/10.1111/jiec.12176

Zoogah, D. B. (2011). The dynamics of green HRM behaviors: A cognitive social information processing approach. *German Journal of Human Resource Management, 25*(2), 117–139.

Applying Circular Economy Thinking to Food Systems in the Hospitality Industry in Nigeria

Adunola Okupe

1 Introduction

Nigeria is an oil-rich dependent nation with an estimated population of about 206.1 million, accounting for 50% of the West African Nations. The country's population is projected to reach 400 million by 2050. 50% of the total population lives in urban centers. Nigeria's GDP as of 2020 is estimated to be $432 billion, which positions the country as the largest economy in Africa. However, the agricultural sector is not well organized and developed. As a result, the country relies on imported processed food and agricultural products estimated at $10 billion. As of September 2021, US food and agricultural exports to Nigeria reached $532 million—wheat importation accounted for 70% (Ebenezer, 2022).

Foodservice is a dominant segment of the hospitality industry representing a significant proportion of the economy. The COVID-19 pandemic highlighted certain risks within the food distribution system

A. Okupe (✉)
Lagos Business School, Lagos, Nigeria
e-mail: aokupe@lbs.edu.ng

M. O. Erdiaw-Kwasie and G. M. M. Alam (eds.), *Circular Economy Strategies and the UN Sustainable Development Goals*,
Sustainable Development Goals Series,
https://doi.org/10.1007/978-981-99-3083-8_12

that led to unprecedented challenges within the hospitality industry; these challenges caused a strain on the operations of the industry as well as a decline in revenue; other impacts include job losses and a reduction in hotel occupancy (Bello & Bello, 2021). In Nigeria, the Federal Government instituted a two weeks nationwide lockdown from the 30th of March 2020 to 13th of April 2020 in Lagos State—the economic capital of the country; Abuja—the Federal Capital Territory; and Ogun State. This was done to curb the spread of the virus, given the number of cases recorded within a short time in the country. The nationwide lockdown resulted in the shutdown of operations in several industries, including the Tourism Industry. The mandated movement restriction greatly impacted the hospitality sector food systems, hotel occupancy rate, etc. (Kimbu et al., 2021). Globally, 70% of the hotel staff were furloughed due to the low occupancy rate of hotel rooms (Alsetoohy et al., 2021). With the gradual off of activities in the industry, stakeholders have had to adapt to the new normal and consider crisis management strategies that can help mitigate some of the risks that may be presented in the future (Dayour et al., 2021). One of the avenues to reconsider is the exposure to the international supply chain for food and beverages and how to identify ways to rethink food systems and distribution in Nigeria's hospitality industry. This chapter starts with an understanding and analysis of the food systems in hospitality, food systems, and sustainability. The impact of the pandemic on the food distribution system in the economy and the role of the circular economy in the food distribution system were reviewed. Findings from the descriptive and exploratory research conducted in West Africa to understand the coping mechanisms adopted by players in the hospitality sector during the pandemic were documented. Recommendations to guide stakeholders in the industry on how the food systems can be redesigned by adopting the circular economy thinking to forestall a future breakdown of activities in the industry.

2 Food Systems in the Hospitality Industry

The Nigerian hospitality industry is undergoing significant development and change with the rising middle class and demand for world-class hospitality services. According to the W Hospitality Group 2022 report on hotel pipeline development in Africa, Nigeria is the third largest hotel development after the Egypt and Morocco market, with 255 and 344 rooms planned to be developed in Lagos and Abuja, respectively, over the next four (4) years. This figure refers to the reported development pipeline activity by international and regional hotel brands and does not include local development activity for hotels and guest houses.

The hospitality industry is a part of food distribution within the food system. Food systems involve all interlinked activities with different actors involved in the production, aggregation, processing, distribution, consumption, and disposal (loss or waste) of food products that come from activities of agriculture (including livestock), forestry, fisheries and food industries, alongside the broader economic, societal, and physical environments in which these activities are taking place (Nguyen, 2018). Furthermore, the food system encompasses the full value chain of all activities involved in food distribution, including human waste management and disposal (Ellen MacArthur Foundation, 2019).

As shown in Fig. 1, the food system involves several activities undertaken by different actors—from production to processing, retailing, and consuming, including storage and disposal, as demonstrated in the diagram. In reality, food systems involve multiple interacting value chains. These require a broad set of supporting services, including physical and market infrastructure, transport, financial services, information, and technology. Furthermore, the incentives and operating conditions for the actors are influenced by the *institutional environment* of policies, rules, and regulations (e.g., food safety and quality, financial, taxation, environment, etc.). Together these institutions create the formal and informal 'rules of the game' that govern how the food system functions.

Food processing in Nigeria is underdeveloped due to poor infrastructure (power, water, roads) and high operating costs. As a result, the country remains dependent on imports to meet the demand for quality processed foods. In 2018, food imports from Nigeria represented 10.9% of the total merchandise imports. Nigeria is a substantial net importer of agricultural products, with total imports of over USD 5.81 billion (N2.21 trillion) in 2018 (Flander Investment and Trade, 2020). Imports are dominated by bulk/intermediate commodities such as wheat, rice, sugar, frozen fish, dairy products, vegetable oil, and other medium and consumer-oriented products. The top partner countries from which Nigeria imports food products are Brazil, China, the United States, the United Kingdom, Indonesia, and South Africa. Nigeria's trade links with the United States remain strong, and South African firms are increasingly establishing a presence in the West African market. Imports from Asia, especially China, have grown markedly in recent years, and China's

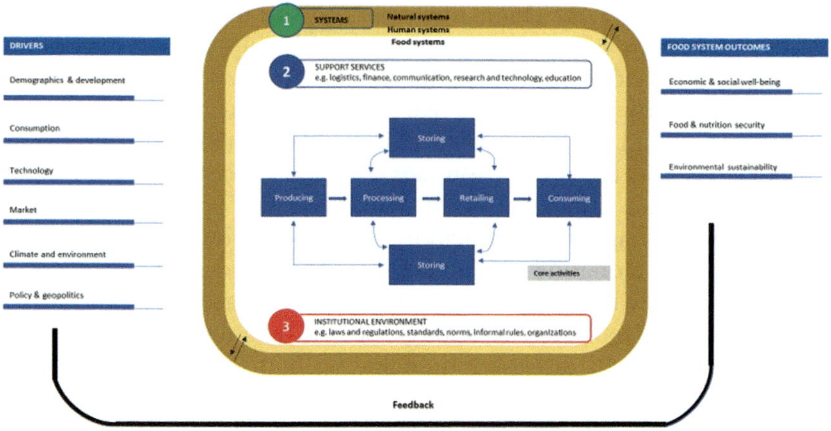

Fig. 1 Food system (*Source* Adapted from Foresight4Food Initiative, 2019)

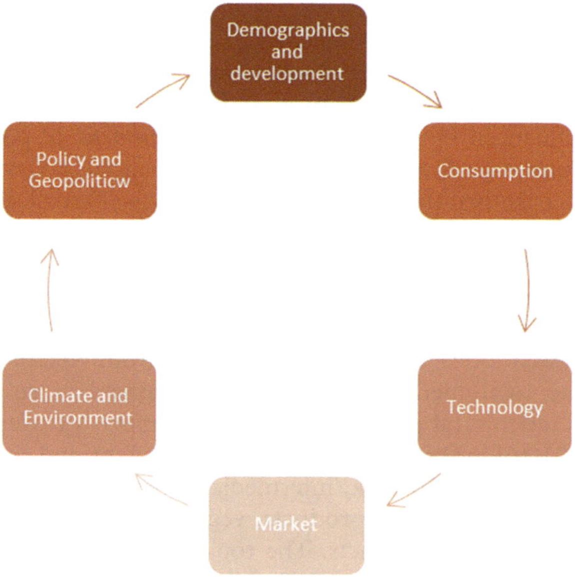

Fig. 2 Drivers of the food system

investment in all sectors of the economy has experienced rapid growth (Flander Investment and Trade, 2020).

The Nigerian hospitality industry is import-dependent, with most of the supplies for food and drinks imported from Brazil, China, the United States, the United Kingdom, and Ireland (World Integrated Trade Solution, 2019). According to Nigeria's Federal Ministry of Agriculture and Rural Development (FMARD), Nigeria spends an estimated US$5 billion on importing food and beverages annually (Vanguard, 2020). This observation is not unique to Nigeria and extends to other parts of the continent. Africa has more than 60% of the world's arable uncultivated land, yet the continent spends about US$50 billion annually on food imports, most of which could be produced in Africa (Gakpo, 2020). Activities surrounding food are increasingly being linked to the sustainability of the hospitality industry because it has far-reaching economic impacts on the actors within the industry. This is the need to consider sustainable approaches to food production and the role of the demand for locally produced/sourced food in addressing some of the sustainability challenges in the food system (Ellen MacArthur Foundation, 2019).

3 Food Systems and Sustainability

Food production is complex, with several actors working together to ensure adequate food provision. Actors include seed and fertilizer firms, storage, processing, large and small-scale retailers, and consumers (Szita & Toth, 2005). Nutrition has been identified to be one of the biggest problems that confront the global system, especially in developing countries. After Southern Asia, Africa is ranked to have the second largest share of the world's undernourished people (36% in 2019). However, models developed pre-pandemic showed that the number of food-insecure persons within the continent was expected to fall, as was the share of the region's food-insecure population (Rosen et al., 2016).

Global food production, distribution, and consumption patterns are highly unsustainable (Alam et al., 2021; Jurgilevich et al., 2016). In addition, about 30–50% of food meant for consumption is wasted within the food system (Stuart, 2009), implying a waste in productivity, energy, and natural resources as well as the cost of wasting food. The amount of food wasted globally varies across countries and usually depends on the size of income; level of economic usually depends on the size of income, level of economic growth, and urbanization (Chalak et al., 2016). Developed countries in Europe, North America, Oceanian countries, Japan, South Korea, and China account for 56% of global food waste (Lipinski et al., 2009). On the other hand, less developed countries account for 44% of global food waste during the post-harvest and processing stages (Ishangulyyev et al., 2019; Gustavsson et al., 2011). The cause for this is attributed to a series of issues ranging from technological limitations, labor and financial restrictions, and lack of adequate infrastructure for transportation and storage (Churchill et al., 2023; Gustavsson et al., 2011). As a result, the African food system experienced significant changes, including a reduction in poor nutrition and an improvement in nutrition outcomes as undernutrition decreased from 33% to around 23% between the years 1990 to 1992 and 2014 to 2016 (May & Mentz-Coetzee, 2021).

Morawicki and Gonzalez (2018) defined food sustainability as providing enough food for the human population now and for future generations. According to the United Nations Global Impact (2008), food sustainability is the idea that food processing and production is done in a way that does not waste natural resources and is available to future generation without being detrimental to the environment or health. Food sustainability is one of the key aspects of world food program in 2019 reports that 1 in 9 people in the world goes hungry every day, and a lot of people suffer from starvation and malnutrition. This is so rampant that the United Nations emphasized an intense change in the global food and agriculture system to tackle world hunger. The cost of food has been increasing over the years, with many central governments swinging into action to subsidize food prices. In 2008, food prices globally reached their highest level in 30 years, which was critical as the poverty rate also increased (UN Global Impact, 2008). These events brought about a catalytic drive toward discourse and action for food sustainability as a shift from this cause would result in global food and nutrition insecurity which transcends to the development of environmental, socio-economic human rights, developmental, security, and political challenges.

There are two elements in the sustainability context of food; the food system and food security are integrated principles that must be aligned to ensure food sustainability. This includes sustainability along the entire value supply within the food chain, which includes producers, processors, distributor, consumer chain, optimal usage, and replacement of resources and wastage, which are crucial principles to ensure a sustainable food system. Furthermore, access to food is one of the fundamental human flights of every human on Earth at all times, as stated by the 1966 International Universal Declaration of Human Rights. Therefore, an adequate food supply must be provided even during a crisis.

4 Impact of COVID-19 on the Food Distribution in the Nigerian Hospitality Industry

The pandemic revealed the fragility of the linear food system. It heightened the effects of food insecurity which were already being experienced in developing countries before the pandemic, with acute hunger and chronic malnutrition as a result of reduced income, natural hazards, political unrest, lack of investment in the agricultural sector, export restrictions, inadequate food supply to cater to the increased global demand, and worsened climate change (Workie et al., 2020). The pandemic affected the global food supply system, starting from the production phase, processing, transportation, and wholesaling, to the retail (distribution) and consumption phases (Food & Agriculture Organisation, 2020). In addition, it revealed the weaknesses in the food system brought about by the disruptions in the global and local supply chain due to lockdowns and the ensuing travel restrictions, which affect the processing and transportation of food crops, resulting in food crops wasting away on farms; and at the same time, people who lived in cities experienced food shortages (Bjornlund et al., 2022).

The halt in operations in the hospitality industry, which were typically large-scale purchases of food products, led to reduced demand for food products, especially imported food products. For example, there was a disruption in the need for fish from large-scale buyers such as restaurants, hotels, and even markets (Fisheries Committee for the West Central Gulf of Guinea, 2020). This disruption in the supply chain of seafood and the reduced demand for fish led to the rise in the price of fish and other aquatic food. Furthermore, the cost of food production and transportation increased, resulting in the rise in staple food prices in the market, leading to heightened food security (WorldFish, 2020). According to the Global Alliance for Improved Nutrition, developing countries such as Rwanda, Tanzania, India, Ethiopia, Indonesia, Kenya, and Mozambique recorded a hike in food prices with the respective percentages of 19.5%, 12.3%, 3.8%, 3.4%, 2.5%, 4.2%, and 10.5%, respectively (Global Alliance for Improved, Nutrition, 2020). In Nigeria, the prices of staple food items increased by about 80–120% (PriceWaterhouseCoopers, 2020). The devaluation of the Naira by 5.5% in 2020 and 7.6% in 2021 (Komolafe, 2020, Onu & Alake, 2021) also revealed the risks inherent in the import-dependent hospitality sector—the lack of access to the ingredients for items offered on menus for hotels, restaurants, and bars, in particular, perishable foods which are in some instances ordered several times a week and flown into the country.

Consumer demand for food changed with more awareness of and appreciation for locally grown food products from a wellness perspective (Alsetoohy et al., 2021), strengthening the case for a sustainable and localized food system.

Certain themes emerge that are to be considered, even as the hospitality industry restarts. Firstly, the exploration of how the hospitality industry in Nigeria can influence food demand and the strategies by which the industry can showcase and present locally grown food items for tourists and visitors. Secondly, how the industry can be an agent for sustainable food systems by redirecting the increase in demand for locally sourced food items toward sustainable food production, processing, and distribution.

5 Circular Economy and Food Distribution System in the Hospitality Industry

The circular economy has hitherto focused on the products side regarding manufactured goods. However, circular economy thinking can be applied to service like the hospitality industry. Food distribution is important as it requires goods (food) to be looped within the service provision context. In a circular food economy, food waste is eliminated by being redesigned into the food production system to ensure that the food waste as waste is a necessary ingredient for the further production of valuable, affordable, and healthy food which becomes accessible to everyone (The Conversation, 2021).

Circular economy thinking advocates that the food system needs to eliminate the linear *take-make-dispose* consumption pattern by adopting strategies that 'close' the loop to reduce waste and reuse waste usage of materials/products. It is design-led, which means that at the stage of the design of the intervention, the loop is closed, which is different from what is observed with sustainability efforts within the food system, which are primarily focused on reducing already produced waste. The hospitality industry is revealed to be one of the most energy-intensive industries due to its multi-usage function and high usage of disposable products, water, and energy. As such need to embrace the practice of circular economy for waste management by adopting strategies to explore the adoption of locally grown food items into their menus (Juliao et al., 2019).

The food system encompasses the entire food production value chain for human consumption. Hence, the system comprises activities from agricultural activities with subsections such as handling, transportation, storage, processing, distribution, consumption, waste management, and disposal (Ellen MacArthur Foundation, 2019). The current food system has been instrumental in growing the population, developing the economy, and increasing urbanization across the globe, yet the myriad of advantages the system brings also comes with its own cost. Therefore, the model must be more fit for a sustainable long practice.

The linear food production model, which extracts finite resources from the earth, is estimated to cost the world a total of 5.7 trillion dollars which is attributed to the way food is handled in the food value chain; hence, the model in itself is wasteful, polluting, and harmful to the natural environment. The drawback of the current linear food production model results in environmental, economic, and health degradation; thus, the model's impact on people is one of the leading causes of some diseases associated with man's health. Furthermore, the devastating impacts of the linear food model result in releasing harmful oxidants like ammonia, which can mix up with other harmful pollutants in the atmosphere, contaminating the environment (Pozzer et al., 2017). The associated risks involved in this model severely impact economies with their associated costs. Following the growth in urbanization, cities offer a unique pathway for transformation geared toward the circular path, as an estimated 80% of food is expected to be consumed in cities by 2050 (Ellen MacArthur Foundation, 2019). Urban food actors, through emerging trends and technological advancement, can influence this shift by getting more value for food consumed, thereby influencing what food is produced and how it is also produced.

In more recent times, consumers are shifting their preference to plant-based protein, a more regeneratively grown food; consumers are also paying more attention to where and how food is grown, which offers an opportunity for patronizing locally grown food items. Technological advancement is opening new possibilities for easy access to healthy meals by consumers.

The circular economy for the food system offers a pathway for designing out waste and pollution in the environment, keeping materials in use, and regenerating the natural system, which improves the food production process and offers an improved access to healthy and nutritious meals. The hospitality industry can reimagine the current food system by adopting the circular economy model to promote a much healthier, diverse, and resilient food system. This can only be achieved through collaboration with actors within the food value chain; hence, the industry can encourage local actors to source food grown regeneratively and locally, make the most of food, and encourage the sales of healthier food in the market.

Through the circular economy, local sourcing of food regeneratively can help the hospitality industry increase its resilience in food supply by creating partnerships with a diverse range of suppliers, hence managing the industry's volatility to shocks. It can also support the expansion of growing a variety of crops fitting for the local environment; the populace can also become encouraged to adopt healthier diets and reduce food waste. It also offers an opportunity to reduce excess packaging, thus reducing the number of actors in the food distribution chain (Barber, 2014).

By making the most of food, societies focus on reducing food waste and designing out the concept of food waste altogether. This entails the transformation of by-products into new reusable products. Hence, urban centers can serve as a hub for redesigning food waste instead of being just a final destination for food. A significant amount of food consumed globally is structured to suit the design of certain food brands, retailers, restaurants, and other food providers. These providers have significantly impacted man's food preferences and consumption patterns, particularly in cities. Hence, these brands can use their influence to educate people about adopting the regenerative food system. Additionally, in discouraging food waste, marketing efforts can be tilted toward positioning delicious food and products as easy and accessible choices.

The circular economy enhances the traditional sustainability approaches based on an eco-efficiency technique through the combination of economic advantages, reducing the costs of input, managing supply risks, and reducing associated externalities (Ellen MacArthur Foundation, 2013) and also ensures the propagation of a greener economy through the promotion of more efficient and ecological use of resources and innovative business models [Stahel, 2016].

6 Circular Economy in the Tourism and Hospitality Industry

A growing body of knowledge recognizes the importance of sustainability to the tourism and hospitality industry (Boley, 2011). While the concept of circular economy has taken cognizance across sectors like the agriculture and construction sector within the Chinese economy, there has been a limited application of the circular economy to the tourism and hospitality sectors (Ming, 2006). Similarly, Pattanaro and Gente (2017) reached the same conclusions, arguing that research work and academic debate about the circular economy within the tourism industry are quite scarce. The application of the circular economy to the tourism industry has inherent benefits present in its application largely because of the multiplier effects it has on an economy and its ability to boost circular flows between suppliers and customers. For example, Ming and Shu (2007) emphasized the application of circular economy in the tourism industry as its application can encourage the sustainable use of resources which could lead to the enhancement of the efficiency of the industry, thus enabling the industry to contribute more to the advancement of sustainable development. Similarly, Green Tourism (2016) opined that encouraging and applying circular economy tourism ensure that tourists/travelers assume a responsible approach during all periods of their stay, from when they prepare for a trip then through when they prepare for a trip to their local experience. In light of showing commitment to the circular economy within the tourism and hospitality industry, hotels can give priority to local, natural, recycled, recyclable, and seasonal products as an avenue of managing waste and maximizing recycled waste that is inherent in hotel operations (Jones & Wynn, 2019). Regarding using this in the hotel business, food systems can be bolstered, hence ensuring its circularity. However, developing the circular economy framework within activities of the tourism industry is quite intricate and hence entails the application of effective policies and advocacy geared toward the advancement of green consumption (Ming & Shu, 2007). Integrating elements of the circular economy into the business models of small and medium-sized enterprises can ensure an easy transition toward a circular economy within the tourism and hospitality sectors (Manniche et al., 2019). The Ellen MacArthur Foundation, Sun, and Mckinsey Center for Business and Environment (2015) emphasized that the concept of circular economy is preset on these principles—that is, preserving and improving natural capital, optimizing outputs from resources in use, and ensuring the system's efficiency by removing negative externalities. Observing these themes would adequately attend to the subject of sustainability and hence complement the need for companies to be sensitive enough to the needs of their stakeholders, as suggested by the stakeholder theory (Jones & Wynn, 2019). Beyond the theoretical understanding of the circular economy, Welford (2016) argued that policies and strategies are imperative in translating theories into practical reality. Tech is especially important in empowering leaders in the industry to make informed and intelligent decisions regarding circular economy and sustainability, which is the case for practical applications within the food system in Africa. Therefore, technology can be useful in addressing food distribution and kitchen resource sharing between restaurants and organizations (e.g., NGOs that feed the poor/those in need of surplus items, etc.)

7 Methodology

This was exploratory and descriptive research conducted in Francophone and Anglophone countries of West Africa for two months to understand hospitality managers' response and recovery options. The research adopted the quantitative data collection method with a sample size of 70. The study population includes hotel managers, food and beverage managers, chefs, hotel owners, and restaurant managers. In addition, self-administered questionnaires were shared online with respondents from the hospitality industry—bars, restaurants, hotels, and cafes.

8 Research Findings

Figure 3 reveals that the challenge in the supply of ingredients, the change in need of customers, and the difficulty in meeting customers' demand were the greatest impacts COVID-19 had on the industry. In addition, about 34% of respondents reported having changes and bottlenecks in the supply of raw materials due to the steps taken by governments to curb the spread of the pandemic, which included partial or total lockdowns in some cases. The lockdowns created bottlenecks in movement, which obstructed the supply of raw materials to businesses in the hospitality sector.

In addition, customers' demands changed due to precautions taken to curb the spread of COVID-19. This made it difficult for businesses in the hospitality sector to meet customer demands meet customer demands

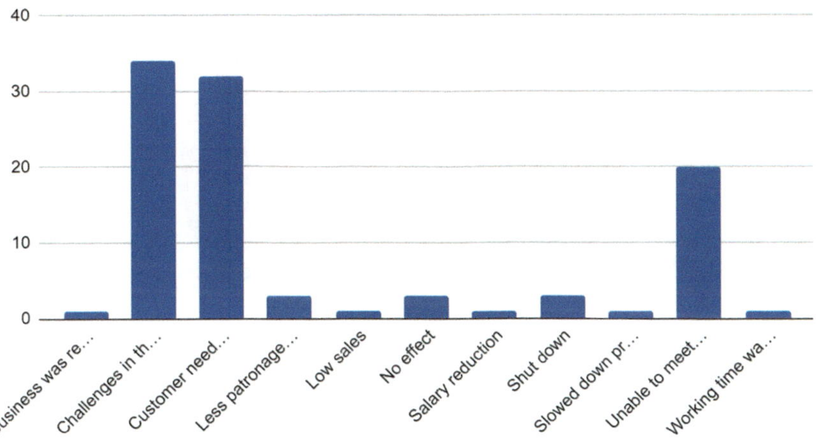

Fig. 3 Covid 19 impact on business

effectively. As a result, there was a deficit in the supply of food items on the menu as demanded by consumers. Furthermore, the pandemic heightened the weaknesses in the food system as there were disruptions in the global and local supply chain because of lockdowns and travel restrictions, which led to food crops wasting away on farms; the same was also the case for people who lived in cities who experienced food shortages.

Figure 4 reveals the pandemic's impact on the industry's service offerings. Approximately 4 out of every 10 businesses pivoted to take away menu; approximately 2 to 3 out of every 10 either closed down or cut down on the number of staff, while 1 out of every 10 had to change menu. Many businesses, especially restaurants in West Africa, pivoted to take away as a strategic means to keep business afloat during the lockdown. The change in the menu resulted from the restrictions on movement, and as such, there were difficulties in getting certain food items. Businesses closed down completely because they had to follow government directives in a bid to reduce the spread of the disease.

Figure 5 reveals that businesses sourced their ingredients from other local vendors, and the sector was still hugely affected because only about 3% of businesses generate their raw materials/ingredients in-house. The 71% that source from other local vendors is indirectly affected since some of those local vendors also make up the 11% of businesses that import ingredients/raw materials from other countries, which were

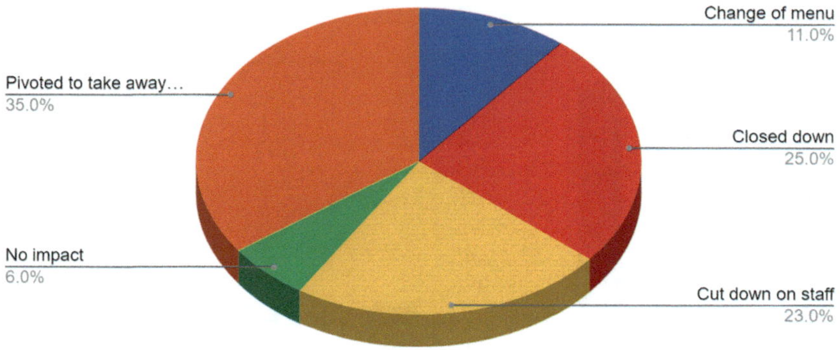

Fig. 4 Covid 19 impact on service offering

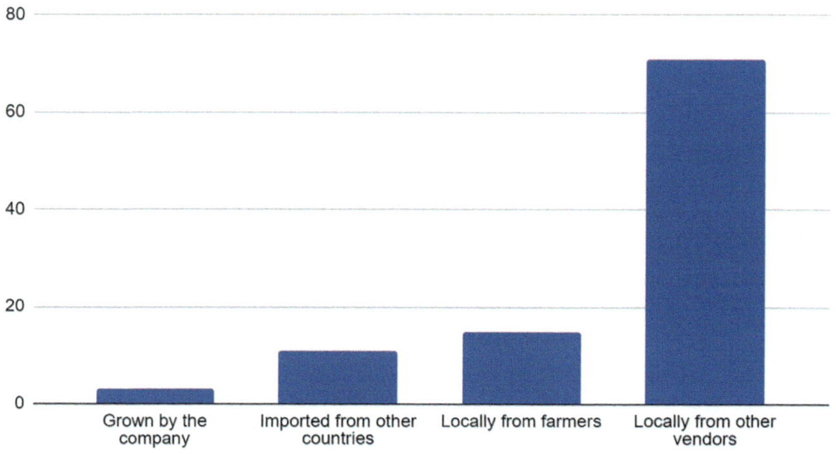

Fig. 5 Source of ingredients during the pandemic

greatly affected due to the global pause on trans-border trade during the period.

Figure 6 reveals that 91% of the respondents opined that they will consider including more locally sourced food items in their menu. The respondents adduced that there are lots of reasons guiding their choices, such as the fact that the preference of customers is subject to change which is the case now; also because including locally sourced food items

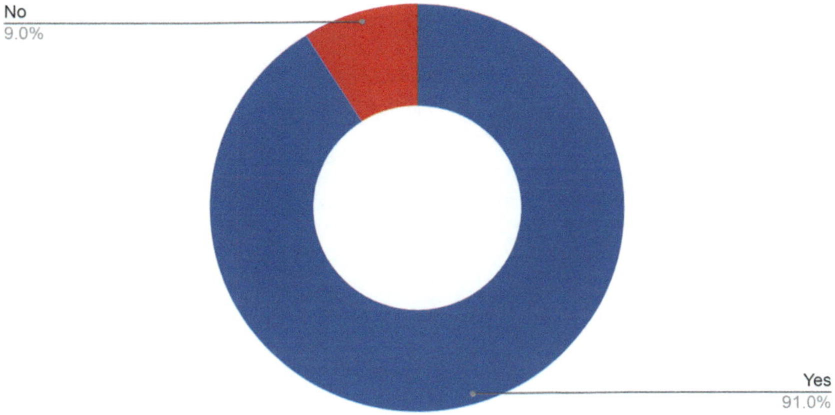

Fig. 6 Business desire to include locally sourced food items in menu

could help them strengthen local engagements because it also helps them cut costs which impact positively on their profits, of the respondents also argue that imported food items are highly inflated, and hence, there is a greater need to patronize local food items because over 70% of their customers are locals.

In addition, about 9% argue that they do not consider including locally sourced food items on their menu for the following reasons, because there was low patronage when other businesses adopted it. Some respondents of the 9% (hotels and restaurants) argue that their business is not interested in running a full-option restaurant because they are an international brand with a lot of patronage from customers who are expatriates, save cost, and have a positive impact on profit. This can be attributed to increased health consciousness and the fact that most customers are indigenes within the country.

Figure 7 reveals that 76% of the respondents responded that there is more demand for locally sourced food on the menu. This was especially true during the COVID-19 pandemic as more customers became conscious of their food with less demand for imported items. However, 24% of the respondents from the hospitality industry also argued that they are yet to see any demand for locally sourced food on their menu.

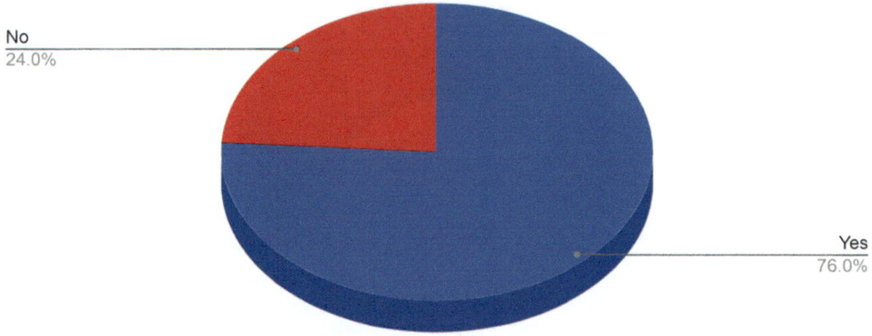

Fig. 7 Customer's demand for locally sourced food

9 Discussion of Findings

The findings from the research revealed the crisis created within the food distribution system of the industry. The crisis can mainly be attributed to the global and local supply chain disruptions because of lockdowns and travel restrictions. The findings of scholars can corroborate this. The pandemic had consequences on the seafood trade such that the industry witnessed a disruption in the demand for fish, particularly from buyers who are patronizers in large scale as well as restaurants, hotels, and even markets (Fisheries Committee for the West Central Gulf of Guinea, 2020). Historical events have shown the highly sensitive nature of the hospitality industry to negative external factors such as war, economic downturn, natural disasters, and wars. These catastrophes have severely impacted the hospitality industry's different sectors (Kim et al., 2005). Some of these events negatively impacted the sector, but none of those crises compared to the kind of impact COVID-19 had on the sector (Moreno et al., 2021). The spread of the coronavirus has altered the global scenario of how things are done (Moreno et al., 2021). The pandemic created crippling effects on many sectors of the economy as well as shocks but especially had a severe impact on the hospitality sector (Moreno et al., 2021; Sigala, 2020). A report from the United Nations World Tourism Organization (UNWTO, 2021) estimated that international tourist arrivals globally declined by 74%, impacting tourism revenue. The effects of the pandemic on the hospitality sector are unique, which necessitates the demand for research on crisis management and planning for the sector (Rivera, 2020).

A crisis management framework adopted by organizations incorporates elements of the 'capacity to prevent, contain, recover and learn' from different aspects of crises (Ghaderi, 2021; Kovoor-Misra, 1996). For example, Pearson and Mitroff (1999) proposed a five-stage crisis management framework: (1) Detection; (2) Preparation and Prevention; (3) Damage Limitation; (4) Recovery; (5) Learning. All five stages are very important for managing a crisis, but the most important is preparation and the ability to limit the damage, as this is necessary to prevent damages that may have consequential impacts. For example, during the surge of the pandemic in Nigeria, COVID-19 affected the hospitality industry so that it could not meet its customers' demand.

The adoption of the circular economy framework is imperative for a country like Nigeria due to the many benefits of its adoption. For Nigeria as a country, the model promotes the design of products and services that will reduce the depletion of natural resources and environmental degradation. Furthermore, the circular economy will help the country reduce its dependence on imported goods as well as stimulate economic growth. Deployment of the circular economy in the food system in Nigeria.

10 Concluding Remarks

This chapter contributes to the limited research gap on the circular economy in the tourism industry, especially in the hospitality sector. It further contributes to the emerging academic and practical discussions about the circular economy as an opportunity for food systems to incorporate locally sourced menus in hospitality operations.

A circular economy provides a framework in which society can create a cross-sectoral policy to support varied initiatives in different 'parts of the circle' to break away from the linear and extractive model to a more sustainable mode of production and consumption. Its concept has become popular among governments and businesses as a way to promote sustainability based on restorative and regenerative production and consumption systems, particularly in North America, Europe, and some Asian countries such as China and Japan, but opportunity exists for this in Africa too.

Transitioning to a circular economy will lead to economic development. Jobs will be created, and the citizens' standard of living will improve. There will be a reduction in the use of natural resources and thus its preservation for the next generation. As a result, we will have a safer environment and reduce the depletion of ozone layers. This is expected to increase life expectancy.

Circular economy sustainability solutions include supporting local food supply chains with less waste, closing nutrient loops, pricing the true cost of resource consumption and losses in natural capital, and creating policy mechanisms to promote recovery and reduce loss of critical raw materials in particular. In addition, a circular food economy through the hospitality sector provides opportunities for tourist experiences to be enhanced through food, trips, and food souvenirs. Digital technology can be harnessed to intensify this enhancement further—food is a strong way through which local culture can be experienced.

11 Actionable Recommendations

The hospitality industry can reimagine the current food system by adopting the circular economy model to promote a much healthier, diverse, and resilient food system. This can only be achieved through collaboration with actors within the food value chain; hence, the industry can encourage local actors to source food grown regeneratively and locally, make the most of food, and encourage the sales of healthier food in the market. The following are recommendations that industry players can consider:

- Engaging in supportive policy frameworks, innovations, financial instruments, and communications to encourage participation by a wider audience will suffice in creating the enabling conditions for a shift in the system.
- Companies play an important role in the circular economy because they are responsible for creating products and services. Public awareness is needed to sensitize people on the importance of the circular economy in hospitality—for food but also in an even wider/broader context—and there is the need to provide circular economy capacity building for the hospitality sector (including private and public sector agents).
- Adopting administration-focused practices, such as green certifications and CSR (corporate social responsibility), is also important contributory factor to positively drive and influence customers' decision since it reduces consumers' perceived uncertainty and risks with food products.

- Adopting green restaurants' strategy in the hospitality industry can promote recycling and composting, water and energy efficiency, and waste management, and it could as well offer the option of locally grown or organic food on the menu while also devoting efforts to the three Rs (reduce, reuse, and recycle) and the two Es (energy and efficiency).

References

Alam, G. M., Erdiaw-Kwasie, M. O., Nagy, G. J., & Leal Filho, W. (Eds.). (2021). *Climate vulnerability and resilience in the Global South: Human adaptations for sustainable futures*. Springer Nature.

Alsetoohy, O., Ayoun, B., & Abou-Kamar, M. (2021). COVID-19 pandemic is a wake-up call for sustainable local food supply chains: Evidence from green restaurants in the USA. *Sustainability, 13*, 9234. Available at: https://doi.org/10.3390/su13169234

Barber, D. (2014). *The third plate: Field notes on the future of food.*

Bello, Y. O., & Bello, M. B. (2021). Introduction to hospitality and tourism management. Ppabulun Global Limited.

Bjornlund, V., Bjornlund, H., & van Rooyen, A. (2022). Why food insecurity persists in sub-Saharan Africa: A review of existing evidence. *Food Security, 14*(4), 845–864.

Boley, B. B. (2011). Sustainability in tourism and hospitality education: Towards an integrated curriculum. *Journal of Tourism and Hospitality Education, 23*(4), 22–31.

Chalak, A., Abou-Daher, C., Chaaban, J., & Abiad, M. G. (2016). The global economic and regulatory determinants of household food waste generation: A cross-country analysis. *Waste Management, 48*, 418–422.

Churchill, S. A., Cai, Y., Erdiaw-Kwasie, M. O., & Pan, L. (2023). Financial development and tourism: A century of evidence from Germany. *Applied Economics, 55*(3), 305–318.

Dayour, F., Adongo, C., Amuquandoh, F. E., & Adam, I. (2021). Managing the COVID-19 crisis: Coping and post-recovery strategies for hospitality and tourism businesses in Ghana. *Journal of Hospitality and Tourism Insight, 4*(4), 373–392.

Ebenezer. (2022). *United States Department of Agricultural Report: Global agricultural information network.* NI2021-0012.

Edem, E. (2020). COVID 19: Prices of foodstuffs skyrocket in Cross River. *Daily Post.*

Ellen MacArthur Foundation. (2013). *Towards the circular economy 2: Opportunities for the consumer goods sector*. https://doi.org/10.1162/108819806775545321
Ellen MacArthur Foundation. (2015). *Growth within a circular economy vision for a competitive Europe*. Ellen MacArthur Foundation, Isle of Wight, UK.
Ellen MacArthur Foundation. (2019). *Cities and circular economy for food: An executive summary*.
Fisheries Committee for the West Central Gulf of Guinea. (2020). *Nigeria: Addressing impact of COVID-19 on fish*.
Flander Investment and Trade. (2020). *Food and beverage industry in Nigeria: Market overview*.
Food and Agriculture Organisation. (2020). Addressing the impacts of COVID-19 in food crises|April–December 2020: FAO's component of the global COVID-19 humanitarian response plan. https://doi.org/10.4060/ca8497en
Foresight4Food. (2019). *The dynamics of food systems—A conceptual model*.
Gakpo, J. O. (2020). COVID-19 reviving Africa's confidence in locally produced food. *Alliance for Science*.
Ghaderi, Z., King, B. E. M., & Hall, C. M. (2021). Crisis preparedness of hospitality managers: Evidence from Malaysia. *Journal of Hospitality and Tourism Insights, 5*(2), 292–310.
Gibson, A. (2020). Supply chain disruptions from COVID-19 will take years to rebalance, particularly for the food industry for food, implications stretch beyond when everyone else reaches a "new normal". Idaho District Export Council.
Global Alliance for Improved, Nutrition. (2020). Annual report 2020/2021. Access this file through https://www.gainhealth.org/sites/default/files/annual_reports/GAIN-Annual%20Report%2020-21_0.pdf
Green Tourism. (2016). *Green ideas for tourism in Europe*. Available at: www.greentourism.eu/
Gustavsson, J., Cederberg, C., Sonesson, U., Van-Otterdijk, R., & Meybeck, A. (2011). *Global food losses and food waste*. Food and Agriculture Organization.
Ishangulyyev, R., Kim, S., & Lee, S. H. (2019). Understanding food loss and waste—Why are we losing and wasting food? *Foods, 8*(8), 297.
Jones, P., & Wynn, M. G. (2019). The circular economy, natural capital, and resilience in tourism and hospitality. *International Journal of Contemporary Hospitality Management*. https://doi.org/10.1108/IJCHM-05-2018-0370
Juliao, J., Gaspar, M., & Tjahjonp, B. (2019). Exploring circular economy in the hospitality industry. In *Innovation, engineering and entrepreneurship* (pp. 953–960). Springer International.
Jurgilevich, A., Birge, T., Kentala-Lehtonen, J., Korhonen-Kurki, K., Pietikainen, J., Saikku, L., & Schosler, H. (2016). Transition towards circular economy in the food system. *Sustainability, 8*(1), 69.

Kimbu, A. N., Adam, I., Dayour, F., & De-Jong, A. (2021). COVID-19-induced redundancy and socio-psychological well-being of tourism employees: Implications for organizational recovery in a resource scarce context. *Journal of Travel Research, 62*(1), 55–74.

Kim, S. S., Chun, H., & Lee, H. (2005). The effects of SARS on the Korean hotel industry and measures to overcome the crisis: A case study of six Korean five-star hotels. *Asia Pacific Journal of Tourism Research, 10*, 369–377. https://doi.org/10.1080/10941660500363694

Komolafe, B. (2020). Exchange rate unification: CBN devalues naira to 380.5 in the official market. *Vanguard*.

Kovoor-Misra, S. (1996). Moving toward crisis preparedness: Factors that motivate organizations. *Technological Forecasting and Social Change, 53*(2), 169–183.

Lipinski, B., Hanson, C., Lomax, J., Kitinoja, L., Waite, R., & Searchinger, T. (2009). *Reducing food loss and waste* (World Resources Institute, UNEP working paper)

Manniche, J., Larsen, K. T., Broegaard, R. B., & Holland, E. (2019). *Destination: A circular tourism economy. A handbook for transitioning toward a circular economy within the tourism and hospitality sectors in the South Baltic region*. http://cirtoinno.eu/ce-handbook

May, J., & Mentz-Coetzee, M. (2021). Re-imagining resilient food systems in the post-COVID-19 era in Africa. *Sustainability, 13*(19), 10752.

Ming, Q. Z. (2006). *The new development conception of tourism circular economy and its systematic operation mode*.

Ming, Q. Z., & Shu, X. L. (2007). The research on the development idea and the running system of the circular economy on tourism. *Human Geography, 22*(3), 78–81.

Morawick, R. O., & Gonzalez, D. J. D. (2018). Food sustainability in the context of human behavior. *The Yale Journal of Biology and Medicine, 91*(2), 191.

Moreno, G. M., Morales, V. J. G., & Rojas, R. M. (2021). Going beyond the curve: Strategic measures to recover hotel activity in times of COVID-19. *International Journal of Hospitality Management, 96*, 102928.

Nguyen, H. (2018). *Sustainable food systems: Concept and framework*.

Onu, E., & Alake, T. (2021). Nigeria devalues naira as part of path to single exchange rate. *Al Jazeerra*

Pattanaro, G., & Gente, V. (2017). Circular economy and new ways of doing business in the tourism sector. *European Journal of Service Management, 21*(1), 45–50.

Pearson, C., & Mitroff, I. (1999). From crisis prone to crisis prepared: A framework for crisis management. *Academy of Management Executive, 7*(1), 48–59.

PriceWaterhouseCoopers. (2020). *Responding to the impact of COVID-19 on food security and agriculture in Nigeria*.

Pozzer, A., Tsimpidi, A. P., Karydis, V. A., de Meij, A., & Lelieveld, J. (2017). Impact of agricultural emission reductions on fine-particulate matter and public health. *Atmospheric Chemistry and Physics, 17*(20), 12813–12826.

Rivera, M. (2020). Hitting the reset button for hospitality research in times of crisis: Covid19 and beyond. *International Journal of Hospitality Management, 87*, 102528. https://doi.org/10.1016/j.ijhm.2020.102528

Rosen, S., Thome, K., & Meade, B. (2016). International Food Security Assessment, 2016–2026. *Economic Research, 2016*, 2026.

Sigala, M. (2020). Tourism and COVID-19: Impacts and implications for advancing and resetting industry and research. *Journal of Business Research, 117*, 312–321. https://doi.org/10.1016/j.jbusres.2020.06.015

Stahel, W. (2016). The circular economy. *Nature, 531*, 435–438. https://doi.org/10.1038/531435a

Stuart, T. (2009). *Waste: Uncovering the global food scandal*. Penguin Books.

Szita, K., & Toth. (2005). *Food systems and sustainability*.

The Conversation. (2021). *A circular food system can withstand crises like COVID-19—And provide delicious meals*.

The Nation. (2020). *COVID-19: Kwara fish farmers seeks exemption from lockdown*.

United Nations Global Impact. (2008). *Food sustainability: A guide to private sector action*.

UNWTO. (2021). *World tourism organization dashboard: "International Tourism and COVID-19"*. https://www.unwto.org/international-tourism-and-covid-19

Vanguard. (2020). *Nigeria spends $5bn on food import, $1.5bn on milk annually*.

Welford, R. (2016). *Corporate environmental management 3: Towards sustainable development*. Routledge.

W Hospitality Group Report. (2022). *2022 hotel chain development pipelines in Africa*.

WorldFish. (2020). *Fish and aquatic food systems COVID-19 updates: Nigeria*.

World Integrated Trade Solution. (2019). *Nigeria food products imports by country in US$ thousand 2019*.

World Tourism Organization, UNWTO. (2020). *UNWTO launches global guidelines to restart tourism*.

Workie, E., Mackolil, J., Nyika, J., & Ramadas, S. (2020). Deciphering the impact of COVID-19 pandemic on food security, agriculture, and livelihoods: A review of the evidence from developing countries. *Current Research in Environmental Sustainability, 2*, 100014.

Zhang, T., Chen, J., & Hu, B. (2019). Authenticity, quality, and loyalty: Local food and sustainable tourism experience. *Sustainability, 11*(12), 3437.

Circular Economy in Turkish Manufacturing Sector: The Roles of Green Manufacturing and Innovation

Nuran Çakır Yıldız, Zehra Binnur Avunduk, and Michael Odei Erdiaw-Kwasie

1 Introduction

Today, manufacturing systems are responsible for 35% of all global electricity use, generating 20% CO2 emissions and accounting for a quarter of all extractions of primary resources (World Economic Forum, 2018). To prevent environmental damage, companies need to transform their

N. Ç. Yıldız · Z. B. Avunduk (✉)
Business and Economics Discipline, Vocational School of Social Sciences, Istanbul University-Cerrahpaşa, Istanbul, Türkiye
e-mail: avundukz@istanbul.edu.tr; zehrabinnur.avunduk@iuc.edu.tr

N. Ç. Yıldız
e-mail: nurcak@iuc.edu.tr

M. O. Erdiaw-Kwasie
Business and Accounting Discipline, Faculty of Arts and Society, Charles Darwin University, Darwin, Northern Territory, Australia
e-mail: michael.erdiaw-kwasie@cdu.edu.au

M. O. Erdiaw-Kwasie and G. M. M. Alam (eds.), *Circular Economy Strategies and the UN Sustainable Development Goals*, Sustainable Development Goals Series,
https://doi.org/10.1007/978-981-99-3083-8_13

production systems circularly. This can be possible by developing creative and innovative methods in product/service, process, marketing, and management systems. Innovation is a critical success factor for transforming manufacturing techniques into greener methods. According to Florida (1996), adopting process innovations creates incentives for adopting environmentally conscious production strategies D4. Therefore, all innovation types (product/service, process, etc.) actualised by firms to realise sustainable production within the scope of a CE are more expressed explicitly with the concept of eco-innovation. For instance, a successful example of eco-innovation in the automobile industry is demonstrated by BMW's automobile recycling strategy in Germany. BMW has initiated a design for disassembly, which may result in the first 100% fully recyclable cars (Hart, 1995; Sezen & Çankaya, 2013).

Turkey, whose greenhouse gas emissions have increased by 130% in the last 30 years, may experience severe losses in its exports to European Union (EU) countries if it does not reduce its emission intensity (World Wide Fund for Nature) because "in December 2019, the EU announced the European Green Deal to create a climate-neutral continent by 2050 (Acar et al., 2022, p. 8162)." "A national circular economy roadmap has not been devised yet in Turkey to date; however, the *Green Deal Action Plan* announced in July 2021 by the Ministry of Industry and Trade has considered the implementation of circularity principles in Turkey. Furthermore, as part of Turkey's efforts to integrate into the EU, legal harmonisation has been undertaken to match the EU's current legislative frame on waste and circularity issues" (KPMG & SKD Türkiye, 2022, p. 6). According to the EU Commission's "Fit for 55" legislative package, a carbon emission limit value will be set for each product imported by the EU, and an additional customs duty of 55–60 Euros on average will be applied for each ton of excess carbon emission for products exceeding these limit values, environmentally friendly (Yılmaz, 2022). This situation may reshape the concept of competitive advantage and closely affect countries exporting to the EU, such as Turkey. This situation may reshape the concept of competitive advantage and closely affect countries exporting to the EU, such as Turkey. This is because low-emission countries can trade with the EU tax-free or with low taxes; additional carbon costs will arise for countries with high emission values, such as China and Turkey.

Emerging countries must contend with many challenges because increased economic activity causes a simultaneous upward movement in energy demand (especially fossil fuels such as oil, gas, coal, etc.) (Shan et al., 2021). Along with the industrial revolution, the intensive increase in production since the nineteenth century, the excessive and wasteful use of natural resources, the mixing of toxic chemicals and greenhouse gases due to production into nature, and the pollution/destruction of wastes have adversely affected the environment and resources (Mao & Wang, 2018). As a result, the earth is under threat of deterioration due to greenhouse gas emissions from the excessive use of fossil fuels (Alam et al., 2021; Ding et al., 2021) and the harmful methods used in the industry. Thus, enterprises face pressure to change production methods from linear to circular, reinforcing the need to evaluate their performance on sustainability issues (Erdiaw-Kwasie et al., 2023; Gupta et al., 2021). The CE is an economical approach that replaces the "end-of-life" concept by recycling, reducing, recovering, and reusing materials in manufacturing processes (Kirchherr et al., 2017). Therefore, CE can be used to minimise the usage of resources, decrease waste generation (Abreu & Ceglia, 2018; Bag et al., 2020; Kalmykova et al., 2018; Schroeder et al., 2019), and reuse wastes by processing and recycling instead of being garbage, etc., can be seen as most appropriate and a holistic way of how to transform the manufacturing methods towards a sustainable manner.

Integrated CE reduces waste, retains value for long periods, reduces the use of primary resources, and provides socio-economic benefits within closed loops (de Jesus & Mendonça, 2018; Hailemariam & Erdiaw-Kwasie, 2022; Kirchherr et al., 2018). Besides promoting sustainable development, CE decouples economic growth from resource depletion and environmental degradation (Babbitt et al., 2018; Hofmann, 2019; Murray et al., 2017). It is widely acknowledged that accelerated transitions from linear to circular paradigms are required (Haas et al., 2015) despite significant efforts and attention given to circularity. The global circularity rate is just 8.6% (Haas et al., 2015). Organisational adoption of CE practices is less well understood in emerging economies (Blomsma et al., 2022; Haas et al., 2015). Kirchherr and van Santen (2019) found that only 5% of articles on circular economies were written by authors from emerging and developing countries.

CE can be effectively categorised into three layers: micro, meso, and macro (Yuan et al., 2006). Enterprises act first by innovating at the micro-level or individual firm-level (Ruggieri et al., 2016), and CE mainly focuses on GM and eco-design strategies and applications. Second, the aim is to develop an eco-industrial network that benefits regional manufacturing systems and the environment at the meso-level. Finally, the macro-level comprises CE in regions, provinces, and cities (Sousa-Zomer et al., 2018; Yuan et al., 2006). This study examines CE in terms of the micro-(firm)level. To tolerate the damage to natural resources and to reduce environmental, economic, and social costs, it is urgently vital for firms to abandon production methods that harm nature and adopt GM (clean production) methods. GM is "the continuous application of an integrated preventive environmental strategy to processes, products, and services to increase overall efficiency and reduce risks to humans and the environment" (UNEP—UN Environment Programme). It includes a condition for achieving environmental improvements in new products/services, process development, and contribution to a more sustainable world (Glavič & Lukman, 2007). Firms are changing their processes from traditional methods to circular models to become sustainable.

Considering the significance that the use of renewable resources, waste recycling, redesign, eco-innovation, remanufacturing, and such production techniques will play in meeting future needs, it is surprising that so little work has been done on understanding the transformation of businesses applying GM and innovation to transition to a CE in emerging country settings. The discussion about CE is still highly shallow, and the implementation could be more (Franco, 2017; Lieder & Rashid, 2016). Thus, this paper aims to contribute to the literature from the perspective of an emerging country by exploring, understanding, and inferring the motivation behind the sustainable—GM practices of Arçelik, which is Turkey's largest white goods company and Europe's second-largest manufacturer of, according to the market share ranking based on quantity. As it is one of the leading firms in sustainability and has succeeded in being carbon neutral in global manufacturing with its carbon credits that is the reason we have chosen the case of Arçelik in this study. Furthermore, this study follows the Natural Resource-Based View (NRBV) theory as it offers a valuable lens for evaluating the context of the circular organisation.

Following the introduction section, the rest of the paper is structured as follows: in Sect. 2, the theoretical background of the field has been revealed by literature review; in Sect. 3, the methodology is described; and in Sect. 4, the research findings are given. Finally, Sect. 5 includes the conclusion of the research, implications, and discussion.

2 Literature Review

In the linear economic approach, starting from the production process planning, businesses continue their activities with a low-cost structure, high sales, and excessive profit motive, ignoring the fact that natural resources, which have a crucial place in production factors, are scarce and can be exhausted. On the other hand, consumers throw these products as garbage when the ability of the products to meet their needs is lost or when their intended use is over. Businesses and consumers who are not yet aware of the importance of sustainability do not question what happens to the products after they are thrown away. In this respect, the linear economy approach is called the "take-make-dispose" system. In contrast to the "take-make-dispose" linear model, a CE is regenerative by design and purpose to decouple growth from the consumption of limited resources (Ellen Macarthur Foundation, n.d.).

The focus and perspective shifted towards a circular economy model from 2005 onwards, mainly due to the realisation of the increasing future supply–demand mismatch and the limited nature of resources (Goyal et al., 2018). CE involves creating a closed-loop ecosystem for effectively utilising resources and conscious consumption. This implies reconfiguring the material flows from a resource-product-waste (linear) approach towards a resource-product-renewed resource (closed-loop) system. This new economic model to use natural resources efficiently creates a waste-free and resilient ecosystem, thanks to adopting the 3Rs principle (reducing, reusing, and recycling). Joshi et al. (2006) propose to fulfil the 6Rs concept (*recovering*, *reusing*, *recycling*, *redesigning*, *reducing*, *and remanufacturing*) to achieve the goal of sustainable production, which consists of six stages of material flow in a product life cycle, compared to the traditional 3Rs concept (Yan & Feng, 2014). Moreover, Kristensen and Mosgaard (2020) classified micro-level CE in nine dimensions beyond 6Rs according to the results of data-driven coding in their research. These are *recycling*, *remanufacturing*, *reuse*, *resource efficiency*, *disassembly*, *lifetime extension*, *waste management*, *end-of-life management*, and *multidimensional indicators*.

The CE approach has gained growing attention in business and academia (Geissdoerfer et al., 2017). However, analysis and discussions about CE development are mostly done from resource deficiency and environmental impact perspectives, leaving the economic benefits of industrial actors and the firm level missing (Lieder & Rashid, 2016).

Nevertheless, particularly at the micro-level, CE implementation is still limited (Ghisellini et al., 2016; Linder & Williander, 2017; Sousa-Zomer et al., 2018). This shortcoming in the CE research is crucial because fundamental activities for successful CE implementation, such as the application of GM, new product/service design, the transformation of processes, business models, usage of renewable energy, restructuring/greening of the supply chain, and choice of materials are in control and therefore finally determined by production firms with the underlying motivation of gaining economic benefits (Lieder & Rashid, 2016). Moreover, the micro-level requires multiple changes, such as the collaborative new business model development (Lieder & Rashid, 2016) and distinctive strategies to develop the circularity of the manufacturing systems (Winkler et al., 2011) (i.e., GM/cleaner production) in the CE adoption.

There is increasing attention being paid to CE adoption in scholarly discourse. Recently, several studies have highlighted the importance of speed in systemic transitions, despite the widespread applications of circular strategies (Blomsma & Brennan, 2017), as well as the rapid adoption of circular approach by many sectors (Bauwens et al., 2022; Haas et al., 2015; Mathivathanan et al., 2022). Therefore, CE research is now focused on creating a circular disruption (Droege et al., 2022; Neligan et al., 2022). The circular disruption process results in a systemic, widespread, and rapid shift from the destructive "take-make-use-dispose" model (Blomsma et al., 2023). It uses circular strategies to reduce resource consumption and structural waste (Bauwens et al., 2022). The effects of a circular rebound are reduced or negated by a circular disruption, as Zink and Geyer (2017) noted. Over the past few years, it has become increasingly apparent that CE adoption and disruption are hindered by barriers (de Jesus & Mendonça, 2018; Kirchherr et al., 2018). As a critical driver of eco-innovative practice in developed economies, organisational factors and CE adoption have been extensively theorised and discussed (Bag et al., 2020; Bossle et al., 2016; Yadav et al., 2020), but little empirical evidence exists for the manufacturing sector in emerging economies.

Eco-innovation has been recognised as a critical factor in developing competitive technologies and new business models that allow "environmental benefits," including greater efficiency in consumption and use of resources. Eco-innovation acts as a catalyst in the transition from a linear to a CE model (Pichlak & Szromek, 2022) and plays a central role in transforming the linear system into an economic system characterised by circular flows of resources (Doranova et al., 2016). Eco-innovation has also been a key economic growth driver (Sezen & Çankaya, 2013).

In the manufacturing system of the firm, GM is one of the essential strategies that can be considered preparatory towards CE (Bilitewski, 2012; Ghisellini et al., 2016). Pollution prevention and zero waste (Veleva et al., 2017), two of the CE's main goals, can only be achieved through GM principles (Bilitewski, 2012). Florida (1996) drew attention to the importance of zero emissions, which emphasises eliminating all environmentally damaging byproducts from the production process within the scope of GM. Florida also stated that the types of pollution prevention in production to process improvement are: *recycling*, *improved process technology*, *new product and process technology*, *closed-loop system*, *conversion to substitute fuels*, and *others*. However, firms' processes to implement a circular business model are still unclear at the micro-level (Urbinati et al., 2017). Jayal et al. (2010) propose adopting the 6Rs approach, i.e., recycle, reduce, reuse, redesign, remanufacture, and recover, which ensures diminished resource usage and energy in the production process. Although some CE implementation strategies involve GM, eco-innovation, and waste management (Ghisellini et al., 2016), further studies are needed to explore the extent to which these strategies are implemented and individual companies' journeys to the CE transition (Franco, 2017). This need creates a research gap in the field.

GM expresses a cumulation of activities in the production area oriented to decrease pollution, reduce the environmental impact of the production process, and adequately manage waste following legal and normative compliance with CE principles. It is strongly connected with greening the production processes, cleaner production, clean technologies, recycling, remanufacturing, and reusing waste, focusing on reducing greenhouse gases. Cleaner production is accounted for processes (saving raw materials and energy, reducing the use of toxic substances, emissions, and waste) and products/services (reducing impacts along the life-cycle) (Afum et al., 2020; Draghici & Ivascu, 2022; Huiling & Dan, 2020). Dheeraj and Vishal (2012) defined GM as adopting reliable, fast, and energy-efficient manufacturing processes and equipment towards reducing waste and enhancing productivity. GM also includes generating processes that utilise inputs with lower environmental impact, as well as processes that are profoundly productive with nearly zero waste and pollution (Ghazilla et al., 2015), perceived as highly efficient production systems and processes that utilise inputs that generate minimal/zero pollution and eliminate waste, thereby significantly reducing the negative impacts. To prevent contamination, some of the GM practices are; decreasing the usage of raw materials, energy, and solid waste, reusing products as well as renewable materials, recycling water and clean energy, redesigning processes and products, and employee training regarding product stewardship practices (Ghazilla et al., 2015). Also, GM provides several benefits: cost reduction, reduced carbon footprint, the efficiency of the number of raw materials, and improved production time, image, and brand (Draghici & Ivascu, 2022). Thus, the essential rationale is that GM provides sustainability by increasing productivity in CE.

3 Methodology

This research adopts the case study method to understand how businesses are transforming their manufacturing methods to become more sustainable and greener by adopting CE principles. The case study approach is defined as "a research strategy which focuses on understanding the dynamics present within single settings," and "case studies typically combine data collection methods such as archives, interviews, questionnaires, and observations" (Eisenhardt, 1989, p. 534). Qualitative research, in terms of its in-depth interviews, enables us to understand the organisation's fundamental issues, motivations, and intentions (Arda et al., 2017). We utilised a case study design to investigate a contemporary phenomenon about the limited extent of information over which the researchers had little control (Yin, 2009).

Researchers generally assume that case studies should merely rely on qualitative data. This is a misconception, as no empirical research methods use a single data type. Data in the case study method are collected by multiple means that may consist of potential qualitative data sources, such as interviews, direct observations, documentation, historical records, and quantitative data sources (e.g., surveys). Multiple data sources enhance data credibility (Patton, 1990). In addition, case study data often provide advantages in integrating perceptual and objective data. This integration is particularly significant in business research. Thus, using a variety of data, including the combination of subjective and more objective factual knowledge, may add much to our understanding of organisational processes and outcomes. Each data source is one piece of the puzzle, with each element contributing to understanding the phenomenon. This convergence adds strength to the results.

In the pre-interview stage, the authors prepared specific questions on sustainability, CE, GM, and innovation (please refer to the Appendix for the questionnaire). The questionnaire was developed in a 3-stage process. Firstly, a theory-based version of the questionnaire was prepared. This version was reviewed by three academicians and high and middle-level managers from different fields in the company. Secondly, the initial questionnaire was modified to accommodate the suggestions and comments of these specialists. The final version of the questionnaire, which contains the feedback from the second interview, was prepared at the last stage. Then, the managers were contacted and requested to meet with a cover letter and a list of interview questions. Next, a set of semi-structured interview questions was prepared as mentioned above to start a research discussion. Finally, primary data were collected through face-to-face interviews with open-ended questions gathering views and opinions from senior middle-level managers and specialists in three departments (marketing, R&D, and sustainability) in Arçelik Global. The work experience years of managers vary from R&D specialists who began to work recently to senior managers with more than 15 years of experience.

In-depth interviews were conducted with 12 experts through a focus group study and took approximately 3 hours. Interviews were conducted in Turkish, then summarised and translated into English. Besides the primary data, supplementary data was collected from secondary sources such as published reports, company websites, newspapers, trade magazines, and other media reports. In the case of study research, documents are typically used to corroborate and augment evidence from additional sources. According to Kotlar and De Massis (2013), these secondary data sources provided a richer context to understand firms' goal-setting. Finally, the qualitative research findings aim to theoretically evaluate Arçelik's strategic practices in terms of NRBV.

4 Research Background and Findings

4.1 Company Profile

Arçelik has been chosen as the sample in this study to examine sustainable manufacturing strategies from a developing country perspective. This single case study was selected because it permitted an in-depth and rich investigation of the study's phenomena compared to multiple case designs (Siggelkow, 2007; Yin, 2009). As an emerging market leader firm Arçelik, controlled by Koç Group, is the only Turkish conglomerate within the Fortune's Top 500 Global Companies list 2021 (Fortune Global 500 2021), and the company "had the highest score in the Household Durables category for the third year in a row in the Dow Jones Sustainability Index (based on the results dated November 2021)." Arçelik's corporate vision is "Respects the Globe, Respected Globally" as it passionately nurtures its global growth with GM, better utilisation of natural resources, and more sustainable business processes. As a result, it "was awarded a Gold Class Sustainability Award for the second time and recognised as an Industry Mover in the 2022 S&P Global Sustainability Yearbook" (S&P Global, 2022). In addition, Arçelik has become one of the first 45 firms in the world entitled to receive the Terra Carta Seal, presented by the UK's Prince of Wales to firms that have committed to struggle against climate change with specific targets within the scope of the Sustainable Markets Initiative. In addition, Arçelik is the only and first firm in the home appliances sector to receive this honour.

Arçelik has over 40000 employees worldwide in 51 countries, 28 production facilities in 9 countries, and 12 brands (Arçelik, Beko, Blomberg, Grundig, Arctic, ElektraBregenz, Defy, Flavel, Altus, Leisure, Dawlance, and Voltas Beko). Arçelik reached a consolidated turnover of more than EUR 6.5 billion in 2021. With 69.5% of its revenues coming from international markets, Arçelik is the R&D leader in Turkey—holding more than 3000 international patent applications to date through the efforts of more than 2000 researchers in 16 R&D and Design Centres in Turkey and 13 R&D Offices across nine countries. "According to the results published by the United Nations' World Intellectual Property Organization, Arçelik is the only Turkish company in the top 200 list of Companies with the Highest Number of International Patent Applications in the last ten years (Arçelik, 2020)."

4.2 *Findings of the Research: Arçelik Case Within the Scope of NRBV*

The qualitative data obtained from the interviews with Arçelik's managers and specialists are summarised below. In addition, by compiling the secondary data obtained from Arçelik's sustainability reports (2019, 2020, 2021), CE, GM, and innovation practices are examined within the scope of NBRV theory.

According to NRBV, there are three primary strategic capabilities; pollution prevention, product stewardship, and sustainable development. Decreasing emissions is the main aim of pollution prevention, whereas product stewardship guides the selection of raw materials and disciplines of product design intending to minimise the negative environmental impact of manufacturing systems. A sustainable development strategy, however, also dictates that efforts be made to sever the harmful links between the environment and economic activity in especially emerging countries (Hart, 1995).

5 Sustainable Development

The Arçelik Sustainability Board, chaired by the Deputy General Manager responsible for Finance and Financial Affairs (Chief Financial Officer-CFO), "was established to determine corporate sustainability and climate change policies and strategies and ensure their integration with company business processes and monitor sustainability performance." The Sustainability Directorate, affiliated with the Quality, Sustainability, and Government Relations (KSRI) Directorate, encourages the dissemination of the sustainability strategy to all business units within the firm as a business model. The department's duties include determining the long-term sustainability strategy of the firm, carrying out projects with the divisions within the firm, representing the firm in local and international organisations, and ensuring the continuity of its successful performance in the sustainability indexes. The Sustainability Board, which acts as the governing body of the Interfunctional Sustainability Working Groups, determines the strategy based on the reports of the Sustainability Working Groups. The Directorate consists of the following sub-units (working groups): Environmental Coordination, Energy Coordination, Green Chemistry Coordination, Sustainable Supply Chain, Occupational Health and Safety, and Human Rights Committee Working Group (Arçelik, n.d.-a).

Sustainable development vision is based on a triple pillar consisting of economic, environmental, and social components. The interviewees' statements are summarised, and their views on sustainability and CE issues are given below. The answer of each interviewee is shown by numbering with the abbreviation "I."

> Initially, the primary purpose is to follow the regulations and not fall behind the regulations…The main motivation here is to create a structure that determines the company's sustainability strategy and establish a relationship with the financial side. This department was established because of the need for a unit to consolidate with other units and create a different perspective. It has turned to how we can differentiate ourselves from the competition, and we greatly benefit from this right now. (I1)

Arçelik has 2030 targets. These are not only on the environmental side but also on the managerial side. The increase in the participation rate of women in the workforce and management can be given as an example.

> Increasing the rate of women on the board of directors from 17% to 25%, the number of all female employees in the company to 30%, and the number of women at the management level to a certain percentage and the proportion of women in turnover-generating positions. We have set some goals, such as increasing the proportion of women in science and technology positions. In other words, we are trying to pioneer the sector once again. (I2)

> We also have 100 female dealer projects to cover the entire supply chain. You know, the dealer ecosystem in Turkey is predominantly male. Unlike a male dealer's application, when a woman applies to become a dealer, she is first provided with a certain amount of initial investment support. In addition, certain pieces of training are provided by making some positive discrimination there, which cannot be done in a male dealership. In addition, training is given on which government incentive mechanisms can benefit from and receive grants and what she can do as a woman entrepreneur. (I3)

> At the same time, training is provided to women entrepreneurs to increase the number of women entrepreneurs in Anatolia with the Union of Chambers and Commodity Exchanges of Turkey. We also have a target of 500 female technicians. Here, too, we aim to increase the number of service technician women. We are a global company, a large structure with 28 factories in 9 countries, more than 40000 employees, our products and services are sold in 150 countries and so on. Therefore, we are planning to carry these and similar examples of good practices that we have done in Turkey abroad as another step. (I1)

It is expected that the methods changed with the GM mission will also affect the CE as an increase in social welfare by saving resources. The responses of the participants regarding this can be summarised as follows.

> We foresee that the steps taken towards a CE will also bring positive social effects. First, with the increasing population and the climate crisis, access to resources will become limited in many areas, especially food. At this point, providing resource efficiency with CE makes it possible to prevent social crises from occurring in geographies with limited opportunities in the world. In addition, it is calculated that the CE will create a significant amount of employment, especially in waste management. The World Bank projects global employment growth of 3% in the CE transformation by 2030. Reducing the manufacturing cost by recycling products and services will also contribute to social welfare. In addition, the CE as a system that excludes the negative effects of linear economic activities on human and natural systems positively impacts public health by reducing pollution and waste. (I4)

> Transitioning to a CE in the manufacturing sector requires significant changes in primary production and supply chain processes. According to the International Institute for Sustainable Development (ILO) report, employment growth in the manufacturing sector is expected by 0.5%, corresponding to approximately 4 million people... However, we can say that there will be changes in the skills and areas required for employees in the coming period. Furthermore, while a significant increase is expected, especially in repair, recycling, and remanufacturing businesses, a shrinkage is expected in areas working with fossil fuels such as coal and oil. (I5)

> The CE will increase the demand for and employment of skilled labour. It is also seen that the productivity of companies investing in CE innovation has increased. With the increase in the transition to the CE, companies will also need to transfer their investments to these areas. As Arçelik, we are working to create a resource efficiency-based circular business model in our sector by taking part in the EU Horizon 2020 project CSERVEES Project. With the transformation within the scope of the project we participated in with our washing machine and electronics businesses, our focus is on improving products according to the CE system and improving designs for recycling products that have expired. (I4)

> With LEOPET, which we have been working on since 2017, we have created a system in which recycled pet bottles are used as new raw material.

> Circularity is important for manufacturing companies to provide environmental and economic outputs. It will positively affect the business with positive economic outputs by reducing resource use, reducing waste, meeting customer and investor demand. (I5)

6 Pollution Prevention

Pollution prevention, "to meet the company's Near Zero Waste Target, Arçelik uses resources more efficiently, prevents and reduces waste resulting from the company's operations, and improves the effectiveness of separating waste as its source (Arçelik A.Ş., 2022, p. 9)." Arçelik also uses CE principles for designing their products' packaging volume and weight to produce minimum waste. They carry out recycling and reusing projects to reduce the environmental effects caused by packaging processes (Arçelik, n.d.-b). Arçelik develops products and services to prevent and reduce plastic pollution that harms the environment, health, and ecological balance. They manufacture white goods product parts from recycled plastics and washing machines that hold microplastics (Arçelik, 2021, p. 89). Arçelik is investing in R&D and developing new technologies to prevent plastics from their products from entering the natural resources. For example:

> Investing in R&D and developing new technologies to prevent plastics from our products from entering the sea...By 2030, improve water quality by eliminating dumping, reducing pollution, minimising the release of hazardous chemicals and materials, halving the proportion of untreated wastewater, and substantially increasing recycling and safe reuse globally. Prevent and significantly reduce marine pollution of all kinds, particularly from land-based activities, including marine debris and nutrient pollution. (I6)

As a signatory of the Business Plastics Initiative, the following are examples of Arçelik's 2023 interim targets (Arçelik, 2021, p. 21):

- Increasing the recycled plastic content in products from 3000 to 15,000 tons per year
- Increasing the recycled plastic parts made from waste fishing nets from 8 to 25 tons per year

- Increasing the recycled plastic parts made from waste PET bottles from 28 to 60 million per year
- Using plastics generated from their Waste Electrical and Electronic Equipment recycling plants in oven, dishwasher, and tumble dryer product categories
- Recycling 600 tons of packaging waste from their manufacturing plants to be used in plastic parts.

7 Product Stewardship

Product stewardship is an umbrella term for a "shared responsibility" approach to managing products at end-of-life, also called "Extended Product Responsibility" (Lewis, 2005). According to The US EPA, "a product-centered approach to environmental protection" calls on those in the product life cycle—manufacturers, suppliers, retailers, users, and disposers—to share responsibility for diminishing environmental impacts of products. It recognises that producers must take on new responsibilities to decrease the ecological footprint of their products. However, manufacturers cannot accomplish real change alone: consumers, retailers, and the existing waste management infrastructure must help provide the most cost-effective and workable solutions (US EPA, n.d.). Arçelik's product stewardship process is shown in Fig. 1.

In the context of product stewardship, the eco-circularity approach aims to eliminate, rather than manage, waste and to redesign the linear model into a circular model in which products and packaging are kept in a system by improving recyclability, efficiency, and repairability. Under the European Green Deal, the EU Commission adopted a new CE Action Plan, focusing on the design and production of a CE to ensure that the resources used are circulated in the economy as often as possible. The implications of the EU Green Deal and the CE Action Plan are significant for Arçelik as a company generating almost half of its revenues from sales in Europe. To proactively manage the transition processes and implications in line with the Action Plan, in 2020, Arçelik was involved in the World Economic Forum Circular Electronics Partnership, which is a pre-competitive industry platform that sets a vision and roadmap committing to a CE for electronics by 2030 with a network of top electronics companies, and group of pioneering global organisations. Along with the partnership, Arçelik was also involved in preparing an

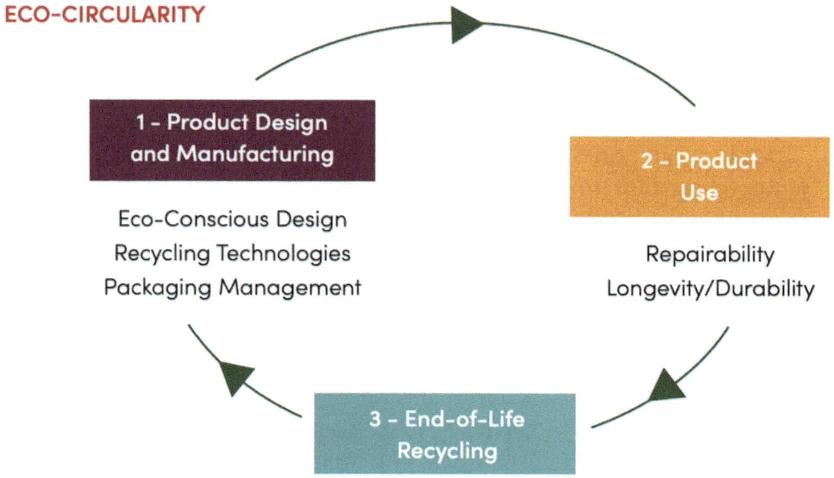

Fig. 1 Product stewardship process in Arçelik sustainability report (*Source* Arçelik [2020, p. 81])

action roadmap released in March 2021, which identifies six pathways to circularity. Furthermore, they actively participated in the working groups focused on these pathways: Circular Design, Take-Back and Collection, and, Recycling and Sourcing.

8 Pro-Activeness

> Since we report to international sustainability indices, some of the issues we constantly see in those indices force us to take us to different places. Because for example, we have been the leader in our category in the Dow Jones Sustainability Index for the last three years. For example, Electrolux was the leader for 13 years before us. Electrolux is a company in Europe where we normally compete very much in products. Especially since the issue has become critical in the retail channel; for example, our European customers are curious about what we do in the field of sustainability and how we differentiate from our competitors. Therefore, maintaining leadership in this and similar indices, acting before other companies, being more proactive, and communicating with our brands put us ahead of our competitors. Even my interaction with retail channel managers has increased a lot lately. Now they have started to talk about sustainability a

lot. Therefore, I can say that staying proactive has turned into a business continuity process. (I10)

Unlike our competitors, we have established a human rights department within the scope required by the sustainability indices. This is an independent department from the human resources department. We also have global business ethics principles. Within these principles and policies, we extended the legally paid maternity leave as a request of the sustainability indexes, unlike our industry. (I11)

9 Green Manufacturing

Examples of clean technologies applied by Arçelik within the scope of GM are given below.

Currently, our company aims to purchase 100% green electricity by 2030. We meet 69% of the electricity we consume from green sources. The share of energy-efficient products in the turnover of the products we sell is approximately 47%. The rate of turnover may decrease continuously because we are constantly growing. In the future, of course, we aim to increase this to at least 50%. In addition, Arçelik used green financing and issued a green financing bill of € 350 million last year. We allocated 189,000.000 € of the 350,000,000 € financing for cleaning production technologies. We carry out hundreds of energy efficiency projects in production every year. This year, I think we have realised more than 220 projects in this way. To reduce water consumption in production processes, we target to increase the use of recycled water to 70% by 2030. We also aim to reduce energy and water consumption in production by 45% per product. (I7)

We have used more than 140,000.000 recycled PET bottles in our products. How was this method discovered? A colleague from the R&D unit has had training. Our colleague learned in this training, that Glass fibre is used in Arçelik's washing machine tubs. There (s)/he learns that the formulation of this PET bottle with glass fibre can be used with a recycled PET bottle. (S)/he goes back to the company and tries it. Later, this became a patent exclusive to Arçelik, and this product, which started in the first washing machine, has now spread to air conditioners and dishwashers. We are expanding its use more and more. (I8)

> The Recycled PET Tub designed by Grundig for washing machines and washer-dryers is produced with recycled PET bottles and, thus, saves energy by reducing plastic waste. Furthermore, the Leopet plastic raw material, obtained from recycled PET bottles, developed by Arçelik's R&D Center and patented by Arçelik, is also used in air conditioners produced by Arçelik-LG. Of the plastic parts in all split air conditioners produced by Arçelik brands, 20% are made of recycled PET bottles, and up to 15 recycled PET bottles are used in each air conditioner. By the end of 2019, 300,000 recycled PET bottles were used in 30,000 air conditioners. (Arçelik, n.d.-c)

Arçelik innovates its processes and products to reach a closed CE model by increasing products' recyclability rates, using recycled material in products and adequately managing the end-of-life processes, including take-back-recycling. They increase product durability, repairability, reusability, and alternative end-of-life stages to contribute to a CE. Arçelik uses recycled raw materials while enhancing overall product recyclability in manufacturing. When products are sold to consumers, Arçelik supports longevity and repairability to minimise product environmental impact while preserving natural resources. Furthermore, the firm develops eco-innovative products by recycling natural resources with its pioneering R&D works. Arçelik CEO Hakan Bulgurlu gave examples of the eco-innovative products they produced: "Eco-friendly innovations with examples such as the washing machine with the world's first synthetic microfiber filtration system that captures up to 90% of the microfibers that would otherwise end up in the ocean, white goods produced with ghost fishing nets, washing machine, and washer dryer tubs and air conditioners manufactured with recycled PET bottles. Moreover, Arçelik took further upcycling efforts and produced small domestic appliances with biocomposite materials. For example, they used coffee grounds in manufacturing new coffee machines and tea fibers in Newline tea makers. These eco-product innovations will be released into the market in 2021" (Arçelik, 2021).

> Our holistic approach considers almost all the plastics in our products to maximise the recycled plastic content. We design recyclable products with reduced environmental impact and share product recyclability rates. The innovative solutions focus on becoming pioneers in the industry. Examples include using PET bottles, waste fishnets, bioplastics, eggshells, tomatoes, and coffee waste in products. Packaging wastes are also turned

> into valuable components to be used in products. The waste dishwasher components from Arçelik's own Waste Electrical and Electronic Equipment Recycling facilities are recycled to be used in Arçelik products again to close the loop in production. (I9)

Some examples of the value generated are provided below:

Using PET bottles in products: Approximately 58 million PET bottles have been recycled since 2017, when the project was first initiated, until the end of 2020. PET bottles are used in washing machines, dishwashers, and air conditioners. The studies are expanded into Koc Holding Group companies. The project was and is still the first in our industry. It was the first project that combined using waste and turning it into a valuable source. It proved that sustainability makes business sense by generating savings for the company and increasing brand value.

Using waste fishnets and textile wastes in products: Eight tons of recycled waste fishing nets and 111.7 tons of recycled industrial thread waste are used in the oven, washing machine, washer-dryer, and dishwasher parts.

"Fiber Catcher: First-of-its-kind microfiber filter integrated inside washing machines and dryers to filter up to 90% of the microplastics. The product will be released in different markets in 2021.

Products made from Organic Composites: The BioFridge uses biobased polyurethane insulation material (Bio-Cool) and biocomposite raw materials containing organic materials such as soy and castor oil. The egg trays are made of 20% eggshell waste and 80% bio-based plastics made from organic resources such as corn starch and sugarcane, and the fan cover is produced with 100% bio-based plastics. The door seal material is also partly made of soybean oil. As a result, the CO2 emission per unit produced is reduced by 6 kg, given biomaterials' 80% lower carbon footprint than petroleum-based traditional materials. The product will be released in different markets in 2021 (Arçelik, n.d.-d)."

10 Open Innovation

Developing collaborations with external stakeholders through a holistic approach enables firms to better address current social, economic, and environmental issues. In the scope of Arçelik's open innovation culture embedded in their business, they collaborate with various stakeholders such as universities, startups, suppliers, and companies across industries.

This creates diverse perspectives and ideas, especially in developing green innovations.

> Arçelik Garage leads design thinking and intrapreneurship mindset within the company, supporting the employees throughout the product development journey, from idea to prototype and product launch. (I12)

Arçelik considers sustainability a business model and emphasises the importance of acting with stakeholders. It leads change with its sustainability practices and continues its journey with a mission that shapes the future. Arçelik implements its sustainability strategy by setting goals in the fields of environmental, social, and governance. Arçelik's sustainable strategic priorities are Transition to Net Zero, Gender Equality and Diversity, Digital Transformation and Innovation, Circular Economy, Supporting Local Communities, Future Fit, Culture, Talent and Organisational Management, Water Management, Ensuring Healthy Lives and Well-Being for People Creating Value in, Supply Chain, Quality, Safety and Customer Management.

As a result, we summarised Arçelik's circular and GM strategies within the theoretical framework of NRBV in Table 1.

11 Discussion and Conclusion

This study contributes to understanding circular business models in Turkey. In recent years, all countries and organisations have developed policies to minimise the effects of global warming. However, if emerging economies such as Turkey cannot reduce emissions, they will not be able to export to EU countries, which may cause significant economic losses. For this reason, it has become necessary for developing economies to understand CE practices, especially at the micro-level.

The pioneering strategies and policies of the white goods giant Arçelik, which operates globally within the Koç Holding group, are discussed using a case study methodology. This paper will contribute to researchers, managers, and policymakers' understanding of how effective GM and innovation strategies are practised. By considering the NRBV of the firm, sustainability initiatives such as CE principles and GM practices are believed to create unique and long-lasting competitive advantages for companies. Arçelik, with its pioneering activities in its sector worldwide,

Table 1 Arçelik's CE and GM strategies within the scope of NRBV

Strategic capability	*Environmental driving force*	*Key resource*	*Competitive advantage*
Pollution Prevention	**Minimize emissions, effluents, & waste** • Total financial savings through waste reduction and water efficiency projects: EUR 1,593,000 • The rate of recycling and recovery in total waste was 96% in the manufacturing plants included in the reporting scope • The waste recycling rate increased to 98% in Turkey operations • The average hazardous waste per product was reduced by 30% in manufacturing in Turkey, Romania Gaesti, Russia, China, and South Africa operations, compared to 2012 • Approximately 1.3 million WEEE units were recycled between 2014 and 2020, a total of 326 GWh of energy was saved, nearly 160,000 tons of CO2 emissions were prevented, and around 6.5 million tons of water was saved	**Continuous improvement** In 2020, the performance management process continued to revolve around continuous improvement and feedback	**Lower costs** By using digital services such as ChatBot and robotic process automation more, productivity has been increased and costs have been reduced Eco-innovative Energy Factory Management System is an EU-funded Horizon 2020 project based on enhanced Life Cycle Assessment and Life Cycle Cost Analysis towards resource efficient manufacturing

(continued)

Table 1 (continued)

Strategic capability	*Environmental driving force*	*Key resource*	*Competitive advantage*
Product Stewardship	**Minimize life-cycle cost of products (in the context of CE):** • PET Tub: 28.2 million recycled PET bottles were used, adding up to 58 million in total since 2017. CO_2 emissions were reduced by 2200 tons and 19,000 GJ of energy was saved • EcoSustain: The proportion of recycled plastic raw material, Eco-Sustain, has reached 31% in Atak vacuum cleaners, 17% in Jaguar vacuum cleaners, and 64%–73% in toast & grills. A total of 190.4 tons of recycled plastics were used in these product categories • 8 tons of recycled waste fishing nets and 111.7 tons of industrial thread waste were transformed into high-performance recycled polyamide compounds which were then used in oven, washing machine and washer-dryer and dishwasher parts • BioFridge: The version with bio-based fan cover and bio-composite with eggshells was released into the market, and the efforts to use bioplastics in SDAs were accelerated. Arçelik used coffee grounds in coffee machines, and tea fibers in tea makers	**Stakeholder integration** Business partners Technology entrepreneurs Suppliers Customers Government, public institutions Local and global NGOs Intergovernmental organizations Universities	**Preempt competitors** Proactive strategy

Strategic capability	*Environmental driving force*	*Key resource*	*Competitive advantage*
Sustainable Development	**Minimize environmental burden of firm growth and development** • *Products that Create Environmental and Social Value:* e.g. 50% of turnover was from energy-efficient products. Compared to 2019, average energy consumption was reduced for tumble dryers in Turkey by 23%, refrigerators, dishwashers, and washing machines by 4%, 3%, and 1%, respectively • *Biodiversity:* e.g. A part of Arctic Ulmi factory is near Natura 2000 area named "Lacurile de pe Valea Ilfovului" for Birds Directive Special Protection Area (SPA), so, according to the National Air Quality Law 104/2011, Arçelik has to measure PM10, NO2 and SO2 quarterly. This is the only production site near to the protected areas and Key Biodiversity Areas where 431.369 m2 of border area is located between Arçelik's production site and the protected area • *Healthy Living:* e.g. Beko —Eat Like A Pro, in collaboration with UNICEF, reached 172,000 children across Latin America region through school-based interventions, and approximately 76.500 teachers were provided with training During COVID-19, an estimated total of 25 million school children, and 3.5 million people were reached through different channels	**Shared vision** • Digitalization and Consumer Experience • Customer Satisfaction and Communication • Open Innovation	**Future position/2030 Targets** ***Increasing resource efficiency in manufacturing*** • Establish renewable energy systems with 15 MW capacity • Purchase 100% green electricity in global manufacturing plants • Make an additional USD 50 million investment in renewable energy and energy efficiency • Reduce energy consumption per product by 45% • Reduce water withdrawal per product by 45% • Increase the water recycling ratio to 70% in all manufacturing plants • Reduce Scope 1–2 GHG emissions by 30% • Increase the waste recycling rate to 99% in global operations • Reach the target of 450 MW Arçelik-branded PV panel sales per year by 2025 **Improving product efficiency and eco-friendly product range** • Increase recycled plastic content to 40% • Increase bio-based material content to 5% • Reduce Scope 3 GHG emissions from use of sold products by 15% • Standardize Arçelik Green Chemistry Management System in products and production globally

(continued)

Table 1 (continued)

Strategic capability	*Environmental driving force*	*Key resource*	*Competitive advantage*
	• *Food Waste:* e.g. Grundig—Respect Food, having met its 2023 targeting advance, helped to create + 450,000 meals and prevented 230 tons of food from being wasted, increasing the total number of prevented food waste to + 500 tons • *Social Investment Programs:* e.g. Over EUR 5.8 million was spent globally for social investment and corporate citizenship activities. 500 Women Technicians Project was launched in Turkey to support women in the workforce • *Products that Create Environmental and Social Value:* e.g. Beko—HygieneShield Range that provides more than 99% virus reduction, including coronavirus, was released in seven key markets including Spain, Germany, the UK, Turkey, the Balkans, Romania, and Poland, also was released into the Turkish market with Arçelik-branded Ultra Hygiene Series • *Products for Disadvantaged Groups:* e.g. Product donations: White goods and small domestic appliances were donated to 550 + hospitals across 20 countries In Turkey, more than 5000 Arçelik- and Beko-branded products were delivered to 202 hospitals across 75 provinces		

Source This Table was created by compiling the Arçelik (2020) sustainability report and interview information

considers sustainability an indispensable business model and competitive strategy to ensure CE with the vision of “Respecting the World, Respected Worldwide.” Arçelik gains a global competitive advantage as it develops a proactive strategy in all aspects of sustainability. Adopting the principles required by the CE worldwide can contribute to the end of the climate crisis and increase development and welfare. Furthermore, Arçelik’s planning to implement its activities to eliminate gender inequality, which is one of its sustainability goals, not only for Turkey but also for countries such as Pakistan and Bangladesh can be considered as steps that contribute to the increase in the welfare level of underdeveloped countries.

Another critical practice of Arçelik is to reduce emission rates throughout the supply chain. In this way, Arçelik encourages its collaborating supplier firms, generally SMEs, to adopt cleaner production methods. Furthermore, Arçelik announced that it would stop working with suppliers not receiving ISO Environment and Energy quality management certificates. This can be considered a strategy that both increases the competition among SMEs and encourages eco-innovative product and process development of SMEs. Considering the cost–benefit analysis for CE, using recycled raw materials in manufacturing can sometimes turn into cost savings and cost increases. At this point, specific percentages of recycled materials can be tested in product design. Countries such as Spain and Italy impose a particular rate of cyclical materials on products. Those not fulfilling this criterion will be subject to tax, which may force companies financially. Before the new regulations, the company’s proactive strategy, using cyclic materials in its manufacturing processes and searching for innovative ways, will bring the business to the breaking point after a while.

Despite the progress made by the case company, certain barriers limit CE implementation and the need for managers’ attention. There is a paucity of technical and scientific knowledge of CE in many manufacturing companies, which makes it challenging for them to implement CE principles. According to managers, CE implementation is inhibited by process digitalisation and design challenge circular. By considering CE challenges in their planning, manufacturing industries can improve their processes in terms of eco-design, eco-innovation, remanufacturing, managing wastes, byproducts, and supply chains. In the manufacturing industry, integrated CE designs can lead to promising solutions, such as monitoring wastes and natural resources, revamping closed-loop supply chains into technological supply networks, and reducing carbon emissions. To implement the eco-design principles, industry experts can rethink their CE business models to balance environmental and economic behaviour. Furthermore, integrated CE allows companies to customise their products by providing extreme customer service by facilitating customer-product communication.

It is crucial to introduce specific incentive mechanisms to the government to companies that have electrical-electronic waste facilities, use high rates of recycled materials in production, or report world-renowned sustainability indices and engage in similar activities. A settlement system can be introduced for businesses that make R&D investments and develop eco-innovation products and processes within the scope of GM. In addition, sustainable financing is a critical issue for the development of SMEs. For the development of sustainable finance, it is crucial to introduce specific tax incentives, in cash or other discounts, both to the credit institutions that use this financing and to the companies that will receive it. Another issue is to raise awareness for consumers to buy green products. Instead of paying high prices for green products, consumers tend to buy the cheapest product that meets their needs by considering their current budget. Businesses and organisations must explain that using water and energy-saving, long-lasting, and higher-quality products creates economic and environmental benefits. Moreover, there are no specific incentives for energy-efficient product sales. For example, introducing a stimulus to the consumer, value-added tax reduction, having a particular incentive mechanism, or introducing green financing practices to the consumers to buy energy-efficient products can increase the sales of green products.

CE practices are significant for manufacturing companies to provide environmental and economic outputs. It will positively affect the companies by reducing resource use, lessening waste, and meeting customer and investor demands. In addition, sustainable and responsible production activities are likely to the positively affect corporate reputation and image. In future studies, the impact of CE activities on investors and consumers can be examined empirically by considering the dimensions of sustainability.

The study also indicates that CE innovation reduces a firm's costs by using fewer resources. When these cost savings are reflected in the prices of the products by the manufacturing firm, the demand for the products is likely to increase in the long run. On the other hand, since CE innovations require incurring high investment costs in the short run, price increases may be possible in the short run to cover these costs. This means that the firm's product demand decreases in the short run, hence its output and employment. It can be argued that U-shaped relationship that exists over time between CE innovations and firm-level employment. Therefore, CE will likely have an employment-enhancing effect in the long run.

As a result, CE and GM applications lead to lower-cost products in the long run; contribute to ensuring sustainability in economic, environmental, and social dimensions; and are most likely to have an employment-increasing effect of meeting the increase in demand for green products. In addition, CE innovations can make a company's products more attractive to the consumer (in terms of better recyclability, energy savings, etc.). Thus, the demand for the products manufactured by the firm and the employment level of the firm increase. But on the other hand, CE innovations increase the ecological value of the company's products, leading to a rise in consumers' willingness to pay. This, in turn, leads to increased demand for the firm's products and firm employment.

In the study, we explore and analyse GM and innovation to implement CE. According to the study, integrated CE strategies have the potential to enhance sustainability's triple bottom line by optimising waste, raw materials, and carbon emissions. Technology resources provide a great deal of data that can be used to make logistics and production decisions. Hence, the manufacturing industry can maximise revenue generation from post-usage products, byproducts, and production waste by recovering resources from different varieties of post-usage products, byproducts, and production waste. The significant barriers require automating the circular supply chain, semantic communication between machines and products, capturing information throughout the system, and improving production and consumption rates to enhance sustainability. Thus, manufacturing industries focus on improving the efficiency, precision, and accuracy of the sustainable supply chain by integrating end-to-end, as in the case of Arçelik.

Appendix: Questionnaire

Sustainable development
When and for what purpose did Arçelik begin its sustainability journey?
What are the main policies you have adopted within sustainability? Can you tell about your practices?
What are the possible social impacts of the transition to a CE?
What is to be expected regarding the social impacts if the CE transition continues rapidly?
If the manufacturing sector goes circular: What is the job impact?
Recent research (Repp et al., 2021) found significant job reductions in emerging economies in the textile industries when circularity increases. What is your opinion on that?
Product Stewardship (CE Practice—R&D, Innovation and GM)
Do you develop new products in which the materials from the collected old products are used by recycling?
Can you give examples of the circular material used products you have developed within the eco or green innovation scope?
What are the clean production technologies you use in your production system?
Is your in-house R&D department developing the cleaner production technologies you use? Are you developing those with your external stakeholders within open innovation?
"How do you apply the 6Rs (recovering, reusing, recycling, redesigning, reducing, and remanufacturing)" within the scope of CE in manufacturing and supply chain management? (How is the waste energy, wastewater, etc., generated during the manufacturing process treated and converted into a clean and reusable form?
Pro-activeness
Do you have a proactive nature as a company? In other words, can we say that you started sustainability attempts as a pioneer, not to follow something?
Pollution Prevention
What kind of activities do you do to prevent environmental pollution?
Do you have any projects you have developed on environmental issues to contribute to society?

References

Abreu, M. C. S. de, & Ceglia, D. (2018). On the implementation of a circular economy: The role of institutional capacity-building through industrial symbiosis. *Resources, Conservation and Recycling*, *138*, 99–109. https://doi.org/10.1016/J.RESCONREC.2018.07.001

Acar, S., Aşıcı, A. A., & Yeldan, A. E. (2022). Potential effects of the EU's carbon border adjustment mechanism on the Turkish economy. *Environment, Development, and Sustainability, 24*(6), 8162–8194. https://doi.org/10.1007/s10668-021-01779-1

Afum, E., Osei-Ahenkan, V. Y., Agyabeng-Mensah, Y., Amponsah Owusu, J., Kusi, L. Y., & Ankomah, J. (2020). Green manufacturing practices and sustainable performance among Ghanaian manufacturing SMEs: The explanatory link of green supply chain integration. *Management of Environmental Quality: An International Journal, 31*(6), 1457–1475. https://doi.org/10.1108/MEQ-01-2020-0019

Alam, G. M., Erdiaw-Kwasie, M. O., Nagy, G. J., & Leal Filho, W. (Eds.). (2021). *Climate vulnerability and resilience in the global south: Human adaptations for sustainable futures*. Springer Nature.

Arçelik. (2019). *Arçelik sustainability report*. Retrieved July 27, 2022, from, https://www.arcelikglobal.com/media/5833/arcelik_master_eng_online_1808.pdf

Arçelik. (2020). *Arçelik sustainability report*. https://www.arcelikglobal.com/media/6347/sustainability_report_2020.pdf

Arçelik. (2021). *Sustainability report*. Retrieved July 27, 2022, from, https://www.arcelikglobal.com/media/6938/arcelik21_sustainability_report.pdf

Arçelik A. Ş. (2022). *Yeşil Tahvil Alokasyon ve Etki Raporu*.

Arçelik A. Ş. (n.d.-a). *Supplier portal*. Retrieved July 27, 2022, from, https://supplier.arcelik.com/en/enviroment

Arçelik. (n.d.-b). *Sustainability management*. Retrieved July 26, 2022, from, https://www.arcelikglobal.com/tr/surdurulebilirlik/yaklasimimiz/surdurulebilirlik-yonetimi/

Arçelik. (n.d.-c). *Waste management*. Retrieved July 27, 2022, from, https://www.arcelikglobal.com/en/sustainability/in-touch-with-our-planet/waste-management/

Arçelik. (n.d.-d). *Recycled pet tub*. Retrieved July 27, 2022, from, https://www.arcelikglobal.com/en/technology/rd/recycled-pet-tub/

Arçelik. (n.d.-e). *Fighting plastic pollution*. Retrieved July 27, 2022, from, https://www.arcelikglobal.com/en/sustainability/in-touch-with-our-planet/fighting-plastic-pollution/

Arda, O. A., Delen, D., Tatoglu, E., & Zaim, S. (2017). An analytic approach to assessing organisational citizenship behavior. *Decision Support Systems, 103*, 9–23. https://doi.org/10.1016/J.DSS.2017.08.004

Babbitt, C. H., Gibson, K. E., Sellers, S., Brozovic, N., Saracino, A., Hayden, A., ... & Zellmer, S. (2018). The future of groundwater in California: Lessons in sustainable management from across the West.

Bag, S., Wood, L. C., Mangla, S. K., & Luthra, S. (2020). Procurement 4.0 and its implications on business process performance in a circular economy.

Resources, Conservation and Recycling, *152*, 104502. https://doi.org/10.1016/J.RESCONREC.2019.104502

Bauwens, T., Vaskelainen, T., & Frenken, K. (2022). Conceptualising institutional complexity in the upscaling of community enterprises: Lessons from renewable energy and carsharing. *Environmental Innovation and Societal Transitions, 42*, 138–151.

Bilitewski, B. (2012). The circular economy and its risks. *Waste Management, 1*(32), 1–2.

Blomsma, F., Bauwens, T., Weissbrod, I., & Kirchherr, J. (2023). The 'need for speed': Towards circular disruption—What it is, how to make it happen and how to know it's happening. *Business Strategy and the Environment, 32*(3), 1010–1031.

Blomsma, F., & Brennan, G. (2017). The emergence of circular economy: A new framing around prolonging resource productivity. *Journal of Industrial Ecology, 21*(3), 603–614.

Blomsma, F., Tennant, M., & Ozaki, R. (2022). Making sense of circular economy: Understanding the progression from idea to action. *Business Strategy and the Environment*.

Bossle, M. B., de Barcellos, M. D., Vieira, L. M., & Sauvée, L. (2016). The drivers for adoption of eco-innovation. *Journal of Cleaner Production, 113*, 861–872.

De Jesus, A., Antunes, P., Santos, R., & Mendonça, S. (2018). Eco-innovation in the transition to a circular economy: An analytical literature review. *Journal of Cleaner Production, 172*, 2999–3018.

De Jesus, A., & Mendonça, S. (2018). Lost in transition? Drivers and barriers in the eco-innovation road to the circular economy. *Ecological Economics, 145*, 75–89.

Dheeraj, N., & Vishal, N. (2012). An overview of green supply chain management in India. *Research Journal of Recent Sciences*, *1*(6), 77–82. www.isca.in

Ding, Q., Khattak, S. I., & Ahmad, M. (2021). Towards sustainable production and consumption: Assessing the impact of energy productivity and eco-innovation on consumption-based carbon dioxide emissions (CCO2) in G-7 nations. *Sustainable Production and Consumption*, *27*, 254–268. https://doi.org/10.1016/J.SPC.2020.11.004

Doranova, A., Roman, L., Bahn-Walkowiak, B., Wilts, H., O'brien, M., Giljum, S., Kong, M. A., & Hestin, M. (2016). *Policies and practices for eco-innovation up-take and circular economy transition EIO bi-annual report*. www.eco-innovation.eu

Draghici, A., & Ivascu, L. (2022). Green manufacturing in the context of circular economy. In *sustainability and innovation in manufacturing enterprises* (pp. 1–15). Springer.

Droege, H., Kirchherr, J., Raggi, A., & Ramos, T. B. (2022). Towards a circular disruption: On the pivotal role of circular economy policy entrepreneurs. *Business Strategy and the Environment*.

Economy Watch. (n.d.). *Emerging and developing economies economic structure*. Retrieved July 14, 2022, from, https://www.economywatch.com/emerging-and-developing-economies-economic-structure

Eisenhardt, K. M. (1989). Building theories from case study research. *Academy of Management Review, 14*(4), 532–550.

Ellen Macarthur Foundation. (n.d.). *The circular economy in detail*. Retrieved July 15, 2022, from, https://archive.ellenmacarthurfoundation.org/explore/the-circular-economy-in-detail

Erdiaw-Kwasie, M. O., Abunyewah, M., Yusif, S., & Erdiaw-Kwasie, A. (2023). Does circular economy knowledge matter in sustainable service provision? *A moderation analysis Journal of Cleaner Production*, *383*, 135429. https://doi.org/10.1016/j.jclepro.2022.135429

Florida, R. (1996). Lean and green: The move to environmentally conscious manufacturing. *California Management Review, 39*(1), 80–105. https://doi.org/10.2307/41165877

Fortune. (n.d.). *Fortune global 500 2021 | Full list of rankings* | Retrieved July 11, 2022, from, https://fortune.com/global500/

Franco, M. A. (2017). Circular economy at the micro level: A dynamic view of incumbents' struggles and challenges in the textile industry. *Journal of Cleaner Production, 168*, 833–845. https://doi.org/10.1016/J.JCLEPRO.2017.09.056

Gao, Y., Li, J., & Song, Y. (2009). Performance evaluation of green supply chain management based on membership conversion algorithm. *2009 Second ISECS International Colloquium on Computing, Communication, Control, and Management, CCCM, 3*, 237–240. https://doi.org/10.1109/CCCM.2009.5267895

Geissdoerfer, M., Savaget, P., Bocken, N. M., & Hultink, E. J. (2017). The circular economy–A new sustainability paradigm? *Journal of Cleaner Production, 143*, 757–768.

Ghazilla, R. A. R., Sakundarini, N., Abdul-Rashid, S. H., Ayub, N. S., Olugu, E. U., & Musa, S. N. (2015). Drivers and barriers analysis for green manufacturing practices in Malaysian SMEs: A preliminary findings. *Procedia CIRP, 26*, 658–663. https://doi.org/10.1016/J.PROCIR.2015.02.085

Ghisellini, P., Cialani, C., & Ulgiati, S. (2016). A review on circular economy: The expected transition to a balanced interplay of environmental and economic systems. *Journal of Cleaner Production, 114*, 11–32. https://doi.org/10.1016/J.JCLEPRO.2015.09.007

Glavič, P., & Lukman, R. (2007). Review of sustainability terms and their definitions. *Journal of Cleaner Production, 15*(18), 1875–1885. https://doi.org/10.1016/J.JCLEPRO.2006.12.006

Goyal, S., Esposito, M., & Kapoor, A. (2018). Circular economy business models in developing economies: Lessons from India on reduce, recycle, and reuse paradigms. *Thunderbird International Business Review, 60*(5), 729–740. https://doi.org/10.1002/TIE.21883

Gupta, H., Kumar, A., & Wasan, P. (2021). Industry 4.0, cleaner production and circular economy: An integrative framework for evaluating ethical and sustainable business performance of manufacturing organisations. *Journal of Cleaner Production, 295*, 126253. https://doi.org/10.1016/J.JCLEPRO.2021.126253

Haas, W., Krausmann, F., Wiedenhofer, D., & Heinz, M. (2015). How circular is the global economy? An assessment of material flows, waste production, and recycling in the European Union and the world in 2005. *Journal of Indian Ecology, 19*(5), 765–777.

Hailemariam, A., & Erdiaw-Kwasie, M. O. (2022). Towards a circular economy: Implications for emission reduction and environmental sustainability. *Business Strategy and the Environment*.

Hart, S. L. (1995). A natural-resource-based view of the firm. *The Academy of Management Review, 20*(4), 986. https://doi.org/10.2307/258963

Hofmann, F. (2019). Circular business models: Business approach as driver or obstructer of sustainability transitions? *Journal of Cleaner Production, 224*, 361–374.

Huiling, L., & Dan, L. (2020). Value chain reconstruction and sustainable development of green manufacturing industry. *Sustainable Computing: Informatics and Systems, 28*, 100418. https://doi.org/10.1016/J.SUSCOM.2020.100418

Jayal, A. D., Badurdeen, F., Dillon, O. W., Jr., & Jawahir, I. S. (2010). Sustainable manufacturing: Modeling and optimisation challenges at the product, process and system levels. *CIRP Journal of Manufacturing Science and Technology, 2*(3), 144–152.

Joshi, K., Venkatachalam, A., & Jawahir, IS. (2006). A new methodology for transforming 3R concept into 6R concept for improved product sustainability. *Proceedings of the IV Global Conference on Sustainable Product Development and Life Cycle Engineering*.

Kalmykova, Y., Sadagopan, M., & Rosado, L. (2018). Circular economy—From review of theories and practices to development of implementation tools. *Resources, Conservation and Recycling, 135*, 190–201. https://doi.org/10.1016/J.RESCONREC.2017.10.034

Kirchherr, J. W., & Van Santen, R. (2019). Research on the circular economy: A critique of the field. *Resources, Conservation and Recycling, 151*.

Kirchherr, J., Piscicelli, L., Bour, R., Kostense-Smit, E., Muller, J., Huibrechtse-Truijens, A., & Hekkert, M. (2018). Barriers to the circular economy: Evidence from the European Union (EU). *Ecological Economics, 150*, 264–272.

Kirchherr, J., Reike, D., & Hekkert, M. (2017). Conceptualizing the circular economy: An analysis of 114 definitions. *Resources, Conservation and Recycling, 127*, 221–232. https://doi.org/10.1016/J.RESCONREC.2017.09.005

Kotlar, J., & De Massis, A. (2013). Goal setting in family firms: Goal diversity, social interactions, and collective commitment to family–centered goals. *Entrepreneurship Theory and Practice, 37*(6), 1263–1288.

KPMG, & SKD Türkiye. (2022). *Preliminary research on circularity potential of five sectors in Turkey.*

Kristensen, H. S., & Mosgaard, M. A. (2020). A review of micro level indicators for a circular economy—Moving away from the three dimensions of sustainability? *Journal of Cleaner Production, 243*, 118531. https://doi.org/10.1016/J.JCLEPRO.2019.118531

Lewis, H. (2005). Defining product stewardship and sustainability in the Australian packaging industry. *Environmental Science & Policy, 8*(1), 45–55. https://doi.org/10.1016/J.ENVSCI.2004.09.002

Lieder, M., & Rashid, A. (2016). Towards circular economy implementation: A comprehensive review in context of manufacturing industry. *Journal of Cleaner Production, 115*, 36–51. https://doi.org/10.1016/J.JCLEPRO.2015.12.042

Linder, M., & Williander, M. (2017). Circular business model innovation: Inherent uncertainties. *Business Strategy and the Environment, 26*(2), 182–196. https://doi.org/10.1002/BSE.1906

Mao, Y., & Wang, J. (2018). Is green manufacturing expensive? Empirical evidence from China. *International Journal of Production Research, 57*(23), 7235–7247. https://doi.org/10.1080/00207543.2018.1480842

Mathivathanan, D., Mathiyazhagan, K., Khorana, S., Rana, N. P., & Arora, B. (2022). Drivers of circular economy for small and medium enterprises: Case study on the Indian state of Tamil Nadu. *Journal of Business Research, 149*, 997–1015.

Murray, A., Skene, K., & Haynes, K. (2017). The circular economy: An interdisciplinary exploration of the concept and application in a global context. *Journal of business ethics, 140*, 369–380.

Neligan, A., Baumgartner, R. J., Geissdoerfer, M., & Schöggl, J. P. (2022). Circular disruption: Digitalisation as a driver of circular economy business models. *Business Strategy and the Environment.*

Patton, M. Q. (1990). *Qualitative evaluation and research methods.* Sage.

Pichlak, M., & Szromek, A. R. (2022). Linking eco-innovation and circular economy—A conceptual approach. *Journal of Open Innovation: Technology, Market, and Complexity, 8*(3), 121. https://doi.org/10.3390/JOITMC8030121

Repp, L., Hekkert, M., & Kirchherr, J. (2021). Circular economy-induced global employment shifts in apparel value chains: Job reduction in apparel production activities, job growth in reuse and recycling activities. *Resources, Conservation and Recycling, 171*, 105621.

Ruggieri, A., Braccini, A. M., Poponi, S., & Mosconi, E. M. (2016). A meta-model of inter-organisational cooperation for the transition to a circular economy. *Sustainability, 8*(11), 1153. https://doi.org/10.3390/SU8111153

S&P Global. (2022). *The sustainability yearbook, rankings.* Retrieved July 11, 2022, from, https://www.spglobal.com/esg/csa/yearbook/2022/ranking/

Schroeder, P., Anggraeni, K., & Weber, U. (2019). The relevance of circular economy practices to the sustainable development goals. *Journal of Industrial Ecology, 23*(1), 77–95. https://doi.org/10.1111/JIEC.12732

Sezen, B., & Çankaya, S. Y. (2013). Effects of green manufacturing and eco-innovation on sustainability performance. *Procedia—Social and Behavioral Sciences, 99*, 154–163. https://doi.org/10.1016/J.SBSPRO.2013.10.481

Shan, S., Genç, S. Y., Kamran, H. W., & Dinca, G. (2021). Role of green technology innovation and renewable energy in carbon neutrality: A sustainable investigation from Turkey. *Journal of Environmental Management, 294*, 113004. https://doi.org/10.1016/J.JENVMAN.2021.113004

Siggelkow, N. (2007). Persuasion with case studies. *Academy of Management Journal, 50*, 20–24.

Sousa-Zomer, T. T., Magalhães, L., Zancul, E., Campos, L. M. S., & Cauchick-Miguel, P. A. (2018). Cleaner production as an antecedent for circular economy paradigm shift at the micro-level: Evidence from a home appliance manufacturer. *Journal of Cleaner Production, 185*, 740–748. https://doi.org/10.1016/J.JCLEPRO.2018.03.006

UNEP—*UN Environment Programme.* (n.d.). Environmental agreements and cleaner production. Retrieved July 13, 2022, from, https://www.unep.org/resources/report/environmental-agreements-and-cleaner-production

Urbinati, A., Chiaroni, D., & Chiesa, V. (2017). Towards a new taxonomy of circular economy business models. *Journal of Cleaner Production, 168*, 487–498.

US EPA. (n.d.). Wastes—Resource conservation—Conservation tools. Retrieved July 26, 2022, from, https://archive.epa.gov/wastes/conserve/tools/stewardship/web/html/basic.html

Veleva, V., Bodkin, G., & Todorova, S. (2017). The need for better measurement and employee engagement to advance a circular economy: Lessons from Biogen's "zero waste" journey. *Journal of Cleaner Production, 154*, 517–529.

Winkler, H. (2011). Closed-loop production systems—A sustainable supply chain approach. *CIRP Journal of Manufacturing Science and Technology, 4*(3), 243–246.

World Economic Forum. (2018). *Readiness for the future of production report.*

World Wide Fund for Nature (WWF). Retrieved July 14, 2022a, from, https://www.wwf.org.tr/?10722/Turkiye-Avrupa-Birligi-Yesil-Mutabakatn-dusuk-karbonlu-ekonomiye-gecis-icin-bir-frsat-olarak-gormeli

World Wide Fund for Nature (WWF). (n.d.). *Türkiye, Avrupa Birliği Yeşil Mutabakatı'nı düşük karbonlu ekonomiye geçiş için bir fırsat olarak görmelidir* | Retrieved July 14, 2022, from,. https://www.wwf.org.tr/?10722/Turkiye-Avrupa-Birligi-Yesil-Mutabakatn-dusuk-karbonlu-ekonomiye-gecis-icin-bir-frsat-olarak-gormeli

Yadav, G., Mangla, S. K., Bhattacharya, A., & Luthra, S. (2020). Exploring indicators of circular economy adoption framework through a hybrid decision support approach. *Journal of Cleaner Production, 277*, 124186.

Yan, J., & Feng, C. (2014). Sustainable design-oriented product modularity combined with 6R concept: A case study of rotor laboratory bench. *Clean Technologies and Environmental Policy, 16*(1), 95–109. https://doi.org/10.1007/S10098-013-0597-3/TABLES/18

Yılmaz, F. (2022). Enerji Yönetimi ve Türkiye: Avrupa Yeşil Mutabakatı Çerçevesinde Bir Değerlendirme. *Akademia Doğa ve İnsan Bilimleri Dergisi, 8*(1), 19–37. https://dergipark.org.tr/tr/pub/adibd

Yin, R. K. (2009). *Case study research: Design and methods* (4th ed.). Thousand Oaks.

Yuan, Z., Bi, J., & Moriguichi, Y. (2006). The circular economy: A new development strategy in China. *Journal of Industrial Ecology, 10*(1–2), 4–8. https://doi.org/10.1162/108819806775545321

Zink, T., & Geyer, R. (2017). Circular economy rebound. *Journal of Industrial Ecology, 21*(3), 593–602.

Corporate Sustainability and Circular Economy in Turkish Service and Industrial Businesses

Melek Yurdakul

1 Introduction

The 'EU green Deal' (European Commission, 2019) adopted by the EU in 2019, which reshapes its economic policies and strategies in the fight against climate change, is an integral part of the 2030 Agenda and the sustainable development goals (SDGs). Accordingly, it aims to reduce greenhouse gas emissions to 55% in 2030 and become carbon net zero by 2050. An economic growth model is needed to keep resource consumption within planetary boundaries, reduce consumption footprint, and increase the material use rate cycle to achieve this. The 'circular economy action plan' prepared in this context offers a future-oriented agenda for a cleaner and more competitive Europe together with economic actors, NGOs, individuals, and consumers. Since the success of transformation requires collective action, the EU strives to lead a circular economy at the global level and to implement the SDGs (European Commission,

M. Yurdakul (✉)
Istinye University, Istanbul, Turkey
e-mail: melek.yurdakul34@gmail.com; melek.yurdakul@istinye.edu.tr

M. O. Erdiaw-Kwasie and G. M. M. Alam (eds.), *Circular Economy Strategies and the UN Sustainable Development Goals*, Sustainable Development Goals Series,
https://doi.org/10.1007/978-981-99-3083-8_14

2020). Through the 'green deal diplomacy' developed for this purpose, it aims to increase the climate action level of international partners to approve and effectively implement the agreements on climate change. Economies growing based on fossil resources, especially developing countries, are essential in combating climate and realizing the agreement's goals. Therefore, developing countries that are responsible for 63% of global greenhouse gas emissions (CGDEV, 2022) have been featured in complementary areas of intervention by EU action on the circular economy (European Commission (c), 2021). In this respect, developing countries must move from the traditional linear economic growth model to a circular one. While CE has received relatively little attention in developing and emerging countries, it has only recently begun to be recognized and applied (Erdiaw-Kwasie et al., 2023; Halog & Anieke, 2021). However, compared with developed countries, it can be said that this interest is some (Elia et al., 2020).

The transition to the circular economy model is also a pretty new issue for businesses (Azizuddin et al., 2021; Ghisellini et al., 2016; Tunn et al., 2019; Upadhyay et al., 2021); especially in developing countries, this is more uncertain (Azizuddin et al., 2021; Halog & Anieke, 2021; Upadhyay et al., 2021). As a requirement of the system approach in the transition of businesses to circularity, all internal and external units and elements must be compatible and supportive. The reluctance of business stakeholders or incompatibilities between different actors may prevent businesses from experiencing uncertainty and failure in the transformation process. For example, the reluctance of regulatory institutions, one of the business stakeholders, to regulate existing laws and regulations designed according to the linear economy model under the requirements of the circular economy causes incompatibilities and disruptions in green transformation (Baah et al., 2022; Kirchherr et al., 2018). The report prepared by the Turkish Industry and Business Association (TUSIAD, 2021), which proposes pioneering policies on CE in Turkey[1] emphasizes the need to update numerous legislations for industrial transformation. At the same time, the report shows that this transformation is necessary for Turkey, a party to international agreements, significantly dependent on foreign energy and intermediate goods, where increasing population,

[1] Turkey is a developing country according to World Bank and United Nations (2022).

growth, inefficient resource management, and practices increase environmental pressures. In this context, it can be said that the lack of a comprehensive roadmap brings uncertainties in circular economy decisions and practices of enterprises. Also, the CE issue is still in its infancy in the literature (De Angelis, 2021; Steward & Niero, 2018). Therefore, this subject is open to contributions both in practice and theory. The first reason for this research is to contribute to the literature and practice by presenting concrete outputs about CE approaches and the implementation of businesses in this uncertain environment. Thus, it aims to draw attention to the CE issue in the relevant sectors, raise awareness, determine its place in business management and practice, and emphasize its relationship with the SDGs.

Turkey adopted the Paris climate agreement in 2021. In addition, Turkey aims to fulfill its obligations under the Kyoto Protocol and fully comply with the EU Climate Acquis. Within the framework of the Paris Agreement realized in this context, Turkey's greenhouse gas reduction target is 21% until 2030, according to the reference scenario, and Turkey has adopted the Net Zero Emission target until 2053 (WWF, 2021). There needs to be a concrete guide to reaching the United Nations' sustainable development goals. Still, there is an emphasis that scientific studies should guide the planning of roadmaps in sustainability transformation (SDG Compass, 2015). The second reason for the research is to measure the business's activities and targets in the context of the United Nations SDGs and reveal sustainability performance in economic, social, and environmental terms. Thus, in roadmaps planning, businesses' SDGs and sustainability performance indicators can guide decision-makers. Accordingly, this study aims to reveal the general picture of the circular economy practices of business in a developing economy and their contribution to sustainable development goals. In addition, determining the sustainability performances of businesses is another purpose of the study. In this sense, it is aimed to contribute to the literature and provide insight to practitioners according to systems and stakeholder theories.

Although the struggle for sustainability in industrial production has continued since the industrial revolution, sustainability in the industry sector is more on the agenda today, where global warming has reached critical levels. Due to global value chains and resource-intensive production, the industry sector has an essential environmental responsibility. The industry is responsible for 40 percent of global carbon dioxide emissions (PwC). Due to this responsibility, the industry sector has a crucial role in achieving sustainability goals and ensuring sustainability. With this understanding, the 'The new Industrial Strategy' (European Commission (d), 2020) developed by the EU aims to fully mobilize the industry in the transition to a climate-neutral and circular economy and to reduce the carbon footprint of the sector with clean technology and new business models. At the same time, ISO (Istanbul Chamber of Industry) and TUSİAD, two leading and important institutions for Turkish industrial businesses, are trying to develop a sustainability vision and guide businesses within the framework of the EU Green Deal and sustainability. In this sense, findings on CE approaches and practices, sustainability, and SDGs performance in the transformation process, which is a new process, may be helpful for sector representatives. The creation of a CE system requires responsible production and consumption. Therefore, the service sector, which has a strategic position between producers and end users, plays a vital role in transitioning to a circular economy (Heyes et al., 2018). Also, today the service sector is one of the fastest-growing sectors (TCTB, 2022), and to talk about CE success, this growth should support the functioning of the circular economy. In this sense, it is necessary to raise awareness and transition to circular business models in the service sector and the industry. Due to the importance of industry and service sectors in the transition to the CE model, they are the focus of the research practice. Since the private sector has a crucial role in achieving the 2030 Sustainable Development Goals and 2050 net zero emissions target, the research is designed based on solutions sought at the business level in a developing economy. The research questions are as follows.

- What are the circular economy approaches in businesses?
- Is the circular economy defined as strategic and management policy?
- What are the practices carried out by businesses within the scope of the circular economy?
- What is the performance of businesses to sustainable development goals?
- What are the economic, social, and environmental sustainability performances of businesses?

All questions were analyzed according to businesses and comparing industry and service sectors. The content analysis method has been used to realize the research purpose and answer the questions. Since sustainability reports are helpful tools for measuring the outputs of corporate sustainability (CS) (Boiral & Henri, 2017; Tiscini et al., 2022; Upadhyay et al., 2021), a report review was preferred to reveal the practices and performances of businesses. This study aims to contribute to the literature and decision-makers determining businesses' current performances and approaches. The service and industrial businesses listed on the stock exchange with independent auditing and international reporting standards were selected as the sample. While 284 companies listed on the stock exchange were determined, it was stated that 25 of these companies had been independently audited and published sustainability reports of international standards. Significant findings were obtained in the analysis based on content analysis.

The next part of the research is the conceptual framework and literature review, the application of the study in the third part, the analysis findings in the fourth part, and the result and discussion in the fifth and sixth parts.

2 Conceptual Framework and Literature

2.1 *Corporate Sustainability and Circular Economy*

Sustainability is asserting the value of longevity and intergenerational justice while recognizing mortality and finitude. In the report Our Common Future, published by the World Commission on Environment and Development, operating under the United Nations in 1987, the concept of sustainability is defined as meeting the needs of the present without compromising the ability of future generations to meet their own needs (WCED, 1987). Systems thinking emphasizes that sustainability should not be defined in terms of a single dimension but rather by the economy, society, and the ecological system (Elkington & Rowlands, 1999). The systems perspective envisages an organization for sustainability that is not used faster than natural resources' renewal, recycling, and regeneration rate (Alam et al., 2022; Marshall & Brown, 2003). Most of the production comes from non-renewable resources; if this does not change, the expected life of our economy is much shorter than that of the universe (Daly, 2006).

Sustainable development addresses conceptualizations of sustainability at the macro level. TBL acts as a proxy for the conceptualization of sustainability and reshapes management discourse by making sustainability part of the business agenda (Elkington & Rowlands, 1999). In this respect, corporate sustainability is a more specific form of sustainable development and, like sustainable development, includes three key dimensions: economic, social, and environmental (Baumgartner & Ebner, 2010). For this reason, in the business and management academia literature, sustainable development is addressed under corporate sustainability (De Angelis, 2018; Erdiaw-Kwasie et al., 2023; Nikolaou & Tsagarakis, 2021). Corporate sustainability emphasizes the role of businesses as members of larger systems and their role in stabilizing or destabilizing these systems. In corporate sustainability, economic, social, and environmental dimensions tend to be considered interdependently. Accordingly, it recognizes that the economy is a part of society and that these two dimensions are part of a more extensive ecological system and define a nested system (Meuer et al., 2020; Montiel, 2008). CS can be defined as meeting the needs of a business's stakeholders without compromising its capability to meet the needs of future stakeholders as well. To this end, businesses must protect and grow their economic, social, and environmental capital bases while actively contributing to sustainability in the political arena (Thomas & Kaj, 2002). Although there has yet to be a consensus on the concept of sustainable development in the CS literature, there are two different perspectives. The first of these is the win–win approach (Porter et al., 1995). Accordingly, economic growth is possible if it is supported environmentally and socially. The second is the reductionist approach at the business level. The assumption states that sustainability is reducible to the business level and corresponds to each sustainable business's sum. One of the critical points to understand in the relationship between sustainability and CS sciences is the vitality of the biosphere. The biosphere does not depend on the existence of economic activities, but the continuity of economic activities depends on the presence of the biosphere. This reality is based on win–win and business-level sustainability assumptions (Isil & Hernke, 2017).

The vitality of the biosphere and the criticality of achieving sustainable development call for solutions that are different from current ways of doing business and approaches. The circular economy offers an effective way for a significant transformation to new principles of circularity from the traditional linear approach to doing business (Lacy et al., 2020). CE, a model that fundamentally changes the way of production and consumption (European Lacy et al., 2020; Parliament, 2022), creates a thriving ecosystem that circulates value throughout the economy and society (Lacy et al., 2020). The circular economy has the potential to help meet UN Sustainable Development Goals, in particular SDG12 (responsible consumption and production), as well as tackle both the causes and effects of climate change (EMF, 2021). The circularity approach means keeping the product components in the economy as long as possible when the product ends its life cycle. Based on value creation, this system brings resources to use continuously through efficiency (European Lacy et al., 2020; Parliament, 2022). An alternative model, CE, assigns a crucial role to businesses, overrides the existing linear model, and addresses ecological and social concerns (De Angelis, 2018). CE is relevant to corporate sustainability in examining the role of businesses in solving sustainability challenges, using natural resources efficiently, and achieving the SDGs (Blinova et al., 2022). CE is an influential model that inspires companies to corporate strategies that foster the development of an economy based on ecological limits (De Angelis, 2018). Accordingly, CS and CE are complementary; both concepts create a synergistic effort to trigger the necessary change to realize sustainable production and consumption (Santos et al., 2017). Also, CE enables businesses to achieve a competitive advantage. It brings new opportunities for companies to enter new markets with innovative products (Lacy et al., 2020). Natural resource expenditures of businesses are reduced, which brings more independence. Increased interaction with consumers strengthens consumer loyalty. Simplified product design enables companies to manage products' life cycles better (Gallaud & Laperche, 2016). CE is important because it minimizes resource use, reduces emissions, increases resource efficiency, offers new market opportunities, and supports sustainable consumption (Tunn et al., 2019).

In the literature, CE is seen as a tool for achieving sustainable development goals (Geissdoerfer et al., 2017). However, while research on the relationship between circular economy and sustainability is ongoing, overcoming the blurring and integration problem between concepts is essential. In different studies, sustainability and CE are often used in similar contexts, thus blurring the conceptual contours. To eliminate this ambiguity, Geissdoerfer et al. (2017) emphasize three aspects of CE's relationship with the concept of sustainability: conditional, beneficial, and trade-off. In the conditional relationship, CE is defined as a condition for achieving sustainability; in the beneficial relationship, CE is defined as helpful; and in the trade-off relationship, CE is defined as benefits and costs.

CE can contribute to different dimensions of sustainability, but the two concepts are not integrated. While the sustainable development agenda focuses on harmony between humans and nature and recognizes economic prosperity as a means to achieve this, CE focuses on technological solutions with the promise of economic growth. Therefore, CE measures must be integrated with sustainable development and climate change; both concepts cannot be achieved without a sustainable circular economy. The integration requires responsibility and effort on the part of each actor in the system (Velenturf & Purnell, 2021).

2.2 Theoretical Framework and Literature Review

The sustainability and CE perspective requires a holistic approach and stakeholder engagement. Thus, the study uses stakeholder and system theories to explain CE and sustainability decisions and activities at the business level. It is inevitable that all actors of the system take action and cooperate in sustainability solutions.

CS encourages managers to make decisions in environmental, social, and economic tensions. Therefore, designing sustainable business strategies and adopting CE brings dynamic and comprehensive decision-making processes. CE is an approach businesses cannot realize on their own; sustainability and CE inherently require a more participatory model (Parker, 2014), and therefore stakeholders are critical dynamics of sustainability transformation. Regulatory changes, changes in producers' business models, and consumer behavior are essential for CE success. The globalized society is adamant in its perception that businesses are responsible to their shareholders and a wider group of stakeholders (Rendtorff, 2019). Therefore, the only way to design and implement CE is to collaborate with internal and external stakeholders. In the literature, the importance of stakeholders in CE transformation is emphasized by different studies (Baah et al., 2022; Bansal et al., 2022; De Angelis, 2018; Geissdoerfer et al. 2017; Ghisellini et al., 2016; Gupta et al., 2019; Marjamaa et al., 2021; Velenturf & Purnell, 2021). The stakeholder theory proposed by Freeman (1984) advocates balancing employee, customer, supplier, and shareholder benefits. The approach is based on the need for decision-makers to consider the interests of both shareholders and stakeholders in implementing new strategies. Since the circular business model is based on a closed-loop system, it requires more networking and collaboration with all actors in the system. Therefore, it is clear that many important decisions need to be made by all parties throughout the process. Integration and coordination of stakeholders are necessary for recycling, refurbishing, reusing, or repairing products. Failure to effectively manage stakeholder relationships makes CE design and implementation difficult or even impossible (Gupta et al., 2019). Stakeholder theory provides a good background for CS and CE design and implementation (Montiel & Delgado-Ceballos, 2014; Soomer, 2012).

Businesses are members of larger groups, and the need to adapt to the expectations of these groups has a significant impact on their sustainability and CE activities. The CE approach draws significantly on systems theory (De Angelis, 2022). Biologist Ludwig von Bertalanffy first proposed the systems approach. This approach considers the business a combination of many systems but examines the operation of all these systems (Koçel, 2014). The systems perspective guides the pursuit of sustainability and is a valuable tool for businesses that want to be sustainable (Marshall & Brown, 2003). Within a systems approach, sustainability activities influence and mobilize the vision and strategies of all existing players. Strongly influenced by stakeholder theory, this translates directly into collaboration. Linkages between different actors in the system, including environmental groups, labor unions, international organizations, government agencies, and community and resource-based industries, provide solutions to sustainability challenges (Prasad & Elmes, 2005). CE promotes the reintegration of the economy into the ecology. In CE, products are designed according to their interaction with economic and ecological dimensions throughout their life cycle. In the closed-loop model, resource stocks interact, and materials are in a flow in this cycle and interconnected. CE emphasizes a system-centered rather than a business-centered view, thus providing a more holistic value-creation mechanism by addressing economic, social, and environmental aspects (De Angelis, 2022). CE consists of subsystems and practices that are interconnected with business strategies. Yang et al. (2020) emphasize three key factors, namely system regulations, internal enablers within the system, and external enablers from the environment, to ensure synergies between subsystems and achieve the goals of a system as a whole.

As a requirement of the systems and stakeholder approaches, all businesses, sectors, and economies are responsible for nature. Weber (2012) examined how Canadian banks integrate environmental risks into corporate loans and where they stand compared to their global peers. In the categorization carried out by the GRI directive, it has been determined that Canadian banks manage their environmental risks to avoid financial troubles. Some of these banks associate sustainability issues with their business strategies. Compared with other countries, it was concluded that Canadian banks are better and more proactive in environmental issues.

Fifka and Drabble (2012) compared the 100 most prominent British and Finnish businesses through sustainability reporting with the GRI standard. The environmental sustainability dimension has a greater weight in the activities of Finnish companies. Of all British businesses, 94% of Finnish businesses publish sustainability reports. 16% of Finnish companies focus on economic, 45% on social, and 39% on environmental issues, while 15% of British businesses focus on ecological, 55% on social, and 29% on economic issues. Since most of the economies are made up of SMEs, it is vital to determine the approach of SMEs to sustainability to spread the sustainability culture. Bos-Brouwers (2010) researched the Dutch Plastics and Rubber industry to gain more insight into sustainability and innovation activities in transforming SMEs. According to the themes prepared by the GRI standard, it has been seen that SMEs have behavioral advantages in compensating for the lack of resources, even if they are less equipped. Leadership style, flexible organization, and motivated employees benefit SMEs more than large organizations.

When the transition from a linear model to a circular-based model takes place, businesses need a feedback process that connects inputs, activities, outputs, and outcomes in a continuous and widespread manner. Barnebe and Nazir (2020) recommend an integrated report to address this need and emphasize compliance with CE. According to Opferkuch et al. (2021), CE remains a voluntary issue for reporting, with the only exception being the GRI frame, which requires companies to report on a qualitative index designed to describe their circularity measures. Inconsistencies between reporting approaches result in companies either not reporting CE issues or simply defining CE practices concerning waste management. This situation causes businesses to associate CE practices only with the environmental dimension. Similar findings were found by Sehnem et al. (2020) in the cosmetics sector in Brazil. Accordingly, while the share of economic indicators was higher in previous reports, it was observed that the environment was more intense in the period analyzed. In addition, symmetry was found between sustainability practices and CE antecedents in the research conducted according to the GRI report analysis. Tiscini et al. (2022) frequently encountered the concepts of GRI, packaging, recycling, GHG emissions, and CO2 emissions in the Italian cosmetics industry in their sustainability report review. Similarly, it was determined that the environmental statements were relatively higher, but the weight of CE, governance, strategy, future-oriented discourses and management issues was insufficient.

A similar finding was reached by Upadhyay et al. (2021). In the research conducted on multinational mining companies, while CE is mentioned at the operational level in mining companies, it is not discussed extensively at the management level. It can be argued that this has not become a mainstream industry subject because of the CE-related policy deficiencies. The reviewed reports determined that the terms of recycling and circular economy were minimal, and there was no increase in these concepts over the years. The analysis findings demonstrate the concern in developing countries. Developing economies are economically dependent on the mining sector. Therefore, the environmental impact created is enormous. However, in Bangladesh, a developing economy, the public has little awareness of the consequences of harmful substances, so it does not pressure businesses or governments about the circular transformation (Azizuddin et al., 2021). At the same time, Azizuddin et al. (2021) propose the CE model as a distinguished model for differentiation. The Paris Agreement has given joint responsibilities to developed and developing countries in reducing carbon emissions. Developing countries must shift their economic growth away from carbon-intensive practices for a responsible future. They should develop a strategy based on the lowest emission or renewable energy investments and maintain growth while reducing emissions (Delgado, 2021). This transformation clarifies the importance of regulators and society (Baah et al., 2022). Sustainability needs to be accepted by consumers to become widespread. For this, Tunn et al. (2019) developed a business model of CE that supports sustainable consumption. Accordingly, the most promising business model for sustainable consumption has been the ones that reduce the overall consumption level and reduce consumer effort.

With the effects of the SDGs and the EU green deal on systems, the issue of transformation has become even more important in businesses. However, the literature reveals that this transformation is very uncertain, especially in the economies of developing countries, and the practices are very inadequate. In this research, SDGs and CE are discussed under the topic of CS at the business level. With this research, sustainability and SDGs performances of Turkish businesses and circular economy practices are presented comparatively.

3 Methodology

3.1 *Sample and Data*

In order to realize the purpose of the research, businesses in the industry and service sector registered in Borsa Istanbul (BIST) were taken as the sample. These businesses represent a significant portion of Turkey's industrial and service sectors. The sample selection selected industrial and service sectors with high-carbon intensity and the potential to reduce this intensity and CE transformation effectively. The industrial sector, which significantly impacts global warming and pollution (EEA, 2020; EPA (a); IPCC, 2014), is responsible for one-fifth of global greenhouse gas emissions. Decarbonization in industrial sectors, which consume 54% of the world's energy resources (WEF, 2022), is very important. Therefore, in the CE transformation, the EU has separately addressed industry, which is responsible for 15% of the EU's emissions, and prepared an industrial policy (European Commission (d), 2020) to support this transformation. In this way, the Commission aims for climate-neutral competitiveness and assigns industry a vital role in achieving this goal.

The transition to climate neutrality is possible with the participation of sectors in the system. Reducing emissions is possible through low-emission technologies and services (European Commission (d), 2020). The system perspective and the need for a closed loop necessitate responsible production and consumption (EMF, 2021; Lacy et al., 2020). Therefore, the service sector, which has a strategic position between producers and end users, is essential in the transition to a circular economy (Heyes et al., 2018). The service sector is a rapidly growing economic sector in developed and developing economies (TCTB, 2022). The absolute amount of emissions caused by the service sector, which covers areas such as transportation, retail, or distribution, is expected to increase with its growing share in the economy. In all sectors, transportation is the most dependent on fossil fuels (IEA, 2022). 21% of global carbon dioxide emissions in 2018 (Ritche, 2020), and 24.5% of these emissions in 2021 came from the transportation sector (EPA(b)). The service sector is at the forefront of the CE transformation due to the need to reduce emissions and its role between industry and the end consumer. Research shows that developing countries are responsible for 63% of current carbon emissions (CGDEV, 2022). The historical link between economic growth and emissions highlights the importance of developing countries in global emission reduction efforts (UNCTAD, 2021). The fact that Turkey is both a developing economy and its largest trading partner is the EU (TUSIAD, 2021) makes the CE issue important in all sectors.

It has been determined that there are 284 businesses in the BIST industry and service sector. The recent 2020-year report published by the businesses was considered in the evaluation, and two essential criteria were followed in the business report. The first of these is that the report has the GRI international standard. GRI standards are an effective way to measure an organization's economic, social, and environmental status. At the same time, the GRI principles are an effective way for businesses to comprehensively share SDG-aligned material issues in their sustainability activities (Boiral & Henri, 2017). Because standards are comparable and have credibility, they help stakeholders and different parties meet expectations, evaluate and make decisions (GRI, 2022). GRI has been preferred because it is the most widely used standard worldwide (KPMG, 2020), is suitable for businesses of all sizes and sectors, and allows comparison (GRI, 2022). The second criterion is that the reports have independent auditing. Twenty-five businesses are determined according to these criteria; 11 in the service sector and 14 in the industry sector. 'CRFSA and TKNSA', 'CIMSA and AFYON', and 'TIRE and OLMKM' coded businesses are evaluated as group businesses because they have published joint reports. Business names are handled in the form of the code they use in the BIST.

3.2 *Measurement of Research Variables and Data Analysis*

The content analysis method was used to measure and analyze research variables. Content analysis is a research technique for the objective, systematic, and quantitative description of the explicit content of communication (Berelson, 1952). The strengths of content analysis are that it provides the opportunity to examine activities that occur over a broad period, that research has no effect on the subjects, and that it is reliable (Babbie, 2007). The steps of content analysis application determined by Babbie's (2007) research were followed. These steps are the selection of the research question, sample selection, conceptualization and development of code categories, data collection coding, description, evaluation, inference, and interpretation. This study is both exploratory and descriptive with a qualitative approach.

The research analysis was carried out in two stages. In the first stage, the weighted scoring system was used in the content analysis to determine the sustainability and SDGs performance of the businesses. In the second stage, a content analysis based on themes was conducted to determine businesses' CE approaches and practices. These steps are explained in detail in the related sub-headings.

3.2.1 *Measurement and Data Analysis of Corporate Sustainability and Sustainable Development Goals*

At this stage of the analysis, data were collected and analyzed to measure the sustainability performance of businesses with the economic, social, and environmental dimensions set forth by Elkington and Rowlands (1999) and to determine the CS disclosure levels. One of the most critical challenges in the field of CS is the lack of a standard measurement method (Montiel & Delgado-Ceballos, 2014). Reporting on CS is an accepted practice by investors, and investors are demanding an increasing amount of information to evaluate businesses. Sustainability reporting is a holistic corporate governance tool and essential for shaping the business's image. The reporting process, which includes stakeholder engagement, keeping stakeholders in the loop, open communication with stakeholders, and external auditing, creates trust. Sustainability reporting is no longer a marginal issue today but an essential element of modern corporate governance (PwC, 2022). Since sustainability reporting is the most important way to measure and share the results of business management's sustainability attitudes and decisions (Boiral & Henri, 2017; Upadhyay et al., 2021), a content analysis method based on report analysis was used in this study to achieve the research purpose.

The sustainability and integrated reports of the businesses were analyzed based on the GRI (2016) standards. The GRI (2018), GRI (2019), and GRI (2020) standards added to the GRI (2016) standard in the following years were also considered. A weighted approach, inspired by the research of Rouf (2011), was used in scoring sustainability disclosures and performances. The standards required to be disclosed by businesses adopting the GRI are accepted as the disclosure index. The GRI index consists of four main themes and one hundred and forty-eight codes. If the business disclosed information in the index, a scoring form was created by giving a '1' point; if not, a '0' point was given. Afterward, the total scores obtained by the businesses were proportioned to the disclosure index, and the sustainability performance scores were reached. SDG performances of businesses have been determined according to 17 goals set by the United Nations. A scoring form was created by scoring the businesses that included the activities or target-based explanations regarding the SDG targets as '1' and the businesses that did not have '0'.

3.2.2 Measurement and Data Analysis of Circular Economy

In this research, the subject of CE was examined based on the content analysis method in the sustainability report. At this stage of the research analysis, data were collected and analyzed to determine the CE approaches and implications of the businesses. Despite the increasing interest in CE, there are deficiencies in measuring and evaluating this concept (Garza-Reyes et al., 2019). Most of the studies in the CE literature are based on theoretical and systematic literature reviews (de Jesus & Mendonça, 2018; de Sousa Jabbour et al., 2019; Halog & Anieke, 2021). Despite emerging interest in integrating CE practices into sustainability processes, research on its role in sustainability reporting remains scant, especially in developing countries (Sanches et al., 2022). The potential of sustainability reporting in legitimating and comparing the sustainability contributions of CE strategies has yet to be discovered (Opferkuch et al., 2021).

A two-stage process was followed in the data collection and analysis related to CE. In the first stage, the CE reporting approach of the businesses was determined. These approaches were designed according to four categories, inspired by the study of Opferkuch et al. (2021).

- Fully integrated: CE is a central topic of business activities, associated with multiple contexts, embraced by senior management, and handled as a business strategy.
- Predominantly integrated: CE is associated with many contents but is not very strong overall.

- Partially integrated: CE is associated with a few content elements and has a weak place in all business activities.
- Not to mention: CE is not included among reported activities and targets.

The reports were scanned with circular economy and circular codes for the CE approaches. The topics and chapters related to these concepts are examined. Then, the content of these topics was evaluated. By not considering the repetitive contents, it was determined how many different contents related to CE were included in the report. Thus, according to the CE content level, which category the businesses are suitable for has been relatively revealed.

After the CE approaches of the businesses were determined, the analysis of CE implementations was started. Businesses with no content related to CE are not included in this analysis. First of all, conceptual themes were created to determine CE implementations. These are circular management, circular implementation, and circular collaboration and membership. The creation of circular management and circular collaboration themes was inspired by the work of Garza-Reyes et al. (2019). The key circular implementation theme is inspired by the work of Stahel (1982).

As in the study of Opferkuch et al. (2021), circular economy and circular codes were used to determine the contents related to the themes. In addition to these codes, the 4R (reuse, repair, remanufacturing, recycle) cyclic framework was determined by Stahel (1982). This model is based on creating new value with no unused and reduced waste. Reuse keeps the product in its original form and used for its original purpose and is an effective way to combat waste. Repair allows for the extension of the use of products as long as possible. It involves repairing, replacing, and upgrading components. Remanufacturing is the redesign and repurposing of products and components. It is done when the product cannot remain in circulation in its current form. Recycling is the return of unused materials to the production process as raw materials when the product cannot be kept in the cycle in its original condition. A product becomes a new product by recycling its basic materials (EMF, 2019). The design phase is crucial in creating value because 80% of the environmental impacts of the products are determined during the design phase (European Commission (b), 2020). The most critical point for successful product transformation is that it is designed with its highest circulation in mind. For example, a product designed to withstand user use is easy to repair, easy to maintain, simple to disassemble, and designed with replaceable components is suitable for conversion (EMF, 2019). EU strives to influence global standards in product sustainability and worldwide product design and value chain management (European Commission (b), 2020). For this reason, the design code was also used to determine the content related to the themes. Reduce, another research code, is another critical element of the CE process, aiming to reduce resource inputs and increase efficiency (Pichlak & Szromek, 2022).

After the themes and codes were determined, report reviews were started. At this stage, report reviews were done in more detail. The topics and sections related to the codes of circular, circular economy, recycle, reuse, repair, remanufacture, design, and reduce are examined. When the business activities related to the circular economy were handled as the top management discourse, strategic, priority topics, sustainability, or environmental management, this data was associated with the circular management theme. At the same time, the collaborations, memberships, and implementations were associated with the relevant theme.

4 Empirical Results

This section includes businesses' sustainability performance, CE approaches and implementations, and SDGs performance analysis findings.

Within the scope of the research, the reports of 284 businesses were examined. 50.35% of these businesses do not publish any sustainability reports. 27.11% of these businesses publish the 'business sustainability statement' required by the SPK (Capital Markets Board of Turkey). Although 12.67% of these businesses have published reports with international standards, they have not received any independent audit. 1.05% of these businesses published a sustainability report without being bound by any standard. Therefore, 25 reports with GRI standards and independent audits were evaluated. The number of businesses included in the evaluation is 8.8% of the total sample.

4.1 Empirical Results of Sustainability Performance

In this section, the research question 'What are the economic, social, and environmental sustainability performances of businesses?' is analyzed by businesses and sectors.

The weighted and average sustainability performances calculated according to the GRI (2016) index and the performance of the businesses in explaining their sustainability activities are given in Table 1. Data is shared at the business level. Businesses are listed according to their sustainability performance in their sector. Accordingly, while the average sustainability performance of the service sector is 0.55, it is 0.47 for the industrial sector. Similarly, the performance of CS disclosure is 0.62 in the service sector and 0.58 in the industry sector. While the highest level of sustainability performance among businesses is 0.79, the lowest performance level is 0.26. The average sustainability performance of all businesses in both sectors is 0.50.

While the social sustainability performance of businesses in the service sector is higher (0.63), the environmental sustainability performance of businesses in the industrial sector is higher (0.60). Economic sustainability performance has the lowest share in both sectors.

The weight of sustainability activities in service and industrial businesses is environmental performance (0.38), social performance (0.37), and economic performance (0.25), respectively. Although there is no significant difference between environmental and social performance, environmental sustainability has a higher share in businesses (Fig. 1).

In addition to the research questions, the businesses were analyzed according to the report type. It was determined that two businesses were the comprehensive option. When the sustainability performance of these businesses was evaluated, it was determined that both the sustainability and CS disclosure performance were higher in the businesses reporting by the GRI-Comprehensive option (Fig. 2). The main difference between the reports stems from the governance dimension.

4.2 Empirical Results of SDGs Performance

In this section, the research question 'What is the performance of businesses to sustainable development goals?' is analyzed by businesses and main sectors.

Table 1 Sustainability performance of businesses and sectors

	Business/Sub-sectors	*Report*	*Economic performance*	*Social performance*	*Environmental performance*	*Sustainability performance*	*Disclosure performance of CS*	*SDGs*
Service Sector	AYDEM/Electric gas and water	SR-Cr	0.76	0.68	0.94	0.79	0.72	11
	DOAS/Wholesale and retail trade	SR-Cm	0.76	0.83	0.72	0.77	0.86	11
	ENKA/Construction	SR-Cr	0.71	0.68	0.72	0.70	0.68	13
	ZOREN/Electric gas and water	SR-Cr	0.59	0.68	0.66	0.64	0.65	4
	TCELL/ Telecommunication	Int-Cr	0.53	0.80	0.50	0.61	0.73	11
	MGROS/Wholesale and retail trade	SR-Cr	0.29	0.80	0.59	0.56	0.62	17
	THY/Transportation	SR-Cr	0.47	0.53	0.38	0.46	0.52	12
	CRFSA and TKNSA Group/Wholesale and retail trade	SR-Cr	0.12	0.43	0.50	0.35	0.53	10
	ENJSA/Electric gas and water	SR-Cr	0.18	0.48	0.38	0.34	0.47	12
	MAVI/Wholesale and retail trade	SR-Cr	0.12	0.38	0.28	0.26	0.44	9
	Average Service Sector		**0.45**	**0.63**	**0.57**	**0.55**	**0.62**	**11**
Industry Sector	MONDI Group/Paper and paper products printing	SR-Cr	0.53	0.60	0.81	0.65	0.69	7
	ARCLK/Electrical devices	SR-Cm	0.24	0.78	0.78	0.60	0.77	12
	BFREN/Electrical devices	SR-Cr	0.41	0.70	0.69	0.60	0.62	9

(continued)

Table 1 (continued)

Business/Sub-sectors	*Report*	*Economic performance*	*Social performance*	*Environmental performance*	*Sustainability performance*	*Disclosure performance of CS*	*SDGs*
CCOLA/Food, beverage, and tobacco	SR-Cr	0.41	0.70	0.59	0.57	0.62	14
CIMSA and AFYON Group/Based on stone and soil	SR-Cr	0.59	0.43	0.53	0.51	0.56	8
PETKM/chemical pharmaceuticals petroleum rubber and plastic products	SR-Cr	0.59	0.38	0.50	0.49	0.53	8
KORDS/Textile	SR-Cr	0.24	0.53	0.66	0.47	0.57	5
ULKER/Food, beverage, and tobacco	SR-Cr	0.18	0.38	0.63	0.39	0.53	11
AKCNS/Based on stone and soil	SR-Cr	0.12	0.40	0.58	0.37	0.51	13
FROTO/Electrical devices and transportation vehicles	SR-Cr	0.18	0.35	0.50	0.34	0.49	14
TOASO/Electrical devices and transportation vehicles	SR-Cr	0.18	0.40	0.34	0.31	0.57	1
AEFES/Food, beverage, and tobacco	SR-Cr	0.00	0.40	0.53	0.31	0.49	13
Average Industry Sector		**0.30**	**0.50**	**0.60**	**0.47**	**0.58**	**9.6**

Note SR: sustainable report, Int: integrated report, Cr: core, Cm: comprehensive. There are two options for the preparation of GRI, core and comprehensive

Fig. 1 Weight of sustainability dimensions in business

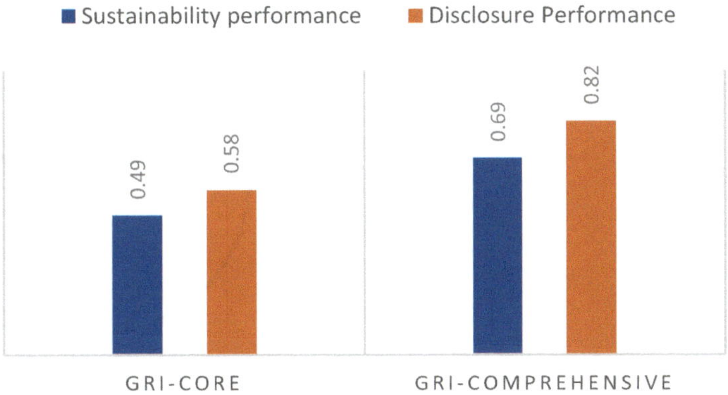

Fig. 2 Sustainability performance by report type

The SDGs performances of each business are shared in Table 1, and the average was determined as 10. 36% of the businesses were below the average SDGs performance. MGROS, which includes all SDGs targets in its targets and activities, performs well. The lowest performance was achieved by TOASO, which included only one of the SDGs in its activities.

In examining the SDGs performances of the businesses, which target was weighted was primarily analyzed. Accordingly, businesses mainly include SDG8, SDG12, SDG9, and SDG13 in their activities and targets. On the other hand, the goals that have the least place in the activities and targets of the businesses are SDG1, SDG2, SDG14, and SDG16, respectively (Fig. 3).

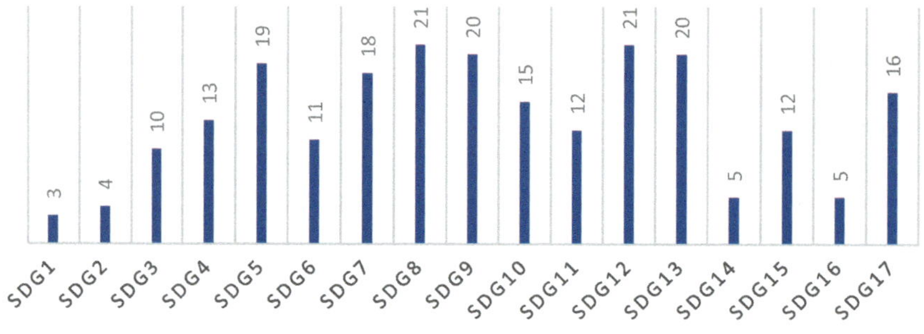

Fig. 3 SDGs weights in businesses

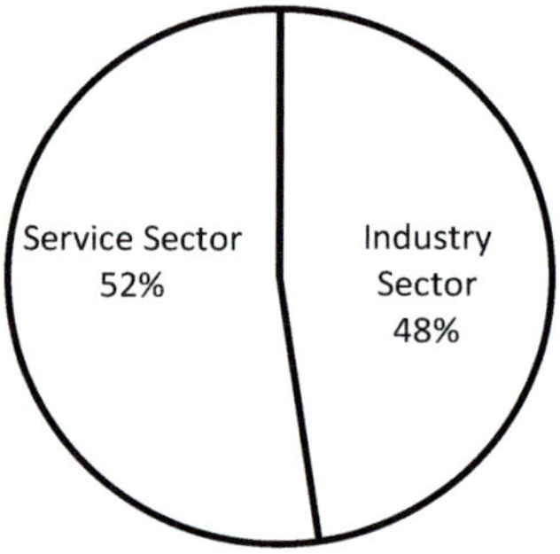

Fig. 4 SDGs performances of main sectors

In addition, the SDGs performance of the Main sectors were compared. Accordingly, it was determined that the SDGs performance share of the Service sector (52%) was higher with a slight difference (Fig. 4).

4.3 Empirical Results of Circular Economy

In this section, the research questions '*What are the circular economy approaches in businesses?*', '*Is the circular economy defined as strategic and management policy?*', *and* '*What are the practices carried out by businesses within the scope of the circular economy?*' are analyzed by businesses and main sectors.

According to the findings of the analysis conducted to determine the CE approaches of the businesses, the circular economy approaches of businesses are mostly similar. When the results in Table 2 are examined, 68% of these businesses partially integrated their CE targets and activities into their reports. The weight of CE in these business activities is deficient. Relevant codes are associated with only a few contents in the reports. The CE-related content of these businesses is mainly related to management and waste management, while content related to supply chain management, raw material management, or R&D is rarely found.

While 23% of these businesses predominantly integrated their CE activities and targets, no CE-related targets and actions were found in a report defined as a group enterprise. Therefore, this business was not included in the CE analysis. In businesses with predominantly integrated reports, CE is addressed within the scope of different topics. These integration areas include strategy, waste management, natural resource management, innovation, efficiency, or R&D. Compared to other businesses, a fully integrated business (ARCLK) has been identified that addresses CE with many contents. In this business report, CE is addressed by top management in several different contexts. These include design and manufacturing processes, responsible production and consumption, supply chain management, innovation, R&D, end-of-life assessment and recycling, product life cycle, packaging management, market risk, policy risk, and company and brand reputation risk.

CE approach weights vary across sectors. Accordingly, the weight of CE partially integrated approaches of businesses in the service sector is 80%, while the weight of industrial sector businesses is 58%. The weight of CE predominantly integrated approach of service sector businesses is 20%, while the weight of CE fully and predominantly integrated approaches of industrial sector businesses is 33%. According to these findings, it is determined that the weight of CE approaches is higher in industrial sector businesses than in service sector businesses.

As a result of examining the CE concept within the scope of business strategy and management policy determined that it is addressed in strategy, top management discourse, or among priority issues in different CE approaches. While CE is addressed as a strategy or policy in all businesses with predominantly and fully integrated approaches, this rate is 47% in businesses with a partially integrated approach. CE is addressed poorly in business implementations, although it is included in priority issues, management, or executive discourse in partially integrated approaches. Waste management constitutes a significant part of these activities.

Table 2 The circular economy in businesses

	Business/Sector	*CE reporting approaches*	*Circular management*	*Circular implementations*	*Circular collaboration and membership*
Service Sector	AYDEM/Electric gas and water	Partially integrated	General Manager Message Sustainability principle	Reduce Recycle	–
	ZOREN/Electric gas and water	Partially integrated	-	Reuse Recycle	R&D and innovation
	ENJSA/Electric gas and water	Partially integrated	Environmental Management Priority issues	Reduce Recycle Repair	Suppliers CE Sub-Working Group
	DOAS/Wholesale and retail trade	Partially integrated	Chairman Message	Reduce Reuse Recycle Repair	Local communities ÇEVKO PETDER AKÜDER
	CRFSA and TKNSA Group/ Wholesale and retail trade	Predominantly integrated	CEO message Sustainability Roadmap Priority issues Sustainable business model	Reuse (re-fill) Reduce Recycle	R&D Innovation and technology Sustainable Development Association (SDA)
	MAVI/Wholesale and retail trade	Predominantly integrated	CEO message Circular strategy Priority issues	Design (durability) Recycle Reuse	Consumer innovation
	MGROS/Wholesale and retail trade	Partially integrated	Priority issues Circular plan	Recycle Reuse Reduce	Suppliers
	TCELL/Telecommunication	Partially integrated	–	Reuse Recycle	SDA

	Business/Sector	*CE reporting approaches*	*Circular management*	*Circular implementations*	*Circular collaboration and membership*
	THY/Transportation	Partially integrated	–	Reuse Recycle Reduce	–
	ENKA/Construction	Partially integrated	–	Design Reduce Recycle Reuse	–
Industry Sector	AKCNS/Based on stone and soil	Partially integrated	Sustainable management	Recycle Reuse	–
	CIMSA and AFYON Group/Based on stone and soil	Partially integrated	Priority issues Environmental management	Recycle Reduce Reuse	–
	AEFES/Food, beverage, and tobacco	Predominantly integrated	Priority issues Global sustainability committee	Reduce Reuse Recycle	Circular Economy Working Group and SDA Circular Economy Platform Enterprise Workshop
	ULKER/Food, beverage, and tobacco	Partially integrated	–	Reduce Recycle Reuse	-
	CCOLA/Food, beverage, and tobacco	Partially integrated	Priority issues	Recycle Reuse Reduce	Local communities, NGOs, industry peers, and consumers UNDP, ÇEVKO The Collect and Recycle Alliance

(continued)

Table 2 (continued)

Business/Sector	*CE reporting approaches*	*Circular management*	*Circular implementations*	*Circular collaboration and membership*
ARCLK/Electrical devices	Fully integrated	Priority issues Roadmap	Design (durability) Reduce Recycle Repair Reuse Remanufacturing	Innovation Suppliers World Economic Forum Circular Electronics Part EU Horizon 2020 Framework Prog AMDEA
BFREN/Electrical devices	Predominantly integrated	Vision Circular economy strategy	Design (durability) Reduce Recycle Repair Reuse Remanufacturing	Sustainable innovation R&D
FROTO/Electrical devices and transportation vehicles	Predominantly integrated	General Manager Message Priority issues	Design Reduce Reuse Recycle	R&D innovation
TOASO/Electrical devices and transportation vehicles	Partially integrated	–	Recycle Reduce Reuse	-
KORDS/Textile	Partially integrated	CEO message	Reuse Recycle Reduce Remanufacturing	PolynSPIRE
PETKM/chemical pharmaceuticals petroleum rubber and plastic products	Partially integrated	–	Recycle Reduce Reuse	R&D and innovation
MONDI Group/Paper and paper products printing	Not mention	–	-	–

In predominantly integrated businesses CE is addressed as consumer engagement, innovation and technology, the 3R (reduce, recycle, and reuse) model, resource efficiency, and lean production with a zero waste approach. The fully integrated business proactively manages the transition process with the action plan. The action plan has three focal points: circular design, take-back, collection, recycling, and sourcing.

According to the findings, most businesses (about 76%) carry out CE implications such as recycling, reducing, and reusing. While reuse, repair, and remanufacturing are prioritized in CE transformation, recycling is the last element of transformation (EMF, 2019). Still, recycling is used more than other concepts in the reports reviewed. Design is critical in the CE cycle (EMF, 2019; European Commission (b), 2020), and design influences the success of the transformation. Despite this, a smaller part of the businesses (about 16%) includes design at the same time, repair, and remanufacturing applications in their CE activities. Systems perspective needs collaborations in transformation. This issue is emphasized in 'A new Circular Economy Action Plan' prepared by the European Commission (b) (2020). According to the data, 14 businesses have further cooperation or membership levels. When cooperating businesses are examined, the issue of innovation stands out. 50% of these businesses collaborate on Innovation. Responsible production and consumption, especially supplier and consumer cooperation (Van Buren et al., 2016), stand out. According to the findings, the cooperation of businesses in this direction is quite low. For sectoral comparison, all collaborations of businesses were collected and weighted according to sectors. Accordingly, all businesses realized a total of 39 collaborations. Industrial businesses realized 61.5% of these collaborations.

Since CE offers the potential to contribute significantly to the realization of the SDGs (EMF, 2021), the circular economy approaches of businesses have been evaluated within the scope of SDGs' performances in addition to the research questions. Accordingly, the share of the fully integrated approach in the SDGs performance is 31%, while the share of not mentioning the business is 18%. Compared to all other approaches, the business with CE fully integrated approach has the highest level of SDGs performance (Fig. 5). In this respect, it can be said that the SDGs performance is related to the weight of CE in business management and practice.

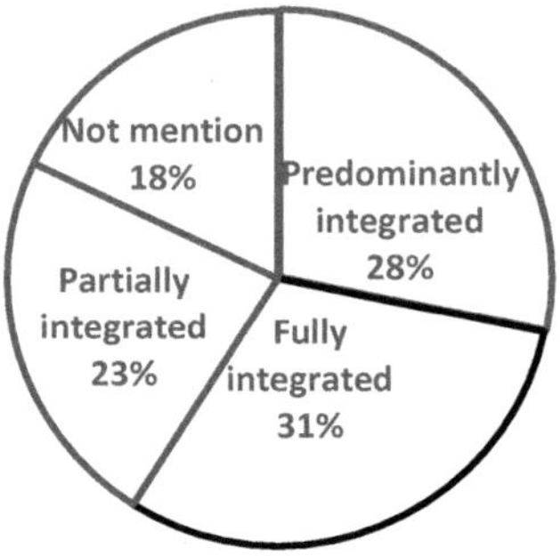

Fig. 5 SDGs performance by CE reporting approach

5 Discussion

This study aims to reveal the general picture of the circular economy practices of business in a developing economy and their contribution to sustainable development goals. The reason for the emergence of this research is to examine the green transformation of businesses in Turkey, which is located in one of the regions expected to be most affected by climate change and has a developing country economy that has a critical importance in reaching the 2030 and 2050 targets. CE, sustainability, and SDGs in industry and service businesses measure this transformation. Research findings are handled separately in terms of theoretical and practical contributions.

5.1 Theoretical Contributions

The analysis findings conducted to answer the research questions are expected to contribute to the literature. Especially in developing countries, the uncertainty of CE transformation (Azizuddin et al., 2021; Halog & Anieke, 2021; Upadhyay et al., 2021) and the relative scarcity of relevant research (Elia et al., 2020) bring uncertainty in the decisions and practices of businesses. This research can contribute to the literature by revealing the current situation and general trends of businesses in terms of approaches and practices of sustainability transformation.

The need for change as a system in the effectiveness and success of the sustainability and CE model requires stakeholder engagement and interaction (Baah et al., 2022; Velenturf & Purnell, 2021). In this sense, reporting can be used to track circularity steps, control and clarify processes and strategies (PwC, 2020), share progress with internal and external stakeholders, and receive feedback (ISO, 2022). Therefore, as part of CS management, sustainability reporting is the best way for businesses to share their CE and sustainability activities with stakeholders (Blackburn, 2007). However, one of the most important findings is that only about 9% of businesses have international standards and independently audited sustainability reporting. Approximately 50% of these businesses do not publish sustainability reports. Most of the remaining businesses have only touched upon very limited sustainability in their annual reports. According to these findings, it can be said that the interaction of businesses with their stakeholders on sustainability and CE is quite low. One reason for this may be that publishing sustainability reports are voluntary. Another reason may be that the weight of sustainability and CE in the decisions and practices of businesses or stakeholder interactions is absent or low (Çalışkan, 2012). Similarly, according to KMPG (2017), Turkish businesses are well below the world average in sustainability reporting.

The concepts of CE and sustainability encompass economic, social, and environmental systems (Lacy et al., 2020). In all business activities, environmental sustainability (38%) and social sustainability (37%) have a higher share than economic sustainability (25%). In the research conducted by Araci and Yuksel (2016) on BIST Sustainability Index businesses in Turkey, economic performance was determined as 41%, social performance as 33%, and environmental performance as 26%. Accordingly, the weight of social and environmental sustainability in business activities has increased over the years, and the share of economic sustainability has decreased. This finding is in line with the results of the analysis conducted by Sehnem et al. (2020) and Tiscini et al. (2022) in Brazil and Italy.

According to the analysis conducted to determine the CE reporting approaches of businesses, 27.5% of businesses have full and predominance approaches, while most have partial approaches (68%). In the research of Opferkuch et al. (2021), the concept of CE was not used in most businesses, while it was partially included in the smaller part. In this respect, it can be argued that the finding of this study is relatively positive. Although, these businesses have associated CE activities and goals with only one or a few contents. It shows that the circular model, expected to bring transformation in realizing sustainability goals, has not yet become a central issue. This finding is supported by previous studies (Azizuddin et al., 2021; Opferkuch et al., 2021; Upadhyay et al., 2021). In the literature,

CE applications are accepted as a potential model for achieving SDGs (EMF, 2021; European Commission (b), 2020). In this research, SDG8 (Decent Work and Economic Growth), SDG12 (Responsible Consumption and Production), SDG9 (Industry, Innovation, and Infrastructure), and SDG13 (Climate Action) are dominant in the goals and activities of businesses. The SDGs performance rate of the fully CE-integrated business (31%) is better than any CE approach and differs significantly from the not mentioned business share (18%). Fully integrated approaches differ in how they integrate CE with the organization's strategy, vision, goals, and practices. So, research findings support the literature (Geissdoerfer et al., 2017). Also, in evaluating the sectoral view, the partially integrated approach is more dominant in the service sector. In contrast, the predominantly and fully integrated approaches have a higher weight in the industrial sector. The reasons for this may be the higher expectations from the industrial sector due to its environmental impacts (European Commission (d), 2020) and the need for the industrial sector to adapt to international trade (TUSIAD, 2021).

There are circular practices at different levels in almost all businesses. Still, circular practices are mostly handled under the heading of waste management in businesses and stand out as reduce, recycle, and reuse. Especially in developing countries, waste is an issue that brings important effects on its own. Therefore, the CE is the best way to deal with waste (Halog & Anieke, 2021). However, addressing CE only as waste management in businesses is insufficient to achieve the SDGs and ensure the desired transformation. The number of businesses that take the circular model as a business strategy and integrate it into the entire product life cycle is limited. It is in line with the findings of the studies of Upadhyay et al. (2021), Opferkuch et al. (2021), and Tiscini et al. (2022).

The transition toward CE adoption is possible with cleaner production and eco-design (Ghisellini et al., 2016). Therefore, the design and production phases are critical points of the circular approach. EU has put design at the center of the CE model in the circular transformation (European Commission (b), 2020). EMF (2021) argues that CE is shaped entirely by design-based principles. A design orientation ensures that pollution and waste are eliminated first (Ghisellini et al., 2016) and contributes directly to regeneration (de Sousa Jabbour et al., 2019). Accordingly, focusing on design helps to prevent the problem at its source. The prevention approach goes down to the source to reduce or eliminate pollution or waste, but an opposite is a control-based approach.

In this approach, the problem is tried to be solved after it has arisen. While the problem solved at its source has the potential to prevent carbon emissions and reduce production costs through efficiency significantly, its control-based approach has no potential to reduce production costs (del Rio Gonzalez, 2005). So, the prevention approach is appropriate for the CE model. The analysis shows that CE design activities remain limited in business applications. Only a few business activities address the CE-related design issue. It is possible to see the traces of green transformation in businesses but reaching the emission target scenarios requires innovative and comprehensive investments.

Top management attitudes and collaborations are emphasized in reaching SDGs and circular transformation (European Commission (b), 2020; United Nations, 2015). The system-wide changes bring multilateral and multi-participatory responsibility. In the systems approach, sustainability is strongly influenced by Stakeholder theory as it affects the vision and strategies of all actors and translates directly into the concept of collaboration (Prasad & Elmes, 2005). Almost half of the businesses have CE memberships or collaborations at different levels. While approximately 13% of these businesses cooperate with suppliers, 8% cooperate with consumers. CE encourages new relationships by reintegrating suppliers and consumers into production cycles (Van Buren et al., 2016). The success of CE depends on the participation of all actors (De Angelis, 2018; Ghisellini et al., 2016), and increasing consumer responsibility further strengthens this virtuous cycle (Ghisellini et al., 2016). In the sectoral comparison of collaborations, industrial businesses (61.5%) interact with more stakeholders than service sector businesses. According to the analysis, although the CE issue was brought to the agenda by the senior management (CEO, General Manager, etc.) in general, the CE issue was either not encountered at all or was only limited in practice. For these businesses, it can be said that CE is not adopted in business strategies and policies and generally remains at the level of discourse.

5.2 *Practical Contributions*

The research findings make it possible to conclude the concepts of CE, sustainability, and SDGs at the business and sector levels. It has been determined that CE is at the beginning level in businesses, the focus of the implementation is on waste management, and it is included in a limited number of business strategies, visions, and plans. The research has the potential to draw attention and raise awareness on the issue as it reveals the general situation of sustainability transformation in the relevant sectors and businesses and offers findings that can be useful for decision-makers.

The average of all businesses' sustainability performance is determined as 0.50. Similarly, the average CS disclosure performance of businesses was determined as 0.60. While the industrial sector remains below this performance, the performance of the service sector is 0.55. Also, the service businesses' SDGs performance is better than industry businesses. In addition, most business has adopted the CE partially integrated approach. Therefore, considering the 2030 and 2050 targets, it can be said that businesses' performance should be improved. In this sense, businesses need to take more action. PwC (2020) recommends six steps for service sector businesses on the path to circularity. These are; identifying opportunities for circularity, clarifying strategy and vision, planning the transformation route, developing collaborations, establishing control mechanisms, and being proactive. ISO (2022) proposes four strategies in its sectoral roadmap for industrial businesses within the climate change framework and CE. These are situation assessments, including stakeholder expectations, local and global trends, determining the vision, strategy, goal setting and implementation, performance monitoring, and reporting. Accordingly, although the importance of reporting on sustainability and CE management is clear, it has been determined that there is no standard related to CE in the reports examined. GRI consists of standards based on the concepts of SDGs and sustainability, so disclosures on CE may lead to different practices. On the other hand, the rate of business publishing sustainability reports is quite low. Sectoral efforts may be beneficial in reviewing the dynamics related to sharing sustainability reports by businesses and raising awareness.

One of the reasons for the significant variation in the sustainability performance of businesses was governance. Since the concept of governance is related to stakeholder engagement, transparency, accountability, fairness to stakeholders, consistency, and allocation of corporate trust (ISO, 2022), it is also addressed in this study. Only 9% of businesses included information on governance in their reports. This finding is in line with the research findings of Tiscini et al. (2022). Parallel to this finding, the collaboration of businesses with stakeholders within the scope of CE is also limited. As in the fully and predominantly integrated approaches, interaction with stakeholders is higher in the industrial sector.

In this respect, stakeholder collaboration varies according to the CE management and naturally according to the approach. Therefore, businesses need to make more efforts to include CE in their strategies, visions, and goals.

Micro efforts are valuable in the transformation process but insufficient. Due to the systems approach, CE includes a wide stakeholder group and imposes responsibilities on different parties. In this sense, sector representatives can draw more attention to this issue and mobilize the sector on measures to increase cooperation and prepare the ground. Macro-level efforts are also necessary for sectors to take action. TUSIAD (2021) recommends that the CE transformation cannot be realized with the current policies and regulations and needs to be updated. Similarly, according to the research of WWF (2021) and Şahin et al. (2021), current practices and policies should be updated; otherwise, Turkey is moving away from its sustainability goals.

6 Conclusion

This research has tried to explain businesses' current situation regarding CE, SDGs, and sustainability with the support of the literature. The research has important outputs representing industrial and service businesses operating in the developing country's economy in terms of its scope. The SDGs and 2050 targets adopted by Turkey bring important responsibilities to businesses. On the other hand, the EU is trying to influence other countries in its green transformation and to participate in this transformation. Significant changes in the sense of businesses' responsibility to sustain life for a while and fulfill our responsibility toward future generations is a need. This change in understanding covers all businesses regardless of the level of economic development (Rendtorff, 2019). In this sense, developing countries need to realize their economic growth based on sustainable values.

Within the scope of this study, it is promising that CE is included in almost all business reports. However, CE is addressed limitedly by most businesses within the waste management framework; this distracts businesses from the focus of the system and the effectiveness of CE implementation. The finding can be explained by the fact that CE is quite new on the business agenda and has not been internalized. In businesses with fully and predominantly integrated approaches, CE is addressed in the management and implementation focus, and stakeholder collaborations are more important. In this sense, the industrial sector is more effective. Although the service sector has higher sustainability performance, its focus on CE management, practices, and cooperation with stakeholders is lower. It shows that the awareness of the service sector on CE is lower than the industrial sector.

On the other hand, the fully integrated approach has the highest share of SDGs performance. Sustainability, which emerges as a necessity for creating value for the system approach and stakeholders, should be included in strategies and targets. More efforts are needed for green transformation.

This research has some limitations. One of these limitations is that businesses may not have reflected all of their CE decisions and practices in their reports due to the lack of a reporting standard on CE reporting. It may limit the representativeness of the findings. Another limitation of this research is that the research analysis covers one year. Since preparing international reports is voluntary, some businesses do not publish reports regularly. Therefore, the change between periods could not be observed. At the same time, examining the research variables outside the framework provided by the report's scope was impossible. Future studies should focus on in-depth examinations of CE approaches based on interviews and field studies to increase the scope and depth of the topic. The research findings are representative of BIST-listed industrial and service sector businesses. Therefore, results cannot be generalized to different sectors and regions. Therefore, future studies may obtain stronger results with data from different developing countries to strengthen the representation of developing countries. In addition, future studies can use quantitative methods to reveal businesses' CE approaches and SDGs performance.

Bibliography

Alam, K., Ali, M. A., Erdiaw-Kwasie, M., Shahiduzzaman, M., Velayutham, E., Murray, P. A., & Wiesner, R. (2022). Impact of ICTs on innovation and performance of firms: Do start-ups, regional proximity and skills matter? *Sustainability, 14*(10), 5801.

Azizuddin, M., Shamsuzzoha, A., & Piya, S. (2021). Influence of circular economy phenomenon to fulfill global sustainable development goal: Perspective from Bangladesh. *Sustainability, 13*(20), 11455.

Baah, C., Afum, E., Agyabeng-Mensah, Y., et al. (2022). Stakeholder influence on adoption of circular economy principles: Measuring implications for satisfaction and green legitimacy. *Circular Economy and Sustainability, 2*, 91–111. https://doi.org/10.1007/s43615-021-00093-2

Babbie, E. (2007). *Tile practice of social research.* Thomson Higher Education.

Barnabe, F., & Nazir, S. (2020). Investigating the interplays between integrated reporting practices and circular economy disclosure. *International Journal of Productivity and Performance Management.*

Bansal, S., Jain, M., Garg, I., & Srivastava, M. (2022). Attaining circular economy through business sustainability approach: An integrative review and research agenda. *Journal of Public Affairs, 22*(1), e2319. https://doi.org/10.1002/pa.2319

Baumgartner, R. J., & Ebner, D. (2010). Corporate sustainability strategies: Sustainability profiles and maturity levels. *Sustainable Development, 18*(2), 76–89.

Berelson, B. (1952). *Content analysis in communication research.* American Psychological Association.

Blackburn, W. R. (2007). *The sustainability handbook: The complete management guide to achieving social, economic and environmental responsibility, UK and US: Earth Scan.* ISBN: 13–978–1–84407–495–2.

Blinova, E., Ponomarenko, T., & Knysh, V. (2022). Analyzing the concept of corporate sustainability in the context of sustainable business development in the mining sector with elements of circular economy. *Sustainability, 14*(13), 8163. https://doi.org/10.3390/su14138163

Boiral, O., & Henri, J.-F. (2017). Is sustainability performance comparable? A study of GRI reports of mining organizations. *Business & Society, 56*(2), 283–317. https://doi.org/10.1177/0007650315576134

Bos-Brouwers, H. E. J. (2010). Corporate sustainability and innovation in SMEs: Evidence of themes and activities in practice. *Business Strategy and the Environment, 19*(7), 417–435.

Çalışkan, A. Ö. (2012). Sürdürülebilirlik Raporlamasi. *Journal of Accounting and Taxation Studies, 5*(1), 41–68.

CGDEV. (2022). https://www.cgdev.org/media/developing-countries-are-responsible-63-percent-current-carbon-emissions. 10.10.22.

Daly, H. E. (2006). Sustainable development—Definitions, principles, policies. In M. Keiner (Ed.). In *The future of sustainability* (pp. 39–53). Springer.

De Angelis, R. (2018). *Business models in the circular economy: Concepts, examples, and theory.* Springer.

De Angelis, R. (2021). Circular economy and paradox theory: A business model perspective. *Journal of Cleaner Production, 285*, 124823. https://doi.org/10.1016/j.jclepro.2020.124823

De Angelis, R. (2022). Circular economy business models: A repertoire of theoretical relationships and a research agenda. *Circular Economy and Sustainability, 2*(2), 433–446. https://doi.org/10.1007/s43615-021-00133-x

De Jesus, A., & Mendonça, S. (2018). Lost in transition? Drivers and barriers in the eco-innovation road to the circular economy. *Ecological Economics, 145*, 75–89. https://doi.org/10.1016/j.jclepro.2017.11.111

de Sousa Jabbour, A. B. L., Luiz, J. V. R., Luiz, O. R., Jabbour, C. J. C., Ndubisi, N. O., de Oliveira, J. H. C., & Junior, F. H. (2019). Circular economy business models and operations management. *Journal of Cleaner Production, 235*, 1525–1539.

del Rio Gonzalez, P. (2005). Analyzing the factors influencing clean technology adoption: A study of the Spanish pulp and paper industry. *Business Strategy and the Environment, 14*(1), 20–37.

Delgado, C. (2021). How developing countries can reduce emissions without compromising growth. https://earth.org/how-developing-countries-can-reduce-emissions-without-compromising-growth/. 10.10.22.

EEA (2020). Turkey–Industrial pollution profile 2020. https://www.eea.europa.eu/themes/industry/industrial-pollution/industrial-pollution-country-profiles-2020/turkey. 29.01.22.

Elia, V., Gnoni, M. G., & Tornese, F. (2020). Evaluating the adoption of circular economy practices in industrial supply chains: An empirical analysis. *Journal of Cleaner Production, 273*, 122966. https://doi.org/10.1016/j.jclepro.2020.122966

Elkington, J., & Rowlands, I. H. (1999). Cannibals with forks: The triple bottom line of 21st-century business. *Alternatives Journal, 25*(4), 42.

EMF. (2019). *Circular economy system diagram*. https://ellenmacarthurfoundation.org/circulate-products-and-materials. 10.12.22.

EMF. (2021). *How the circular economy tackles climate change*. Completing the Picture — How the circular economy tackles climate change | Shared by Climate Change (thirdlight.com). 17.10.22.

Erdiaw-Kwasie, M. O., Abunyewah, M., Yusif, S., & Erdiaw-Kwasie, A. (2023). Does circular economy knowledge matter in sustainable service provision? A moderation analysis. *Journal of Cleaner Production*, 135429.

Erdiaw-Kwasie, M. O., Alam, K., & Kabir, E. (2017). Modelling corporate stakeholder orientation: Does the relationship between stakeholder background characteristics and corporate social performance matter? *Business Strategy and the Environment, 26*(4), 465–479.

European Commission (a). (2019). *The European green deal, COM/2019/640 final, Brussels*. resource.html (europa.eu). 16.10.22.

European Commission (b). (2020). *A new circular economy action plan for a cleaner and more competitive Europe, COM/2020a/98 final, Brussels*. https://eur-lex.europa.eu/resource.html?uri=cellar:9903b325-6388-11ea-b735-01aa75ed71a1.0017.02/DOC_1&format=PDF. 16.10.22.

European Commission (c). (2021). *Results and Indicators for development circular economy*. Unit D4/October/2021. circular_economy_results_chain_final_0.pdf (europa.eu). 16.10.22.

European Commission (d). (2020). *A new industrial strategy for Europe, COM/2020c/102 final, Brussels*. https://eur-lex.europa.eu/legal-content/EN/TXT/PDF/?uri=CELEX:52020DC0102&from=IT. 16.10.22.

European Parliament. (2022). *Circular economy: Definition, importance and benefits*. Circular economy: definition, importance and benefits | News | European Parliament (europa.eu). 16.10.22.

EPA (a). *Global emissions by economic sector*. https://www.epa.gov/ghgemissions/global-greenhouse-gas-emissions-data. 29.01.22.

EPA (b). *Transport*. https://www.epa.ie/our-services/monitoring--assessment/climate-change/ghg/transport/. 10.10.22.

Fifka, M. S., & Drabble, M. (2012). Focus and standardization of sustainability reporting—A comparative study of the United Kingdom and Finland. *Business Strategy and the Environment, 21*(7), 455–474.

Freeman, R. E. (1984). *Strategic management: A stakeholder approach* (p. 276s). Pitman.

Gallaud, D., & Laperche, B. (2016). *Circular economy, industrial ecology, and short supply chain*. John Wiley & Sons.

Garza-Reyes, J. A., Salomé Valls, A., Peter Nadeem, S., Anosike, A., & Kumar, V. (2019). A circularity measurement toolkit for manufacturing SMEs. *International Journal of Production Research, 57*(23), 7319–7343. https://doi.org/10.1080/00207543.2018.1559961

Geissdoerfer, M., Savaget, P., Bocken, N. M., & Hultink, E. J. (2017). The circular economy–A new sustainability paradigm? *Journal of cleaner production, 143*, 757–768. https://doi.org/10.1016/j.jclepro.2016.12.048

Ghisellini, P., Cialani, C., & Ulgiati, S. (2016). A review on circular economy: The expected transition to a balanced interplay of environmental and economic systems. *Journal of Cleaner Production, 114*, 11–32.

GRI. (2016). *GRI standards 2016*. Amsterdam, ISBN: 978-90-8866-095-5. https://www.globalreporting.org/standards/media/1036/gri-101-foundation-2016.pdf. 10.10.22.

GRI. (2018). *GRI standards 2018*. Amsterdam, ISBN: 978-90-8866-094-8. https://www.globalreporting.org/standards/media/1913/gri-standards-glossary.pdf. 10.10.22.

GRI. (2019). *GRI standards 2019*. Amsterdam, ISBN: 978-90-8866-131-0. https://www.globalreporting.org/standards/media/2482/gri-207-tax-2019.pdf. 10.10.22.

GRI. (2020). *GRI standards 2020*. Amsterdam, ISBN: ISBN: 978-90-8866-094-8. https://www.globalreporting.org/standards/media/2594/gri-standard-glossary-2020.pdf. 10.10.22.

GRI. (2022). *A short introduction to the GRI standards*. Amsterdam. https://www.globalreporting.org/media/wtafl4tw/a-short-introduction-to-the-gri-standards.pdf. 10.01.22.

Gupta, S., Chen, H., Hazen, B. T., Kaur, S., & Gonzalez, E. D. S. (2019). Circular economy and big data analytics: A stakeholder perspective. *Technological Forecasting and Social Change, 144*, 466–474. https://doi.org/10.1016/j.techfore.2018.06.030

Halog, A., & Anieke, S. (2021). A review of circular economy studies in developed countries and its potential adoption in developing countries. *Circular Economy and Sustainability, 1*(1), 209–230.

Heyes, G., Sharmina, M., Mendoza, J. M. F., Gallego-Schmid, A., & Azapagic, A. (2018). Developing and implementing circular economy business models in service-oriented technology companies. *Journal of Cleaner Production, 177*, 621–632. https://doi.org/10.1016/j.jclepro.2017.12.168

IEA. (2022). *Transport: Improving the sustainability of passenger and freight transport*. https://www.iea.org/topics/transport. 09.10.22.

IPCC. (2014). *Climate change 2014: Mitigation of climate change*. Cambridge University Press. https://www.ipcc.ch/site/assets/uploads/2018/ 02/ipcc_wg3_ar5_full.pdf. 6.09.2020.

Isil, O., & Hernke, M. T. (2017). The triple bottom line: A critical review from a transdisciplinary perspective. *Business Strategy and the Environment, 26*(8), 1235–1251.

ISO. (2022). *EU green deal*. https://www.iso.org.tr/surdurulebilirlik/video-ab-yesil-mutabakat-iklim-degisikligiturkiyeye-etkileri.html. 13.12.22.

Kirchherr, J., Piscicelli, L., Bour, R., Kostense-Smit, E., Muller, J., Huibrechtse-Truijens, A., & Hekkert, M. (2018). Barriers to the circular economy: Evidence from the European Union (EU). *Ecological Economics, 150*, 264–272.

Koçel, T. (2014). İşletme Yöneticiliği, Genişletilmiş 15. Baskı, İstanbul, Beta Yayınları.

KPMG. (2020). *The road ahead*. https://assets.kpmg.com/content/dam/kpmg/xx/pdf/2017/10/kpmg-survey-ofcorporate-responsibility-reporting-2017.pdf. 14.12.22.

Lacy, P., Long, J., & Spindler, W. (2020). *The circular economy handbook: Realizing the circular advantage*. Palgrave Macmillan.

Marjamaa, M., Salminen, H., Kujala, J., Tapaninaho, R., & Heikkinen, A. (2021). A sustainable circular economy: Exploring stakeholder interests in Finland. *South Asian Journal of Business and Management Cases, 10*(1), 50–62. https://doi.org/10.1177/2277977921991914

Marshall, R. S., & Brown, D. (2003). The strategy of sustainability: A systems perspective on environmental initiatives. *California Management Review, 46*(1), 101–126.

Meuer, J., Koelbel, J., & Hoffmann, V. H. (2020). On the nature of corporate sustainability. *Organization & Environment, 33*(3), 319–341.

Montiel, I. (2008). Corporate social responsibility and corporate sustainability: Separate pasts, common futures. *Organization & Environment, 21*(3), 245–269.

Montiel, I., & Delgado-Ceballos, J. (2014). Defining and measuring corporate sustainability: Are we there yet? *Organization & Environment, 27*(2), 113–139.

Nikolaou, I. E., & Tsagarakis, K. P. (2021). An introduction to circular economy and sustainability: Some existing lessons and future directions. *Sustainable Production and Consumption, 28*, 600–609.

Opferkuch, K., Caeiro, S., Salomone, R., & Ramos, T. B. (2021). Circular economy in corporate sustainability reporting: A review of organizational approaches. *Business Strategy and the Environment, 30*(8), 4015–4036.

Parker, J. (2014). *Critiquing sustainability, changing philosophy*. Routledge.

Pichlak, M., & Szromek, A. R. (2022). Linking eco-innovation and circular economy—A conceptual approach. *Journal of Open Innovation: Technology, Market*, 8, 121. https://doi.org/10.3390/joitmc8030121

Porter, M., & Van der Linde, C. (1995, September–October). Green and competitive: Ending the stalemate. In F. M. Wubben, Emiel (Ed.), *The Dynamics of the eco-efficient economy: Environmental regulation and competitive advantage* (pp. 33–55).

Prasad, P., & Elmes, M. (2005). In the name of the practical: Unearthing the hegemony of pragmatics in the discourse of environmental management. *Journal of Management Studies, 42*(4), 845–867.

PwC. (2020). *Taking on tomorrow: The rise of circularity in energy, utilities and resources*. Döngüsel ekonominin yükselişi | PwC Türkiye. 15.12.22.

PwC. (2022). *Sustainability in industrial production*. https://www.pwc.de/en/sustainability/sustainability-in-industrial-production.html. 28.07.22.

Rendtorff, J. D. (2019). *Philosophy of management and sustainability: Rethinking business ethics and social responsibility in sustainable development*. Emerald Group Publishing.

Ritche. (2020). *Cars, planes, trains: where do CO2 emissions from transport come from?* https://ourworldindata.org/co2-emissions-from-transport. 9.10.22.

Rouf, M. A. (2011). The corporate social responsibility disclosure: A study of listed companies in Bangladesh. *Business and Economics Research Journal, 2*(1), 19–32.

Şahin vd. (2021). Türkiye'nin Karbonsuzlaşma Yol Haritası: 2050'de Net Sıfır, İstanbul Politikalar Merkeci (İPC), Sabancı Üniversitesi. https://ipc.sabanciuniv.edu/Content/Images/CKeditorImages/20211026-23103436.pdf. erişim tarihi 24.03.22.

Sanches, J. R., Trevisan, A. H., Seles, B. M. R. P., Castro, C. G., Piao, R. S., Rozenfeld, H., & Mascarenhas, J. (2022). Sustainable circular economy strategies: An analysis of Brazilian corporate sustainability reporting. *Sustainability., 14*(10), 5808. https://doi.org/10.3390/su14105808

Santos, B., Azevedo, S., & Matias, J. (2017). The relation between corporate sustainability and the circular economy. In S. G. Azevdo & J. C. O. Matias (Ed.), *Corporate sustainability: The new pillar of the circular economy* (1–14). Nova.

SDG Compass. (2015). *The guide for business action on the SDGs*. Erişim 29.01.2020. https://sdgcompass.org

Sehnem, S., Pandolfi, A., & Gomes, C. (2020). Is sustainability a driver of the circular economy? *Social Responsibility Journal., 16*(3), 1747–1117. https://doi.org/10.1108/SRJ-06-2018-0146

Sommer, A. (2012). *Managing green business model transformations*. Springer Science & Business Media.

Stahel, W. R. (1982). *The product life factor. An inquiry into the nature of sustainable societies: The role of the private sector* (Series: 1982 Mitchell Prize Papers), NARC, 74–96.

Stewart, R., & Niero, M. (2018). Circular economy in corporate sustainability strategies: A review of corporate sustainability reports in the fast-moving consumer goods sector. *Business Strategy and the Environment, 27*(7), 1005–1022. https://doi.org/10.1002/bse.2048

TCTB. (2022). Hizmet Ticaret İstatistikleri. https://ticaret.gov.tr/hizmet-ticareti/hizmet-ticaret-istatistikleri. 13.10.22.

Thomas, D., & Kai, H. (2002). Beyond the case for corporate sustainability. *Business Strategy and the Environment*, 11.

Tiscini, R., Martiniello, L., & Lombardi, R. (2022). Circular economy and environmental disclosure in sustainability reports: Empirical evidence in cosmetic companies. *Business Strategy and the Environment, 31*(3), 892–907.

Tunn, V. S., Bocken, N. M., van den Hende, E. A., & Schoormans, J. P. (2019). Business models for sustainable consumption in the circular economy: An expert study. *Journal of Cleaner Production, 212*, 324–333.

TUSİAD. (2021). Avrupa Yeşil Mutabakatı Döngüsel Ekonomi Eylem Planı Türk İş Dünyasına Neler Getirecek?, TÜSİAD-T/2021–06/621. https://tusiad.org/tr/yayinlar/raporlar/item/10790-avrupa-yesil-mutabakati-dongusel-ekonomi-eylem-plani-turk-is-dunyasina-neler-getirecek. 13.12.2022.

UNCTAD. (2021). *Carbon emissions anywhere threaten development everywhere*. https://unctad.org/news/carbon-emissions-anywhere-threaten-development-everywhere 10.10.22.

United Nations. (2015). *Paris agreement*. 27.07.2022. https://unfccc.int/process-and-meetings/the-paris-agreement/the-paris-agreement

United Nations. (2022). *World economic situation and prospects 2022.*. 4.12.22. https://www.un.org/development/desa/dpad/wp-content/uploads/sites/45/WESP2022_ANNEX%20.pdf

Upadhyay, A., Laing, T., Kumar, V., & Dora, M. (2021). Exploring barriers and drivers to the implementation of circular economy practices in the mining industry. *Resources Policy, 72*, 102037.

Van Buren, N., Demmers, M., Van der Heijden, R., & Witlox, F. (2016). Towards a circular economy: The role of Dutch logistics industries and governments. *Sustainability, 8*(7), 647.

Velenturf, A. P., & Purnell, P. (2021). Principles for a sustainable circular economy. *Sustainable Production and Consumption*, *27*, 1437–1457. https://doi.org/10.1016/j.spc.2021.02.018

WCED. (1987). *Report of the world commission on environment and development: Our common future*. 26.07.2022. https://sustainabledevelopment.un.org/milestones/wced

Weber, O. (2012). Environmental credit risk management in banks and financial service institutions. *Business Strategy and the Environment, 21*(4), 248–263.

WEF. (2022). *Reducing the carbon footprint of the manufacturing industry through data sharing*. https://www.weforum.org/impact/carbon-footprint-manufacturing-industry/. 10.10.22.

World Bank. https://data.worldbank.org/income-level/low-and-middle-income?view=chart. 12.12.22.

WWF. (2021). İklim Değişikliğinin Akdeniz'deki Etkileri. 27.07.2022. https://www.wwf.org.tr/ne_yapiyoruz/iklim_degisikligi_ve_enerji/iklim_degisikligi/

Yang, Y., Chen, L., Jia, F., & Xu, Z. (2020). Complementarity of circular economy practices: An empirical analysis of Chinese manufacturers. *International Journal of Production Research, 57*(20), 6369–6384. https://doi.org/10.1080/00207543.2019.1566664

Circular Value Chains: Circular Strategies and Managerial Perceptions of Supply Chain Professionals from Turkey

Halim Kazan and *Adil Ünal*

1 Introduction

A supply chain is a network of manufacturers and service providers who work together during the transformation and transportation of products from raw material form to the final product (Bozarth & Handfield, 2008: 4). Association for Supply Chain Management (ASCM) defines supply chain management as "The design, planning, execution, control, and monitoring of supply chain activities to create net value, building a competitive infrastructure, leveraging worldwide logistics, synchronising supply with demand, and measuring performance globally" (Pittman & Atwater 2019: 185). According to Bozarth and Handfield, supply chain

H. Kazan (✉)
Faculty of Economics, Istanbul University, Istanbul, Türkiye
e-mail: halim.kazan@istanbul.edu.tr

A. Ünal
Department of Business Administration, Faculty of Political Sciences, Istanbul University, Istanbul, Türkiye
e-mail: aunal@istanbul.edu.tr

M. O. Erdiaw-Kwasie and G. M. M. Alam (eds.), *Circular Economy Strategies and the UN Sustainable Development Goals*,
Sustainable Development Goals Series,
https://doi.org/10.1007/978-981-99-3083-8_15

management is the active management of supply chain activities and relationships to maximise customer value and provide sustainable competitive advantage (Bozarth & Handfield, 2008: 8). In the words of Eliyahu M. Goldratt in his book "The Goal", the mission or purpose of supply chain management can be explained as "reducing inventory (stock) levels and operating expenses simultaneously while increasing sales" (Hugos, 2006: 9).

In the literature, some studies analyse the supply chain at different levels. In this direction, explanations are made at the level of bilateral relations (dyadic level), between supply chain members, at the chain level, which includes more than one bilateral relationship, and at a network level consisting of different operations. Within bilateral relations, analyses are carried out over two companies such as supplier-manufacturer and producer-distributor. In the chain approach, the set of relations from the supplier's supplier to the customer's customer and analyses are based on upstream and downstream relations at the network level (Croom et al., 2000).

The field of supply chain management, by its nature, covers intra-business and inter-business relations. Therefore, supply chain management research requires understanding inter-organisational and intra-organisational systems. The supply chain's holistic, organisational and inter-organisational focus extends systems thinking to the theory of "systems of systems" (SoS). Modern supply chains include interrelated and intra-firm processes of many globally complex SoS (Randall & Mello, 2012: 864).

Bechtel and Jayaram discussed the development of supply chain understanding and classified supply chain schools of thought (Hieber, 2002). The authors focused on developing the supply chain field in four main axes: supply chain awareness school, connection/logistics school, information school and integration school. In addition to these schools of thought, some studies have developed approaches that deal with supply chains in terms of sustainability, business continuity, reverse supply chain, closed-loop supply chain, digitalisation and circular supply chain (CSC) since the 2000s (Table 1).

In traditional linear supply chains, consumers or industrial buyers of products are considered in a passive position and are on the chain as the last link of the supply chain. Reverse supply chain flows from buyers to the focus company or other supply chain members are limited. Therefore, it can be argued that businesses should position consumers or industrial

Table 1 Supply chain schools of thought

Authors	*Main focus of supply chain management definition*
	Chain Awareness School
Erdiaw-Kwasie et al. (2023), Jones and Riley (1985), Houlihan (1988), Langley and Holcomb (1992), Cavinato (1992), Novack and Simco (1991), Stevens (1990), Lee and Billington (1992)	"Material flow process from suppliers to end users"
	Linkage/Logistics School
Scott and Westbrook (1991), Turner (1993)	"Supply chain connects all links of the production and supply process, from the raw material to the end customer"
	Information School
Johannson (1994), Towill et al. (1992)	"Connection and flow of information between supply chain management and the various members of the supply chain"
	Integration School
Ellram and Cooper (1990), Cooper and Ellram (1993), Hewitt (1992), Fung et al. (2008)	"Integrative philosophy for managing the stock points and total flow of a distribution channel from the supplier to the end user"
	Supply Chain Sustainability School
Carter and Rogers (2008), Seuring and Müller (2008)	"Strategic, transparent integration and achievement of an organisation's social, environmental and economic goals in the systematic coordination of key business-to-business processes to improve the long-term economic performance of the focus company and its supply chains"
	Supply Chain Business Continuity School
Kildow (2011), Ho et al. (2015)	"Supply chain managers need to understand the fundamentals of business continuity and identify, assess, mitigate and monitor unexpected macro and micro-level events or conditions that may adversely affect any part of a supply chain"

(continued)

Table 1 (continued)

Authors	*Main focus of supply chain management definition*
	Reverse Supply Chain Management School
Prahinski and Kocabasoglu (2006)	Reverse supply chain management (RSCM) is the effective and efficient management of the activities required to retrieve a product from a customer and either dispose of it or recover its value
	Closed-loop Supply Chain Management School
Govindan et al. (2015)	"Closed-loop supply chain management (CLSCM) is the design, control, and operation of a system to maximise value creation over the entire life cycle of a product with the dynamic recovery of value from different types and volumes of returns over time"
	Digitalisation School
Alam et al. (2022), Liotine (2019)	"A digital supply chain is a chain that uses data and systems to connect operations and business functions with domestic and foreign trade partners and relevant stakeholders"
	Circular Supply Chain School
Batista et al. (2018)	"The coordinated forward and reverse supply chains via purposeful business ecosystem integration for value creation from products/services, by-products, and useful waste flows through a prolonged life cycles that improve the economic, social and environmental sustainability of organisations"
Farooque et al. (2019)	"Circular supply chain management (CSCM) is the integration of circular thinking into managing the supply chain and its surrounding industrial and natural ecosystems"

Source Adapted from Bechtel and Jayaram (1997: 17)

buyers not as the end of linear supply chains but as the centre of dynamic value creation and maintenance networks to transform their supply chains (Hazen et al., 2022). However, there is a reverse flow of materials in the closed-loop and open-loop supply chain understanding. Products can flow as a closed-loop input from industrial buyers and consumers within their supply chain or as an open-loop input within other supply chains and industries.

The circular supply chain approach adopts a more innovative and integrated approach. According to Farooque et al. (2019), circular supply chain management systematically restores technical materials and regenerates biological materials towards a zero-waste vision through system-wide innovation in business models and supply chain functions from product/service design to end-of-life and waste management, involving all stakeholders in a product/service life cycle including parts/product manufacturers, service providers, consumers and users (Fig. 1).

Montag (2022) discusses different understandings of the supply chain regarding value focus and waste elimination focus (Fig. 1). According to the author, circular supply chains focus more strongly on value creation and waste removal (Montag, 2022: 30) (Fig. 2).

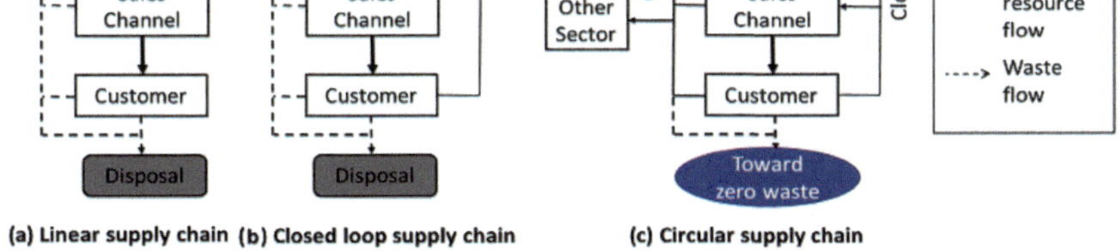

Fig. 1 Linear, closed-loop and circular supply chains (*Source* Farooque et al. [2019])

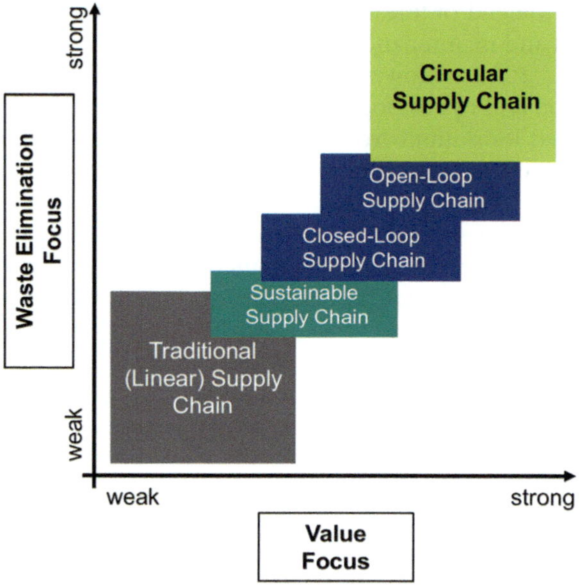

Fig. 2 Value and waste elimination focus of CSCs (*Source* Montag [2022])

2 Value Chain vs. Supply Chain

According to Holweg and Helo (2014), when it comes to designing and managing the supply chain, we come across two fields of study. The authors note a difference in understanding between the "value chain" view, which considers the value creation and allocation aspect of the supply chain, and the operational "supply chain", which develops strategies for designing and operating efficient systems within and between businesses. While the value chain view focuses on financially capturing the maximum added value, the supply chain view focuses on operational efficiency (Holweg & Helo, 2014: 230).

Michael Porter classified value chain activities in his book "Competitive Advantage" in 1985. According to him, the value chain of a business and its operation reflects the history of the business, its strategy and the way it implements strategies. Porter defined value chain activities as primary and support activities (Porter, 1985: 36). These activities are presented in Fig. 3.

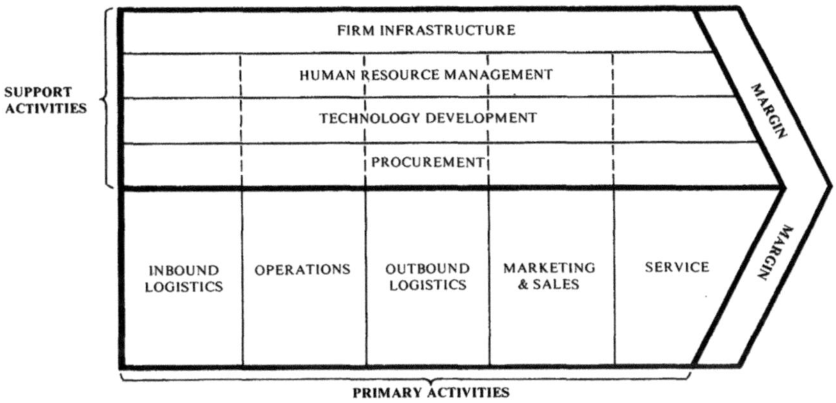

Fig. 3 Primary and support activities in the value chain (*Source* Porter [1985])

Value chain activities are the main building blocks of competitive advantage. Comparing the competitors' value chains and the firm's value chain reveals the differences determining the competitive advantage (Porter, 1985: 39). Some authors emphasise that Porter's value chain approach should be updated with a new concept. Monczka et al. (2016) state that the focus should be on the concept of "extended value chain" (extended business) because the value chain only focuses on the inside of the business, ignoring suppliers and customers (Monczka et al., 2016: 14).

In the extended value chain approach, the activities presented in horizontal flows represent the operational flows of the value chain, in other words, the supply chain operations (Fig. 4). Therefore, we can think of the concept of the supply chain as a subset of the value chain. So, for example, all employees in the organisation are part of the value chain. But the same is not true for the supply chain (Monczka et al., 2016: 14).

Some normative process models classify and categorise supply chain activities in the literature. A normative model consists of a predefined set of alternatives. In addition, normative models describe how an object in the model can be viewed and defined and how it should behave. In this respect, it sets a norm for the chosen field (Poluha, 2016: 74). The two most accepted normative process models are the "Supply Chain Operations Reference" (SCOR) model, published by ASCM in its 14th version

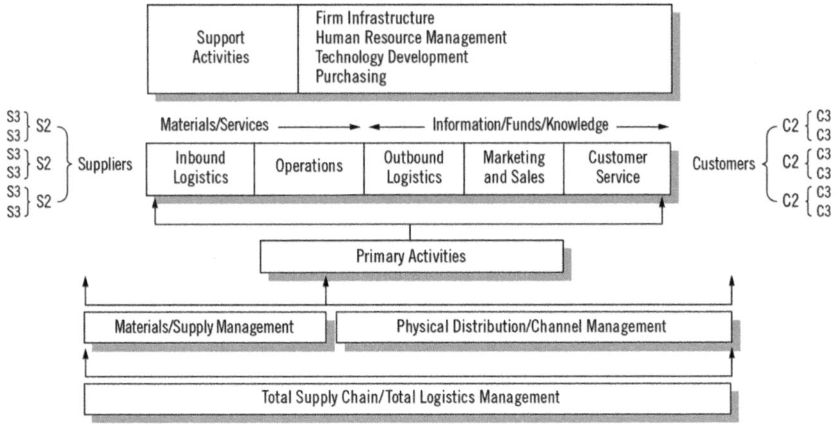

Fig. 4 Extended value chain (*Source* Monczka et al. [2016])

as of 2022 and the model created by the Global Supply Chain Forum (GSCF) within the Ohio State University.

A process reference model describes, characterises and evaluates a complex management process. Once the challenging process is "captured" in a process reference model, it can be clearly defined, consistently studied, and redesigned to gain a competitive advantage (Stewart, 1997: 63).

2.1 SCOR Model

SCOR model is the first cross-industry framework for evaluating and improving supply chain performance and management across the enterprise. It was developed in 1996 in partnership with Advanced Manufacturing Research (AMR) due to the integrated supply chain benchmarking research of Pittiglio Rabin Todd and McGrath (PRTM) (Stewart, 1997: 62). The model is updated regularly and adapted to modern business practices with the contributions of ASCM member companies.

With the 14th version of the SCOR model, ASCM radically changed its view of the supply chain and the schematic representation of the model. In addition, it introduced new process definitions compared to previous versions. For example, the "Plan, Source, Make, Deliver, Return and Enable" processes have been replaced by "Plan, Order, Source, Transform, Fulfil, Return and Orchestrate" processes (ASCM, 2022).

The illustration of the model shows a paradigm shift from a linear supply chain to a circular supply chain understanding. Finally, in Fig. 5, the use of SCOR in model diagrams until 2022 is presented.

The model's name was changed to "SCOR Digital Standard (SCOR-DS)" with SCOR v13. The SCOR model has been developed to describe the business activities associated with all phases of satisfying customer demand (ASCM, 2022). The SCOR-DS v14 model is presented in Fig. 6.

The graphic illustrating the SCOR-DS model is a double infinity diagram, representing today's supply chain's looped, continuous, and connected nature and the seven critical processes within every supply chain. The SCOR-DS graphic displays the balance of "Supply and Demand" in a horizontal infinity loop and "synchronise and Regenerate" in a vertical infinity loop (ASCM, 2022).

In SCOR-DS, circular activities are handled within the Return process. The return process includes processes that coordinate the reverse flow of products and circular activities such as product disposition or transformation processes (ASCM, 2022).

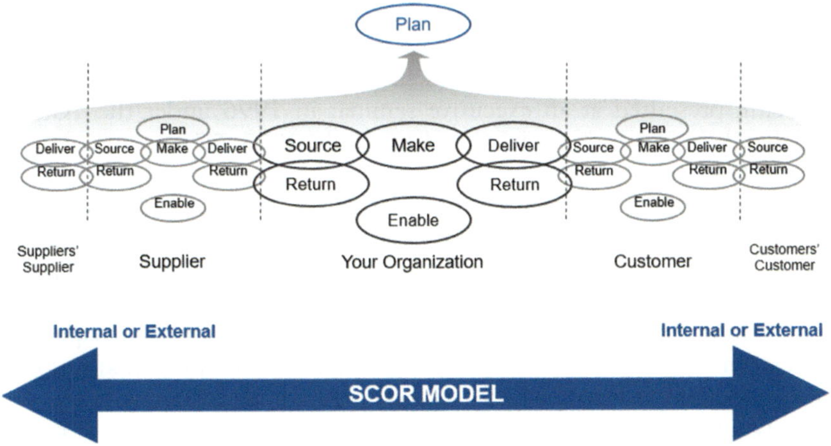

Fig. 5 SCOR model 1996–2020 (*Source* ASCM, SCOR v13 [2020])

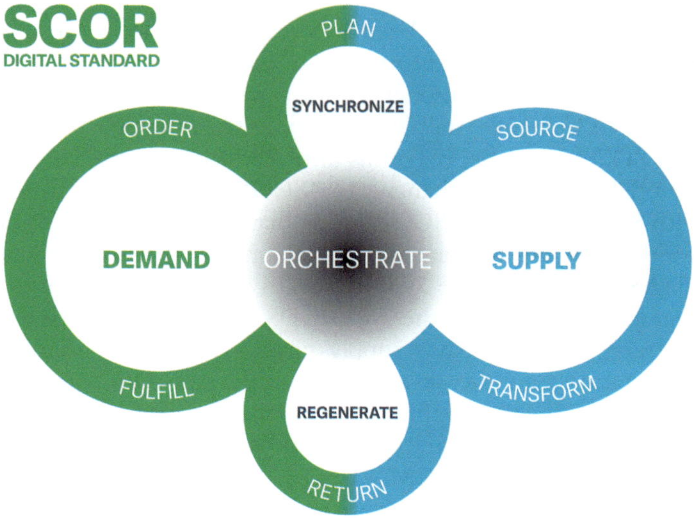

Fig. 6 SCOR-DS model v14.0 (*Source* ASCM, SCOR Digital Standard v14.0 [2022])

2.2 *GSCF Model*

In 1994, executives of a group of multinational companies developed a definition of the supply chain. The structure was brought to the literature by being presented at an executive seminar in 1996 under the "Global Supply Chain Forum" model (Lambert et al., 2005: 27).

The GSCF model deals with the process from the supplier's supplier to the customer's customer and includes 8 basic processes with a holistic approach (Fig. 7). All of these processes are cross-functional and inter-firm.

In the model, circular activities are expressed in "Returns Management". In this process, activities include determining returns management targets and strategies, developing avoidance, gatekeeping and disposition guidelines, expanding the returns network and flow options, and identifying secondary markets (Lambert et al., 2005: 49).

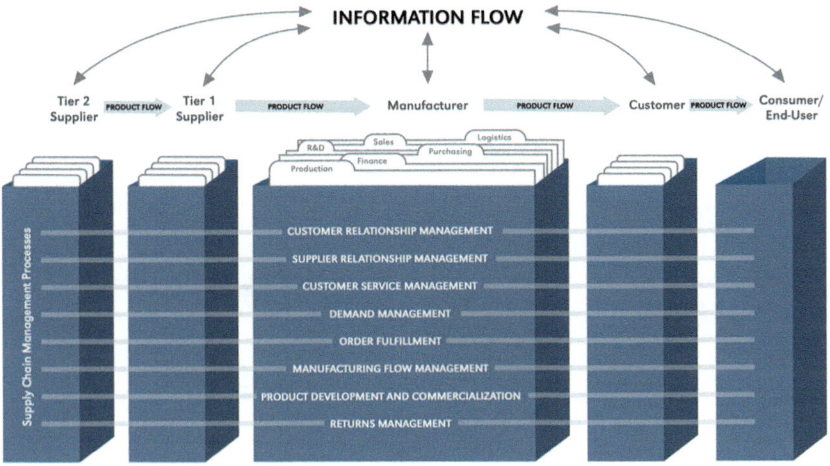

Fig. 7 GSCF model (*Source* Lambert [2008])

3 Circular Economy and Supply Chain

The circular economy refers to a cradle-to-cradle approach in which materials and products are designed to be reused for as long as possible and then reintroduced differently when reuse is no longer possible. Designing for the circular economy requires a holistic perspective that covers all aspects of design and development and considers many supply chain actors far beyond traditional supply chains (Fig. 8) (Bals et al., 2022: 3).

According to Bressanelli et al. (2018), a circular economy is defined as "an economic system restorative and regenerative by design, implemented by one or more supply chain actors through one or more of the four building blocks (circular product design, servitised business models, reverse logistics and enablers) to replace the end-of-life concept with reducing, alternatively reusing, recycling and recovering materials in production, distribution and consumption processes, for both technical and biological materials, to accomplish sustainable development" (Bressanelli et al., 2018).

The transition to a circular economy requires government, industry and society collaboration (Erdiaw-kwasie et al., 2023; Miles & Gold, 2021). There needs to be more than the circularity of the supply chain as an individual effort. With a circular economy, manufacturers need

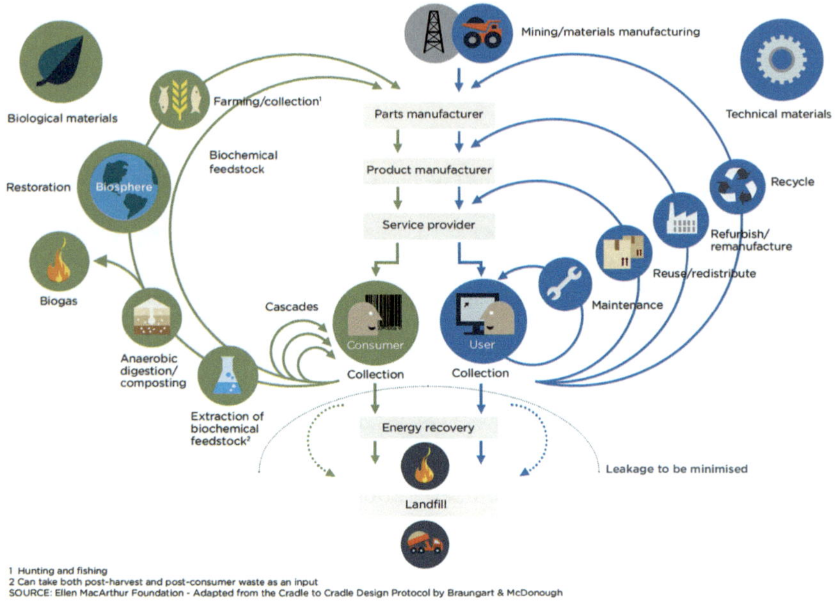

Fig. 8 The circular economy—an industrial system that is restorative by design (*Source* Ellen Macarthur Foundation [2014])

to rethink how they design products and use resources more sustainably, focusing on binding regulations throughout the entire life cycle of their products. Product designers and engineers are encouraged to focus on innovative, resource-efficient design that allows products to be easily disassembled and repaired. In addition, businesses should be encouraged to look at new business models that prioritise circular activities, such as avoiding consumption and switching to a rental model while selling products (Jones, 2021: 42). Complex supply chains require the integration of circularity into managerial thinking models and strategic decision-making in the context of the circular economy while operating (Eisenreich et al., 2022). This mental shift can enable managers to create circular value chains by engaging in shared value creation by interacting with different supply chains.

4 Methodology

As emphasised above, the transition to a circular economy will only be possible by renewing the value chain activities of enterprises with circular strategies or by switching directly to circular business models. Therefore, this study aims to determine the circular strategies of the enterprises operating in Turkey and managers' perceptions regarding the concept of a circular value chain.

To achieve this objective, a two-stage qualitative research methodology was preferred. In the first stage, the case study method was used due to the exploratory nature of the subject. 5 businesses included in the case study were determined by convenience sampling method. Three of these businesses have been selected to reveal the subject's reflections in practice due to their circular business models. The other two businesses act as catalysts in the circular economy. Information on these businesses is presented in Table 2.

Hit Grill and Hookah Charcoal Inc. It operates in the province of Düzce. The other four organisations are based in Istanbul.

In the second stage, data were collected through semi-structured interviews from 15 manufacturing enterprises operating in Istanbul using convenience sampling. Finally, the collected data were analysed by the content analysis method. Information about the participants is presented in Table 3.

During the interviews, 2 open-ended questions were asked to the participants. These questions are listed as follows:

Table 2 Case companies overview

Case company	*Industry*	*Business type*
Nivogo	Textile and Apparel	Retailer and Wholesaler
Hit Grill and Hookah Charcoal Inc	Charcoal Production	Manufacturer
AJ International Group	Textile and Apparel	Waste Separator and Wholesaler
DCube	Consultancy	CE Project Advisor
Business 4 Goals (B4G)	Consultancy	Non-Governmental Organisation

Table 3 Participant profiles

No.	*Participant's department*	*Industry*	*Number of employees*
1.	Procurement	Pharmaceutical	1000+
2.	Supply Chain	Apparel	1000+
3.	Procurement	Pharmaceutical	1000+
4.	Logistics	Textile & Apparel	1000+
5.	Logistics	Machine	500+
6.	Sales	Pharmaceutical	150+
7.	Retail	Food	100+
8.	Category Procurement	Textile	1000+
9.	Procurement	Energy	1000+
10.	Supply Chain	Energy	1000+
11.	Logistics	Textile	1000+
12.	Supply Chain	Ceramic	1000+
13.	Procurement	Machine	50+
14.	Supply Chain	Household Appliances	1000+
15.	Quality	Farm Equipments	250+

1. Could you give some brief information about the supply chain understanding of your business? Then, you can exemplify the approaches you apply while conducting your operations (circularity, sustainability, buying and using recycled materials, waste management, environmental management systems, etc.).
2. What does the circular economy mean to you?

5 Findings

5.1 Nivogo—Circular Transformation Movement

The Nivogo brand, a combination of the words Nivo (new) and go (to take action) is a circular transformation movement, representing the "new fashion" in the face of fast fashion rapidly consuming the world. The journey of Nivogo started in February 2021 with a team of 3 in a 400 m^2 apartment in Kağıthane and continues today in Eastern Europe's largest Renewal Centre with an area of 5000 m^2 and a monthly renewal capacity of 100,000 products powered by circularity and cutting-edge technology.

Nivogo extends the life of products taken from users and brand partners and brings them back into the circular economy. To become a circular alternative to fast fashion by reducing the damage caused by 85 billion fashion waste, they have saved more than 3 billion litres of water by renewing more than 500 thousand products they received from their brand partners in the last 15 months. In addition to the Renewal Centre, Nivogo is expanding its influence by including individual users in the circular fashion movement. Nivogo recently opened Türkiye's first circular store to sell its renewed products at Akasya Mall in Istanbul (Fig. 9).

Among the products which Nivogo took from their brand partners and users, there are defective, lack barcode, customer returned, overstock and old seasoned products, most of which are still usable. These products start their circular transformation journey by going through a 6-stage renewal process at the renewal centre (Fig. 10).

The foundations of Nivogo's unique business model were laid by thoroughly analysing global companies' positive and open-ended aspects.

Fig. 9 Nivogo's circular store

Fig. 10 Operations at Nivogo's renewal centre

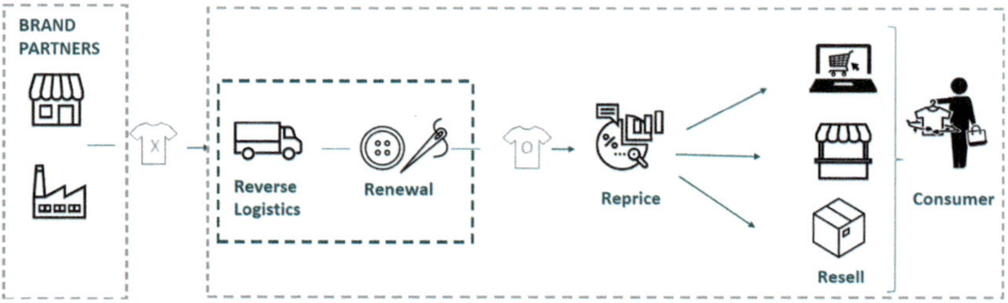

Fig. 11 Nivogo's 4R model

As a result, these companies demonstrated that the circular business model scales and transforms users' shopping habits for a sustainable and better world. Based on these teachings, Nivogo designed a unique business model called "Nivogo's 4R". Nivogo 4R model includes the stages "Reverse Logistics, Renewal, Re-price, Re-sell" (Fig. 11).

In the first stage, Nivogo takes products from their "Circular Transformation Brand Partners" with reverse logistics. After the product's arrival, experts examine the product's current status by identifying an artificial intelligence-based NivoKod for each of them to determine and follow the stations where the processes need to be applied. Finally, the products that have been identified with NivoKod are sent to the renewal stations, including repairing, cleaning, and ironing (Fig. 12).

Fig. 12 Reverse logistics and renewal process

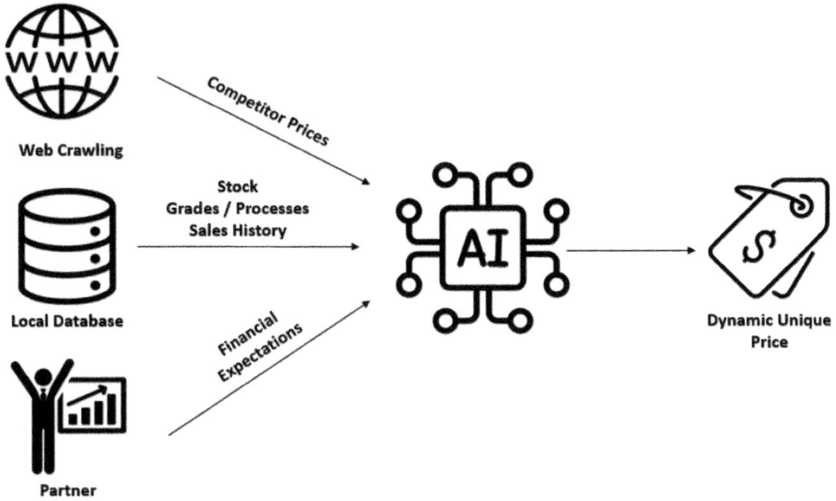

Fig. 13 Re-pricing process

After the renewal stage, with the artificial intelligence-based smart re-pricing technology Nivogo developed, special and fair pricing is made for the renewed products according to the parameters such as the first version, the renewal processes applied, the final version, the quality control score, the year, the season, etc. (Fig. 13).

At the re-sell stage of the 4R Model, products that have been renewed, sanitised and re-priced with AI-based pricing technology, are packaged and transferred to sales channels by completing their circular transformation journey (Fig. 14).

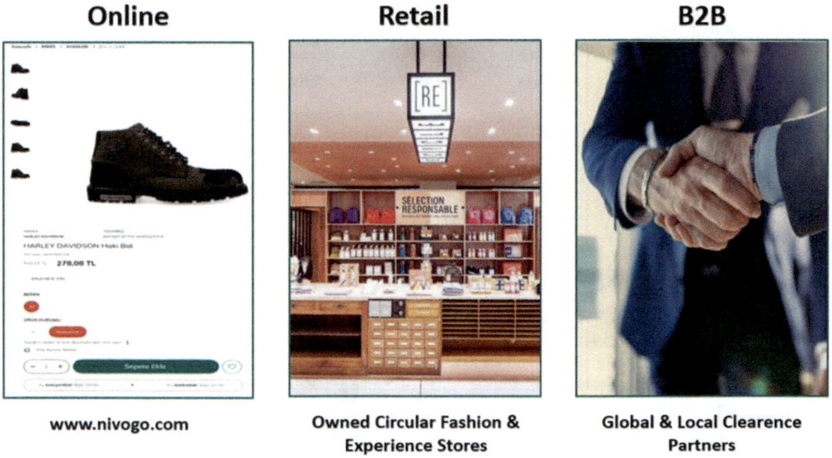

Fig. 14 Sales channels for re-sell

5.2 *Hit Grill and Hookah Charcoal Inc.—Circularity-Driven Production Using Hazelnut Shells*

Turkey is the world's largest producer and exporter of hazelnuts, accounting for about 70% of world production and around 75% of world exports (USDA, 2022).

Hit Grill and Hookah Charcoal Inc. is a business established in Düzce, one of the regions where hazelnuts are grown the most, to bring nutshell wastes into the economy. The enterprise plans to convert around 40,000 tons of hazelnut shell waste annually at its facility (Fig. 15).

Hazelnut shell wastes obtained from Düzce region are subjected to a carbonisation process in the pyrolysis machine. Roasting takes place by heating the stainless steel in an oxygen-free environment in the pyrolysis machine. The first run is made with natural gas. After 500 degrees, the gas produced by the burning of the hazelnut shell is released, and the system is provided to burn with its own gas without allowing it to cool. Thus, the system is maintained with lower energy consumption. The system is carried out with automation technology with a fully intelligent system. The carbonised shells are crushed and turned into powder. The carbon powder output is then mixed with edible starch to form a paste. The

Fig. 15 Production facility—hazelnut shell warehouse

output, which is made into dough, is cut into small pieces. Finally, it is turned into charcoal or hookah charcoal (Fig. 16).

With 7500 calories, it provides burning for around 6 hours. The resulting coal has no soot, odour, flame or smoke. During carbonisation, 15% natural vinegar is formed from the hazelnut shell. It has been determined that the hazelnut shell vinegar produced in the process increases the soil and crop yield in the agricultural sector. It is sold to greenhouses located in the Antalya region of Turkey. Thus, it is seen that the enterprise provides input to enterprises from other sectors with an open-loop supply chain understanding. In addition, the enterprise produces waste material input. Another waste but valuable material generated during the process is offered for use in the agricultural industry.

As of 2022, with the increase in input costs, economic sustainability problems exist. The hazelnut shell purchased by the company over 450 Turkish Liras/Ton in the previous year increased to 2,300 Turkish Liras/Ton in January 2022. The main reason for this is the Ukraine-Russia war. In addition, especially the natural gas-energy crisis in Europe has increased the demand for hazelnut shells. In this case, since hazelnut shell is a product with high heat calories, they demand it from Turkey for direct burning. This leads to an increase in the company's costs as a chain. The

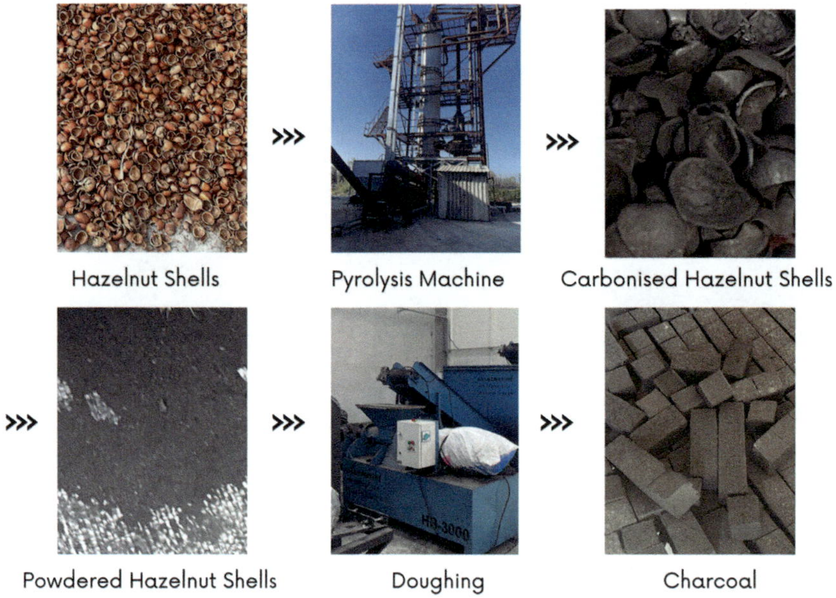

Fig. 16 Production process

company states that they are planning R&D studies to produce activated carbon, used in many sectors, instead of charcoal from hazelnut shells if operating costs continue to rise.

5.3 AJ International Group

AJ International Group is a business established in Turkey in 2018. The business has international operations. It carries out clothing collection activities in 61 provinces in Turkey. AJ continues its activities by collecting idle textile products in clothes piggy banks, re-selling the usable ones to the users and recycling the unusable ones for producing fibre, felt and cotton.

The enterprise categorises the clothing products obtained from the clothes piggy banks placed at the relevant points under the coordination of the municipalities in the sorting facilities in Istanbul and Mersin. The garments, categorised according to availability, are packaged in bulk after the pressing process (Fig. 17).

Fig. 17 Journey of waste clothes

All products that have been sorted are exported to countries such as Afghanistan, India and Pakistan in large packages. As a result, there is an increase in the amount of waste clothes collected during wardrobe renewal periods. The products are recycled in the relevant country or meet with the final buyers at local markets. With the recycling activities of the enterprise, the basic raw material of regenerated fabrics is formed.

5.4 *DCube—Sustainability Through Circular Economy*

DCube, founded in 2018, is one of the first companies in Turkey which solely focus on circular economy solutions and business models to achieve sustainable development. DCube works towards sustainable development through:

- Facilitating circularity through raising awareness for behavioural change;
- Capacity development for enabling innovative, regenerative solutions and creating circular value chains and business models;
- Providing hands-on consultancy for groups such as (but not limited to) entrepreneurs, businesses, industrial zones and local governments to achieve circularity;

- Investing in the startup community to advance circular economy principles;
- Connecting clean technologies and circular solutions with the Turkish industry to facilitate digitalisation and green economy transition of energy and material-intensive sectors;
- Advocating for designing and adapting necessary circular economy policies and action plans.

In 2018, the company started its operations with a project financed by Zafer Development Agency. The project took place in Uşak with the coordination of the Uşak Chamber of Commerce and Industry. The projectre focus on understanding the current situation of the regenerative fibre industry in Uşak and creating circular strategy. Besides, in 2020, DCube collaborated with Eastern Anatolia Development Agency to analyse the circular economy transition potential of the city of Van.

During the field visits and workshops with the representatives of the industries in Uşak and Van, the team realised that there is a strong misunderstanding of recycling as being the most effective solution to reach sustainability and very limited knowledge of what circular economy is. Therefore, this was a verification that raising awareness for behavioural change plays a key role in the circularity journey of Turkey.

Following this project and in line with this verification, in 2020, the DCube team developed a "Circular Economy Guidance for Businesses" to increase understanding of circular business models. The guidance was created with the sponsorship of B4G (Business for Goals), a platform founded by UNDP, TÜSIAD (Turkish Business and Industry Association) and TÜRKONFED (Turkish Enterprise and Business Confederation).

In addition to this contribution to Turkish circular economy literature, DCube created different contents for capacity building to explain circular economy to different target groups. As a result, beginning from 2019 until now (November 2022), DCube team has reached around 1500 people, including students, entrepreneurs, businesses, NGOs and local governments only through mentoring, workshops and lectures.

5.5 *Business 4 Goals (B4G) Platform*

Business for Goals Platform is a joint action platform established in 2019 under the leadership of TÜRKONFED, TÜSİAD and UNDP to build the bridge between the business world and Sustainable Development Goals (SDGs) and strengthen the role of the business world in sustainable development. The platform is open to the participation of all large, medium and small-sized enterprises that want to integrate the SDGs into their corporate policies and strategies, all public institutions that observe public–private sector cooperation, local administrations, academia and non-governmental organisations within the framework of the principle of inclusiveness.

The platform is among Turkey's first and only Circular Economy Week organisers. Thanks to the dialogue environment created within the scope of Turkey Circular Economy Week and its structure that brings together the public, civil society, academia and business world. The B4G platform aims to raise awareness about the circular economy model, to increase the contribution of the private sector in the fight against climate change, encourage sustainable production and consumption processes, and follow the reflections of the EU Green Deal and the EU Circular Economy Action Plan in Turkey.

5.6 *Circular Strategies and Managerial Perceptions of Supply Chain Professionals*

This section presents the findings related to the qualitative research carried out by the in-depth interview method. This section presents the qualitative research findings carried out by the in-depth interview method.

5.6.1 *Supply Chain Approaches in Terms of Circularity*

The participants' understanding of circularity differs by industry. Some participants emphasised supply chain strategies through their knowledge of sustainability. Most examples seem to be in the form of waste recycling or related to closed-loop supply chain operations. Table 4 presents the approaches businesses implement in the supply chain regarding circularity.

Table 4 Supply chain approaches in terms of circularity

No.	*Industry*	*Emphasised approach*
1.	Pharmaceutical	In 2006, we started to implement the reputation management model. Thus, the issue of sustainability has entered the company's strategies
2.	Apparel	The optimal ratio of recycled and organic cotton-containing products is determined for buying. In addition, efforts are being made to increase the number of certified suppliers that can produce these products
3.	Pharmaceutical	In 2010, the carbon footprint report began to be published
4.	Textile & Apparel	10,000 boxes are shipped daily from our warehouses to our stores. We repair and reuse the boxes in need of repair. When they become unusable, they are sold to recycling businesses
5.	Machine	Recycled products are not used in manufactured products
6.	Pharmaceutical	Productions comply with international standards such as EMA and FDA
7.	Food	Attempts are made specifically for our company to produce cosmetic products, ceramic products and fertilisers from the coffee pulp
8.	Textile	We are trying to reduce the damage we cause in jeans washing and grinding processes. We sort our waste at the source
9.	Energy	An average of 750 tons of manure per day is brought to the facility from the surrounding dairy farms, and electrical energy is produced with the gas released after processing. When the fertilisers remaining in the tanks for about a month are discharged, high-yield organic fertiliser is obtained for use in the fields
10.	Energy	We aim to contribute to the energy transition based on renewable energy in fuel use with our EV Charger (Electric Vehicle Charging Stations) products
11.	Textile	We are planning to switch to the use of plastic boxes. Thus, the boxes we use to ship products to our stores will be in a continuous cycle
12.	Ceramic	By increasing recovery rates by over 70%, we save water and energy consumption required for production processes and contribute to the sustainability of raw materials
13.	Machine	The company uses natural gas at its 3 co-generation facilities to produce the electrical energy it needs. In this way, energy savings and efficiency are achieved
14.	Household Appliances	Products are transported on recyclable second-hand plastic pallets

(continued)

Table 4 (continued)

No.	*Industry*	*Emphasised approach*
15.	Farm Equipments	In Çerkezköy plant, rainwater is stored and reused. In some locations, energy use is provided by solar panels. The total CO_2, CH_4 and N_2O emissions generated in all our shipments are calculated and followed

5.6.2 *Circular Economy Understanding of Supply Chain Professionals*

Participants were asked what the circular economy means to them, and it was aimed to determine their current perceptions. These perceptions are presented at Table 5.

6 Conclusion

The perceptions of supply chain professionals regarding the concept of circular economy indicate that they are familiar with it. However, it is seen that they have a limited and different understanding of circularity activities in terms of supply chain strategies. Participants focused heavily on closed-loop supply chain activities. Purchasing and using recycled regenerated materials for production is common in the textile and apparel industry. In indirect purchases of materials such as office supplies and packaging materials, the supply of products made with recycled materials is preferred. All businesses have goals and practices regarding evaluating waste in a way that does not harm the environment.

Since the business subject of case studies is directly carrying out circular activities, awareness of the concept is high. Businesses apply unique strategies in their own fields. Nivogo, Hit Grill and Hookah Charcoal Inc. and AJ International Group are examples of circularity-oriented business models. Practical applications and strategies of businesses contribute to business practices.

Generally, it can be said that the catalyst support of organisations that will establish symbiotic industrial relations between enterprises and the government's economic incentives are important for transforming supply chains into circular value chains in Turkey.

Table 5 Circular economy perceptions of supply chain professionals

No.	*Industry*	*Circular economy approach*
1.	Pharmaceutical	It is an IC model that aims to protect the environment in the process, from using raw materials in production to generating waste
2.	Apparel	It is the return of all products to the production and consumption cycle after use
3.	Pharmaceutical	The circular economy is a model that will support the economy in times of crisis. In addition, it is the benefit of waste to the environment and economy
4.	Textile & Apparel	It means recycling. 90% recycled material is used in our company boxes and packaging materials
5.	Machine	It can be expressed as ING of inert materials as raw materials or by recycling and reusing them as raw materials or revising them
6.	Pharmaceutical	It is a concept that deals with transformation and recycling rather than the process of production, use and destruction in the industrial economy
7.	Food	It creates an economy for transforming consumable commodities after the consumption of commodities
8.	Textile	It means everything that is less harmful to nature in every field or nature-friendly can be used more instead of disposable and can be reused instead of thrown away
9.	Energy	We carry out this as a business line with our biogas production facility in our company
10.	Energy	The circular economy is to maintain the usability of the enterprises' products, protect the product value and ensure the effective use of resources
11.	Textile	I can express it as a systematic approach that will add value to businesses and consumers by adopting a systemic approach on the production and consumption side and increasing the efficiency of the use of products
12.	Ceramic	It is the recycling of leftovers from produced or consumed products and the disposal of waste products without harming the environment
13.	Machine	The circular economy refers to the use of recycling and recycling rather than the waste or disposal process
14.	Household Appliances	Instead of removing the wastes generated in production from the system, a circular economy is to be used in the business again and to contribute to the economic structure of the business by obtaining from the wastes
15.	Farm Equipments	It uses raw materials and energy resources throughout the supply chain as little as possible and suitable for reuse

References

Alam, K., Ali, M. A., Erdiaw-Kwasie, M., Shahiduzzaman, M., Velayutham, E., Murray, P. A., & Wiesner, R. (2022). Impact of ICTs on innovation and performance of firms: Do start-ups, regional proximity and skills matter? *Sustainability, 14*(10), 5801.

ASCM, Supply Chain Operations Reference Model—SCOR Digital Standard v13.0. (2020). https://www.ascm.org/globalassets/ascm_website_assets/docs/intro-and-front-matter-scor-digital-standard.pdf. 10 November 2022.

ASCM, Supply Chain Operations Reference Model—SCOR Digital Standard v14.0. (2022). https://www.ascm.org/globalassets/ascm_website_assets/docs/intro-and-front-matter-scor-digital-standard2.pdf. 10 November 2022.

ASCM, SCOR DS v14. (2022). Return process. https://scor.ascm.org/processes/return/R1. 10 November 2022.

Bals, L., Tate, W. L., & Ellram, L. M. (2022). Introduction to circular economy supply chains: From supply chains to value systems. In L. Bals, W. L. Tate, & L. M. Ellram (Eds.), *Circular economy supply chains* (pp. 3–27). Emerald Publishing.

Batista, L., Bourlakis, M., Smart, P., & Maull, R. (2018). In search of a circular supply chain archetype—A content-analysis-based literature review. *Production Planning & Control, 29*(6), 438–451.

Bechtel, C., & Jayaram, J. (1997). Supply chain management: A strategic perspective. *The International Journal of Logistics Management, 8*(1), 15–34.

Bozarth, C. C., & Handfield, R. B. (2008). *Introduction to operations and supply chain management*. Pearson Prentice Hall.

Bressanelli, G., Perona, M., & Saccani, N. (2018). Challenges in supply chain redesign for the Circular Economy: A literature review and a multiple case study. *International Journal of Production Research*, 1–28

Carter, C. R., & Rogers, D. S. (2008). A framework of sustainable supply chain management: Moving toward new theory. *International Journal of Physical Distribution & Logistics Management, 38*(5), 360–387.

Cavinato, J. L. (1992). A total cost/value model for supply chain competitiveness. *Journal of Business Logistics, 13*(2), 285–301.

Cooper, M. C., & Ellram, L. M. (1993). Purchasing and logistics strategy. *The International Journal of Logistics Management, 4*(2), 13–24.

Croom, S., Romano, P., & Giannakis, M. (2000). Supply chain management: An analytical framework for critical literature review. *European Journal of Purchasing and Supply Management, 6*(1), 67–83.

Eisenreich, A., Füller, J., Stuchtey, M., & Gimenez-Jimenez, D. (2022). Toward a circular value chain: Impact of the circular economy on a company's value chain processes. *Journal of Cleaner Production, 378*, Article 134375.

Ellen Macarthur Foundation. (2014). Towards the Circular Economy Vol 3, Accelerating the scale-up across global supply chains, https://emf.thirdlight.com/file/24/cDm30tVcyxPQsxcD10AcOo2GK/Towards%20the%20circular%20economy%20Vol%203%3A%20Accelerating%20the%20scale-up%20across%20global%20supply%20chains.pdf. 9 November 2022.

Ellram, L. M., & Cooper, M. C. (1990). Supply chain management, partnerships and the shipper-third party relationship. *The International Journal of Logistics Management, 1*(2), 1–10.

Erdiaw-Kwasie, M. O., Abunyewah, M., Yusif, S., & Erdiaw-Kwasie, A. (2023). Does circular economy knowledge matter in sustainable service provision? A moderation analysis. *Journal of Cleaner Production, 383*, 135429.

Farooque, M., Zhang, A., Thurer, M., Ting, Q. U., & Huisingh, D. (2019). Circular supply chain management: A definition and structured literature review. *Journal of Cleaner Production, 228*, 882–900.

Fung, V. K., Fung, W. K., & Wind, Y. J. R. (2008). *Competing in a flat world: Building enterprises for a borderless world*. Wharton School Publishing.

Govindan, K., Soleimani, H., & Kannan, D. (2015). Reverse logistics and closed-loop supply chain: A comprehensive review to explore the future. *European Journal of Operational Research, 240*(3), 603–626.

Hazen, B., Confente, I., Pellathy, D., & Russo, I. (2022). From end-of-the-road to critical node: The role of end-user "consumers" in shaping circular supply chain management. In L. Bals, W. L. Tate, & L. M. Ellram (Eds.), *Circular economy supply chains* (pp. 151–165). Emerald Publishing.

Hewitt, F. (1992). Supply chain integration. *Annual Conference Proceedings. Council of logistics management* (pp. 334–341). Oak Brook, IL.

Hieber, R. (2002). *Supply chain management: A collaborative performance measurement approach*. vdf Hochschulverlag AG.

Ho, W., Zheng, T., Yildiz, H., & Talluri, S. (2015). Supply chain risk management: A literature review. *International Journal of Production Research, 53*(16), 5031–5069.

Holweg, M., & Helo, P. (2014). Defining value chain architectures: Linking strategic value creation to operational supply chain design. *International Journal of Production Economics, 147*(Part B), 230–238.

Houlihan, J. B. (1988). International supply chains: A new approach. *Management Decision, 26*(3), 13–19.

Hugos, M. (2006). *Essentials of supply chain management*. John Wiley & Sons Inc.

Johannson, L. (1994, Summer). How can a TQEM approach add value to your supply chain? *Total Quafity Environmental Management, 3*(4), 521–530.

Jones, S. M. (2021). *Advancing a circular economy—A future without waste?* Palgrave Macmillan.

Jones, T. C., & Riley, D. W. (1985). Using inventory for competitive advantage through supply chain management. *International Journal of Physical Distribution & Materials Management, 15*(5), 16–26.

Kildow, B. A. (2011). *A supply chain management guide to business continuity*. New York: American Management Association.

Lambert, D. M. (2008). *Supply chain management: Processes, partnerships, performance* (3rd ed.). Supply Chain Management Institute.

Lambert, D. M., Garcia-Dastugue, S. J., & Croxton, K. L. (2005). An Evolution of Process Oriented Supply Chain Management Frameworks. *Journal of Business Logistics, 26*(1), 25–51.

Langley Jr, C. J., & Holcomb, M. C. (1992). Creating logistics customer value. *Journal of Business Logistics, 13*(2), 1–27.

Lee, H. L., & Billington, C. (1992, Spring). Managing supply chain inventory: Pitfalls and opportunities. *Sloan Management Review, 33*(3), 65–73.

Liotine, M. (2019). Unlocking digital innovation: Guiding principles for driving digital technology in the supply chain. In A. M. Pagano & M. Liotine (Eds.), *Technology in supply chain management and logistics* (pp.143–167), USA.

Miles, A., & Gold, S. (2021). Mapping the social dimension of the circular economy. *Journal of Cleaner Production, 321*, 128960.

Monczka, R. M., Handfield, R. B., Giunipero, L. C., & Patterson, J. L. (2016). *Purchasing and supply chain management*. Cengage Learning, USA.

Montag, L. (2022). Circular economy and supply chains: Definitions, conceptualisations, and research agenda of the circular supply chain framework. *Circular Economy and Sustainability*, 1–41.

Novack, R. A., & Simco, S. W. (1991). The industrial procurement process: A supply chain perspective. *Journal of Business Logistics, 12*(1), 145–167.

Prahinski, C., & Kocabasoglu, C. (2006). Empirical research opportunities in reverse supply chains. *Omega, 34*(6), 519–532.

Pittman, P.H., Atwater, J. B. (2019). *APICS dictionary the essential supply chain reference* (16th ed.). Association for Supply Chain Management (ASCM).

Poluha, R. G. (2016). *The quintessence of supply chain management*. Springer.

Porter, M. (1985). *Competitive advantage*. The Free Press.

Randall, W. S., & Mello, J. E. (2012). Grounded theory: An inductive method for supply chain research. *International Journal of Physical Distribution & Logistics Management, 42*(8/9), 863–880.

Scott, C., & Westbrook, R. (1991). New strategic tools for supply chain management. *International Journal of Physical Distribution & Logistics Management, 21*(1), 23–33.

Seuring, S., & Müller, M. (2008). From a literature review to a conceptual framework for sustainable supply chain management. *Journal of Cleaner Production, 16*, 1699–1710.

Stevens, G. C. (1990). Successful supply-chain management. *Management Decision, 28*(8), 25–30.

Stewart, G. (1997). Supply-chain operations reference model (SCOR): The first cross-industry framework for integrated supply-chain management. *Logistics Information Management, 10*(2), 62–67.

Towill, D. R., Naim, M. M., & Wikner, J. (1992). Industrial dynamics simulation models in the design of supply chains. *International Journal of Physical Distribution & Logistics Management, 22*(5), 3–13.

Turner, J. R. (1993, December). Integrated supply chain management: What's wrong with this picture? *Industrial Engineering, 25*(12), 52–55.

USDA. (2022). Tree nuts annual, October 2022. https://apps.fas.usda.gov/newgainapi/api/Report/DownloadReportByFileName?fileName=Tree%20Nuts%20Annual_Ankara_Turkey_TU2022-0041.pdf. 11 November 2022.

Exploring Factors Affecting the Adoption of Green Process Management Model in the Software Industry: Progress Towards Sustainability and Circularity

Changiz Valmohammadi and Farkhondeh Mortaz Hejri

1 Introduction

In response to global warming and rising energy costs, governmental agencies and private enterprises have expanded their environmental protection investments (Alam et al., 2022; Marquis et al., 2011). As a result, a great global movement has emerged to use information and communication technology (ICT) in an environmentally responsible manner. The title "Green Information and Communication Technology" was chosen to signify the usage of ICT to reduce energy consumption

C. Valmohammadi (✉)
Department of Industrial Management, Faculty of Management, South Tehran Branch, Islamic Azad University, Tehran, Iran
e-mail: valmohammadi@yahoo.com

F. M. Hejri
Department of IT Management, Faculty of Management, South Tehran Branch, Islamic Azad University, Tehran, Iran

M. O. Erdiaw-Kwasie and G. M. M. Alam (eds.), *Circular Economy Strategies and the UN Sustainable Development Goals*,
Sustainable Development Goals Series,
https://doi.org/10.1007/978-981-99-3083-8_16

and waste in the environment. In addition, it indicates the manufacturing of green hardware devices and components, cutting-edge techniques for building energy-efficient software, and a collection of tools to help attain the primary goal of "saving the world" (Garcia et al., 2016). Every aspect of life, including new technology, has become environmentally conscious. Mankind recognises the importance of reducing energy consumption and carbon footprint. On a global scale, innovations are being developed to address these issues. Significant research and development programmes and substantial budgeting for projects to develop eco-friendly technologies demonstrate the global focus on this issue (Calero et al., 2019). Users are increasingly utilising software applications due to the growing number of mobile devices, including smartphones, tablets, and laptops, as well as internet users, high accessibility provided by cloud computing, and the Internet of Things (IoT). Accordingly, the impact of software development on greenhouse gas emissions has increased. In the meantime, engineering equipment, procedures, and IT services harm the environment due to their high energy consumption (Anthony et al., 2018).

As demands for corporate software increased, it has become a significant concern in recent years. Green software development is a relatively new research area in green computing, which requires more attention from academicians (Calero et al., 2022; Yahaya et al., 2019). Technologies, such as software and applications used by the community, and their usage are developing more than ever, which affect modern human lifestyle. Besides, technological signs of progress increased energy consumption rates (Mishra & Mishra, 2021). Manotas et al. (2016) argue that there are two main problems, including the lack of need for more knowledge and tools for developing energy-efficient software. From a technical point of view, examples of successful tools and experiences are available. However, not all of these tools have yet been presented in a consistent and integrated framework and cannot provide an integrated view for software developers and designers. Furthermore, the study of the relationship between software features and natural resources demand stemming from the structure and use of ICT systems has not received much attention so far (Alam et al., 2021; Hilty & Aebischer, 2015). Previous studies have focused more on the greenness of software products, and less attention has been paid to the green features in software processes (Ibrahim, 2021). Therefore, this study aims to explore how identified factors impact the green software development processes to facilitate the progress towards enhancing the suitability and circular economy of the software industry using the interpretative structural modelling (ISM) approach in the context of Iran.

2 Context

Sustainability has become the foundation of IT; it commits that IT will support our future and that the industry will improve. Using IT to create more eco-friendly systems has been considered to reduce energy consumption (greening of IT). As Calero et al. (2019), and Calero and Piattini (2017) mention on the hard side, such as green data centres, green hardware, etc., necessary attention has been made, so now it is necessary to pay careful attention to the soft (software-related) aspect of the issue. Green software engineering is an important aspect of the software engineering process in this era. Previously, software engineers focused on hardware or software development, regardless of sustainability. Duboc et al. (2019) believe that engineers focus more on the technical aspects of a software system, and any concerns beyond the rapid release of software are usually less priority. Sustainability is important for all software systems because using each new system depends on becoming part of the technical infrastructure. Its frequent usage may cause new burdens on social ecosystems (Raisian et al., 2016; Saputri & Lee, 2021). As hardware becomes more powerful, the impact of software behaviour on energy consumption increases significantly (Calero et al., 2022).

The development of integrated software systems has made daily human life highly dependent on technology and increased energy consumption (Saputri & Lee, 2021). Sustainability is an important issue for future jobs and especially for software engineers, because of its impact on society (Mishra & Mishra, 2021). Green business process management includes a sustainable approach to managing business processes, which can help to create and undertake environmentally sustainable operations (Maciel, 2017). In different industries, the importance of business processes with a green approach has been considered, and by defining the corresponding business strategies, the first steps have been taken to improve the environmental impact on processes. However, there are still no clear solutions for achieving the green stage and sustainability in the software development life cycle (Raisian et al., 2016).

In most cases, the green process of IT has led to improved resource efficiency. Reducing costs in the company's structure and strengthening its competitive position compared to competitors and organisations provide new services in today's society and unique benefits for current products, leading to a better image of the company compared to its competition. This will increase the worthiness of the organisation and thus increase efficiency (Alam et al., 2022; Garcia et al., 2016).

According to new technologies, green software can fit the software development process in complex software settings where software systems are green (Jannat, 2016). Therefore, developing concepts, methods, and tools that lessen environmental software impact, software products, and resulting processes is essential. However, using energy-efficient software development tools and methodologies has rarely been explored. Furthermore, because of Not bof the abundance of energy-efficient low-priced hardware components, designers have not paid enough attention to the energy-saving potentials of software (Mahmoud & Ahmad, 2013).

3 Literature Review

3.1 Business Process

Process management has gained increasing importance in various economies. The importance of well-designed processes in optimising limited resources is well established. In addition, processes can be redesigned to use available resources better. However, to transition to a sustainable economy, we must learn how to redesign many aspects of modern life, using business process management as a key tool. Therefore, we require a set of appropriate processes to apply business process management to sustainability issues (Vom Brocke et al., 2012). Furthermore, business practices are a subset of business processes in most organisations, being in place to help create value in the form of profits, credits, and other incentives. Therefore, organisations' appearance, performance, and utility are naturally described by their design and implementation of such processes (Seidel et al., 2012).

Thus far, business process management has not seriously considered environmental sustainability as a change driver or a goal. Thus, the present understanding of business process management must be developed to incorporate the notion of "green business process management" to encompass sustainability as a goal and a catalyst for change in business process management. Furthermore, to better define, implement, and use green business process management in organisational activities consistent with sustainability goals, we need to know how to describe, implement, and apply green business process management in organisational actions. Accordingly, green business process management is all about understanding, documenting, modelling, analysing, simulating, executing, and constantly changing the process with a focus on the environmental effects of these processes (Seidel et al., 2012).

3.2 Software Development Process

The software development process, commonly referred to as the software development lifecycle (SDLC) methodology or model, is a framework for organising, planning, and controlling the development of information systems. Over time, various frameworks have emerged, each with strengths and weaknesses (Spenser, 2010). Software development is the process of writing and maintaining codes. In addition, it encompasses all activities ranging from the first concept of the intended software to its final version, which sometimes occurs in a planned and systematic manner. Hence, software development can encompass any actions resulting in software products, including research, new developments, prototyping, modifications, reuses, reengineering, or maintenance.

This process aims to satisfy user needs while ensuring proper system performance. Thus, it must include procedures for validation (outputs meeting requirements) and verifiability (output performance being accurate) purposes. Manuals, programme documentation, and test items are part of the materials produced throughout the development process, together with computer programmes. For example, directions for software use are detailed in user manuals. In addition, test cases and programme documentation help developers become familiar with the future maintenance and deployment of software (Pressman & Maxim, 2016). Suppose there is no precise software process for an individual or organisation, in that case, it will be defined in terms of the ongoing project, organisational culture, and competencies of individuals involved in the project. Software development is often confused with programming. However, software development is concerned with the commercial production of software, optimal use of existing components, and strategies for organising, expediting, and improving the quality of software projects (Birrell, 1985). The environmental impacts of software are categorised into three categories. The first level includes the ICT supply effects which directly impact the use of energy or natural resources, such as hardware needs, implementation and performance requirements, software product packaging, network bandwidths, etc. Impacts resulting from using software services are referred to as second-level impacts or impacts of information technology use. Third-level impacts, also known as systematic ICT impacts, result from multiple conflicting systems interacting to lead to reactive consequences. For example, if some natural resources are utilised to manufacture a specific type of software, but the same number of resources can be used to produce more diversified software, demands for these natural resources will increase (Naumann et al., 2011).

3.3 Green Software

Sustainable software engineering (SSE), or green software engineering (GSE), has gained prominence in the software engineering community over the past few years. This is because many academicians know the direct and indirect effects of software programmes on the system and environmental energy consumption. As a result, software engineering methodologies are being developed, with green software developments being promoted by software engineers and developers. The major goal of these endeavours is to be green and sustainable at every level of the software development process; furthermore, their other goal is to consider it as a software quality feature or a non-functional aspect of software programmes.

Many research projects on green computing use the term "sustainability" interchangeably with "greenability". Software engineering sustainability, for instance, was considered to be a part of software sustainability in a book written by Calero and Piattini (2017). According to these authors, software sustainability can be applied to various items, including software systems, software products, web applications, data centres, etc. The term "sustainable software" has been defined in two ways. First, software-assisted greening has been around for a while now. Software greening generally refers to environmental activities that any software addresses, including energy management software. In addition, green software refers to how software is developed to produce a more sustainable product (called GSE). Engineering principles are applied to environmental issues using GSE methods. Accordingly, the software is developed, operated, and maintained in a green manner, resulting in green software.

Each step of the SDLC is supported and performed by specified activities. However, according to a study by Shenoy and Eeratta (2011), SDLC implementation may currently cause issues with green environment upgrades. The problem is that several major decisions made at these stages, such as paper use, e-waste production, electricity use, increased carbon effects of transportation and travel, air conditioning, and the like, may directly or indirectly harm the environment. Thus, human, economic, and energy resources are required for operations in the software life cycle. These factors will define the aspects of software stability (Calero & Piattini, 2017). In addition, software development and maintenance endeavours affect a software development community sociological and psychological features and its people. This category addresses labour rights, mental health, social protection, social fairness, and viability, among other topics. Conversely, economic stability is determined by how the software life cycle process safeguards stakeholder investment, ensures profitability, lowers risks, and preserves assets. Finally, environmental sustainability is defined as the impact of a software product's development, maintenance, and use on energy consumption and resource use (Calero & Piattini, 2017).

According to Jetley et al. (2013), the software industry needs to incorporate top-notch software engineering methodologies into its software development strategy to improve quality and costs. However, this requires a comprehensive set of software industry processes and tools and software engineering methods, tools, and techniques that are limited thus far. This paper discusses some of the issues the software industry faces when applying software engineering concepts and procedures to application development. Furthermore, this research stresses the need for more research and initiatives to promote the software industry in adapting software engineering, methodologies and concepts.

Dick et al. (2013) used agile methodologies in their GSE research. Accordingly, they provided a methodology integrating green computing aspects of software engineering approaches into agile green methods. In addition, they introduced a model that incorporated green computing features into software engineering methods and agile methods for developing green and sustainable software. Ardito and Morisio (2014) surveyed information system guidelines and data to see how they would reduce energy consumption. The authors, for instance, provided energy efficiency guidelines for infrastructures, applications, operating systems, hardware, and networks. Finally, Betz and Caporale (2014) focused on the design of sustainable software systems. Accordingly, they considered sustainability management as one of the most pressing issues today, which explained why private and public organisations were so interested in "sustainable" practices and solutions. In addition, their research highlighted the scarcity of existing sustainable development methods and solutions. Besides, the authors introduced a conceptual model for integrating sustainability features into business development processes to accomplish this goal. Furthermore, this study suggested that stability needs to be measured during the SDLC to integrate stability features into software engineering.

Lago et al. (2015) discussed the opportunities and challenges of sustainable software development. Software, according to these authors, is critical to societal success. Thus, environmental sustainability has become an integral component of software system operation and development. According to Lago et al. (2015), key challenges in producing green and sustainable software include software power shortages, green software, supervision of software energy consumption in practice, sustainable architectural design, green software, and cloud-based ready-made software. According to Pinto and Castor (2017), green software is concerned with lowering energy consumption through software analysis and optimisation. Accordingly, they added that "even tiny inefficiencies in applications compound throughout the system, with a major impact on battery life". Yahaya et al. (2019) conducted an empirical study in Malaysia on the software process regarding stability factors contributing to the development of green software. Green IT is becoming more prevalent in GSE, according to evidence. Besides, green software design and implementation are demanding and challenging from a managerial standpoint, including complicated equations between several actors. Although numerous tools and methods are available, they have not yet been well integrated into a single framework to provide software developers and designers with a unified perspective. Against this background, the present study introduced a new model for green software development from a process perspective.

In Iran, research has been done on green information technology, and the importance of this issue and effective factors in the country have been studied. Green management has also been considered in various industries. Taghva et al. (2019) provided a model for developing organisational sustainability through green information technology. The results of their research indicate that the dimensions of green information technology are as follows: green readiness, information and communication technology, as an empowerment (low carbon), green actions and activities, green information technology cycle management, green organisations and data centres, green information technology monitoring that can be used for the sustainability of organisations. Khadivar et al. (2018) presented a dynamic system model to study the component relations and determine the appropriate strategy to reduce energy consumption and eventually increase the level of green information technology maturity in the organisation. This study aims to analyse a complex system focused on reducing energy consumption and thus increasing the maturity of green information technology and then modelling its dynamic behaviour using Vensim software. Also, Shahbandarzadeh and Kabgani (2016) state that the level of acceptance and movement towards using green information technology among industries and industry managers is gradual. And budget constraints put managers under pressure to invest smartly in green information technology.

4 Circular Economy

The concept of circular economy (CE) as a potential guide towards a sustainable economic system has been noticed by industry and academia. The circular economy, like ICT systems, affects all aspects of modernity. Cutting-edge technologies such as big data, cloud computing, cyber-physical systems, the internet of things, virtual and augmented reality, and blockchain have an important role in accepting the concepts and strategies of the circular economy in the direction of reduction, reuse, recycling, and restoration (Demestichas & Daskalakis, 2020). The results show that solutions related to data collection and analysis, particularly to the Internet of things (IoT), blockchain, digital platforms, artificial intelligence algorithms, and software tools, are among the most popular solutions proposed by researchers. Results also suggest that greater emphasis is placed on the "reduce" component of the CE, although ICT solutions for the other "R" components, as well as holistic ICT-based solutions, do exist as well.

Specific important challenges impeding the adoption of ICT solutions for the CE are also identified and reviewed, with consumer and business attitudes, economic costs, possible environmental impacts, lack of education around the CE, and the lack of familiarisation with modern technologies being found among the most prominent ones (Demestichas & Daskalakis, 2020). The software has become an indispensable part of industrial production and thus influences the life cycle of manufacturing systems, as many of these systems have to be replaced or evolved due to changing requirements. Software adaptation through continuous evolution extends the service time of these systems and thus saves valuable resources (Kutscher et al., 2020). An alternative is reengineering the existing software, which enables the recycling of systems but causes major challenges. First, the existing software must be analysed efficiently, and only knowledge implicitly present in the source code must be obtained. Second, the modified software needs to be functionally verified and validated. Finally, it must be configured and commissioned on the production system. The software has an end-of-life. This ageing process results from environmental transformations (Grottke et al., 2008). In industry, this change is induced by new technologies, business models, and changing requirements, i.e. Industry 4.0 (Anderl, 2015). Although, since the 1990s, Product Service Systems (PSS) have been heralded as one of the most effective instruments for moving society towards a resource-efficient, circular economy (Tukker, 2015), PSS has also become embedded in a wider range of science fields such as manufacturing, ICT, business management, and design.

One definition of PSS is 'a mix of tangible products and intangible services designed and combined to fulfile the customer needs jointly (Tukker & Tischner, 2006). From the mid-1990s, PSS became a popular subject for researchers engaged with sustainability and business alike. Experts in the realm of sustainability argue that if one focused on the end user's needs or the service a user wants rather than the product, it would become much easier to design need-fulfilment systems with radically lower impacts. In a product-oriented business model, firms are incentivised to maximise the number of products sold. This is their principal method of boosting turnover, increasing market share, and generating profits. However, in service-oriented business models, in theory, the incentive differs. Firms then make money, by offering services and consumables, which are key in creating the cost factors. Hence, firms will be incentivised to prolong the service life of products, to ensure that they are used efficiently and to reuse parts as far as possible after the end of the product's life cycle. These elements could lead to a depreciation of material flows in the economy while maximising service output or user satisfaction.

In line with realising the global transformation towards digitising businesses and smart initiatives, automation, and robotisation and given the intensive competitive business environment, manufacturing organisations are shifting from traditional product-based business models towards developing and implementing service-oriented business models (Martin et al., 2018).

5 Methodology

5.1 Statistical Population and Sample

The statistical population of this study consisted of academic and industry specialists. Playing a key role, as identified by others, theoretical understating of the subject, diversity, and agreement to participate were the criteria for selecting participants in this study. Accordingly, significant persons were identified in the disciplines of information systems, business processes, networking and hardware, and business strategy for this study. These individuals had an important role in organisational software engineering and information technology. Besides, academic specialists were among the selected participants as they were familiar with the research model concepts and dimensions. The industry experts included senior and middle managers, consultants, and university professors. Additionally, people of various levels of knowledge, experience, and organisational affiliations were employed in this study. The interviewees were contacted before the interviews to ensure that everything was in place. Table 1 shows the profile of the respondents.

5.2 Data Collection Methods and Tools

Desk research was used to gather information about the literature and research background. Besides, semi-structured interviews were conducted to gather experimental data. The interviewees were directly contacted using this tool, which allowed for a more in-depth assessment of their perceptions, attitudes, interests, and aspirations. Interviewing is used to delve into complex issues, follow up on answers, determine root causes, and ensure that subjects comprehend questions. In in-depth semi-structured interviews, interviewees are free to describe how they perceive, act, and behave as much as possible (Kallio et al., 2016). The interview

Table 1 Profile of the respondents

Characteristics	*Frequency*	%
Education		
Bachelor	2	20
Master	5	50
Ph.D.	3	30
Major		
Information Technology	3	30
computer	4	40
other	3	30
Work Experience		
5–10	3	30
10–15	3	30
15–20	2	20
Up to 20	2	20

questions were extracted from the first stage of the theoretical framework. Table 2 presents the 11 factors extracted from the extant literature.

5.2.1 The Interpretive Structural Modelling Approach

The theory of ISM is based on discrete mathematics, graph theory, social sciences, group decision-making, and computer assistance. The procedures of ISM were developed through individual or group mental models that were used to calculate binary matrices, also called relation matrices. Besides, ISM was proposed by Warfield (1974) to present the relations of the criteria. ISM is a well-established method for recognising relationships among specific elements defining a problem. It originates as an interactive group learning process, yet individuals can also use it. In this process, a set of directly or indirectly linked elements are structured into a systematic model. Moreover, ISM is utilised to understand the relationships among the barriers and to develop insights into a collective understanding of these relationships (Huang et al., 2005).

This method identifies effects and explains the direction among the system's attributes. In addition, it establishes relationships among specific attributes to define a problem through their dependence and driving power (Mathiyazhagan et al., 2013). Due to its capability, ISM is a popular tool among academicians for analysing interrelationship attributes.

Table 2 The extracted factors

Factor (selected code)	*ID*	*Reference*
Governance	C1	Anthony et al. (2018), Maciel (2017), Gallotta et al. (2016), Nowak et al. (2011), Post et al. (2011)
Emerging trends	C2	Baiyere et al. (2020), Srivastava et al. (2020), Jnr et al. (2018), Kawdawatta and Marcelline (2018), Michanan et al. (2017), López-Pintado et al. (2017), Zhu et al. (2015), Djemame et al. (2014)
Infrastructure	C3	Mohammed et al. (2019), Anthony et al. (2018), Maciel (2017), Mohankumar and Kumar (2016), Gandomi and Amin (2014), Ghamkhari and Mohsenian-Rad (2013), Bianzino et al. (2010)
Strategy	C4	Rashid and Khan (2018), Anthony et al. (2018), Maciel (2017), Mohankumar and Kumar (2016), Ardito and Morisio (2014)
Green readiness	C5	Molla and Cooper (2014), Molla et al. (2011), Campbell et al. (2019), Nhamo (2013), Molla and Cooper (2010)
Tool	C6	Chowdhury and Hindle (2016), Noureddine and Rajan (2015), Li et al. (2015), Pathak et al. (2012)
Method	C8	Afum et al. (2021), Rashid and Khan (2018), Dick et al. (2013)
Stakeholders	C9	Saputri and Lee (2021), Roscoe et al. (2019), Anthony et al. (2018), Karita et al. (2019), Kern et al. (2019), Jnr et al. (2018), Gallotta et al. (2016), Manotas et al. (2016)
Policies and laws	C10	Shahzad et al. (2020), Ibrahim (2021), Anthony et al. (2018), Maciel (2017), Laskurain et al. (2017), Rashid and Khan (2018), Singh and Gond (2017), Ardito and Morisio (2014), Kern et al. (2019), Dick et al. (2013)
Assessment and monitoring	C11	Anthony et al. (2018), Maciel (2017), Gallotta et al. (2016), Mohankumar and Kumar (2016), Ardito and Morisio (2014), Kern et al. (2019), Betz and Caporale (2014), Nowak et al. (2011)

In reality, the ISM method consists of a series of repeated interrogations, usually using questionnaires, of a group of individuals whose opinions or judgements are of interest. After the initial interrogation of each individual, each subsequent interrogation will be accompanied by information regarding the preceding round of responses, usually presented anonymously. Thus, the individual is encouraged to reconsider and, if appropriate, to change their previous response in light of the responses of other group members (Seuring & Mueller, 2008).

The objective of this research was achieved by integrating MICMAC and ISM to produce validated frameworks for strengthening enablers and reducing barriers in different groups. The integrated methodological approach is used because MICMAC is flexible in evaluating the level of interactions between variables on a suitable scale. Based on the interrelationships, ISM, being an established technique in the literature, helps to build hierarchical structures (Dubey et al., 2016). Many researchers employed the ISM method in their studies; for instance, Valmohammadi and Dashti (2016) studied barriers to e-commerce implementation in the context of Iran. However, no study was found to have applied this approach to designing the green process management model in software development. Therefore, this approach was utilised to achieve the objective of the study.

5.2.2 Interpretive Structural Modelling Steps

Interpretive structural modelling was performed using six steps as follows:

1. Formation of the structural self-interaction matrix (SSIM): In this step, enablers were identified and added to the SSIM. Besides, the enablers' dimensions and their comparison results were included in this matrix. In addition, the types of the relations among the enablers in this matrix were determined by relations V, X, O, and A, as follows:
   - V: Barrier i helps to achieve barrier j, but j does not lead to i.
   - A: Barrier i does not help to achieve barrier j, but i leads to j.
   - X: Barriers i and j help to achieve each other.
   - O: Barriers i and j are unrelated.
2. Formation of the initial reachability matrix (RM): This matrix transformed SSIM matrix relation symbols to values 0 and 1 (Table 3 shows the rules).

Table 3 Converting conceptual relations to numbers (Thakkar et al., 2007)

Conceptual symbol	*i to j*	*j to i*
V	1	0
A	0	1
X	1	1
O	0	0

3. Formation of the final reachability matrix: This matrix was created by combining several relations among variables. Secondary relations among the indicators' aspects were managed in this matrix. For example, if dimension I led to dimension J, and dimension J led to dimension K, the dimension I would lead to dimension K, according to the secondary relation. In addition, if this were not the case in the reachability matrix, the matrix would be changed, with missing relations replaced. As a result, certain 0 elements would turn into 1, as represented by the symbol 1*. The final matrix was produced by recognising secondary relations and altering the received matrix.
4. Determining the level and priority of variables: The variables were levelled once reachability and antecedent sets for each element and the common set were determined. A reachability set was created for each element in which the final reachability matrix rows appeared as 1. Besides, a necessary set was one in which all columns numbered 1. In addition, a united collection was obtained by subscribing to these two collections. The first degree of precedence was given to elements whose common and reachability sets were the same. In addition, all elements' levels were determined by eliminating these elements and continuing the process with the remaining ones.
5. Producing an interpretive structural modelling (ISM) based on the final reachability matrix and levels specified.
6. Analysis of the driving power and dependence (the MICMAC method): The amount of driving was determined by the sum of the rows of values in the final reachability matrix for each element, while the amount of dependence was determined by the column sum (Ghasemi & Valmohammadi, 2021). Accordingly, four groups of elements were identified based on these two factors under the categories of autonomous, dependent, connected, and independent. The first group was characterised by low driving power, dependence, and separation from other factors. The second group had little driving power but much dependence. The third group had much power, yet it depended on other elements. Any actions applied to these factors would change other factors. The fourth group had strong driving power but low dependence. In fact, they were known as essential factors falling into one of the two groups of independent or related.

6 Results

The self-interaction matrix was established for the first time in this study. The symbols described in step one were employed in the self-interaction matrix. As Table 4 shows, expert judgement was used to produce this matrix. Next, the initial reachability matrix was created using values 0 and 1, followed by transverse relationships and the final reachability matrix (Table 5). In the original matrix, all entries in this table were 1* and 0. In the fourth step, the reachability set and the antecedent were extracted from the final reachability matrix, with the criteria graded. See Table 6.

6.1 ISM Formation

The ISM was drawn after each indicator's level was determined, with the final reachability matrix considered. Figure 1 depicts the final model. Accordingly, this model has four levels, the first and fourth being the most influential and the most permeable, respectively.

6.2 Driving and Dependence Power Analysis (MICMAC)

Using MICMAC analysis, the criteria were grouped based on the driving and dependent powers of each of the enablers (Table 3, Fig. 2). Accordingly, the criteria namely governance (C1), strategy (C4), stakeholders

Table 4 Structural self-interaction matrix (SSIM)

	C1	*C2*	*C3*	*C4*	*C5*	*C6*	*C7*	*C8*	*C9*	*C10*	*C11*
C1		O	O	O	V	O	O	O	O	O	O
C2			O	O	A	O	V	O	O	O	O
C3				O	A	O	V	O	O	O	O
C4					V	O	O	O	O	O	O
C5						V	V	V	A	A	A
C6							V	O	O	O	O
C7								A	O	O	O
C8									O	O	O
C9										O	O
C10											O
C11											

Table 5 Final reachability matrix

	C1	*C2*	*C3*	*C4*	*C5*	*C6*	*C7*	*C8*	*C9*	*C10*	*C11*	*Driving power*
C1	1	1*	1*	0	1	1*	1*	1*	0	0	0	7
C2	0	1	0	0	0	0	1	0	0	0	0	2
C3	0	0	1	0	0	0	1	0	0	0	0	2
C4	0	1*	1*	1	1	1*	1*	1*	0	0	0	7
C5	0	1	1	0	1	1	1	1	0	0	0	6
C6	0	0	0	0	0	1	1	0	0	0	0	2
C7	0	0	0	0	0	0	1	0	0	0	0	1
C8	0	0	0	0	0	0	1	0	0	0	0	2
C9	0	1*	1*	0	1	1*	1*	1*	1	0	0	7
C10	0	1*	1*	0	1	1*	1*	1*	0	1	0	7
C11	0	1*	1*	0	1	1*	1*	1*	0	0	1	7
Degree of dependence	1	7	7	1	6	7	11	7	1	1	1	

(C9), policy (C10), and assessment and monitoring (C11) were considered independent. These criteria had low dependence and high driving force, indicating that they had a high impact and low permeability. On the other hand, the dependent criteria were the emerging trends (C2), infrastructures (C3), tools (C6), green software development process (C7), and method (C8). Accordingly, these criteria had a high degree of dependence, low driving power, high permeability, and minimal impact on the system. The other criteria included interface factors with a high degree of dependence and driving power; in other words, they had a strong influence and high permeability, meaning that any minor changes in these criteria would cause significant changes.

7 Discussion

According to our findings, green governance is one of the important driving forces in managing green processes in software development. This finding is in line with the result of Hardin-Ramanan et al. (2018), who suggested a model for the governance of green information technology that reflects the responsiveness, mechanisms, and drivers of green IT in large Mauritian organisations. In the same vein, Jr et al. (2017) argue that the IT industry needs to adopt a green approach to its governance process; thus, a green IT governance framework can help reduce

Table 6 Determining levels of indicators

	Reachability set	*Antecedent set*	*Intersection set*	*Level*
C1	C1-C2-C3-C5-C6-C7-C8	C1	C1	4
C2	C2-C7	C1-C2-C4-C5-C9-C10-C11	C2	2
C3	C3-C7	C1-C3-C4-C5-C9-C10-C11	C3	2
C4	C2-C3-C4-C5-C6-C7-C8	C4	C4	4
C5	C2-C3-C5-C6-C7-C8	C1-C4-C5-C9-C10-C11	C5	3
C6	C6-C7	C1-C4-C5-C6-C9-C10-C11	C6	2
C7	C7	C1-C2-C3-C4-C5-C6-C7-C8-C9-C10-C11	C7	1
C8	C7-C8	C1-C4-C5-C8-C9-C10-C11	C8	2
C9	C2-C3-C5-C6-C7-C8-C9	C9	C9	4
C10	C2-C3-C5-C6-C7-C8-C10	C10	C10	4
C11	C2-C3-C5-C6-C7-C8-C11	C11	C11	4

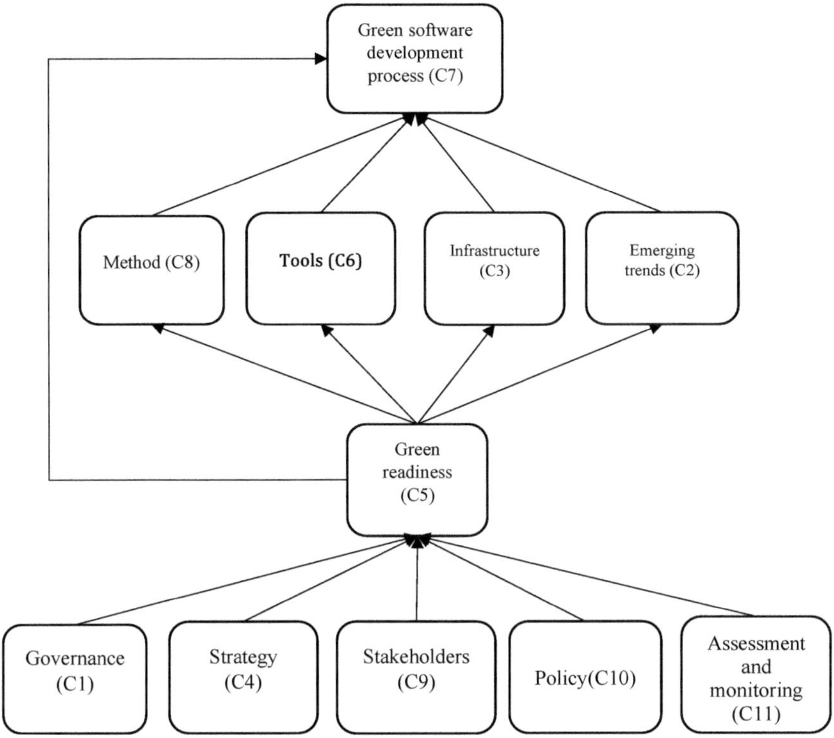

Fig. 1 The levels of structural model

environmental risks in an environmentally friendly manner. Therefore, they proposed a green information technology governance framework to reduce environmental risks.

Organisations need a green strategy fitting green governance, which must be aligned with the strategies of the business and the organisation. Accordingly, adopting a clear green strategy is vital to obtaining long-term and short-term goals. Organisations must develop specific strategies for carbon emissions and energy and waste management. Carbon emission management focuses on managing and reducing carbon emissions by organisations. This will include the use of information technology systems specifically designed to reduce carbon emissions and programmes for measuring software systems' carbon emissions and reducing them

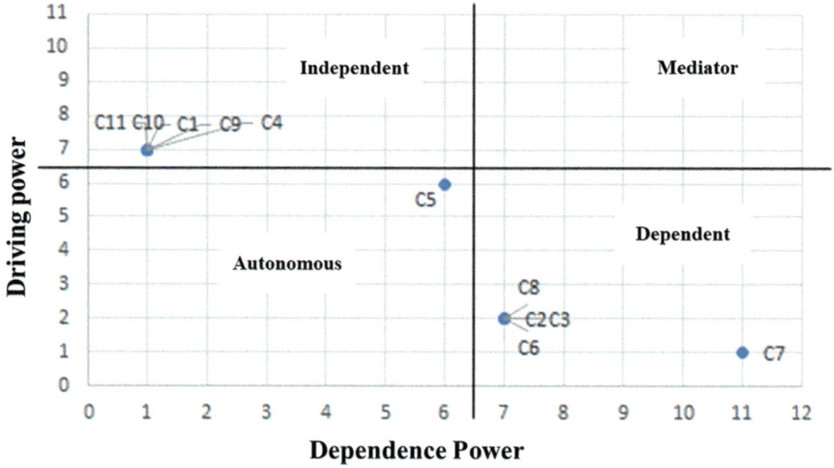

Fig. 2 Driving power and dependence

in the software development process. Energy and waste management is another essential issue considered in organisational strategies. Organisations need comprehensive, coordinated, and well-managed policies and rules. The framework of the green IT policy should be fit to ensure that green IT becomes a business work plan which ultimately leads to the least risk-faced projects. Strategies must be developed to be consistently implemented in various fields, including risk management, energy costs, resource consumption, and environmental effects of carbon emissions. In Dezdar's study (2017), the green policy factor was considered the acceptance factor for green information technology.

Assessment and monitoring of the use of resources and the environmental effects of carbon emissions on a continuous basis and at specified intervals is another important factor which needs specific attention by the software industry. For instance, at the beginning of a software development project, measuring environmental risks is one of the important factors which must be considered. Another factor that organisations should pay attention to is measuring energy costs in such a way that energy costs are considered and monitored separately. In this regard, the research of Noureddine et al. (2012) can be used to implement a time-energy consumption monitoring framework to help developers identify points with high energy consumption. In another study, Duarte et al. (2019) introduced a framework for evaluating software energy costs based on model analysis. Software is modelled as label transfer systems in this model, and the energy costs are interpreted using existing tools. Graph-based algorithms are then applied to scroll the models to obtain information about the software consumption associated with software behaviour, such as the least/most expensive software implementation, specific implementation cost, and average software implementation cost. In their research, they concluded that developers could create more energy-efficient software and consider potential exchanges for time, space, and energy costs when producing new versions of their systems.

Stakeholders are among the most important factors influencing the process of green software development. In software development, stakeholders include senior executives, project managers, engineers, software developers, and software users (Hirasawa, 2020). Having a green attitude towards creating a desire for change and a commitment to change is one of the prerequisites for the process of green software development. The green attitude indicates an organisation and employees' attitude towards environmental sustainability, which is the primary factor for implementing long-term green business process management. Promoting environmental awareness is an easy way to become an environmental observer, which results in improving environmental behaviours. Many resources are available for increasing awareness of environmental aspects. In addition, administrators, developers, and users of the training software can be instructed through group learning, informative and inspiring seminars, environmental books, brochures, and social networks as Karita et al. (2019) surveyed on the software industry's awareness of green and sustainable software engineering. New research results confirm the original survey's evidence, showing that software engineering sustainability is a novel issue for software specialists. However, experts show interest in this issue, and there is a general understanding that sustainability should be considered a qualitative feature. Among the observations made, they developed an initial theory indicating that software specialists are aware of this issue, even subconsciously in the "green in software" state. This study provided evidence about how the industry understands sustainability practices in software development. Although the community of software engineers has increased its interest in green and sustainable software engineering, the software industry in this area has not adequately considered it. Accordingly, green and sustainable practices have not been fully understood. Green readiness is on the second level of the model that affects the tools, methods, and infrastructures used and emerging trends. These four criteria are at the third level. It is worth mentioning that software and hardware make sense together. The use of correct infrastructures is effective in the green software development process. Cloud computing, virtualisation, and green data centre could be used to achieve this goal. In computer science, virtualisation technology has led to great development. The virtual implementation of hardware devices with the same functionality has brought many benefits. Software is virtual, and it is not physical. Therefore, virtualisation is often in software, which runs on a specific piece of hardware.

Various approaches have been introduced to software development. Software development using agile and Lean methods is one of the effective methods in green software development. Various studies examined the agile approach to green software development. In this regard, Rashid and Khan (2018), in their research, addressed agile methods employed by global software development manufacturers to develop green and sustainable software.

Various tools can be used to facilitate the process of green software development. For example, one can refer to automation tools in various processes of the software life cycle (testing, developing, and monitoring [energy measurement]). Fatima et al. (2020) systematically studied private tool support for developing the green android operating system. Macro et al. (2019) introduced a new multi-objective optimisation tool for green infrastructure planning. So, it is inferred that to enhance the practices of circularity, which can lead to the fortification of sustainability practices of companies engaged in software development, paying necessary attention and following the criteria and the interaction identified in this study is vital. As discussed above, all of the identified criteria play an important role in managing the economical consumption of tangible and intangible resources, such as natural resources, energy, the total time spent on software development projects, etc.

8 Conclusion

Green development is one of the most important innovations of the new millennium that will bring about great changes in human society. Sustainability and sustainable development are important issues in all industries, including IT and software systems development. Since the development of integrated software systems has made daily human life highly dependent on technology. Sustainability has been considered for all software systems because each new system becomes dependent as it becomes part of the technical infrastructure. Its continued use may place new burdens on ecological systems. Processes can be redesigned to reduce resource and energy use, and sustainability as an important dimension by considering carbon emissions, renewable energy consumption, waste generation, and other environmental performance criteria can help facilitate a circular economy and establish a sustainable organisation, particularly in the realm of the software industry.

In this model, we presented a set of factors using qualitative content analysis, extracted from texts and specialised interviews and classified into four levels based on the ISM method. By drawing the network of interactions, the factors, as well as their roles and functions, were determined. The final model consists of 4 levels. The first and fourth levels were determined as the most influential and permeable, respectively. Accordingly, the criteria, namely, governance, strategy, stakeholders, policy, and assessment and monitoring, were determined independently. These criteria have low dependence and high driving force, indicating that they have a high impact and low permeability. On the other hand, the emerging trends, infrastructures, tools, green software development process, and method were determined as dependent criteria. These criteria have a high degree of dependence, low driving power, a high level of permeability, and a minimal impact on the system. The other criteria included interface factors with a high degree of dependence and driving power; in other words, they had a strong influence and high permeability, meaning that any minor changes in these criteria would cause significant system changes. In summary, the obtained results might help software industry managers and experts apply the necessary measures to implement a sustainable software development project successfully.

References

Afum, E., Gao, Y., Agyabeng-Mensah, Y., & Sun, Z. (2021). Nexus between lean operations, eco-product innovativeness, social, green and business performances: Empirical evidence from Ghanaian manufacturing SMEs. *Journal of Manufacturing Technology Management, 32*(8), 1557–1577.

Alam, G. M., Erdiaw-Kwasie, M. O., Nagy, G. J., & Leal Filho, W. (Eds.). (2021). *Climate vulnerability and resilience in the global south: Human adaptations for sustainable futures*. Springer Nature.

Alam, K., Ali, M. A., Erdiaw-Kwasie, M., Shahiduzzaman, M., Velayutham, E., Murray, P. A., & Wiesner, R. (2022). Impact of ICTs on innovation and performance of firms: Do start-ups, regional proximity and skills matter? *Sustainability, 14*(10), 5801.

Anderl, R. (2015). Industrie 4.0—Technological approaches, use cases, and implementation. *at-Automatisierungstechnik, 63*(10), 753–765.

Anthony, B. J., Majid, M. A., & Romli, A. (2018). An empirical study on predictors of green sustainable software practices in Malaysian electronic industries. *Journal of Information and Communication Technology, 17*(2), 347–391.

Ardito, L., & Morisio, M. (2014). Green IT—Available data and guidelines for reducing energy consumption in IT systems. *Sustainable Computing: Informatics and Systems, 4*, 24–32.

Baiyere, A., Salmela, H., & Tapanainen, T. (2020). Digital transformation and the new logics of business process management. *European Journal of Information Systems, 29*(3), 238–259.

Betz, S., & Caporale, T. (2014, December). Sustainable software system engineering. In *2014 IEEE Fourth International Conference on Big Data and Cloud Computing* (pp. 612–619). IEEE.

Bianzino, A. P., Chaudet, C., Rossi, D., & Rougier, J. L. (2010). A survey of green networking research. *IEEE Communications Surveys & Tutorials, 14*(1), 3–20.

Birrell, N. D. (1985). *A practical handbook for software development*. Cambridge University Press.

Calero, C., & Piattini, M. (2017). Puzzling out software sustainability. *Sustainable Computing: Informatics and Systems, 16*, 117–124.

Calero, C., Mancebo, J., García, F., Moraga, M. Á., Berná, J. A. G., Fernández-Alemán, J. L., & Toval, A. (2019). 5Ws of green and sustainable software. *Tsinghua Science and Technology, 25*(3), 401–414.

Calero, C., Moraga, M. Á., & García, F. (2022). Software, sustainability, and UN sustainable development goals. *IT Professional, 24*(1), 41–48.

Campbell, L. K., Svendsen, E., Sonti, N. F., Hines, S. J., & Maddox, D. (2019). Green readiness, response, and recovery: A collaborative synthesis. General Technical Reports NRS-P-185. US Department of Agriculture, Forest Service. 358 p., 185, 1–385.

Chowdhury, S. A., & Hindle, A. (2016, May). Characterising energy-aware software projects: Are they different? In *Proceedings of the 13th International Conference on Mining Software Repositories* (pp. 508–511).

Demestichas, K., & Daskalakis, E. (2020). Information and communication technology solutions for the circular economy. *Sustainability, 12*(18), 7272.

Dezdar, S. (2017). Green information technology adoption: Influencing factors and extension of theory of planned behavior. *Social Responsibility Journal, 13*(2), 292–306.

Dick, M., Drangmeister, J., Kern, E., & Naumann, S. (2013). Green software engineering with agile methods. In *2nd International Workshop on Green and Sustainable Software (GREENS)* (pp. 78–85).

Djemame, K., Armstrong, D., Kavanagh, R., Ferrer, A. J., Perez, D. G., Antona, D., & Blasi, L. (2014). Energy efficiency embedded service lifecycle: Towards an energy efficient cloud computing architecture. In *CEUR Workshop Proceedings* (Vol. 1203, pp. 1–6). CEUR Workshop Proceedings.

Duarte, L. M., da Silva Alves, D., Toresan, B. R., Maia, P. H., & Silva, D. (2019, September). A model-based framework for the analysis of software energy consumption. In *Proceedings of the XXXIII Brazilian Symposium on Software Engineering* (pp. 67–72).

Dubey, R., Gunasekaran, A., Papadopoulos, T., Childe, S. J., Shibin, K. T., & Fosso Wamba, S. (2016). Sustainable supply chain management: Framework and further research directions. *Journal of Cleaner Production, 142*(2), 1119–1130.

Duboc, L., Betz, S., Penzenstadler, B., Kocak, S. A., Chitchyan, R., Leifler, O., Porras, J., Seyff, N., & Venters, C. (2019). Do we really know what we are building? Raising awareness of potential sustainability effects of software systems in requirements engineering. In *IEEE International Requirements Engineering Conference*. IEEEComputer Society.

Fatima, I., Anwar, H., Pfahl, D., & Qamar, U. (2020). Tool support for green Android development: A systematic mapping study. In *5th International Conference on Software Technologies—ICSOFT* (pp. 409–417).

Gallotta, B., Garza-Reyes, J. A., Anosike, A., Lim, M. K., & Roberts, I. (2016). A conceptual framework for the implementation of sustainability business processes. In *Proceedings of the 27th Production and Operations Management Society (POMS)*.

Gandomi, Z., & Amin, S. (2014). Towards green computing application for measuring the sustainability of data centers: An analytical survey. In *The International Conference on Electrical, Electronics, Computer Engineering and their Applications, EECEA2014* (pp. 86–96).

Garcia, S., Cintra, Y., Rita de Cássia, S. R., & Lima, F. G. (2016). Corporate sustainability management: A proposed multi-criteria model to support balanced decision-making. *Journal of Cleaner Production, 136*, 181–196.

Ghamkhari, M., & Mohsenian-Rad, H. (2013). Energy and performance management of green data centers: A profit maximisation approach. *IEEE Transactions on Smart Grid, 4*(2), 1017–1025.

Ghasemi, B., & Valmohammadi, C. (2021). Identification and prioritisation of critical success factors of knowledge management implementation using a novel hybrid group decision-making model. *VINE Journal of Information and Knowledge Management Systems*.

Grottke, M., Matias, R., & Trivedi, K. S. (2008, November). The fundamentals of software aging. In *2008 IEEE International Conference on Software Reliability Engineering Workshops (ISSRE Workshop)* (pp. 1–6). IEEE.

Hardin-Ramanan, S., Chang, V., & Issa, T. (2018). A Green information technology governance model for large Mauritian companies. *Journal of Cleaner Production, 198*, 488–497.

Hilty, L. M., & Aebischer, B. (2015). ICT for sustainability: An emerging research field. In *ICT innovations for sustainability* (pp. 3–36). Springer.

Hirasawa, N. (2020). Usability in software development process—Proposal of society/stakeholder centered design (SCD).

Huang, J.-J., Tzeng, G.-H., & Ong, C.-S. (2005). Multidimensional data in multidimensional scaling using the analytic network process. *Pattern Recognition Letters, 26*(6), 755–767.

Ibrahim, S. A. (2021). Green software process assessment: The theoretical framework. *Turkish Journal of Computer and Mathematics Education (TURCOMAT), 12*(3), 2011–2016.

Jannat, U. K. (2016). Green software engineering adaption in requirement elicitation process. *5*, 94–98.

Jetley, R., Nair A., Chandrasekaran, P., & Dubey A. Applying software engineering practices for development of industrial automation applications. In *2013 11th IEEE International Conference on Industrial Informatics (INDIN)* (pp. 558–563).

Jnr, B. A., Majid, M. A., & Romli, A. (2018). An empirical study on predictors of green sustainable software practices in Malaysian electronic industries. *Journal of Information and Communication Technology, 17*(2), 347–391.

Jr, B. A., Majid, M. A., & Romli, A. (2017). A green information technology governance framework for eco-environmental risk mitigation. *Progress in Industrial Ecology, an International Journal, 11*(1), 30–48.

Kallio, H., Pietilä, A. M., Johnson, M., & Kangasniemi, M. (2016). Systematic methodological review: Developing a framework for a qualitative semi-structured interview guide. *Journal of Advanced Nursing, 72*(12), 2954–2965.

Karita, L., Mourão, B. C., & Machado, I. (2019, September). Software industry awareness on green and sustainable software engineering: A state-of-the-practice survey. In *Proceedings of the XXXIII Brazilian Symposium on Software Engineering* (pp. 501–510).

Kawdawatta, T., & Marcelline, Z. (2018, October). Reducing emissions through digital transformation. In *2018 National Information Technology Conference (NITC)* (pp. 1–3). IEEE.

Kern, E., Guldner, A., & Naumann, S. (2019). Including software aspects in green IT: How to create awareness for green software issues. In *Green IT engineering: Social, business and industrial applications* (pp. 3–20). Springer.

Khadivar, A., & Abdian, S. (2018). Modeling the maturity of green information technology with the approach of systems dynamics. *Organizational Resource Management Research, 8*(2), 73–91.

Kutscher, V., Ruland, S., Müller, P., Wasser, N., Lochau, M., Anderl, R., & Hähnle, R. (2020). Towards a circular economy of industrial software. *Procedia CIRP, 90*, 37–42.

Lago, P., Koçak, S. A., Crnkovic, I., & Penzenstadler, B. (2015). Framing sustainability as a property of software quality. *Communications of the ACM, 58*(10), 70–78.

Laskurain, I., Ibarloza, A., Larrea, A., & Allur, E. (2017). Contribution to energy management of the main standards for environmental management systems: The case of ISO 14001 and EMAS. *Energies, 10*(11), 1758.

Li, F., Qanbari, S., Vögler, M., & Dustdar, S. (2015). Constructing green software services: From service models to cloud-based architecture. In *Green in software engineering* (pp. 83–104). Springer.

López-Pintado, O., García-Bañuelos, L., Dumas, M., & Weber, I. (2017, September). Caterpillar: A blockchain-based business process management system. In *BPM* (Demos).

Maciel, J. C. (2017). The core capabilities of green business process management—A literature review. In *Proceedings of the Wirtschaftsinformatik (WI)* (pp. 1526–1537).

Macro, K., Matott, L. S., Rabideau, A., Ghodsi, S. H., & Zhu, Z. (2019). OSTRICH-SWMM: A new multi-objective optimisation tool for green infrastructure planning with SWMM. *Environmental Modelling & Software, 113*, 42–47.

Mahmoud, S. S., & Ahmad, I. (2013). A green model for sustainable software engineering. *International Journal of Software Engineering and Its Applications, 7*(4), 55–74.

Manotas, I., Bird, C., Zhang, R., Shepherd, D., Jaspan, C., Sadowski, C., & Clause, J. (2016, May). An empirical study of practitioners' perspectives on green software engineering. In *2016 IEEE/ACM 38th International Conference on Software Engineering (ICSE)* (pp. 237–248). IEEE.

Marquis, C., Zhang, J., & Zhou, Y. (2011). Regulatory uncertainty and corporate responses to environmental protection in China. *California Management Review, 54*(1), 39–63.

Martin, S. L., Javalgi, R. R. G., & Ciravegna, L. (2018). Service advantage built on service capabilities: An empirical inquiry of international new ventures. *Journal of Business Research, 88*, 371–381.

Mathiyazhagan, K., Govindan, K., NoorulHaq, A., & Geng, Y. (2013). An ISM approach for the barrier analysis in implementing green supply chain management. *Journal of Cleaner Production, 47*, 283–297.

Michanan, J., Dewri, R., & Rutherford, M. J. (2017). GreenC5: An adaptive, energy-aware collection for green software development. *Sustainable Computing: Informatics and Systems, 13*, 42–60.
Mishra, A., & Mishra, D. (2021). Sustainable software engineering: Curriculum development. In C. Calero, M. A. Moraga, & M. Piattini (Eds.), *Software sustainability. Springer*. ISBN978-3-030-69970-3.
Mohammed, M. A., Mohammed, I. A., Hasan, R. A., Ţăpuş, N., Ali, A. H., & Hammood, O. A. (2019, October). Green energy sources: Issues and challenges. In *2019 18th RoEduNet Conference: Networking in Education and Research (RoEduNet)* (pp. 1–8). IEEE.
Mohankumar, M., & Kumar, M. A. (2016). Green database design model in software development life cycle phase. *Indian Journal of Science and Technology, 9*(30).
Molla, A., & Cooper, V. (2010). Green IT readiness: A framework and preliminary proof of concept. *Australasian Journal of Information Systems, 16*(2).
Molla, A., & Cooper, V. (2014). Greening data centers: The motivation, expectancy and ability drivers.
Molla, A., Cooper, V., & Pittayachawan, S. (2011). The green IT readiness (G-readiness) of organisations: An exploratory analysis of a construct and instrument. *Communications of the Association for Information Systems, 29*(1), 4.
Naumann, S., Dick, M., Kern, E., & Johann, T. (2011). The GREENSOFT Model: A reference model for green and sustainable software and its engineering. *Sustainable Computing: Informatics and Systems, 1*(4), 294–304.
Nhamo, G. (2013). Green economy readiness in South Africa: A focus on the national sphere of government. *International Journal of African Renaissance Studies-Multi-, Inter- and Transdisciplinarity, 8*(1), 115–142.
Noureddine, A., & Rajan, A. (2015, May). Optimising energy consumption of design patterns. In *2015 IEEE/ACM 37th IEEE International Conference on Software Engineering* (Vol. 2, pp. 623–626). IEEE.
Noureddine, A., Bourdon, A., Rouvoy, R., & Seinturier, L. (2012, September). Runtime monitoring of software energy hotspots. In *2012 Proceedings of the 27th IEEE/ACM International Conference on Automated Software Engineering* (pp. 160–169). IEEE.
Nowak, A., Leymann, F., Schumm, D., & Wetzstein, B. (2011, August). An architecture and methodology for a four-phased approach to green business process reengineering. In *International Conference on Information and Communication on Technology* (pp. 150–164). Springer.
Pathak, A., Hu, Y. C., & Zhang, M. (2012, April). Where is the energy spent inside my app? Fine grained energy accounting on smartphones with Eprof.

In *Proceedings of the 7th ACM European Conference on Computer Systems* (pp. 29–42).

Pinto, G., & Castor, F. (2017). Energy efficiency: A new concern for application software developers. *Communications of the ACM, 60*(12), 68–75.

Post, C., Rahman, N., & Rubow, E. (2011). Green governance: Boards of directors' composition and environmental corporate social responsibility. *Business & Society, 50*(1), 189–223.

Pressman, R. S., & Maxim, B. R. (2016). Software engineering: A professional approach.

Raisian, K., Yahaya, J., & Deraman, A. (2016). Current challenges and conceptual model of green and sustainable software engineering. *Journal of Theoretical & Applied Information Technology, 94*(2).

Rashid, N., & Khan, S. U. (2018). Agile practices for global software development vendors in the development of green and sustainable software. *Journal of Software: Evolution and Process, 30*(10), e1964.

Roscoe, S., Subramanian, N., Jabbour, C. J., & Chong, T. (2019). Green human resource management and the enablers of green organisational culture: Enhancing a firm's environmental performance for sustainable development. *Business Strategy and the Environment, 28*(5), 737–749.

Saputri, T. R. D., & Lee, S. W. (2021). Integrated framework for incorporating sustainability design in software engineering life-cycle: An empirical study. *Information and Software Technology, 129*, 106407.

Seidel, S., Recker, J., & vom Brocke, J. (2012). Green business process management. In *Green business process management* (pp. 3–13). Springer.

Seuring, S., & Mueller, M. (2008). From a literature review to a conceptual framework for sustainable supply chain management. *Journal of Cleaner Production, 16*(15), 1699–1710.

Shahbandarzadeh, H., & Kabgani M. H. (2016). Quantitative analysis of the opportunities and challenges of utilising green information technology in the organisation. *Intelligent Business Management Studies, 4*(14), 97–116.

Shahzad, M., Qu, Y., Zafar, A. U., Rehman, S. U., & Islam, T. (2020). Exploring the influence of knowledge management process on corporate sustainable performance through green innovation. *Journal of Knowledge Management*.

Shenoy, S., & Eeratta, R. (2011, December). Green software development model: An approach towards sustainable software development. In *India Conference (INDICON), 2011 Annual IEEE*.

Singh, S., & Gond, S. (2017). Green computing and its impact. In *Nature-inspired computing: Concepts, methodologies, tools, and applications* (pp. 1628–1642). IGI Global.

Spenser, M. (2010). Understanding the software development process. https://www.projectsmart.co.uk.

Srivastava, A., Gupta, M. S., & Kaur, G. (2020). Energy efficient transmission trends towards future green cognitive radio networks (5G): Progress, taxonomy and open challenges. *Journal of Network and Computer Applications*, 102760.

Taghva, M. R., Zohrabi, M., & Dehdashti Shahrokh, Z. (2019). A Structural model for green information technology and organizational sustainability. *BI Management Studies, 7*(27), 5–28.

Thakkar, J., Deshmukh, S., Gupta, A., & Shankar, R. (2007). Development of a balanced scorecard: An integrated approach of Interpretive Structural Modeling (ISM) and Analytic Network Process (ANP). *International Journal of Productivity and Performance Management, 56*(1), 25–59.

Tukker, A. (2015). Product services for a resource-efficient and circular economy—A review. *Journal of Cleaner Production, 97*, 76–91.

Tukker, A., & Tischner, U. (2006). Product-services as a research field: Past, present and future. Reflections from a decade of research. *Journal of cleaner production, 14*(17), 1552–1556.

Valmohammadi, C., & Dashti, S. (2016). Using interpretive structural modeling and fuzzy analytical process to identify and prioritise the interactive barriers of e-commerce implementation. *Information & Management, 53*(2), 157–168.

Vom Brocke, J., Seidel, S., & Recker, J. (Eds.). (2012). *Green business process management: Towards the sustainable enterprise*. Springer Science & Business Media.

Warfield, J. N. (1974). Toward interpretation of complex structural models. *IEEE Transactions on Systems, Man, and Cybernetics, 5*, 405–417.

Yahaya, J., Raisian, K., Ibrahim, S. R. A., & Deraman, A. (2019). *Green software process based on sustainability dimensions: The empirical investigation*.

Zhu, C., Leung, V. C., Shu, L., & Ngai, E. C. H. (2015). Green internet of things for smart world. *IEEE Access, 3*, 2151–2162.

Capital Maintenance and Depreciation Accounting Convergence: A New Model for Sustainability and Circularity in Capital Maintenance

Yaw Ndori Queku

1 Background Development

In studying the historical perspective of capital maintenance, it is evident that the theory of depreciation accounting has presented a framework of financial capital maintenance by charging depreciation against revenue to restrict distributable profit and maintain financial capital. However, the current depreciation accounting rubrics have failed to maintain physical capital. The broad phenomenon of capital or asset depreciation is an envelope or umbrella that shelters a wide variety of array of heterogeneous concepts (Carsamer et al., 2021; Escribá-Pérezy et al., 2019; Nobes, 2015). The meaning of capital upon which depreciation evolves is often associated with a specific discipline or field of discussion.

In economics and taxation, capital is associated with wealth and benefits. In accounting, the concept of capital has evolved. Originally, it was a

Y. N. Queku (✉)
Cape Coast Technical University, Cape Coast, Ghana
e-mail: quekuisaaccliford@yahoo.com

M. O. Erdiaw-Kwasie and G. M. M. Alam (eds.), *Circular Economy Strategies and the UN Sustainable Development Goals*, Sustainable Development Goals Series,
https://doi.org/10.1007/978-981-99-3083-8_17

credit concept; however, it often appears as fixed capital, working capital and intellectual capital (Cohen & Harcourt, 2003; Nobes, 2015; Seidu et al., 2022). The commonality of these expressions is that capital in accounting relates to assets. These dynamics in capital or asset reflect the complexity of expenditure classification where the traditional view of asset is linked to 'capital expenditure' and differentiated from 'revenue expenditure', which is splendidly oxymoronic because the former is about buying or acquisition of assets (James & Nobes, 2013). Given that this chapter's central theme is depreciation and capital maintenance rather than the generic capital and expense, there is no space to discuss interesting technicalities such as recognition criteria of expense and assets and the evolution of capital.

Nevertheless, it is important to note that in determining residual income upon which shareholders or equity participants have claims, the corresponding expenses must be matched against the revenue (Barker, 2010; Queku, 2015). Before modern-day accounting, the matching principle traditionally applied to revenue and revenue expenditures and excluded the allocation of the cost of the productive capacity or physical assets (fixed assets) used to generate such income (Napier, 1997; Reid, 1987). This created deficiency in the stewardship accounting and overestimation of residual income as not all resources used in the production of the said income are compensated for usage (Brief, 1966; Morris, 1986).

Capital or asset depreciation accounting was therefore seen as a means to address this deficiency and maintain the value of capital (i.e. production capacity of an asset) (Liapis & Kantianis, 2015; Rihll, 2013). Thus, the primary basis for depreciation accounting is to make residual income more realistic and maintain capital (financial and physical capacity). This suggests that depreciation accounting is primarily a tool for achieving capital maintenance. However, the accounting rules and principles for depreciation have made depreciation a mere notional expense, and a technical accounting requirement, and therefore, it is viewed widely as a non-cash item. In essence, the current accounting rubrics for depreciation emphasise financial capital maintenance, which makes good in any capital erosion by not paying dividends until notional non-cash expense (allocated cost of fixed asset) is deducted from profit. The non-cash characteristics of depreciation neglect physical capital maintenance as physical capital maintenance focuses on asset restoration and replacement, which requires actual cash consideration rather than a mere non-cash expense.

This breeds what we term 'parallelism within capital depreciation and maintenance ecosystem'. The original theory of parallelism is often accredited to Gottfried Wilhelm Leibniz (Hergenhahn, 2009). The theory basically assumes a dualist mind–body interaction. It maintains that physical and mental events run on a parallel course of existence but do not causally interact with one another (Broad, 1925). Relatedly, although there is an interaction of some sort between capital depreciation and capital maintenance through financial capital maintenance, they run in parallel via physical capital maintenance. This parallelism of depreciation accounting and physical capital maintenance has contributed to depleted assets and threatened the going concern of businesses due to the overconcentration of non-cash preservation of capital. This practice has been followed and applied for ages as the basis for maintaining capital without question (Escribá-Garcia, 2020; Nobes, 2015; Escribá-Pérezy et al., 2019); however, with the growing concern for financial sustainability, firms are expected to demonstrate beyond preserving equity to showcase the ability to meet future physical, and operational capacity through extended life, reuse, asset replacement and restoration as echoed in both sustainability and circular economy literature (Circle Capital Equipment Coalition, 2021; Economy, 2021). This would require a framework to converge the theoretical depreciation and capital maintenance accounting. This chapter, therefore, seeks to present a model of convergence of depreciation accounting and capital maintenance to achieve sustainability and circularity in Capital Maintenance. The rest of the chapter is organised as follows: depreciation accounting: theory and practice; depreciation and capital maintenance: historical perspective and controversies; depreciation accounting and capital maintenance: convergence and parallelism; the threat of depreciation-capital maintenance parallelism: business sustainability and circular economy perspective; alternative model for convergence and sustainable capital maintenance and sustainable capital maintenance model: a panacea for circular economy transition and financing.

2 Depreciation Accounting: Theory and Practice

The theory of depreciation accounting evolved historically from the broad capital theory (Boucekkine et al., 2011). Following the historical trends, although the foundational elements which have emerged from the capital, such as income and investment, have seen strong coherent and consensual assumptions leading to the development of powerful theories by the neoclassical theorists (Abel, 1990), there is no consensus for developing a coherent theory of depreciation (Escribá-Pérezy et al., 2019). For instance, market observable events and transactions evoked the dynamics of capital elements such as savings, investment and income (Bischo & Kokkelenberg, 1987).

However, depreciation accounting is associated with unobservable market events, making the available records unreliable. These unobservable transactions relate to a collection of notional, implicit benefits and costs originating from a composite of two distinctive sources: deterioration (physical and economic) and obsolescence (Bitros, 2010; Blanchard et al., 1993). The deterioration is inherent characteristics or intrinsic dynamics of non-current assets associated with usage, age and maintenance of assets resulting in physical wear and tear (output decay) and economic depletion (input decay). Obsolescence is associated with technological dynamics and structural changes that are extrinsic to non-current assets.

Although there is no widely known theory supporting the practice of depreciation accounting, some researchers have built on the foundational understanding of the causes or reasons for the assignment of depreciation (i.e. deterioration and obsolescence) to develop hypotheses and models to explain the dynamics (Bitros, 2010; Boucekkine et al., 2011; Liapis & Kantianis, 2015; Mukoyama, 2008). This has generated diverse proportions. Some studies have followed the vintage capital model (Johansen, 1959; Solow, 1960), while others have explored the proportionality theorem (Jorgenson, 1963). The proponents of the vintage capital model view depreciation as a natural instrument that focuses on accounting for obsolescence (Johansen, 1959). This model is concerned with the rate of replacing obsolete capital arising from technical progress in new capital or equipment. This model nevertheless over-concentrates on the obsolescence of capital equipment but disregards economic deterioration. In effect, the depreciation rate of capital is primarily determined according to the extent to which a new capital asset replaces the old or obsolete capital asset due to technical progress.

The proportionality theorem provides the framework for depreciation accounting based on physical deterioration (wear and tear). Following the pioneering work of Jorgenson (1963), the depreciation rate of capital assets is proportional to the capital stock. The proponents of this theorem recommend using a constant rate in exponential form to represent the depreciation rate or pattern of capital assets (Escribá-Pérezy et al., 2019). Although these theories vary in assumptions and frameworks for determining capital depreciation, they are still common. These theories provide a framework to determine the depreciation rate of capital either via the extent of replacement, constant rate, accelerated rate or double-declining balance.

Alternative theoretical perspectives for depreciation include Time Series Depreciation versus Cross-Section Depreciation. The time series depreciation measures depreciation as the rate of change in the individual asset's value over time. The underlying economic theory of time series depreciation is traced to the work of Hotelling (1925), who viewed depreciation as the rate at which an asset decreases in value concerning time. The cross-section depreciation (CSD) estimates depreciation according to vintage, similar to the vintage capital model. CSD measures depreciation as the difference between asset values of different vintages within the same time horizon. The underlying assumption of this depreciation framework is based on the productive asset efficiency (Hulten & Wycoff, 1996; Yoshida, 2020). This seeks to maintain the production efficiency of productive assets.

These theoretical perspectives of depreciation (Time Series Depreciation versus Cross-Section Depreciation) do not generally coincide. The departure in their estimation widens, especially when asset revaluation is expected from obsolescence or other antecedents of depreciation. The time series depreciation framework recognises such revaluation as an integral part of the depreciation accounting, whereas the framework of cross-section depreciation implicitly recognises it as if such revaluation is a capital loss.

Others have also followed conditional and unconditional conservatism frameworks in discussing the theoretical perspective of depreciation (Mora & Walker, 2015). Proponents of the unconditional conservatism perspective of depreciation argue that depreciation is determined ex-ante, but it often uses a rate higher than economic depreciation (Feltham & Ohlson, 1996; Mora & Walker, 2015). The term 'unconditional' captures the fact that the conservative view estimates the depreciation without considering how the assets are used or consumed in production. Thus, how the asset is used or applied in the business is inconsequential in the determination of depreciation of capital. Unconditional conservatism views often result in an understatement of net assets; however, it does not necessarily understate net income. Over the asset's economic useful life, the depreciation amount is fixed, but the full cost will eventually be allocated as long as the asset is in use. Thus, the unconditional conservative-based depreciation only alters the timing for charging depreciation. Consequently, income is lower in the early stage of the asset's life and higher at the later stage. The conditional conservatism perspective requires depreciation to be adjusted to reflect the bad and good aspects of the asset (Pope & Walker, 2003). It has been acknowledged that the scope of conditional conservatism is dependent on the extent of application of unconditionally conservative (Pope & Walker, 2003; Roychowdhury & Martin, 2013).

The absence of a widely accepted and unified theory of depreciation shows the deep-seated controversies about accounting for depreciation. The only agreement clear from all the theoretical perspectives is the purpose of depreciation: allocate the cost of capital assets to make residual income more realistic and maintain equity and maintain the productive capacity of capital through asset replacement.

In practice, the accounting for depreciation has also seen different methods of estimation. These methods include the straight line, reducing balance, sum of the year digit and production unit methods. International Accounting Standard Board (IASB)'s IAS 16 defines depreciation as 'the systematic allocation of the depreciable amount of an asset over its useful life'. This systematic allocation is based on the choice of depreciation method and rate as explained within the theoretical framework. The accounting practice of depreciation has evolved within the international accounting arena and most importantly, for countries that have followed Britain's accounting rubrics.

Historically, in its effort to regularise depreciation accounting in practice, the Institute of Chartered Accountants in England and Wales (ICAEW) in 1945 issued a recommended depreciation method emphasising the straight-line method (Zeff, 2014). The straight-line method requires a constant depreciation amount over the asset's life span. Subsequently, Accounting Standards Committee (ASC) issued a directive in 1977. This became the first foray of ASC into accounting for non-current assets and was popularised as Statements of Standard Accounting Practice (SSAP) 12 in 1977. Among other things, SSAP 12 specifically required that a building is recognised as a depreciable asset, a practice that was not universal by then.

Currently, the practice of depreciation is regulated largely by the standards issued by IASB, such as IAS 16, issued in 1993 but has since received amendments up to May 2020 (IASplus, 2022). IAS 16 recognises that depreciation allocates the asset's cost to determine realistic residual income into equity but echoes the fact that depreciation is a mere technical accounting requirement. Thus, depreciation accounting primarily maintains equity or financial capital but does not compensate for asset value loss. In effect, depreciation accounting is theoretically and practically connected to capital maintenance. The very essence of depreciation accounting is to achieve sustainability or improve the going concern of firms through capital maintenance. The depreciation reflects the systematic allocation of the cost of assets to limit distributable profits to maintain the financial capital of the entity. The latter part of the chapter will discuss how depreciation converges with capital maintenance to achieve sustainability and improve going concern assumption.

3 Depreciation and Capital Maintenance: Historical Perspective and Controversies

As noted earlier, depreciation accounting primarily exists because of the need to achieve the going concern or sustainability assumption of firms by emphasising capital maintenance. The concept of capital maintenance is broadly viewed as the deliberate employment of all relevant resources to preserve the operating capacity of capital goods or assets. The idea of capital maintenance has almost always existed since the formalisation of accounting rubrics and financial reporting, however, it has largely been neglected in the literature (Jianu et al., 2017; Pakšiová & Oriskóová, 2020). Nevertheless, the relevance of capital maintenance is fundamental as it is tied to firms' going concern or sustainability as both residual income and production capacity are bound up to it. This emphasises why clarity is needed about capital maintenance.

As far back as the thirteenth century, when double-entry bookkeeping (DEB) was invented, it was well-established among merchants that residual income is calculated within the context of capital maintenance (Lee, 1977; Nobes, 2015). Thus, residual income becomes real when all commitments relating to maintaining a firm's ability to use the capital to regenerate future income are accounted for. This is what we term the 'Residual View' of capital maintenance. Although profits were seldom estimated on an annual basis (Yamey, 1949), profit was still determined as an increase in capital. The maintenance of capital limits profit distributions through depreciation accounting as it affects the quantum of the distributable profit. Thus, depreciation accounting becomes an important mechanism to limit distributable profit and enhance capital maintenance.

These historical understandings of capital maintenance and profit relationship existed without legal ramification (Nobes, 2015). In the nineteenth century, stakeholders and accountants began regularising these historical understandings, and consequently, these understandings worked down into almost all financial regulations, laws, and accounting standards today. In the United Kingdom (UK), the 'Companies Act' of 1844, Act 1980 and ACT 1981, Germany Law and the EU's Second Directive on company law, all draw from these historic fundamentals of the capital maintenance-depreciation nexus (Haller, 1992; ICAEW, 2010; Nobes, 1983). This stage of the historical development is the 'Legal Capital View'. These laws and legal pronouncements gave legal backing to prevent distribution that could deplete capital as a pretence of profit (Garcia, 2020; Lutter, 2006). During this stage, accounting principles and standards for recognising residual income or profit were not well developed. Profit followed a stock-based approach, measured as an increase in net assets using either current or liquidated values (Richard, 2015).

The Great Depression also set a new stage for historical development. It affected major facets of global economies, including professional practices such as accounting. As a result, economic principles became important in the accounting profession, including the view of capital maintenance. Thus, in the 1930s, the view of capital maintenance took a shift from the pure legal view to a more dynamic view where there was a clear separation of capital and income (Hicks, 1949; Schmalenbach, 1959). The great contribution of Hicks in the separation of profit has made researchers often recognise Hicks as the father of profit in modern times. At this stage of development, profit was measured as a flow concept from the difference between revenues and expenses and the residuals accumulated in income surplus or retained earnings. The clear separation of residual income and capital at this stage requires that capital is maintained when dividends are only paid out of retained earnings with the distribution of capital surplus and share capital strictly prohibited.

During this historical era, the matching and realisation accounting principle became very dominant. To avoid the accumulation of capital as a pretence of net income into retained earnings and consequently distributing same to deplete capital, the cost of capital assets (non-current assets) needs to be systematically allocated and proportionate share for the period included in the expenses to be matched against the corresponding revenue to determine profit. Thus, depreciation accounting began to be formalised as a tool to implement the matching principle. Moreover, unrealised capital gains were prohibited from being included in retained earnings in line with the realisation principle. In effect, principles of matching and realisation became the cornerstone. Accounting for depreciation and estimation of the residual income, which was important tools for maintaining capital, were still based primarily on historical cost accounting.

Fast forward to the 1970s, accounting began to witness a breakthrough through standardisation and harmonisation (IASB, 2010; Queku, 2015). This era gave birth to IASB and, consequently IASB conceptual framework, International Financial Reporting Standards (IFRSs) and International Accounting Standards (IASs). In the 1970s and 1980s, the world saw sharp price-level changes and this challenged the soundness of capital maintenance based on the dynamic view. Therefore, price-level adjustments were introduced as an alternative approach to dynamic capital maintenance. This inflationary view was recognised in the IASB conceptual framework and other accounting standards.

The inflationary view operationalises capital maintenance into two types: financial capital maintenance and physical capital maintenance. These types of capital maintenance have been articulated by the accounting standard-setters globally. Financial capital maintenance may also be subdivided into *money and real financial capital maintenance*. *Money capital maintenance* follows historical cost accounting where capital is maintained when in the absence of additional capital injection and withdrawal, capital at close is equal to opening capital. Thus, any capital increase thereafter is residual income. *Real capital maintenance* follows current purchasing power accounting where capital is maintained when in the absence of additional capital injection and withdrawal, closing capital is equal to opening capital adjusted for general price-level changes (Garcia, 2020). It is important to state that financial capital maintenance (money or real) concentrates on equity holders' interest with very little regard for the operating capacity of the business, going concern and sustainability. It is therefore called 'Proprietary Capital Maintenance'. *Physical capital maintenance* follows current cost accounting and measures capital as the production capacity of an entity's assets in terms of the volume of production of goods and services. Thus, capital is maintained only when resources are allocated and available to

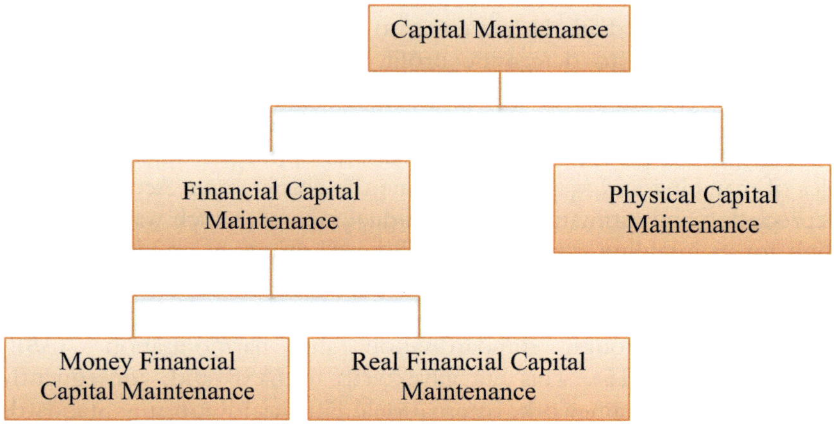

Fig. 1 Structure of capital maintenance (*Source* Author's construct)

increase the physical capital to maintain its volume of production of goods and services. Figure 1 summarises the structure of capital maintenance.

There is, therefore, a sharp difference in maintaining capital in terms of financial and physical capital. The choice in practice is dependent on the accounting regulation and standards in force, 'Choice of Nominal versus Physical'. In the UK, SSAP 7 focused on financial capital maintenance, while SSAP 16 provided guidelines for physical capital maintenance (Nobes, 2015). Similarly, IASB conceptual framework discussed these two streams of capital maintenance, however, the framework allows firms to choose the type of capital to maintain. The US accounting standards are also skewed in favour of financial capital maintenance. In practice, most of the firms have favoured financial capital maintenance. Table 1 presents the accounting framework and the bases for maintaining capital.

In the 1980s, international trade and liberalisation deepened financial integration. This new reality contributed to financialisation in the advanced economies and introduced more complexity in capital maintenance. According to Garcia (2020), this led to a change in profit determination, especially for listed firms. In the US, the Financial Accounting Standards Board (FASB) introduced a new concept of 'comprehensive income' in 1984. Comprehensive Income measures income without consideration for the principle of realisation (FASB, 1984). IASB also followed this new trend in its 2001 conceptual framework

Table 1 Framework and bases for capital maintenance

Capital maintenance	*Accounting framework*	*Basis for capital maintenance*
Financial Capital Maintenance		
Money Capital Maintenance	Historical Cost Accounting	$CC = OC$
Real Capital Maintenance	Current Purchasing Power	$CC = OC_a$
Physical Capital Maintenance	Current Cost Accounting	$PC_C = PC_s$

Source Author's Construct
Note CC, OC, OC_a, PC and PC_o denote Closing Capital, Opening Capital, Opening Capital Adjusted for general price-level changes, Production Capacity at Close (PC_c) and Production Capacity at Start (PC_s)

and introduction of fair value measurements in some accounting standards. The conservative flow-based income measurement began to fade away for a market-based measurement. In its fair value measurement and comprehensive income, the IASB framework and its standards somewhat endorsed the dynamic view of the 1930s by allowing some changes in assets to be recognised in 'Retained Earnings' and others in 'Other Comprehensive Income (OCI)' (Biondi, 2011; IASB, 2013; Queku, 2020). Similar to the dynamic view, to maintain capital, IASB and its associated standards prohibit the distribution of the accumulated gains in the OCI. The measurement approach was mainly a stock-based approach with fair value rather than with conservative assessment.

A critical consideration of these historical developments suggests that behind these evolutions of capital maintenance is the recognition of residual income and capital to the extent that capital is maintained. From the invention of DEB to the fair value measurement regime, the nature of depreciation accounting has been a key determinant in all directions of capital maintenance. However, a more fundamental controversy is the treatment of depreciation: *Should depreciation be accounted for as a mere technical requirement for accounting numbers or treated as a more complex phenomenon for sustainable capital maintenance and going concerned*? This question is becoming louder and louder in recent times owing to the call for sustainability reporting and circular accounting (Cohn, 2021; Garcia, 2020; Circle Economy, 2019, 2020, 2021; Jianu et al., 2017; PACE, 2019). Achieving this sustainability and integrating a circular accounting framework into the traditional accounting practice widens the

Table 2 Evolution of capital maintenance

Fundamental view	*Time span*	*Basis for change*	*Focus and approach*
Residual Income View	13th Century	Preserving capital	Income in the context of capital
Legal Capital View	19th Century	Legal justification	Stock-based approach with conservative assessment
Dynamic View	1930	Great Depression	Flows-based approach with cost allocation
Inflation View	1970	Price-Level Changes	'Nominal' vs 'Physical' capital maintenance, etc
Net Assets View	1990	Financialisation	Stock-based approach with fair value
Sustainable View	2010	Business sustainability, going concern problem and circular accounting	Stock and Flows-based approach

Source Author's construct developed from the literature (Garcia, 2020; Lee, 1977; Nobes, 2015)

debates and controversy. The reason is that sustainability and circular accounting requires simultaneity in maintaining financial and physical capital (i.e. credit and debit capital). This raises a further question: is it credit or debit capital that needs to be maintained: equity capital or net assets maintenance? Should there be a convergence of depreciation, and financial and physical capital maintenance rather than parallelism? This is the new stage of capital maintenance. This new era requires consideration for both stock and flow approaches to adapt to emerging complexities. These questions will be addressed in the latter sections of the chapter. Nonetheless, this debate or controversy seems to have disappeared from the depreciation and capital maintenance complementarity research agenda. The historical development is summarised in Table 2.

4 Depreciation Accounting and Capital Maintenance: Convergence and Parallelism

The connection between depreciation and the pace of capital maintenance in the short and long run can be readily seen from the decomposition dynamics of capital maintenance (Financial and physical capital maintenance). Although the accounting treatment for depreciation has evolved and is currently still dynamic, as evident in the numerous alternative depreciation methods, the non-cash systematic allocation chargeable against income remains undisputable (Liapis & Kantianis, 2015; Queku, 2020). The IASB conceptual framework and the individual accounting standards have reflected capital maintenance in various ways and demonstrated convergence between capital maintenance and depreciation accounting (i.e. including amortisation and impairment).

The depreciation (Depn) and impairment requirements of IAS 16 (Property, Plant and Equipment) and IAS 38 (Intangible Assets) provide evidence of convergence as they are related to maintaining capital. Depreciation (including amortisation and impairment) is debited to income to reduce profit, restricting dividends arising from limited distributable profits, and thereby maintaining original financial capital in nominal terms. The following scenario with numbers could clarify the role of depreciation in capital maintenance.

5 Scenario 1

A firm begins with stated capital of GHS300 million, a truck of GHS200 million and cash of GHS 100 million. At the beginning of the period, an item of inventory is purchased for GHS100 million. Assuming the item is sold for GHS150 million, the truck has a useful life of four (4) years, and any profit is distributed as a dividend to shareholders (See Table 3). The structure of the Profit statement with and without depreciation accounting is as follows:

Under 'A' 'Without Depreciation', we observe that closing capital and reserve still stand at GHS300 million. It will be mistaken to assume that capital is maintained without depreciation because the wasting cost of the truck has not been accounted for. So in real terms, the capital is being depleted in the going concern. Without depreciation, equity participants have full claims on the GHS50 million as dividends, threatening capital when distributed in full. One could argue that to limit distributable profit and restrict dividends, the firm must account for the wasted asset (truck). As it can be observed in 'B', with depreciation, the dividend is restricted, and capital nominally increased to GHS350. Further, a lack of depreciation (including amortisation and impairment) could mean that high dividends would be paid, and consequently, liquidity problems may be experienced due to high cash outflow (Garcia, 2020).

Table 3 Profit statement with and without depreciation accounting

	A *Without depreciation* *GHS million*	*B* *With depreciation* *GHS million*
Sales	150	150
Cost of Sales	(100)	(100)
Operating Profit	50	50
Depreciation	0	(50)
Net Profit	50	0
Dividend	50	0
Closing Capital and Reserve	300	350

There is therefore a window of convergence between depreciation and capital maintenance. However, this convergence seems limited when capital maintenance is decomposed into its constituents: financial capital maintenance (FCM) and physical capital maintenance (PCM). The current practice of depreciation is a notional deduction chargeable to income reflected in non-cash adjustment. This suggests that it is more connected to financial capital maintenance and departed from physical capital maintenance. Physical capital maintenance requires maintenance of the production capacity of an entity's assets in terms of the volume of production of goods and services.

Referring to Scenario 1, although nominal financial capital is maintained through the introduction of depreciation accounting, the production capacity of the truck would still be eroded with time. Although some have argued that the re-measurement model in IAS 16 and IAS 38 implicitly care for physical capital maintenance as upward revaluation would increase depreciation allocation (Nobes, 2015), the mere re-measurements of assets only deepen the convergence of real financial capital maintenance as it is based on changes in the market condition (i.e. changes in general price level). These re-measurement models still follow non-cash adjustment principles where gains arising from an increase in asset value are recorded in other comprehensive income (OCI) which is prohibited for distribution as dividends and depreciation chargeable on the new revalued amount.

How do these effects maintain the physical capacity of the asset? Obviously, these have no implications for maintaining the physical productivity of the asset. In fact, even these re-measurement models sometimes allow recognition of an increase in the value of an asset in profit, making them distributable. Depreciation alone is not a tool to cure physical capital maintenance controversy. Although depreciation-physical capital maintenance seems to move in the same direction, there is no convergence in sight. As depreciation rises, the need to enhance physical capital maintenance becomes urgent but these mechanisms do not converge in current accounting practice. This is what we term '*Parallelism of Depreciation and Physical Capital Maintenance*'. This parallelism is a fundamental issue arising from accountants' preferred definition of 'Capital'.

In the proprietary view, capital is referred to as equity (Van Mourik, 2010). Van Mourik (2010) concluded from reviewing extensive literature that this proprietary view prevails in all accounting practices, including in IFRS and any other Generally Accepted Accounting Principles (GAAP). This might explain why the practice of capital maintenance is skewed towards financial (nominal capital) rather than physical capital, which is required to preserve the entity's productive capacity in the going concern. Nevertheless, the emerging concept of OCI introduced by IASB views capital as an increase in the net assets (Garcia, 2020) as emphasised by the re-measurement models of several individual IFRSs/IAS, including IAS 16 and IAS 38. Of course, this feature deepens the controversy about what constitutes capital and it also shows the lack of a clearer meaning to the concept of capital within the IFRS framework. Thus, it would be difficult to achieve convergence in capital maintenance in the current form of depreciation coupled with controversy about what constitutes capital. This is illustrated in Fig. 2.

The figure depicts depreciation (Depn) and financial capital maintenance (FCM) lying in the same line but departs from physical capital maintenance. This is because any additional depreciation limits the amount of distributable profit in magnitude, restricts the same quantum in dividend and enhances original nominal financial capital. One can argue, therefore, that holding other factors constant, Depn and FCM have a point of convergence at the beginning of the economic life of the depreciable asset as depicted by 'X'. Furthermore, as discussed earlier, introducing re-measurement models based on market information provides the basis for preserving productive capacity through investment in an asset replacement. Thus, the re-measurement or revaluation, NOT

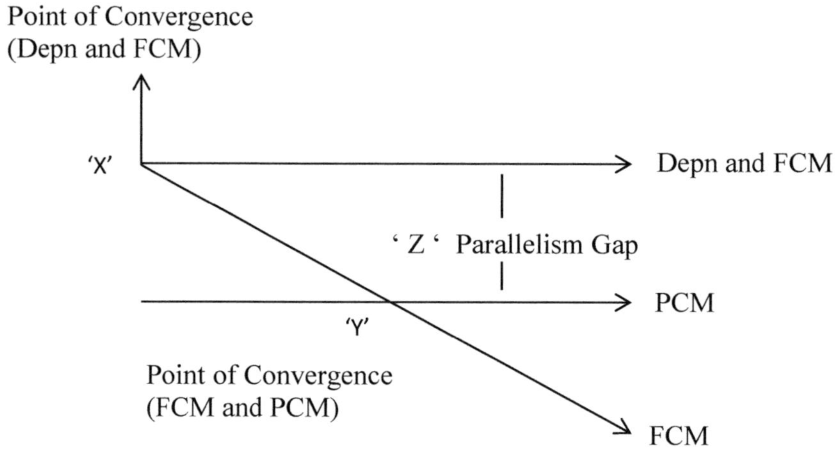

Fig. 2 Depreciation-capital maintenance: Convergence and parallelism (*Source* Author's construct)

the depreciation, connects FCM and PCM at some point in time. This becomes realistic and clear when the productive capital is maintained through asset replacement and restoration. At this stage, the market prices used in the re-measurement may approximate the restoration or replacement cost, hence the convergence depicted in 'Y'.

However, so long as depreciation accounting remains a mere technical requirement for limiting distributed profit without cash allocation or investment to capture the preservation of the productive capacity of assets, parallelism will exist, as illustrated by 'Z'. This parallelism could also threaten business sustainability unless the restricted meaning of capital is relaxed. Piketty (2014), in his study rejected the restrictive meaning of capital and argued that capital is 'the total of nonhuman assets that can be owned or exchanged on some market' (Piketty, 2014, p. 46). This enhanced meaning of capital is consistent with the contribution of Fisher and could provide the foundation for achieving sustainable capital maintenance and circularity accounting. The enhanced definition of capital would require resources to be allocated to increase the physical capital to maintain the volume of the productive capacity of assets. This is important since IASB is inconclusive about capital and recognises it as either debit (increase in net asset) or credit (increase in equity). Accordingly, there is an opportunity to present an alternative framework to develop sustainable and circularity capital maintenance to converge depreciation and all constituent of capital maintenance.

6 Threat of Depreciation-Capital Maintenance Parallelism: Business Sustainability and Circular Economy Perspective

A major question is: how serious is the threat of depreciation-capital maintenance parallelism?

A sustainable business focuses on the going concern of the enterprise through stabilisation, growth and development, including diversification and revitalisation of the enterprise (Pakšiová & Oriskóová, 2020; Sebestova & Nowakova, 2013). Achieving this business sustainability goal requires maintenance of capital and assets (Markovic et al., 2013; Tumpach & Baštincová, 2014). Therefore, it is counterproductive when the capital maintenance model over-concentrates on smoothening equity through financial capital maintenance where depreciation (including amortisation and impairment) is used to preserve capital by limiting distributed profit and increasing distributed retention. This suggests the need to follow the capital maintenance model to maintain financial and physical capital concurrently. Thus, parallelism of depreciation accounting and physical capital maintenance threatens long-run sustainability goals and circularity accounting. This may be illustrated in Scenario 2.

7 Scenario 2

Assuming Haulage Company operates the business of transporting Bauxite from Awaso to Takoradi Port eight (8) times a day using 4 Trucks acquired on the same day at the beginning of the business with the useful economic life of four (4) years for GHS50 million each. Assuming the company has decided to use straight-line depreciation method with no residual value and distributes all residual profit as dividends (see Tables 4 and 5 for resultant financials). The summary of the financial statements of the Company for the Four Years after maintaining financial capital are as follows:

Table 4 Statement of profit or loss and other comprehensive income (Current Accounting Treatment)

Haulage Company Limited	*Year 1* GHSm	*Year 2* GHSm	*Year 3* GHSm	*Year 4* GHSm
Sales	200	220	242	266
Cost of Sales	100	105	110	116
Operating Profit	100	115	132	150
Depreciation	(50)	(50)	(50)	(50)
Net Profit	50	65	82	100
Dividend	(50)	(65)	(82)	(100)
Retained Earnings	–	–	–	–

Table 5 Statement of financial position (Current Accounting Treatment)

Haulage Company Limited	*Opening*	*Year 1* GHSm	*Year 2* GHSm	*Year 3* GHSm	*Year 4* GHSm
Fixed Asset	200	150	100	50	-
Cash:					
Actual	100	100	100	100	100
Notional[a]		50	100	150	200
	300	**300**	**300**	**300**	**300**
Equity	300	300	300	300	300
	300	**300**	**300**	**300**	**300**

[a]This denotes depreciation-induced cash. It arises from depreciation. However, in practice, no cash is set aside for depreciation, so it is denoted as notional cash

This scenario shows that when depreciation is charged against income and subsequently all residual profits are distributed as dividends, the financial capital (equity) is maintained. This is a clear evidence of convergence between depreciation and financial capital maintenance. However, since this convergence does not account for physical capital maintenance, business sustainability is threatened as the Haulage Company may not be able to operate after the fourth year. The reasons are as follows:

i. The value of the Truck is nil, suggesting that holding other factors constant, there is no Truck to transport the Bauxite. All four (4) trucks need to be replaced.
ii. The cash balance is GHS300 million (actual and notional), which cannot be used to replace all four trucks and still continue to operate even if it is assumed that the current market price for the trucks is the same as they were acquired four years ago. Of course, this is unlikely given price-level changes reflected in the growth of cost of sales over the past four years. For Haulage Company Ltd to continue to operate into the foreseeable future (i.e. year 5), it would require a total cash outlay of about GHS327 (see Table 6).
iii. Further threat is the liquidity problem. Using this cash balance will mean no cash will be available to meet other unobservable business running costs. therefore, the company will require at least ghs127 million as working capital.

Thus, as long as convergence is not achieved for financial and physical capital maintenance, business sustainability is unlikely. Therefore, a major question is how have many businesses continued to exist and operated into the future, even though this parallelism still exists. There are at least three possible reasons:

i. The threat of parallelism is minimised when the quantum of sales is very high. In addition, the high volume of sales provides readily available cash to meet current obligations, especially when most of the sales are on a cash basis or receivables are received timely.

Table 6 Analyses of operational outlook for year 5

	Year 4 (Baseline) *GHSm*	*Year 5* *GHSm*
Operational Analyses		
Acquisition of Trucks	–	200
Working Capital @110% of Year 4	116	127
Required Cash for Operation		**327**
Cash Available		300
Additional Cash Required		**27**

ii. Availability of credits and loanable funds cushion sustainability under the threat of parallelism.
iii. In practice, not all distributable profits are actually distributed. This also improves liquidity position.

As illustrated in this section, moving towards full convergence through sustainability and circularity of capital maintenance is significant because it positions business operations to reflect the economic reality of the business environment (Garcia, 2020; Jianu et al., 2017; Paksiova, 2017; Pakšiová & Oriskóová, 2020). We, therefore, argue that to achieve business sustainability, there should be a paradigm shift in the current practice of capital maintenance to sustainability and circularity of capital maintenance where firms distribute profit to the extent that equity is maintained (financial capital maintenance) and resources are left and allocated to cover restoration of the non-current assets (physical capital maintenance).

Furthermore, business sustainability threats could also spill over to circular economy financing. *The circular economy is basically a system which seeks to achieve a state where there is no waste and products of today are not discarded but become raw materials for tomorrow*. The circular economy model affects all facets of life and professions, including the accounting and finance fraternity. The Platform for Accelerating the Circular Economy (PACE) advocates a circular economy for capital equipment (Circle Economy, 2019; PACE, 2019). The objectives of a circular economy for capital equipment are more aligned with physical capital maintenance. As explained throughout this chapter, physical capital relates to maintaining the productive capacity of non-current assets. This could have embedded circular economy principles for capital equipment (Circle Capital Equipment Coalition, 2021; Economy, 2019).

Of course, the traditional circular strategies for 'products that flow' may not be effective or appropriate for capital equipment. Capital equipment will require circular strategies, which will lead to the extended life of the asset, shared use, return for reuse and slowing down degradation and deterioration (Circle Economy, 2020, 2021; Capital Equipment Coalition, 2021). These circular economy strategies would also contribute to physical capital maintenance as they could enhance the productive capacity of the related assets. Achieving this would require active financing and investment. The current capital maintenance, which focuses primarily on technical accounting requirements of non-cash adjustment, threatens business sustainability and the circular economy of capital equipment due to a lack of active funding/cash.

Thus, the summary of the answer to the question raised at the beginning of this section is that: ***depreciation-capital maintenance parallelism can threaten business sustainability and integration of circular model in accounting.***

This leads to the next question: How do we address these threats of parallelism to achieving business sustainability and the circularity of capital maintenance? This will require a new approach to capital maintenance. The next section of the chapter proposes an alternative model for sustainable capital maintenance complementary to business sustainability and circular economy.

8 Alternative Model for Convergence, Sustainability and Circularity of Capital Maintenance

Physical capital maintenance requires investment decisions and continuous allocation for future financing to the extent that the productive capacity is either maintained or improved. This investment would require sufficient cash to support the outlay of capital maintenance. The expected investment outlay to maintain physical capital is equivalent to the replacement or improvement cost of the physical asset. For technical consideration, we follow Revsine (1974). We follow these fundamentals to develop a model for capital maintenance that would deepen business sustainability and a framework for integrating circular accounting.

We assume that financial capital (equity) and physical assets should be maintained to achieve full convergence of capital maintenance and depreciation. Thus, we operationalise capital as both equity (credit) and net asset (debit) (Nobes, 2015). We have already discussed that the current depreciation accounting treatment (non-cash consideration) is enough to maintain only financial capital and achieve convergence. However, more is required to maintain physical capital (cash investment consideration). We also argue that maintaining physical capital will require cash investment tied to depreciation. This would be achieved by treating depreciation as both a non-cash item (i.e. meeting the traditional accounting requirement) and cash investment to meet the requirement of physical capital maintenance. Therefore, at full convergence, full capital maintenance or sustainable capital maintenance is the function of depreciation (D) and cash investment or reserve (I). Unlike traditional accounting for depreciation, this proposed model requires active rolling over investment to maintain the physical capacity of the asset. This is expressed as:

$$CM\binom{i}{t} = f(D, I) \tag{1}$$

The traditional depreciation methods in accounting largely depend on only deterioration (physical wear and tear), which is enough for financial capital maintenance as it is a mere technical accounting requirement. However, for physical capital maintenance (PCM), the estimation of depreciation should account for both deterioration (physical wear and tear) and economic depletion and obsolescence (technological dynamics and structural changes). This assumption is important as depreciation accounting is often associated with unobservable market events, making the available records not sufficiently appropriate. These unobservable events or inputs include the use of broad assumptions, assumptions relating to the entity or class of asset of the entity rather than the asset's unique property (observable input), the historical experience of similar assets instead of the identical, same or actual experience of the specific asset and generic principle. Therefore, the depreciation amount to be used in the capital maintenance will be adjusted (D^*) to reflect these dynamics. Therefore, we substitute D^* for D and recalled Eq. (1) as Eq. (2):

$$CM\binom{i}{t} = f\left(D^*, I\right) \tag{2}$$

Adjusted depreciation is therefore given as:

$$D\binom{i}{t} = D_i(1 + a) \tag{3}$$

where 'a' is the depreciation-adjusted rate based on unobservable and observable market information. Although the 'a' is determined at the time the asset is ready for use, it should be reviewed periodically to reflect changes in observable and non-observable inputs. For instance, assuming four (4) trucks are expected to be used to transport goods throughout the year with an expected daily trip per truck of two (2). However, suppose two trucks break down and the remaining two double their trips to fill the gap, in that case, there is an observable event for an upward review of the depreciation-adjusted rate ('a') due to deterioration and economic depletion changes.

As noted earlier, the cash investment/reserve (I) outlay is expected to be sufficient enough to restore or improve the productive capacity of the asset. Thus, the equivalent cash investment or reserve fund should approximate the asset's expected replacement or restoration cost, i, at the end of the useful life, n. This replacement or restoration cost should also be adjusted by certainty equivalent factor (u) so that firms could be indifferent in the choice of replacement and restoration. The certainty equivalent factor (u) reflects the market information of the asset and the expected rate of return on investment (investment for asset restoration) so as to reasonably avoid over or under-investment to restore or improve the productive capacity of the asset. Thus, 'u' may be adjusted upward or downward depending on the expected restoration, improvement or replacement cost of the asset.

Moreover, since the replacement or restoration is a future commitment, the cost should be discounted to its present value using firm-level or project-specific cost of capital. Therefore, we express the cash investment as:

$$I\left(^{i}_{t}\right) = \frac{R_n(1+u)}{(1+r)^n} \tag{4}$$

where R_n denotes the replacement or restoration cost at the expected replacement or restoration date, n

r is the discount rate.

It is important to note that sustainable investment is incremental after the initial investment outlay. Therefore, capital is fully maintained at t, when CM_t–CM_{t-1} is invested for capital replacement or restoration. Besides the incremental investment, the investment returns or valuation gain or loss on the sustainable investment contributes to the period end reported sustainable asset. Therefore, the final model for sustainability and circularity capital maintenance without parallelism (SCM_p) is expressed as:

$$SCM^{i}_{t} = \sum_{t=1}^{n} D_1(1+a) + \frac{R_n(1+u)}{(1+r)^n} - CM_{t-1} \tag{5}$$

Consistent with Garcia (2020), the practical implementation of the SCM_p model will require financial investment (Asset restoration) and capital maintenance reserve in addition to the usual depreciation accounting. Thus, with this model, full convergence will still be achieved when depreciation accounting still follows its non-cash treatment.

9 Impact of the Sustainable Capital Maintenance Model on Financial Statement

The SCM model affects both the statement of profit and loss and other comprehensive income (or statement of comprehensive income) and the statement of financial position. The investment return or gain or loss on the valuation of the sustainable investment is reported in the statement of profit and loss and other comprehensive income. However, since all the returns remain holding gain for the period the related asset is in use, it is not recognised in income or retained earnings until the asset is put out of use. Thus, all gains or losses on the sustainable asset are recognised rather in other comprehensive income (OCI) and transferred to a special account called 'Sustainable Capital Maintenance Reserve'. This proposed treatment would also eliminate the possibility to cause volatility in net profit arising from unpredictable market conditions of the sustainable asset (Barth, 2006; Queku, 2020; Song, 2015).

In the statement of financial position, the accounting for sustainability and circularity of capital maintenance would affect assets through the recognition of a sustainable asset or investment and additional cash injection. Capital Maintenance Fund/Reserve would also be recognised in equity but remain non-distributable.

To illustrate the impact of the SCM model on the financial statement, we will revisit Scenario 2 and add further assumptions consistent with the model as follows:

i. It assumed that the asset replacement or restoration cost is GHS 200 million.
ii. It is further assumed that the depreciation-adjusted rate is 5% at the end of the year, consistent with changes in the cost of sales and remains at that absolute amount throughout the life of the trucks.
iii. An equivalent certainty factor of 5% and an investment return of 10% are also assumed.
iv. It is also assumed that any additional cash introduced to finance sustainable capital has an opportunity cost of return similar to the investment return of the sustainable asset (i.e. 10% in this scenario).
v. Project finance cost for such assets is assumed to be 15%.

Substituting these parameters into the SCM model in Eq. (5), the financial statement in Scenario 2 may be restated as in Tables 7 and 8. It can

Table 7 Statement of profit or loss and other comprehensive income (Proposed Sustainable Capital Maintenance Model)

Haulage Company Limited				
	Year 1 *GHSm*	*Year 2* *GHSm*	*Year 3* *GHSm*	*Year 4* *GHSm*
Sales	200	220	242	266
Cost of Sales	100	105	110	116
Operating Profit	100	115	132	150
Depreciation	(50)	(50)	(50)	(50)
Net Profit	50	65	82	100
Dividend	(50)	(65)	(82)	(100)
Retained Earnings	–	–	–	–
Sustainable asset return[a]	19	21	25	30

[a]This is the return on asset. It is non-distributable so long as the asset is in active use
It is, therefore not recognised in income/retained earning

be observed that the results in Table 5 (i.e. under traditional depreciation accounting and Table 7 (under sustainability and circularity of capital maintenance) are almost the same except for sustainable asset return (SAR). SAR represents the market return of the investment for capital maintenance. This return is not distributable until the end of the asset's useful life. Therefore, it is not recognised in income or as part of retained earnings.

10 Sustainability and Circularity of Capital Maintenance Model: A Panacea for Circular Economy Transition and Financing

As discussed earlier, sustainability and the circularity of capital maintenance without parallelism is the foundation for business sustainability and circular accounting. This is seen in a firm ability to either replace or restore the productive capacity of its capital. According to the Coalition of Circular Accounting (CCA) 'Circular accounting describes the practice of measuring, analysing, and reporting on a company's financial and non-financial performance to truly reflect the value and impact of circular businesses on all relevant stakeholders'. However, some have largely neglected accounting in the circular economy framework, arguing that accounting application for circular business or circular accounting

Table 8 Statement of financial position (Proposed Sustainable Capital Maintenance Model)

Haulage Company Limited					
	Opening	*Year 1 GHSm*	*Year 2 GHSm*	*Year 3 GHSm*	*Year 4 GHSm*
Fixed Asset	200	150	100	50	-
Sustainable Asset (Investment)		210	251	300	358
Cash	100	(41)	29	55	78
Additional Cash		41	-	-	-
	300	**360**	**381**	**406**	**436**
Equity	300	300	300	300	300
Additional Injection (Equity/Debt)		41	41	41	41
Capital Maintenance Fund/Reserve		19	40	65	95
	300	**360**	**381**	**406**	**436**

may amount to fitting a square peg into a round hole (Barker et al., 2020; Cohn, 2021).

Factually, the current accounting framework and technical reporting requirements are ill-equipped to truly drive and capture the value of circular businesses (Thurum, 2017; WEF, 2021). For instance, non-cash depreciation does not provide for asset restoration and consequently creates environmental waste in the form of scraps. Moreover, accounting is often viewed as a technical issue and fits well in the wider economic system rather than driving its dynamics such as a circular economy. In fact, circular accounting is not common in accounting practice among firms and accountants. Its benefits are also not well-understood. Therefore, the integral role of financial accounting in driving and managing the circular transition has remained under-appreciated (WEF, 2021).

Nevertheless, circular economy (CE) presents an innovative model for developing a business strategy for restoring and preserving business and productive capital (Planing, 2018), which accountants have struggled to address for centuries. In advancing circularity accounting (circular economy for accountants), one critical consideration is how it impacts capital maintenance or maintenance of the productive capacity of capital equipment. According to PACE (2019), with capital equipment, the priority should be reuse, refurbish, repurpose and remanufacture to retain the highest value of the capital equipment. Three principles have been formulated to achieve circularity of capital equipment (Capital Equipment Coalition, 2021, WEF, 2021):

i. Products and components follow circular and digitisation principles for minimal resource consumption and increased reuse strategies.
ii. Value retention is maximised by optimising the product and component utilisation rate and useful life, with the help of servitisation and digitalisation.
iii. End-of-use equipment and components are returned for reuse via efficient reverse logistics.

Of particular interest to accountants, is the emphasis on Lifecycle Extension strategies (LCES) in the CE model (Korhonen et al., 2018). The impact of CE model, in particular for LCES has somehow refreshed capital maintenance through digitalisation (Antikainen et al., 2018; Neligan, 2018; Sarc et al., 2019). For instance, some of the fundamental LCES in CE such as Deposit models (DM), Lease and Rent models (LRM), Sell-and-Buy-Back model (SBBM) and Product-as-a-Service (PaaS) may be seen as drivers for capital maintenance and sustainable productive capacity of the asset. Although these models' technical details are outside this chapter's scope, they are the interface for integrating CE into traditional accounting to achieve sustainability and the circularity of capital maintenance.

The CE through LCES creates a new supply chain structure where manufacturers, service providers and end users (firms) could rearrange contractual arrangements where manufacturers and/or service providers would be ready to either buy back or upgrade or refurbish capital assets at the end of its useful life (Bressanelli et al., 2018; Fontana et al., 2021; Grieves & Vickers, 2017) and may resell the asset to another customer. Therefore, circular accounting recognises that the inherently long life of capital assets could be extended even further, slowing down wear and tear and degradation, emphasising shared use and reuse to eliminate avoidable waste and enhance sustainable, and productive capacity (Circle Economy, 2021). This CE strategy and practice would minimise the cost of capital maintenance (i.e. productive capacity) and increase profit margins.

Nevertheless, since circularity is a strategic issue, companies and accountants, for that matter, would adopt CE models only when such models demonstrate value-for-money and positive synergy with the rest of the reporting value chain. For capital assets, the relevance of circularity accounting would be based on reliable data on asset performance. This makes the concept of digital twins very crucial. *A digital twin is a virtual representation or a digital replica of a physical product or equipment that collect, process, store and synchronise data with information systems* (Demestichas et al., 2020; Grieves, & Vickers, 2017) *which can be used for managing asset circularity.*

This digitalisation process (digital twin) of CE allows for a better understanding of the exogenous properties of a product (i.e. physicochemical properties and dynamics of the process) and endogenous properties (i.e. the behaviour of the various components that compose the equipment, the use of the product, maintenance history and performance rate, etc.) (Fontana et al., 2021). Thus, a digital twin is particularly important for integrating CE into accountancy practice and reporting strategies, especially with respect to equipment life cycle extension. Reliable data about the productive capabilities of the capital asset, equipment health, maintenance history, etc., would smoothen asset circularisation and capital maintenance. Thus, digitally enabled services, equipment utility, maintenance and shared access of the CE model that see beyond a one-off sale to focus on asset functionality instead of ownership of asset offer innovative ways for keeping capital assets in use efficiently and for longer periods.

Achieving a successful integration of the CE model for capital maintenance and sustainable, productive capacity requires active financing and modification of the current framework of capital maintenance. Thus, although circular accounting is one of the surest ways to achieve sustainable capital maintenance, circular accounting is possible only when traditional accounting practices are either modified or extended. Unfortunately, as discussed earlier traditional accounting for capital assets such as depreciation cannot accommodate the investment outlay required for adopting circular accounting. The current depreciation accounting follows the linear approach where an asset is acquired, used, depreciated throughout its useful economic life, and eventually discarded mostly as waste. This traditional depreciation accounting approach does not consider a circular model's fundamental principles (Circle Economy, 2021).

However, the proposed sustainability and circularity of capital maintenance present a different picture showcasing accounting as both an enabler and a driver of circular business. However, this proposed sustainability and circularity of capital maintenance requires full convergence of depreciation and capital maintenance where physical capital (asset) is maintained through restoration investment, thereby aligning asset accounting to a circular model. The investment and financing parameter of the proposed model provides a window for a circular economy for capital equipment as they are more aligned (Capital Equipment Coalition, 2021; Jowitt et al., 2020). The embedded principles of the circular economy for capital equipment require investment in asset restoration (Circle Economy, 2019; PACE, 2019). This proposed model emphasises this. Thus, the modified approach of depreciation used in developing this model, where cash investment for restoration is mandatory, is viewed as a new initiative in aligning traditional accounting with circular accounting.

The circular model seeks to optimise the use and inventory of capital equipment through circular strategies to minimise and reduce environmental waste, address the scarcity of resources, and develop sustainable value chains (Circle Capital Equipment Coalition, 2021; Economy, 2019). Maintaining physical capital through replacement and restoration, as discussed in the proposed model, would lead to an extended life of the asset, shared use, return for reuse and slowing down degradation and deterioration (Circle Economy, 2020). Moreover, circular accounting requires a huge capital outlay, especially the circular economy for capital equipment which has often become a barrier (Capital Equipment Coalition, 2021; PACE, 2019; Ranta et al., 2018). However, with asset restoration investment proposed in this model coupled with its convergence with full capital maintenance, firms could achieve sustainable capital maintenance and circular economy concurrently through embedded strategies of circular and traditional accounting. Therefore, capital maintenance's proposed sustainability and circularity is a panacea for circular accounting.

11 Chapter Summary

This chapter discussed historical development in depreciation literature from theoretical and practical perspectives. The chapter has also presented the evolution of capital maintenance since the thirteenth century during the invention of Double Entry Book Keeping through the legal view, dynamic view, inflation view and net assets view to the current stage where the emphasis is on sustainable capital maintenance. Depreciation-capital maintenance nexus was discussed with a point of convergence articulated, and the threat of parallelism was identified. The chapter further highlighted how sustainable capital maintenance without parallelism, as proposed in this chapter, could deepen business sustainability and provide a window for integrating circular economy models into accounting practice through circular accounting.

References

Abel, A. B. (1990). Consumption and investment. In B. M. Friedman & F. H. Hahn (Eds.), *Handbook of monetary economics* (pp. 725–778). Elsevier Science.

Antikainen, M., Uusitalo, T., & Kivikytö-Reponen, P. (2018). Digitalisation as an enabler of circular economy. *Procedia CIRP, 73*, 45–49.

Barker, R. (2010). On the definitions of income, expenses and profit in IFRS. *Accounting in Europe, 7*(2), 147–158.

Barker, R., Eccles, R. G., & Serafeim, G. (2020). *The future of ESG is … accounting?* Retrieved from Harvard Business Review.

Barth, M. E. (2006). Including estimates of the future in today's financial statements. *Accounting Horizons, 20*(3), 271–285.

Biondi, Y. (2011). The pure logic of accounting: A critique of the fair value revolution. *Accounting, Economics and Law, 1*(1) 2011, Art. 7.

Bischo, C. W., & Kokkelenberg, E. C. (1987). Capacity utilization and depreciation-in-use. *Applied Economics, 19*(8), 995–1007.

Bitros, G. C. (2010). The theorem of proportionality in contemporary capital theory: An assessment of its applicability. *Journal of Economic and Social Measurement, 35*, 1–31.

Blanchard, O., Rhee, C., & Summers, L. (1993). The stock market, profit and investment. *Quarterly Journal of Economics, 108*, 115–136.

Boucekkine, R., De La Croix, D., & Licandro, O. (2011). Vintage capital theory: Three break breakthroughs.

Bressanelli, G., Adrodegari, F., Perona, M., & Saccani, N. (2018). Exploring how usage-focused business models enable circular economy through digital technologies. *Sustainability, 10*, 639.

Brief, R. P. (1966). The origin and evolution of nineteenth-century asset accounting. *Business History Review, 40*(1), 1–23.

Broad, C. (1925). The mind and its place in nature.

Capital Equipment Coalition. (2021). Workshop on circular economy in capital equipment. An initiative by PACE.

Carsamer, E., Abbam, A., & Queku, Y. N. (2021). Bank capital, liquidity and risk in Ghana. *Journal of Financial Regulation and Compliance*. https://doi.org/10.1108/JFRC-12-2020-0117

Circle Economy (2019). The circularity gap report 2019. Closing the circularity gap in a 9% world. https://assets.website-files.com/5d26d80e8836af2d12ed1269/5dea43f562f8ac3e3113fe51_ad6e59_ba1e4d16c64f44fa94fbd8708eae8e34_compressed.pdf

Circle Economy. (2020). The circularity gap report 2020. When circularity goes from bad to worse: The power of countries to change the game. https://pacecircular.org/sites/default/files/2020-01/Circularity%20Gap%20Report%202020.pdf

Circle Economy. (2021). Financial accounting in the circular economy: Redefining value, impact and risk to accelerate the circular transition.

Cohen, A. J., & Harcourt, G. C. (2003). Whatever happened to the Cambridge capital theory controversies? *Journal of Economic Perspectives, 17*(1), 199–214.

Cohn, M. (2021). Sustainability accounting: 'the biggest disrupter.' Retrieved from Accounting Today.

Demestichas, K., & Daskalakis, E. (2020). (2020) Information and communication technology solutions for the circular economy. *Sustainability, 12*, 7272.

Escribá-Pérezy, F. J., Murgui-Garcíaz, M. J., & Ruiz-Tamaritx, J. R. (2019). Capital stock and depreciation: Theory and an empirical application discussion paper 2019-4.

Feltham, G. A., & Ohlson, J. A. (1996). Uncertainty resolution and the theory of depreciation measurement. *Journal of Accounting Research, 34*(2), 209–234.

Financial Accounting Standards Board (FASB). (1984). Recognition and measurement in financial statements of business enterprises, statement of financial accounting concepts No. 5.

Fontana, A., Barni, A., Leone, D., Spirito, M., Tringale, A., Ferraris, M., Reis, J., & Goncalves, G. (2021). Circular economy strategies for equipment lifetime extension: A systematic review. *Sustainability, 13*, 1117. https://doi.org/10.3390/su13031117

Garcia, C. (2020). Accounting and law for equity capital and corporate sustainability. *European Company Law, 17*(5), 176–182.

Grieves, M., & Vickers, J. (2017). Digital twin: Mitigating unpredictable, undesirable emergent behavior in complex systems new findings and approaches. In F.-J. Kahlen, S. Flumerfelt, & A. Alves, (Eds.), *Transdisciplinary perspectives on complex systems new findings and approaches* (pp. 85–113). Springer International.

Haller, A. (1992). The relationship of financial and tax accounting in Germany: A major reason for accounting disharmony in Europe. *International Journal of Accounting, 27*(4), 310–323.

Hergenhahn, B. R. (2009). *An introduction to the history of psychology*. Wadsworth Cengage Learning.

Hicks, J. R. (1949). *Value and capital*. Oxford University Press.

Hotelling, H. (1925). A general mathematical theory of depreciation. *Journal of the American Statistical Association, 1925*, 340–353.

Hulten, C. R., & Wycoff, F. C. (1996, January). Issues in the measurement of economic depreciation: Introductory remarks. *Economic Inquiry, XXXIV*(1), 10–23.

IASB. (2010). *The conceptual framework for financial reporting*. International Accounting Standards Board.

IASB. (2013). *A review of the conceptual framework for financial reporting*. International Accounting Standards Board.

IASplus. (2022). IASB issues amendments to IAS 16 regarding proceeds before intended use. https://www.iasplus.com/en/news/2020/05/ias-16. Accessed 27 Nov 2022.

ICAEW. (2010). Guidance on the determination of realised profits and losses in the context of distributions under the Companies Act 2006, Tech 02/10. Institute of Chartered Accountants in England and Wales; and Edinburgh: Institute of Chartered Accountants of Scotland.

James, S. R., & Nobes, C. W. (2013). *The economics of taxation*. Fiscal Publications.

Jianu, I., Jianu, I., & Turlea, C. (2017). Measuring the company's real performance by physical capital maintenance. *Economic Computation and Economic Cybernetics Studies and Research, 51*(1), 37–57.

Johansen, L. (1959). Substitution versus fixed production coefficients in the theory of economic growth. *Econometrica, 29*, 157–176.

Jorgenson, D. W. (1963). Capital theory and investment behavior. *American Economic Review, 53*(2), 247–259.

Jowitt, S. M., Mudd, G. M., & Thompson, J. F. H. (2020). Future availability of non-renewable metal resources and the influence of environmental, social, and governance conflicts on metal production. *Commun Earth Environ, 1*, 13.

Korhonen, J., Honkasalo, A., & Seppälä, J. (2018). Circular economy: The concept and its limitations. *Ecological Economics, 2018*(143), 37–46.

Lee, G. A. (1977). The coming of age of double entry: The Giovanni Farolfi ledger of 1299–1300. *Accounting Historians Journal, 4*(2), 79–95.

Liapis, K. J., & Kantianis, D. D. (2015). Depreciation method and life-cycle costing (LCC) methodology. *Procedia Economics and Finance, 19*(2015), 314–324.

Lutter, M. (2006). *Legal capital in Europe, European company and financial law review* (Special Vol. 1). https://doi.org/10.1515/9783110926583

Markovic, P., Grancicova, K., & Juhaszova, Z. (2013). Financne rozhodovanie o dlhodobej alokacii kapitalu. Management trends in theory and practice: Scientific papers EDIS. University Publishing House, University of Zilina.

Mora, A., & Walker, M. (2015). The implications of research on accounting conservatism for accounting standard setting. *Accounting and Business*

Research, 45(5), 620–650. https://doi.org/10.1080/00014788.2015.1048770

Morris, R. D. (1986). Lee v. Neuchatel Asphalte Company (1989) and depreciation accounting: Two empirical studies. *Accounting and Business Research, 17*(65), 71–81.

Mukoyama, T. (2008). Endogenous depreciation, mismeasurement of aggregate capital, and the productivity slowdown. *Journal of Macroeconomics, 30*(1), 513–522.

Napier, C. (1997, September 17). The British aristocracy, capital and income, and nineteenth century company accounting. In *Paper Presented at the Accounting, Business and Financial History Conference*.

Neligan, A. (2018). (2018) Digitalisation as enabler towards a sustainable circular economy in Germany. *Intereconomics, 53*, 101–106.

Nobes, C. W. (2015). Accounting for capital: The evolution of an idea. *Accounting and Business Research, 45*(4), 413–441.

Nobes, C. W. (1983). The evolution of the harmonising provisions of the 1980 and 1981 companies acts. *Accounting and Business Research, 14*(53), 43–53.

PACE. (2019). Circular value creation—Lessons from the capital equipment coalition. Davos. https://pacecircular.org/sites/default/files/2019-03/Circular_value_creation_2019_v7_WEB.pdf

Pakšiová, R., & Oriskóová, D. (2020). Capital maintenance evolution using outputs from accounting system. *Scientific Annals of Economics and Business, 67*(3), 311–331.

Paksiova, R. (2017). The critical analysis of profit for its allocation decision-making. *Scientific Annals of Economics and Business, 64*(1), 41–56. https://doi.org/10.1515/saeb-2017-0039

Piketty, T. (2014). *Capital in the twenty-first century (tr. A. Goldhammer)*. Belknap Press.

Planing, P. (2018). Towards a circular economy—How business model innovation will help to make the shift. *International Journal of Business and Globalisation, 20*, 71.

Pope, P., & Walker, M. (2003). *Ex-ante and ex-post accounting conservatism, asset recognition and asymmetric earnings timeliness*. Working Paper, Lancaster University and University of Manchester.

Queku, I. C. (2015). International Financial Reporting Standards (IFRS) in practice and fundamental accounting concepts and conventions: A review approach. *Journal of Advances in Business and Management Research (JABMAR), 1*(1), 19–25.

Queku, Y. N. (2020). Value relevance of fair value measurement and stock price predictability: Incremental effect and synergetic analysis of listed banks in Ghana. *ADRRI Journal of Arts and Social Sciences, 17*(7(5)), 85–111.

Ranta, V. L., Aarikka-Stenroos, P. R., & Mäkinen, S. J. (2018). Exploring institutional drivers and barriers of the circular economy: A cross-regional comparison of China, the US, and Europe. *Resources, Conservation and Recycling,* 135, 70–82.

Reid, J. M. (1987). Judicial views on accounting in Britain before 1889. *Accounting and Business Research, 17*(67), 247–258.

Revsine, R. (1974). *Replacement cost accounting: A Theoretical foundation.* Touche Ross Publications.

Richard, J. (2015). The dangerous dynamics of capitalism: From static towards futuristic IFRS accounting. *Critical Perspectives on Accounting, 30*(2015), 9–34.

Rihll, T. E. (2013). Depreciation in vitruvius. *The Classical Quarterly, 63*, 893–897.

Roychowdhury, S., & Martin, X. (2013). Understanding discretion in conservatism: An alternative view point. *Journal of Accounting and Economics, 56*(2/3, suppl. 1), 134–146.

Sarc, R., Curtis, A., Kandlbauer, L., Khodier, K., Lorber, K. E., & Pomberger, R. (2019). Digitalisation and intelligent robotics in value chain of circular economy oriented waste management—A review. *Waste Management, 95*, 476–492.

Schmalenbach, E. (1959). *Dynamic accounting.* Gee and Company.

Sebestova, J., & Nowakova, K. (2013). Dynamic strategy for sustainable business development: Mania or hazard? *Amfiteatru Economic, 15*(34), 442–454.

Seidu, B. A., Queku, Y. N., Tackie, G., & Mensah, J. (2022). Working capital-targeting and firms life cycle: Analysis of persistence and speed of adjustment. *African Journal of Business and Economic Research, 17*(3), 117–143.

Solow, R. M. (1960). Investment and technological progress. In K. J. Arrow, S. Karlin, & P. Suppes (Eds.), *Mathematical methods in the social sciences* (pp. 89–104). Stanford University Press.

Song, X. (2015). Value relevance of fair values—Empirical evidence of the impact of market volatility. *Accounting Perspectives, 14*(2), 91–116.

Thurum, R. (2017). *The blueprint for new accounting: Laying the foundations for future-fit reporting* (pp. 1–120, Rep). Retrieved from Reporting 3.0 website.

Tumpach, M., & Baštincová, A. (2014). Cost and benefit of accounting information in slovakia: Do we need to redefine relevance? In O. Deev, V. Kajurová, & J. Krajíček (Eds.), *European Financial Systems 2014: Proceedings of the 11th International Scientific Conference* (pp. 655–676). Masaryk University.

Van Mourik, C. (2010). The equity theories and financial reporting: An analysis. *Accounting in Europe, 7*(2), 191–211.

World Economic Forum (WEF). (2021). What is sustainability accounting? What does ESG mean? We have answers. Retrieved from World Economic Forum website.

Yamey, B. S. (1949). Scientific bookkeeping and the rise of capitalism. *The Economic History Review, 1*(2/3), 99–113.

Yoshida, J. (2020). The economic depreciation of real estate: Cross-sectional variations and their return implications. *Pacific-Basin Finance Journal, 61*(C).

Zeff, S. (2014). Principles before standards: The ICAEW's 'N Series' of recommendations on accounting principles 1942–1969. An initiative from the ICAEW Financial Reporting Faculty.

Pathways Towards a Circular Economy in Ghana: The Contribution of Waste Transfer Stations and the Informal Waste Collectors in Solid Waste Management

Kofi Dua Agyei, Bob Offei Manteaw, Andy Bonaventure Nyamekye, Kirk Bright Enu, and Thomas Yeboah

1 Introduction

Many cities worldwide face increasing challenges in providing sustainable solid waste management (SWM). Environmental problems which result from poor SWM continue to increase in volume, affecting air quality, water resources, and human health (Mudu et al., 2021). There is a direct link between waste and sanitation, which, if not managed properly, may affect the natural environment and human health (Cobbinah et al.,

K. D. Agyei (✉) · B. O. Manteaw · K. B. Enu
Centre for Climate Change and Sustainability Studies, University of Ghana, Accra, Ghana
e-mail: kdagyei@outlook.com

A. B. Nyamekye
Citigroup, 388 Greenwich St, New York City, NY 10013, USA

M. O. Erdiaw-Kwasie and G. M. M. Alam (eds.), *Circular Economy Strategies and the UN Sustainable Development Goals*, Sustainable Development Goals Series,
https://doi.org/10.1007/978-981-99-3083-8_18

2015; Yeboah-Assiamah, 2015). Recent increases in population and especially in urban communities, seem to have highlighted how challenging waste management is to city authorities in developing countries, especially in sub-Saharan Africa. Thus, the lack of waste management facilities and infrastructure has significantly worsened the effectiveness of waste management in these regions (Mudu et al., 2021; Oteng-Ababio et al., 2013; Peiris & Dayarathne, 2022).

The multiple challenges in reaching desired targets in solid waste management in developing countries can also be attributed to fragmented responsibilities and limited integration of policy goals towards socio-ecological benefits, a mismatch between national-level policy decisions and local-level action plans, and the lack of benchmarking measures or performance criteria to evaluate the environmental, economic, and social impacts of the existing management practices (Peiris & Dayarathne, 2022). These factors have resulted in low waste collection rates ranging from 20 to 80% in some urbanised communities in developing regions, whereas collection rates in developed countries are close to 90% (Chang & Pires, 2015; UNEP, 2005; Washburn, 2012). Thus, collection services in least-developed countries were registered at 39% in 2018 and 51% for countries in the lower category of middle-income states (Breukelman et al., 2019). The low collection rates in developing countries make it evident that most communities are inundated with piles of waste, which has a high propensity to negatively affect sanitation and hygiene.

Ghana is no exception to this bizarre phenomenon as its large urban communities, such as Accra and Kumasi, are overwhelmed with uncollected solid waste. For instance, Accra generates about 2500 tons of waste per day, out of which over 55% remains uncollected (Kortei & Quansah, 2018). It is, therefore, clear from such scenarios that SWM as a socio-ecological challenge is a function of poor and inefficient collection and management regimes, particularly in sub-Saharan Africa, where most countries lack the requisite infrastructure to facilitate efficient management processes (AfDB, 2012; Emenike et al., 2013).

Effective waste management is necessary for Africa's sustainable development aspirations. Thus, a proper waste management system will help

---

T. Yeboah
Bureau of Integrated Rural Development (BIRD), Kwame Nkrumah University of Science and Technology, Kumasi, Ghana

to create sustainable cities, enhance human health, and mitigate greenhouse gas emissions. As a result, most governments in sub-Saharan Africa have made waste management a policy priority. They have set conscious processes to explore innovative approaches to position their economies on the path of circularity to increase resource productivity and waste recovery. This new orientation would enable these economies transition from the old "make-consume-dispose" linear economic model to the "recovery-reuse-recycle-repurpose-remake" model of the circular economy (CE). The circular economy helps to close and narrow material loops between consumption and disposal by advocating for the reuse, redesign, and recycling of materials and products for as long as possible as a step to reducing waste generation. It reflects a structural shift that promotes long-term stability, optimises the use and circulation of materials and resources, and yields economic, environmental, and societal benefits (Sharma et al., 2021). The driving force of the CE principle is a circular business model (CBM), which shows how organisations capture and retain value (Kirchherr et al., 2017a). But current practices and policies to facilitate a transition to CE in Ghana remain challenging across sectors (Debrah et al., 2022). Nonetheless, one cardinal means by which the circular economy principle can effectively cut down on waste depends on how it is actualised in infrastructures (Savini, 2021).

For the city of Accra to address all these solid waste management challenges, it has recently introduced what is known as a "waste transfer station" as part of efforts to add technological innovation to enhance the efficient delivery of waste management services (Manteaw & Boachie, 2019). Transfer stations are light-industrial facilities where municipal solid waste is unloaded from multiple smaller waste collection vehicles such as garbage trucks, reloaded and consolidated into larger transport vehicles such as rail cars to a landfill or material recovery site (Ağaçsapan & Çabuk, 2020; Bovea et al., 2007). Transfer stations are essentially transportation infrastructure and not waste disposal sites. Therefore, they help cut down on transport costs to disposal sites, thereby incentivising more waste to be collected (Cui et al., 2011).

There are two waste transfer stations for municipal solid waste in Ghana, located at Teshie and Achimota. Many hope introducing these innovative initiatives into the country will allow solid waste to be handled appropriately and deepen circularity in urban communities. However, the significance of these facilities has not been fully assessed as to their effect on properly managing solid waste in the city (Manteaw & Boachie, 2019).

This chapter argues that transfer stations, as new technology, could help to manage solid waste in the city and set it on the path to circularity, particularly with the involvement of the informal tricycle waste collectors (known in local communities as the "aboboyaa") who have become dominant actors in the solid waste management chain.

Thus, to deepen and broaden the growing concept of the circular economy and suggest a pathway to help to improve waste management practices in Ghana, the chapter seeks to assess the potential impact of the waste transfer station in promoting circular economy strategies that lead to sustainable development outcomes in urban communities through the lens of the technological innovation theory. Specifically, it seeks to explore how the transfer station has contributed to realising circular economy principles and the extent to which it has affected the informal waste collectors to promote good solid waste management practices.

2 Literature Review

2.1 Solid Waste Management in Ghana

Solid waste management has evolved to be an urban phenomenon in Ghana. It has thus shifted from being the sole responsibility of various households and currently falls within the sphere of operations of local governments (Mumuni, 2016; Oteng-Ababio, 2012). Waste has become an unavoidable product of society and, in modern times, has changed in nature and form and hence needs to be appropriately managed. Ghana has, however, failed to process the large amount of waste generated over the years into valuable forms that could be utilised for generating energy, raw materials for other production purposes, and creating wealth (Amoah & Kosoe, 2014; Oteng-Ababio et al., 2017).

It is estimated that about 200 kg of waste per capita is generated annually in Ghana (Duku et al., 2011). But waste generation is not uniform across the country. The two most populous cities, Accra and Kumasi, have waste generation rates above the national average of 0.51 kg per person in a day (Miezah et al., 2015). The high waste generation rates could be attributed to the rapid rate of urbanisation and population increase in these areas. It is worth highlighting that only some of the waste generated is taken out of the system. Indeed, Kortei and Quansah (2018) state that 55% of the 2500 metric tons of waste generated daily in the Accra Metropolis remain uncollected. The inexhaustive collection of solid waste from various generation sources has had minimal impact on reducing the volumes of waste generated in the country. This situation explains the adoption of environmentally unsustainable practices such as the throwing of wastes onto the streets, inappropriate public areas, lowlands, drainage channels, and rivers which defeats the fundamental principles of a circular system (Amoah & Kosoe, 2014).

Interestingly, efforts towards SWM in the country are focused mainly on solid waste collection and reflect the government's move to privatise solid waste collection in the late 1990s (Oteng-Ababio, 2010). Thus, to some extent, the government's partnership with the private sector has helped to improve efficiency as more resources are mobilised, risks are decreased, and service delivery is enhanced. In this view, local authorities and private waste service providers often undertake waste management services at a fee to households and businesses (Abunyewah et al., 2022; Oteng-Ababio, 2012). However, the involvement of private formal waste companies did little to curb the solid waste collection inefficiencies in the system as there remained piles of uncollected waste in various generation sources (Amoah & Kosoe, 2014). Oteng-Ababio et al. (2013, p. 5) note that "public–private partnerships have managed to improve the coverage in terms of the collection but have fallen well short of standards for efficiency and reliability."

Against this background, informal waste collectors emerged in Ghana's solid waste management processes as a direct response to the perceived shortfalls in the waste management sector. The quest to find solutions to waste collection lapses has increasingly given credence to the activities of the informal solid waste collectors, a sector whose tremendous role and complementary efforts in solid waste management have been overlooked (Akorsu et al., 2019; Medina, 2007; Oteng-Ababio, 2011). These informal waste actors are individuals or groups engaging in unregistered and unregulated waste management activities. Their work includes waste collection, transportation, separation, recycling, and disposal (Akorsu et al., 2019). This chapter's data collection and analysis focus on those who use tricycles to collect and transport solid waste from places that formal waste contractors cannot access. This group of informal waste actors typically falls into three different classes depending on the ownership arrangements for the vehicle used. They are owner-drivers, worker-drivers, and those who either worked on hire-purchase arrangements or paid a determined daily fee to the vehicle owner. With no formal contract with their clients, they collect waste from households, business enterprises, and marketplaces for onward disposal (Akorsu et al., 2019).

The informal waste actors are not legally recognised; thus, their activities are not registered, regulated, or licensed. Their primary function is to complement the role played by formal solid waste management companies (Oduro-Appiah et al., 2019). Indeed, Ghana's informal waste sector operations have become pronounced. A sizeable number of people in the middle- and low-income suburbs subscribe to their services, as asserted by Oteng-Ababio (2010). He indicates that about 60% of the middle-income class patronised their services in the Accra Metropolis.

Nevertheless, as crucial as they may be in SWM, the informal waste actors are given little recognition in Ghana. Thus, the conversation on partnership in SWM is often viewed concerning only formal sector companies neglecting the informal sector (Akorsu et al., 2019; Oteng-Ababio, 2010). An observation made by Akorsu et al. (2019) indicates that policy emphasis on SWM is skewed towards the plight of private formal and not informal sector operators, who are dominant in the waste management sector. Despite the low recognition, the sector keeps growing and serves as a source of livelihood, especially for the poor urban population. These people work under deplorable conditions, and city authorities and residents often blame them for blurring the beauty of cities.

To improve solid waste management in the country, Ghana has recently introduced two transfer stations for solid waste, one at Teshie and the other at Achimota. The transfer stations facilitate waste collection services and enable collection actors to access them easily with their collection vehicles, mostly skip trucks and motorised tricycles. This novel technology is believed to revolutionise SWM in the country, especially collection services rendered in Accra. Hence, curbing the practice of direct transportation of collected waste from generation sources to final disposal sites (Manteaw & Boachie, 2019).

2.2 *The Concept of Circular Economy and Waste Management*

The steadily rising consumption patterns and diminishing natural resources have prompted many local governments and private businesses to implement and strengthen waste management strategies to achieve greater resource efficiency. Despite significant improvements in waste management efforts over the past decades, numerous opportunities remain. In recent years, the circular economy concept has become more popular for efficiently managing societal resources (Luttenberger, 2020). This paradigm proposes converting the current linear system, based primarily on linear resource flows, into closed-loop resource flows capable of maintaining product value over time. Geissdoerfer et al. (2017) refer to the concept as a regenerative system that minimises waste and resource input, emissions, and energy leakage by slowing and narrowing energy and material loops. De Jesus et al. (2019) state that CE pertains to eco-innovation set on a pathway that fosters system redesign rather than improving resource use and entails a shift to technological innovation that considers sustainability issues and brings about positive ecological effects.

Luttenberger (2020) views CE to imply waste prevention, reuse, and recycling, and serves as a powerful bridging concept to encourage the fundamental links among resource use, waste, and emissions and contributes to integrating economic and environmental policies. In this sense, applying the concept to waste management enhances the use value of materials by creating a closed-loop economy through strategies that enable a second life for waste products. Achievable by employing material recovery, repair, and recycling when the end-of-life is irreversibly reached. This process helps to reintegrate the recovered materials into the value chain and address the impacts at the initial phase of the product life cycle (Nußholz, 2017; Romero-Hernández & Romero, 2018).

It has to be stated that CE is a recent, emerging, and variously contested concept (De Jesus & Mendonça, 2018; Kirchherr et al., 2017a; Korsunova et al., 2022). Kirchher et al. (2017a) state that various stakeholders usually employ fast-developing concepts, which can blur the concept since they frequently operate in different worldviews. As such, there is no commonly accepted definition as witnessed in the case of CE. Kirchherr and Piscicelli (2019) observe the various meanings given to the concept in the different disciplines. They note that human geographers have taken an interest in the CE concept as a vision to manage interactions between the environment and economies. Earth scientists are most interested in the technical aspects of its implementation at the micro level. On the other hand, environmental scientists view the CE as a promising paradigm to bring down companies' environmental impacts, and innovation scientists see the CE as a systemic innovation.

Reike et al. (2018) also note the differences in conceptualising the R-strategies of the CE among different scholars across disciplines. Even though the 3Rs form an accepted notion of CE in theory and practice, there are varying numbers of Rs aside from the known (Reduce, Reuse, and Recycle) and different meanings assigned to them. This variation reinforces the divergent conceptualisations of the CE principle. Nußholz (2017) adds that even though most of these strategies in the concept can also be understood as measures to extend the useful life of resources, most scholars classify this set of strategies differently and use different terminologies. Also, the defining boundaries underlying the categorisations can differ even if the same terminology is used. This is evident in Morseletto's (2020) classification of the framework into ten common circular economy strategies (i.e. reduce, rethink, refuse, recover, reuse, recycling, repurpose, refurbish, repair, remanufacture), which was further organised into three groups approach. These are the useful application of materials, extending the life span of products and their parts, and more innovative product manufacturing and use.

Again, Korsunova et al. (2022) point out that current literature on informal practices in low-income contexts often erroneously labels all CE practices as "recycling," which obscures the benefits of practices that retain the value of materials and goods better than recycling. Citing Nzeadibe's (2009) work, they describe how locals recover waste materials and reuse them to make new products or add value to the materials before they are resold. These practices aim to keep materials in use for a more extended period and can be classified under Reike et al. (2018) shorter, higher priority loops in the hierarchy of waste prevention and value reutilisation systematise under the 10R classification. Kirchherr et al. (2017a) note that CE is mainly to reduce, reuse, recycle, and recover but that, in reality, much policy has been oriented towards promoting recycling. In order to cohere ideas that surround the concept, there ought to be further theoretical development of the concept to help strengthen it and thus circumvent diverging views. The theoretical development of CE needs to be transparent to improve the current understanding of the concept in the discourse (Kirchherr et al., 2017a).

Strong theoretical development is essential as circular economy models enable the smooth transition towards sustainability and strive for business models that integrate the three pillars of sustainable development. These models employ a symbiotic approach to recover materials from waste, design long-lasting products, and extend the service life of systems. Sharma et al. (2021) believe the concept can provide a healthy balance of economic performance, social inclusion, and environmental resilience. However, there are calls for more holistic approaches to explore and understand how CE fulfils the three dimensions of sustainability, as over-simplistic approaches may lead to unintended consequences (Camilleri, 2019). Particularly so as it is seen to satisfy the environmental and economic dimensions of sustainability. Kirchherr et al. (2017a) state that it is particularly worrying that understanding the CE concept only entails one or two of the three dimensions of sustainable development, a move which can lead to unsustainable CE implementation. On the other hand, Repp et al. (2021) state that transitioning to a circular economy has an ethical dimension since its social effects could contribute to or question social sustainability, which many scholars often discount.

The transition to a CE would require contributions from various stakeholder groups. Although the role of the private sector has been particularly highlighted in recent years, the approach has been mainly established in the formal private sector through governance and innovation programmes in different parts of the world. Yet the accounts of how the vast informal private stakeholder group contributes to retaining the value of materials in circulation have received little attention. They are often detached from business and policy discussions on transitions towards a CE. These informal practices are often overshadowed by calls for an improved formal organisation of waste management (Korsunova et al., 2022). Even though there is a general agreement that the informal sector in developing countries will be an integral part of the CE under any scenario, the conceptual foundations of the CE are still rooted in the realities of the formally organised sectors. The informal workers constitute a significant stakeholder group in the municipal solid waste management system. Their presence allows for carefully sorting and recovering materials that would otherwise end up in unsanitary landfills (Sandhu et al., 2017).

Given the debates surrounding the concept, Kirchherr et al. (2017a, p. 229) define CE as an "economic system that replaces the "end-of-life" concept with reducing, alternatively reusing, recycling and recovering materials in production/distribution and consumption processes. It operates at the micro level (products, companies, consumers), meso level (eco-industrial parks), and macro level (city, region, nation, and beyond), to accomplish sustainable development, thus simultaneously creating environmental quality, economic prosperity and social equity, to the benefit of current and future generations." Even though Geissdoerfer et al. (2017) disagree with this definition as they point out some loopholes, the CE concept offers significant benefits to present and future generations if its implementation considers sustainability concerns (Korsunova et al., 2022).

For CE to become functional, it must be backed by a business model to assess how companies can transition into and adopt circularity principles. Kirchherr et al. (2017a) assert that circular business models are the core of the CE principle. They are the driving force in the shift towards CE, and understanding the concept that lacks business models is one with no direction. Circular business models (CBMs) incorporate strategies that describe how organisations retain value by slowing, closing, and narrowing resource loops. Nußholz (2017) defines it as "how a company creates, captures, and delivers value with the value creation logic designed to improve resource efficiency through contributing to extending the useful life of products and parts (e.g., through long-life design, repair and remanufacturing) and closing material loops" (p. 12). They aim to prevent the loss of valuable materials, protect environmental quality, create jobs, and promote social progress, potentially resulting in economic growth (Bocken et al., 2016). Geissdoerfer et al. (2020) state that the business model is composed of recycling measures (cycling), use phase extensions (extending), a more intense use phase (intensifying), and the substitution of products by service and software solutions (dematerialising). They are tools for value creation that change the way business is undertaken and drive it towards sustainable development (Fehrer & Wieland, 2021). These business models are often propelled by customers who consume products or services the business offers. The consumer perspective may thus be seen as the flip side of the CE business model coin (Kirchherr et al., 2017a).

Meanwhile, implementing CE strategies is not straightforward often comes with setbacks. Kirchherr et al. (2017b) outline four barriers that impede the successful implementation of a circular economy, indicating technological factors as a key barrier. Romero-Hernández and Romero (2018) add that many companies lack innovative ideas beyond standard waste management practices. That is, waste initiatives primarily focus on compliance and may not offer sufficient conditions to drive business value from waste. Sharma et al. (2021) note that transitioning into a circular economy (especially in the waste sector) requires innovative strategies and effective policies. Indeed, implementing circular strategies often requires more holistic and radical changes as rapid urbanisation puts pressure on limited natural resources. Such a shift can change the conventional linear flow of materials where potentially valuable materials end up at disposal sites to a system that recovers and retains products for a considerable time. At the same time, it could create employment opportunities, minimise environmental impacts, and improve livelihoods. There is a need for investments in innovation that capitalise on circular strategies to help realise these goals. Technological advancements present an opportunity to move towards the circular economy by integrating zero waste strategies, where the R-components feature strongly (Lacy et al., 2020). Indeed, this paves the way to innovate long-term actionable targets for sustainability, judging that waste management was considered an environmental problem in the past and has been identified as a socio-economic and financial problem by many cities (Peiris & Dayarathne, 2022). The next section explores how technological innovation in waste management can enhance sustainable development outcomes by promoting circular strategies in a system.

2.3 Integrating Innovative Technologies to Achieve Sustainable Development.

We employ the technological innovation perspective for this research, which is premised on Schumpeter's innovation studies—developed initially to explain economics concepts and better appreciate the dynamics that bring about economic growth (Godin, 2008). The framework makes it possible to investigate how specific innovations can catalyse development. Schumpeter defined technological innovation as the set-up of a new production function (Schumpeter, 1939, p. 87) and stated that varying the technique or method of production should not be misconstrued to mean innovation since other external factors could cause such variations. He, however, did not give much detailed analysis of any of the processes of the concept he introduced in economic theory (Godin, 2008).

But in recent decades, researchers have begun to investigate the processes of technological change rather than simply seeing it as a "black box" of growth factors. They have thus come across a vibrant phenomenon that is far too intricate to fit into heuristic theories (Omri, 2020). Corea (2000) notes that technological innovation has a self-perpetuating momentum that gradually increases through limitless chains, radically altering the social structure that nurtures it and is oriented towards problem-solving. In this view, technological change can thus be considered a learning process that entails discoveries, adoption, and experimentation. Technological innovation is a technique to ensure long-term economic growth by allowing for more efficient processes, improved resource use, and less environmental impact through waste minimisation or effect control (Constantinescu & Frone, 2014). As indicated earlier, the essence of technological innovation as a tool to transform societies and economies is not novel in economic theory. The relationship of incorporating technology into an economy is still one of the well-known themes in the area of economics (Fan et al., 2018).

Nonetheless, due to the increase in resource constraints and environmental issues, contemporary researchers have given attention to innovative technologies as a pathway for overcoming these challenges and achieving sustainable processes (Amri et al., 2018). That is, the concept of sustainability has gradually caused a shift in the focus of studies to how innovative technologies must consider the coordination of the environment and economy (Omri, 2020). For instance, the recent attempts to evaluate how technological innovation could improve environmental quality have brought on a surge in technologies that help to deal with the numerous global challenges that confront us. These may include alterations in the fuel mix, using energy-efficient technologies, and setting up end-of-pipe technologies. To this end, we consider technological change pivotal in dealing with environmental problems, of which waste management is no exception (Weitzman, 2017). Numerous channels can be employed to explain and better understand the importance of technological innovation in combating environmental challenges. For instance, Amri et al. (2018) note that incorporating innovative technologies into the energy sector can help reduce emissions of greenhouse gases such as carbon dioxide. They, therefore, advocate for Tunisia to adopt more energy-efficient technologies to reduce the adverse effect of unsustainable energy usage. Using country-level data for 51 countries between 1971 and 2000, Kumar and Managi (2010) note that adopting environmentally friendly technology innovation reduces negative environmental impacts, particularly in advanced economies.

It is important to note that technological innovation not only brings about economic and environmental gains but also fulfils the social dimension of sustainability, as it has the propensity to boost human development. Technological innovation can help improve health and well-being while ensuring equitable access to developmental benefits that arise. These widespread benefits include access to better services and improved life expectancy. It is, therefore, seen as an instrument for advancing human potential (Omri, 2020; UNDP, 2020). This aligns with Hartmann's (2012) view of technological innovation, as he states that it creates a constantly changing environment that aids the development of human skills.

This means technological innovation is necessary to propel economic and social systems. However, as it becomes more widely used, it can cause other system variables to decline or unanticipated effects to emerge (Anadon et al., 2016; Schumpeter, 1939). Wang and Lin (2013) caution that although investment in innovation is profitable, it is necessary to consider the challenges and risks it brings, even in developing countries. This is particularly true as the existing theory of technological innovation has been explored primarily in the cases of developed economies. It, therefore, raises some degree of uncertainty about whether or not similar arguments can be made for developing countries where technological innovation is usually believed to emanate from foreign sources rather than local ingenuity (Amri et al., 2018; Constantinescu & Frone, 2014; Kumar & Managi, 2010).

Our research examines the possibility of technological innovation to cause meaningful gains in sustainable development outcomes by promoting circular strategies. This view is against the background that technological innovation is central to the global sustainable development agenda. Thus, innovation, which is also part of the Sustainable Development Goals (SDGs) is seen as a channel for realising the other goals (Anadon et al., 2016). Also, developing countries are noted to lack the technical know-how to fight environmental issues, and as such innovative technologies are pathways to cut down on pollution and waste (Constantinescu & Frone, 2014).

In adopting this framework, the paper is directed towards what many expect to be a favourable impact of innovative technologies on sustainable development outcomes. In this sense, we argue that the Achimota waste transfer station could help meet set targets in our quest to achieve sustainable development as it can help promote a circular waste management system. That is, it could create a pathway to enhance efficiency and cause positive impacts on the environment and society. Consequently, it can serve as a transformative tool to move away from the business-as-usual scenario to create a socio-economic system based on the concept of the CE (De Jesus & Mendonça, 2018).

The introduction of the transfer station, as noted by (Manteaw & Boachie, 2019), could be critical in SWM in Ghana. However, positive interaction among actors in the waste collection chain can help bring out its usefulness in creating a circular waste economy. As depicted in Fig. 1, a circular waste business model brings out the complexities of the transition to a circular economy and emphasises team efforts and collective entrepreneurship (Fehrer & Wieland, 2021). It does emphasise the systemic and symbiotic perspective of CBMs as a response to the circular economy's complexity, which builds on collective action and collaboration of various actors (waste management business experts, informal waste workers, academics, and actors from relevant government agencies and local authorities) in realising desirable outcomes. This perspective is actor-centric, highlighting that sustainable waste management is co-created through service provision and mutual exchange of interacting stakeholders in the SWM sector. Where the practices of actors are guided by institutional arrangements that, in turn, connect the stakeholders to form self-adjusting, complex adaptive ecosystems (Fehrer & Wieland, 2021), consequently, the dynamic assemblage of heuristic rules and norms can influence innovation through the performative practices of waste management actors. To this end, the model redirects focus from monetary value creation which is mostly the ultimate goal of businesses, to the shaping of exchange practices and perceptions of all stakeholders, which can be of circular importance.

Thus, creating a circular waste economy in Accra would require institutional change resulting from the diverse activities of the various waste management actors in the city. This can give rise to adaptive tensions that, in effect, point out challenges that need to be solved by a new set of practices. Overcoming these waste management challenges cannot be achieved by a single actor. Hence all actors can engage in institutional work by identifying and bringing others on board through knowledge sharing and communicating their ideas of value co-creation practices.

Specifically, the mutual exchange of waste management actors in translating, interpreting, and modifying interconnected and overlapping institutional arrangements is critical for driving environmental, social, and economic sustainability. This needs to include perspectives of informal waste workers and the public (waste generators) at the micro level to industries and social movements at the meso-level and macro-level societal institutions. For instance, collecting and recovering valuable materials by the informal waste workers at the micro level may seem insignificant. But such practices make it clear that the waste industry may shift towards a circular economy that aligns with the global fight against environmental

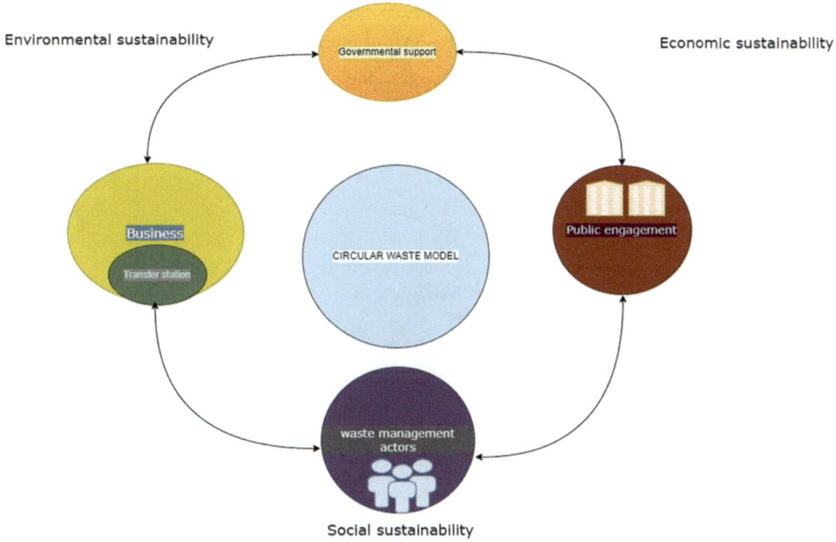

Fig. 1 Conceptual framework

challenges. Essentially, the waste management actors become key players in the waste management system based on the idea of social-collaborative loops. We, therefore, can assert that a circular waste business model that functions well is geared towards a broader institutional change process that hinges on effective communication and knowledge sharing to achieve desired targets.

3 Methodology

3.1 Profile of Study Area

The study was conducted in the Okaikwei North Municipal Assembly (ONMA) in Greater Accra region, Ghana. It was formerly a sub-metropolitan council of the Accra Metropolitan Assembly (AMA) but was elevated to municipal status on 14 November 2017 and was inaugurated on 15 March 2018. The municipal assembly was founded by the Local Government Act 2016 (Act 938) with Legislative Instrument (LI) 2307, making it one of the 29 MMDAs in the Greater Accra region (ONMA, 2020). The number of people living in the municipality based on the

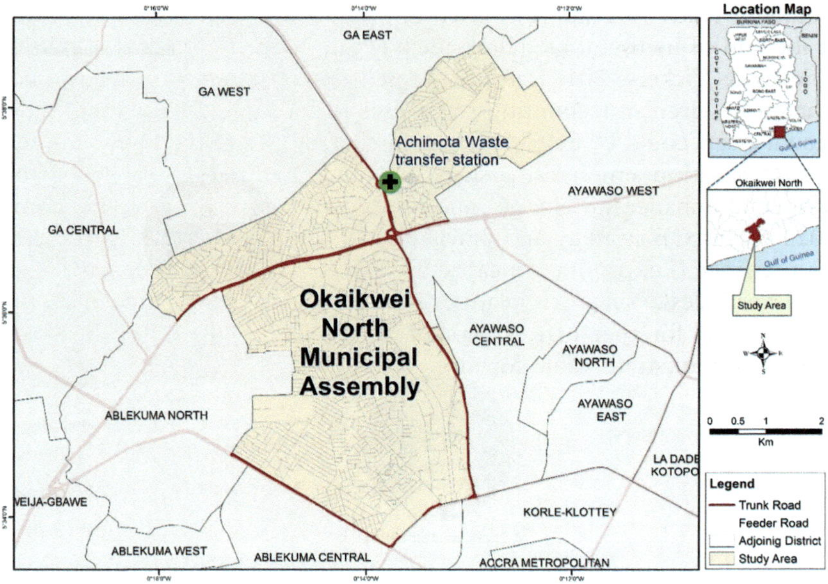

Fig. 2 Map of the study area showing the Actimota waste transfer station (*Source* Authors construct)

2020 population and housing census was 160,446, out of which 78,421 were males, and 82,025 were females (GSS, 2021). The municipality hosts diverse people from all over the country, even though it is known to be a Ga community. The settlement pattern is mainly nucleated, where buildings are clustered together with no clearly defined pattern (Coker, 2016) (Fig. 2).

3.2 Methods

The qualitative case study research approach was employed to collect and analyse data for this study. The choice of the case study design was appropriate because the transfer station is recent waste management technology in Ghana, and very little is known about it. Again, the approach is flexible and helped the researchers to make adjustments, particularly to the instrument used during the study, and to establish a chain of evidence to explain how the Achimota waste transfer station has shaped the waste management processes in the city in recent times (Yin, 2018).

The study adopted a purposive sampling technique in selecting four key informants, including officers of the environmental health and waste department of the ONMA, the Achimota transfer station, and executive members of the informal waste collector's association who coordinates the group's operations in the municipality. They were purposively sampled since they possess requisite knowledge in solid waste management. In order to give informal waste collectors operating in the ONMA an equal chance of being selected, a simple random technique was used to select members to participate in the study. Overall, the concept of saturation was applied to include eleven (11) waste collectors to participate in the study (Boddy, 2016; Fusch & Ness, 2015).

The data collection for the study took place between April and July 2020. Primary data was gathered through in-depth interviews, observations, and field notes. The primary data was complemented with reviews of secondary documents and data, including official waste management documents from the environmental health department of ONMA and the Achimota transfer station, to facilitate data analysis. Before conducting the interview, the researcher met with or called the potential respondents to solicit their support and participation in the study.

Interviews, which solicited information on the impact of the Achimota transfer station in SWM in the city, were audio-recorded after consent was sought from the various respondents. The data gathered was thoroughly scanned, transcribed, grouped into qualitative themes, and analysed, after which conclusions were drawn.

3.3 *Ethical Consideration*

The study objectives and procedures were explained to all interviewees, after which each interviewee gave oral informed consent before the start of the interviews. In addition, study respondents were made aware that their participation was voluntary and that they had the right to refuse to participate or withdraw from the study at any time. The identity of study participants was withheld in the analysis to ensure anonymity and confidentiality.

4 Results

4.1 *Overview of the Achimota Transfer Station*

The Achimota waste transfer station (Zoompak Gh. Ltd), a private-owned facility, was commissioned in May 2017 as the second ultra-modern waste facility in the country. It was established by collaborating with the Jospong Group of Companies (Ghana) and the Campak Group of Turkey. The Achimota transfer station is an engineered facility used to temporarily store, transport, and safely dispose of solid waste. It is equipped with an automation system that calculates the net tonnage of waste before and after waste is tipped into the long trailers. The transportation facility was put up to receive all types of municipal solid waste from solid waste collectors. The waste is compacted into trailers to transfer to the Adepa Landfill (a final disposal site) at Nsawam in the Eastern region. There has been a steady increase in its usage by informal waste collectors even though the facility was initially built to accommodate the formal waste collection companies who mostly dispose of their waste at the Accra Compost Plant or use the landfill sites.

Given this changing trend, a new mechanism was implemented to make the facility user-friendly for waste collectors in the informal sector. The facility's management installed a hydraulic lift platform that will enable them to empty waste into the trailer without difficulty, a process which takes 5–10 minutes. The transfer station is designed to have five shutes through which waste from the waste collection vehicles is channelled into the trailers and has a daily capacity of 1200 tons. The transfer station has a parking space (informal hub) where tricycle waste collectors park their vehicles as they await their turn to use the station (Fig. 3).

Fig. 3 Tricycle waste collectors make their way into the Achimota transfer station (*Source* Authors fieldwork)

As depicted in Fig. 1, tricycle waste collectors enter the Achimota waste transfer station to dispose of their waste. Outside the facility is a weighbridge where the collected waste is weighed, and the appropriate fee is charged before one gets to use the facility. Thus, the collected waste is weighed when collectors believe the waste they carry falls short of the already predetermined flat fee of GH¢ 36 (equivalent to US$ 6.5).

4.2 *Impact of Achimota Transfer Station on Solid Waste Management*

4.2.1 *Screening of Collected Solid Waste*

Before the transfer station was introduced, waste collectors disposed of the waste directly at final disposal sites. This direct connection between collection points and the final disposal sites offered no room for screening and sorting the collected waste. As a result, all the various types of solid waste were taken to disposal sites. These included materials that can be reused, repurposed, recycled and other hazardous wastes that are harmful to human health and the environment. The Achimota transfer station, which has a parking space, has enabled waste workers to remove these materials from the waste stream even before waste collectors access the station to dispose of collected waste. An Official at the transfer station stated:

> The parking space helps the waste collectors retrieve useful and/or hazardous and other dangerous materials from the waste they collect before they enter the facility to tip into the long trailers. This practice is essential as hazardous waste that would otherwise be taken to landfill sites is removed.

An informal waste collectors association executive member also lauded the transfer station's significant change to solid waste management. The member added that:

> The transfer station offers an opportunity for waste pickers to take out reusable/recyclable as well as unwanted materials from the collected waste.

Thus, this technological innovation has helped to promote circular strategies that lead to desirable sustainable development outcomes as only non-hazardous solid wastes with no potential value are transported to final disposal sites. Furthermore, the recovery of reusable/recyclables and other materials from the waste stream creates an opportunity to establish a circular economy, as it has helped drive a closed-loop material flow system in Accra.

4.2.2 Enhanced Waste Collection

It was common for waste collectors to travel several kilometres to dispose of solid waste, affecting the quantities of waste collected in the municipality. Again, the distance travelled made them tired, resulting in dumping at unapproved places at certain times, further compounding the city's bizarre waste situation. However, introducing the transfer station has enhanced the waste collection rate in the city as waste collectors travel less distance between waste generation sources and the disposal point. Consequently, material recovery from collected waste has increased, and this, to a large extent, has improved the waste management situation in the municipality. A tricycle waste collector stated:

> The Achimota transfer station has helped to bring down the distance we travelled to the Kpone landfill site and Mallam dumpsite. We can empty our waste and collect some more because we are not easily tired.

The travel time from waste generation sources is reduced, enabling more waste to be collected out of the system. An officer at the municipal environmental health directorate noted:

> The Achimota transfer station has contributed to improving solid waste collection in the municipality because it is closer to sources of waste generation. The waste collectors can collect more waste and recover many secondary products that can be used for other things. Collection rate within the municipality has improved from 75 to 83%.

The introduction of the transfer station has brought about changes in SWM in the city, and it has helped in gradually creating a circular system where recovered materials stay in the system for a considerable period. The following section looks at the impact of the Achimota transfer station on informal waste collection.

4.3 *The Impact of the Achimota Transfer Station on Informal Waste Collectors*

4.3.1 *Improved Income Levels and Employment Opportunities*

Undoubtedly, the rapid rate of urbanisation in Accra has made finding available lands for siting waste disposal facilities challenging, which is why several disposal sites are located in distant places. In this regard, it becomes extremely costly to use standard vehicles like motorised tricycles (aboboyaa) to transport waste to these places, as waste collectors have to travel several kilometres on a single trip to dispose of collected waste. The Achimota waste transfer station has helped reduce long-distance travel to disposal sites such as the Kpone landfill site and the Mallam dumpsite. As a result, waste collectors can cut down on fuel and other transportation costs. An official at the Achimota transfer station states that:

> Since they are not travelling long distances to dispose of waste, then it means that they are cutting down on transportation costs and thereby reducing fuel consumption which helps them to save more.

An executive member of the informal waste collectors added:

> The tricycle waste collectors are not supposed to travel long distances as we witness, so the transfer station has helped them to save fuel costs and thereby increase their profit margins.

The facility has improved turnaround time, allowing waste collectors to collect more generated waste and increasing their income levels. Again, not all the waste they collect is discarded. The waste is screened at the facility before it ends at the disposal sites. Apart from the human and environmental benefits of undertaking this process, there is an enormous economic boon in retrieving recyclables and other reusable materials from the waste stream compared to the conventional methods of direct disposal after collecting waste (Fig. 4).

Taking out reusable materials and recyclables has created employment opportunities for waste pickers who try to recover materials with a potential value from the collected waste as tricycle operators await their turn to use the transfer station. It thus serves as a source of income-generating opportunity for informal waste workers as they sell these

Fig. 4 Secondary materials recovered at the Achimota Transfer Station (*Source* Authors fieldwork)

recovered products to suitable buyers. An official at the transfer station stated:

> They have their way of doing the sorting. They bring out the plastics, the rubbers, the glasses, the metals, and all other useful things before the waste collectors get inside the station. It creates jobs for waste pickers and helps them to make more money as they sell to suitable buyers.

The extra incomes the waste workers earn are used to sort out essential expenditures they incur during their daily operations. They do not necessarily have to rely on the money they make through house-to-house collections. A waste collector stated:

> The extra money earned from the recovery of materials is used for daily feeding and sometimes fuel purchase when the money is huge. So the money earned through the main activity is used to settle daily tricycle rates and cater for other family needs.

Another advantage the waste collectors get from using the transfer station as they sort and retrieve recyclables is the opportunity to merge their waste with other waste collectors in situations where the volume of collected waste falls short of the user fee charged. The parking space created gives them an excellent opportunity to reduce costs as two or more tricycle operators can merge their waste and share the user fee charged for accessing the transfer station. This act would be difficult to undertake anywhere else.

4.3.2 Reduced Risks

It must be said that some of the tricycles used for solid waste collection are in bad condition and are not roadworthy. They sometimes break down on the highways as they try to make their way to these disposal sites posing severe risks to themselves and other road users. The regular breakdown of tricycles contributes to traffic congestion and environmental health conditions resulting from undisposed waste on the road. They are exposed to severe risk when attempting to fix a broken-down tricycle along the road. Again, waste collectors expressed their displeasure at the little attention they received from other road users during their operations. This situation was more pronounced a few years ago when they had no option but to use the highways to reach final disposal sites. This increased their risk level, and many suffered road accidents during their duties.

> Our lives were at risk as the other road users saw us as a nuisance and did not accord us any respect. I have a friend who has been involved in an accident on the Tema motorway twice. The last accident he suffered was fatal and resulted in severe head injury. (An informal waste collector).

An executive member of the informal waste collectors at the Achimota transfer station shared a tragic story.

> The risk associated with a tricycle using the highway is very high. Imagine a tricycle sharing the same road used by a tipper truck. Three of our people have lost their legs in road accidents alone this year.

These ordeals highlight how the Achimota transfer station has positively impacted the work of the tricycle waste collectors. One could expect that the facility would be highly patronised by the waste collectors and, by so doing, safeguard their lives and that of other road users. The section that follows explores challenges associated with using the transfer station.

4.4 *Challenges Associated with the Achimota Transfer Station*

The introduction of the Achimota transfer station came in handy when the tricycle waste collectors were being hounded by city authorities and law enforcers day after day for using the highways and dumpsites such as the one located at Mallam. On that score, one would be right to think of an increase in the usage of the transfer station. Interestingly, the situation is different as some challenges with the running of the transfer station have affected the facility's patronage. Indeed, checks from documents at the facility revealed a drop in patronage by tricycle waste collectors from an average of 280 to 155 tricycles per day between 2017 and 2020. Thus, regardless of the convenience of using the Achimota transfer station, most tricycle waste collectors are still inclined to utilise the final disposal sites. The underlying factors accounting for the low patronage of the facility include:

4.4.1 *Poor Service Delivery*

When it was commissioned, the Achimota transfer station had seven (7) trailers. But this number has recently reduced to just two (2). The decline in the number of trailers is due to frequent breakdowns that take a long time to be fixed. In addition, tricycle waste collectors alleged that some of these trailers were taken to other facilities without any justification, affecting the facility's smooth running. An officer at the facility revealed a trailer could take the waste of thirty (30) to thirty-five (35) tricycles at a time. It, therefore, suggests that with just two trailers remaining at the facility, it becomes challenging to serve more than seventy (70) tricycle waste collectors at a time. A frustrated tricycle waste collector stated:

> The number of trailers at the Achimota transfer station has reduced from seven to just two. Even procuring spare parts to fix a broken trailer is a problem. We have no idea whether it is the responsibility of the company's management, the president of Zoomlion or the government to fix broken trailers.

The two remaining trailers cannot serve the hundreds of waste collectors visiting the facility daily. An average of seventy (70) tricycle waste collectors can be served at any given time. The rest usually have to wait until the trailers return from the Adepa landfill site to have their turn. Another waste collector indicated that:

> Waste collectors that visit the facility sometimes number more than three hundred (300). One receives quick service when he is lucky to be among the first to get a number. The rest would have to wait several hours before their turn.

The problem of the unavailability of trailers at all times to enable waste collectors to empty the waste they collect often results in long queues, which sometimes extend across the street. This worrying trend has caused many waste collectors to abandon the idea of using the Achimota transfer station entirely, as it has reduced their turnaround time. And even though visiting disposal sites have its challenges, many of them are unperturbed as they cannot keep up with the delays they are subjected to whenever they visit the transfer station. A tricycle waste collector explained that:

> Nowadays, you can get here as early as 8 AM to empty your waste so that you can go and collect some more, but they will tell you there are no trailers available. Sometimes, we can stay here till evening, and the queue will even extend across the streets.

It is important to state that the indiscriminate waste dumping by some collectors is noted to be due to these delays. Indeed, the waste collectors work under deplorable conditions with no protective gear. The operational lapses have caused many to resort to using other disposal sites instead of the transfer station, and these are the supposed culprits responsible for the inappropriate dumping. An executive member of the waste collectors stated that:

> We are not happy with the services they give our people. Their working conditions are bad, meaning they need to be served quickly to change into something better. The delays are too much, and because there is a transfer station, they cannot easily use the landfill sites.

The executive asserted that most of the indiscriminate dumping of waste witnessed resulted from waste collectors being denied access to disposal sites since there is a transfer station. Another reason that has caused the problem to fester is the fact that most waste collectors have daily fixed rates to pay to vehicle owners, so there is a need for them to have regular and prompt services so that their work is not impeded. The executive intimated that:

> They have daily fixed rates to make and families to cater for, but the situation has restricted the number of trips they can have in a day. Previously, some could make as many as three or four daily trips. One would consider himself fortunate to have two trips under the present circumstance.

A waste collector, expressing his dissatisfaction with the services rendered at the transfer station, placed the blame at the feet of the facility's managers. The waste collector stated that:

> Sometimes, the officers take our monies in the morning when we visit the station without telling us the situation at hand. We can wait till evening only to be told to wait till the next morning before we can empty our waste

> because there are no trailers. Some of our colleagues, out of frustration, take away their tricycles and dump in unapproved places just to create space for the next day's work.

Observations on field visits to the Achimota transfer station revealed a fluctuating trend in the number of "aboboyaa" (motorised tricycles) present at the facility. The number was high on some days and very low on other days. An inquiry into this trend revealed that tricycle waste operators patronise the station mostly on rainy days as it becomes difficult for them to use the landfills and other dumpsites since they easily get stuck. They would instead go directly to the disposal site when there are no rains. The poor services rendered at the Achimota transfer station have caused many tricycle waste collectors to use the disposal sites as an alternative.

In contrast, others tend to dump in unapproved places. This situation has hindered waste material recovery as waste collectors use disposal sites. As such, they cannot screen and recover materials from the waste collected.

4.4.2 Higher Fees Charged

Most of the tricycle waste collectors operating in the municipality were under the daily fixed rate ownership arrangement. That is, they paid a fixed rate to the owners of the tricycles at the end of the day's work. The rate paid to owners mainly depends on the newness of the vehicle used to gather waste. Aside from this fixed fee, they also bear other operational costs, such as the transfer station user fees. It indicates that the money they earn daily depends on the number of trips, the services they receive, and the amount paid at the transfer station. Considering the fees charged at some disposal sites and what they paid at the Achimota transfer station, they are convinced that the charges at the facility are high and they are being exploited. On the other hand, officials at the transfer station believed that the fee was moderately priced and that the waste collectors had no issues with it. The tricycle waste operators expressed varied opinions. A tricycle waste collector stated that:

> We admit that the Achimota transfer station has brought enormous benefits. Our major concern is the high fees charged. We do not get anything from the work we do nowadays.

The tricycle waste collectors recounted the fees they previously paid at the various disposal sites, which ranged between fifteen to twenty Ghana Cedis (US$2.7–US$3.6). They were, therefore, of the view that the emergence of the Achimota transfer station would help reduce long-distance travel and, most importantly, help to relieve the pressure on their pockets. But, to their surprise, they are made to pay GH ¢ 36 (US$6.5) for each visit, and the station's officials have even attempted to increase this fee. Another tricycle waste collector added that:

> We thought the transfer station would be cheaper than we used to pay, but this is not the case. They even attempted to increase the fees from the current GH¢ 36 to GH¢ 45 (equivalent to US$ 6.5 to 8.2) per trip, but we opposed it. Most of the tricycles we use for this business are not our own. We pay a fixed rate daily to owners and meet other basic and family needs, so this fee is too high for us.

They were particularly concerned about the changing regulations restricting their operation regarding where to dispose of waste. Yet, they have received little help from the local authorities in resolving their issues. These challenges have made many waste collectors turn away from using the Achimota transfer station to other alternative sources. Indeed, the high charges at Achimota transfer station are cited to be among the causes of the unabated dumping of solid waste in unapproved places in the city. It is believed that some recalcitrant waste collectors who, due to the high fees charged, still make their way to the other disposal sites are responsible for the unsavoury act of dumping waste at inappropriate places.

5 Discussion and Implications

This paper aims to add to the current debate on the concept of a circular economy by assessing the potential impact of the waste transfer station in promoting circular strategies that bring about intended sustainable development outcomes in urban communities through the lens of the technological innovation theory. We summarise the study's key findings before we touch on its implications. Firstly, the transfer station has reduced waste volumes by promoting material recovery process necessary for transitioning to a circular economy. The technology has enabled informal waste workers to realise circular strategies that lead to desirable urban sustainable development outcomes and to reduce negative environmental impacts. Indeed, CE aims to establish a closed loop of materials that stay in a system for a longer period. As highlighted by earlier studies (Beğen, 2002; Nhubu et al., 2019; Zemanek et al., 2011), we find that the transfer station has enhanced the recovery of potentially valuable materials that can be recycled or reused, or repurposed, which is achieved through the screening and sorting of collected waste by the informal waste collectors. Thus, the innovative technology has deepened the principle of circularity in waste management as every little bit of valuable materials is salvaged for a different use (see: Doe et al., 2022), providing the needed condition to safeguard the environment.

Secondly, it is mainly argued that the CE aims to sustain economic prosperity through environmentally beneficial approaches. At the same time, social effects are worth highlighting. Leveraging the transfer station to implement a CE in the SWM sector can create employment opportunities. The results show the creation of employment avenues for waste pickers (a group in the informal waste worker category in the SWM chain). This group of workers or "scavengers" remove all valuable materials from collected waste for resale or repairs as waste collectors wait to use the transfer station. These circular practices serve as livelihood strategies for the urban poor in a developing country (Oduro-Appiah et al., 2019; Sandhu et al., 2017). The "informal packing hub" at the transfer station offers an accessible channel to attract many people into recovering and screening waste to be reused, recycled, or repurposed. As a result, they meet and interact with each other and offer legitimacy to their work by reducing risks and harassment (Korsunova et al., 2022). Contrary to our findings, Repp et al. (2021) indicate that a circular shift in the clothing sector in Europe could cause a significant decline in employment in emerging economies. It does bring out important dynamics of a move to CE in the various stages of development. However, Doe et al. (2022) corroborate our findings by stating that implementing a CE in the coconut waste value chain could create employment for the youth. In a related study, Yaduvanshi et al. (2016) note the importance of circular strategies in providing employment opportunities for the urban poor.

Thirdly, the results show that transitioning to CE through technological innovation improved waste workers' income levels and increased their purchasing power. It is interesting to know that there were sustained income levels for the waste workers during the COVID-19 pandemic, as against the (World Bank, 2021) assertion the pandemic would affect the economies of developing countries. The pandemic was projected to cause a decline in the income of labour workers and other economic indicators. It was further estimated that the pandemic would cause 62% of the global labour workers in the informal sector to lose their livelihoods (Sharma et al., 2021). Thus, the informal waste workers earned additional incomes by engaging in circular strategies as collected waste was taken to the transfer station. In line with our findings, Goyal et al. (2018) demonstrate how the uptake of circular strategies such as reuse offers income for people in small communities in India, especially women. The women reused more than half a million kilograms of waste clothing to create mattresses.

Lastly, the informal waste workers complained about service quality, fees charged, as noted by Atta-Agyem (2013), and lack of support as challenges that mitigate their operations. This has reduced waste collection services and, in effect, potentially valuable materials that can be recovered and put back into the system. Services deemed circular can offer extra benefits due to superior environmental sustainability as a key attribute. However, businesses tend to focus on incremental processes to reap efficiency gains, which fall short of the fundamental changes required. They are often caught up in business-centric and profit-oriented value creation ideas unsuited to addressing sustainability issues (Fehrer & Wieland, 2021). It is also important to note that the uptake of circular strategies does not automatically result in increased efficiency gains and would require setting specific targets (Nußholz, 2017). Realising a CE in the waste management sector in Accra would require coordinated action from enhanced knowledge and information sharing of waste-related matters and the implementation of CE, dialogue and cooperation between all the key stakeholders (informal waste workers, waste experts and managers, waste management researchers, and local authorities) in the waste management space. These could play a pivotal role by providing useful information to establish circular waste business concepts to retain the product's or material's value. Though it takes time to build confidence in circular business models such as the circular waste business model, the process can be facilitated by developing cooperation and information sharing that leads to a systemic change among key stakeholders (Salmenperä et al., 2021). Thus, the lack of cooperation between stakeholders is a major barrier to implementing CE in developing countries (Korsunova et al., 2022).

Again, managerial experience represents a hurdle in handling the relatively high number of waste collectors utilising the transfer station. This is because the facility was initially set up and designed to be used by formal waste contractors. Therefore, the organisational processes to accommodate the informal waste collectors are inadequate to utilise such an innovation fully. Furthermore, building and scaling an innovative offering that responds to evolving trends around circularity would require fundamentally different managerial strategies that are agile and not profit-oriented (Kuhlmann et al., 2022).

As major stakeholders in SWM processes, local authorities need to bring forth favourable regulations and incentives if private businesses are to spearhead the move towards implementing CE in the waste management system. This is so because private companies tend to focus on short-term gains and rewards that are difficult to sustain, which causes them to lose interest entirely in implementing the CE (Kirchherr et al., 2017a). Such reflections must consider the public and informal waste workers' responsibility as contributing to CE and critical enablers of any circular waste business model. Hence, excluding them could lead to adopting a business-oriented logic regarding CE, which risks developing unviable business models (Repo & Anttonen, 2017). As (Fehrer & Wieland, 2021) noted, putting in place a good business model is not achieved by a single actor. The systemic alignment processes that shape business models can only be comprehended when viewed from multiple system levels.

Having outlined possible pathways to transitioning to a CE in the waste sector, we argue that incorporating technological innovation in solid waste management could be leveraged as an opportunity to transition towards full adoption of CE and would be a tool central to achieving sustainable development. Adopting innovative technologies in waste management fulfil all three dimensions of sustainable development. We demonstrate that achieving sustainable development targets, particularly in a developing country context, can be realised by transitioning to a CE. That, in turn, can offer employment opportunities, inclusion and reduced risk for informal waste workers (social sustainability), improved income for waste collectors (economic sustainability), and enhanced waste collection services that promote material recovery (environmental sustainability). The finding is contrary to an earlier study (Omri, 2020) that suggested technological innovation is compatible with the three pillars of sustainable development in only the advanced countries, where technological innovation simultaneously improves the economic, environmental, and social dimensions of sustainable development. In contrast, it only affects the economic and environmental dimensions in the case of middle-income countries and has no impact on these dimensions in the case of low-income countries. Per our findings, we can state that technological innovation plays a pivotal role in the sustainable development aspiration of developing countries.

6 Conclusions

The contribution of innovative technologies to sustainable development outcomes remains to be determined, particularly in developing countries. The study aimed to contribute to the debate, focusing on the waste management sector, by assessing the potential impact of the Achimota transfer station in Accra. The findings revealed that the waste transfer station could be instrumental in solid waste management by promoting circular strategies that lead to intended sustainable development outcomes in the city. The study revealed that technological innovation could be key in achieving sustainable development in developing countries. However, operational gaps associated with the daily running of the Achimota waste transfer station could make transitioning to CE and attain sustainable development an elusive dream. This challenge is surmounted by having a circular waste model where all stakeholders cooperate and collaborate through knowledge and information sharing, effective communication, and different managerial mechanisms to transform the waste management system from a linear to a circular economy. The study concludes that the Achimota transfer station contributes substantially to solid waste management in Accra. Technological innovation can help informal waste workers recover and retain the value of materials for as long as possible while offering opportunities for improved well-being and increased income.

The limitation of the research is derived chiefly from the methodologies used in that the study was undertaken during the peak of the COVID-19 pandemic. Given the many restrictions imposed, the researchers adopted a small sample size and techniques which impacted the recruitment of stakeholders involved in the waste management chain for the study (mainly waste management experts and practitioners). We would encourage future research to employ different methodological approaches in addressing the above limitations to test the validity of our results. We recommend future research to broaden the discussion in understanding the impacts of technological innovation towards the transition to CE in the waste sector, including public perception, understanding, and engagement. We also recommended future studies examine how technological innovation applies to other sectors of developing economies in shaping CE practices and fulfilling the dimension of sustainability. This would help broaden the discussion, improve learning, and deepen understanding of the contribution of technology as a tool for achieving sustainable development through CE implementation in developing countries.

References

Abunyewah, M., Okyere, S. A., Diko, S. K., Kita, M., Erdiaw-Kwasie, M. O., & Gajendran, T. (2022). Flooding in informal communities: Residents' response strategies to flooding and their sustainability implications in Old Fadama, Accra. In *Disaster risk reduction for resilience* (pp. 435–461). Springer.

African Development Bank Group. (2012). Urbanization in Africa. Available at www.afdb.org/en/blogs/afdb-championing-inclusive-growth-across-africa/post/urbanization-in-Africa-10143/

Ağaçsapan, B., & Çabuk, S. N. (2020). Determination of suitable waste transfer station areas for sustainable territories: Eskisehir case. *Sustainable Cities and Society*, *52*(September 2019), 101829. https://doi.org/10.1016/j.scs.2019.101829

Akorsu, A. D., Britwum, A. O., Boampong, O., & Tachie, B. Y. (2019). Waste workers matter. Series no. 1. School for Development Studies, University of Cape Coast.

Amoah, S. T., & Kosoe, E. A. (2014). Solid waste management in Urban Areas of Ghana: Issues and experiences from Wa. *Journal of Environment Pollution and Human Health, 2*(5), 110–117.

Amri, F., Bélaïd, F., & Roubaud, D. (2018). Does technological innovation improve environmental sustainability in developing countries? Some evidence from Tunisia. *Journal of Energy and Development*, *44*(1/2), 41–60. https://www.jstor.org/stable/26863980

Anadon, L. D., Chan, G., Harley, A. G., Matus, K., Moon, S., Murthy, S. L., & Clark, W. C. (2016). Making technological innovation work for sustainable development. *Proceedings of the National Academy of Sciences of the United States of America, 113*(35), 9682–9690. https://doi.org/10.1073/pnas.1525004113

Atta-Agyem, F. (2013). Integration of informal solid waste collectors into mainstream solid waste management in Adentan Municipality [Masters Thesis, KNUST].

Beğen, N. N. (2002). Optimal locations of landfills and transfer stations in solid waste management [Doctoral dissertation, Bilkent University].

Bocken, N. M., De Pauw, I., Bakker, C., & Van Der Grinten, B. (2016). Product design and business model strategies for a circular economy. *Journal of Industrial and Production Engineering, 33*(5), 308–320.

Boddy, C. R. (2016). Sample size for qualitative research. *Qualitative Market Research: An International Journal, 19*(4). https://doi.org/10.1108/QMR-06-2016-0053

Bovea, M. D., Powell, J. C., Gallardo, A., & Capuz-Rizo, S. F. (2007). The role played by environmental factors in the integration of a transfer station in a municipal solid waste management system. *Waste Management, 27*(4), 545–553.

Breukelman, H., Krikke, H., & Löhr, A. (2019). Failing services on urban waste management in developing countries: A review on symptoms, diagnoses, and interventions. *Sustainability, 11*(24), 6977.

Camilleri, M. A. (2019). The circular economy's closed loop and product service systems for sustainable development: A review and appraisal. *Sustainable Development, 27*(3), 530–536.

Chang, N. B., & Pires, A. (2015). Systems analysis for the future of solid waste management: Challenges and perspectives. In *Sustainable solid waste management* (pp. 849–894). https://doi.org/10.1002/9781119035848.ch24

Cobbinah, P. B., Erdiaw-Kwasie, M. O., & Amoateng, P. (2015). Africa's urbanisation: Implications for sustainable development. *Cities, 47*, 62–72.

Coker, L. B. (2016). *Perceptions of parents, teachers and students towards sexual and reproductive health education in junior high schools in Okaikoi North sub-metro, Accra, Ghana* [Masters thesis, University of Ghana].

Constantinescu, A., & Frone, S. (2014). The role of technological innovation in sustainable economic development. *Journal of Knowledge Management*, 17. http://spicks2014.scientificpapers.org/download/13/

Corea, S. (2000). Cultivating technological innovation for development. *The Electronic Journal of Information Systems in Developing Countries, 2*(1), 1–15. https://doi.org/10.1002/j.1681-4835.2000.tb00009.x

Cui, L., Chen, L. R., Li, Y. P., Huang, G. H., Li, W., & Xie, Y. L. (2011). An interval-based regret-analysis method for identifying long-term municipal solid waste management policy under uncertainty. *Journal of Environmental Management, 92*(6), 1484–1494.

Debrah, J. K., Teye, G. K., & Dinis, M. A. P. (2022). Barriers and challenges to waste management hindering the circular economy in sub-Saharan Africa. *Urban Science, 6*(3), 57.

De Jesus, A., & Mendonça, S. (2018). Lost in transition? Drivers and barriers in the eco-innovation road to the circular economy. *Ecological Economics, 145*, 75–89.

De Jesus, A., Antunes, P., Santos, R., & Mendonça, S. (2019). Eco-innovation pathways to a circular economy: Envisioning priorities through a Delphi approach. *Journal of Cleaner Production, 228*, 1494–1513.

Doe, B., Aboagye, P. D., Osei-Owusu, P. K., Amoah, T., Aidoo, A., & Amponsah, N. Y. (2022). Towards circular economy and local economic development in Ghana: Insights from the coconut waste value chain. *Circular Economy and Sustainability*, 1–26.

Duku, M. H., Gu, S., & Hagan, E. B. (2011). Biochar production potential in Ghana—A review. *Renewable and Sustainable Energy Reviews, 15*(8), 3539–3551.

Emenike, C. U., Iriruaga, E. T., Agamuthu, P., & Fauziah, S. H. (2013). Waste management in Africa: An invitation to wealth generation. In *Proceedings of International Conference on Waste Management and Environment.*

Fan, H., Ismail, H. M., & Reza, S. M. (2018). Technological innovation, infrastructure and industrial growth in Bangladesh: Empirical evidence from ARDL and granger causality approach. *Asian Economic and Financial Review, 8*(7), 964–985. https://doi.org/10.18488/journal.aefr.2018.87.964.985

Fehrer, J. A., & Wieland, H. (2021). A systemic logic for circular business models. *Journal of Business Research, 125*, 609–620.

Fusch, P. I., & Ness, L. R. (2015). Are we there yet? Data saturation in qualitative research. *Qualitative Report, 20*(9), 1408–1416.

Geissdoerfer, M., Savaget, P., Bocken, N. M., & Hultink, E. J. (2017). The circular economy–A new sustainability paradigm? *Journal of Cleaner Production, 143*, 757–768.

Geissdoerfer, M., Pieroni, M. P., Pigosso, D. C., & Soufani, K. (2020). Circular business models: A review. *Journal of Cleaner Production, 277*, 123741.

Ghana Statistical Service (GSS). (2021). Ghana 2021 population and housing census: General report volume 3A. Ghana Statistical Service.

Godin, B. (2008). In the shadow of Schumpeter: W. Rupert Maclaurin and the study of technological innovation. *Minerva, 46*(3), 343–360. https://doi.org/10.1007/s11024-008-9100-4

Goyal, S., Esposito, M., & Kapoor, A. (2018). Circular economy business models in developing economies: Lessons from India on reduce, recycle, and reuse paradigms. *Thunderbird International Business Review, 60*(5), 729–740.

Hartmann, D. (2012). *Sen meets Schumpeter: Introducing structural and dynamic elements into the human capability approach* (FZID Discussion Papers), pp. 1–30.

Kirchherr, J., Reike, D., & Hekkert, M. (2017a). Conceptualising the circular economy: An analysis of 114 definitions. *Resources, Conservation and Recycling, 127*, 221–232.

Kirchherr, J. W., Hekkert, M. P., Bour, R., Huijbrechtse-Truijens, A., Kostense-Smit, E., & Muller, J. (2017b). Breaking the barriers to the circular economy.

Kirchherr, J., & Piscicelli, L. (2019). Towards an education for the circular economy (ECE): Five teaching principles and a case study. *Resources, Conservation and Recycling, 150*, 104406.

Korsunova, A., Halme, M., Kourula, A., Levänen, J., & Lima-Toivanen, M. (2022). Necessity-driven circular economy in low-income contexts: How informal sector practices retain value for circularity. *Global Environmental Change, 76*, 102573.

Kortei, N. K., & Quansah, L. (2018). Plastic waste management in Ghana—A complete failure and the consequences. *Allied Sciences* (September 2016). Retrieved from https://www.researchgate.net/publication/308889973_Plastic_waste_management_in_Ghana

Kuhlmann, M., Bening, C. R., & Hoffmann, V. H. (2022). How incumbents realise disruptive circular innovation-Overcoming the innovator's dilemma for a circular economy. *Business Strategy and the Environment*.

Kumar, S., & Managi, S. (2010). Environment and productivities in developed and developing countries: The case of carbon dioxide and sulfur dioxide. *Journal of Environmental Management, 91*(7), 1580–1592. https://doi.org/10.1016/j.jenvman.2010.03.003

Lacy, P., Long, J., & Spindler, W. (2020). Introduction: The path to transformation is circular. In *The circular economy handbook* (pp. 1–14). Palgrave Macmillan.

Luttenberger, L. R. (2020). Waste management challenges in transition to circular economy–case of Croatia. *Journal of Cleaner Production, 256*, 120495.

Manteaw, B. O., & Boachie, J. (2019). Africa's urban waste management and sanitation challenges: Are transfer stations the solution? *Environment and Pollution, 9*(1), 1. https://doi.org/10.5539/ep.v9n1p1

Medina, M. (2007). Waste picker cooperatives in developing countries. In *Membership-based organisations of the poor* (pp. 125–141). Routledge.

Miezah, K., Obiri-Danso, K., Kádár, Z., Fei-Baffoe, B., & Mensah, M. Y. (2015). Municipal solid waste characterisation and quantification as a measure towards effective waste management in Ghana. *Waste Management, 46*, 15–27.

Morseletto, P. (2020). Targets for a circular economy. *Resources, Conservation and Recycling, 153*, 104553.

Mudu, P., Nartey, B. A., Kanhai, G., Spadaro, J. V., & Fobil, J. (2021). *Solid waste management and health in Accra, Ghana*. World Health Organization. Licence: CC BY-NC-SA 3.0 IGO.

Mumuni, I. (2016). *Examining the roles and challenges of informal waste pickers in the solid waste management of the Tamale Metropolis of Northern Ghana* (Master's thesis). International Institute of Social Studies.

Nhubu, T., Muzenda, E., Mbohwa, C., Patel, B., & Mateescu, C. (2019). *Suitability of municipal solid waste transfer stations in Harare, Zimbabwe*. 2019

7th International Renewable and Sustainable Energy Conference (IRSEC) (pp. 1–5). IEEE.

Nußholz, J. L. (2017). Circular business models: Defining a concept and framing an emerging research field. *Sustainability, 9*(10), 1810.

Nzeadibe, T. C. (2009). Solid waste reforms and informal recycling in Enugu urban area, Nigeria. *Habitat International, 33*(1), 93–99. https://doi.org/10.1016/j.habitatint.2008.05.006

Oduro-Appiah, K., Afful, A., Kotey, V. N., & De Vries, N. (2019). Working with the informal service chain as a locally appropriate strategy for sustainable modernisation of municipal solid waste management systems in lower-middle income cities: Lessons from Accra Ghana. *Resources, 8*(1), 12.

Okaikwei North Municipal Assembly (ONMA). (2020). About us. Retrieved from http://www.onmaonline.com/about-us/. Accessed 3 January 2020.

Omri, A. (2020). Technological innovation and sustainable development: Does the stage of development matter? *Environmental Impact Assessment Review, 83*(April), 106398. https://doi.org/10.1016/j.eiar.2020.106398

Oteng-Ababio, M. (2010). Private sector involvement in solid waste management in the Greater Accra Metropolitan Area in Ghana. *Waste Management & Research, 28*(4), 322–329.

Oteng-Ababio, M. (2011). Missing links in solid waste management in the Greater Accra Metropolitan Area in Ghana. *GeoJournal, 76*(5), 551–560. https://doi.org/10.1007/s10708-010-9363-9

Oteng-Ababio, M. (2012). The role of the informal sector in solid waste management in the Gama, Ghana: Challenges and opportunities. *Tijdschrift Voor Economische En Sociale Geografie, 103*(4), 412–425. https://doi.org/10.1111/j.1467-9663.2011.00690

Oteng-Ababio, M., Melara Arguello, J. E., & Gabbay, O. (2013). Solid waste management in African cities: Sorting the facts from the fads in Accra, Ghana. *Habitat International, 39*, 96–104. https://doi.org/10.1016/j.habitatint.2012.10.010

Oteng-Ababio, M., Owusu-Sekyere, E., & Amoah, S. T. (2017). Thinking globally, acting locally: Formalising informal solid waste management practices in Ghana. *Journal of Developing Societies, 33*(1), 75–98.

Peiris, M. T. O. V., & Dayarathne, G. L. N. (2022). Application of life cycle framework for municipal solid waste management: A circular economy perspective from developing countries. *Circular Economy and Sustainability*, 1–20.

Reike, D., Vermeulen, W. J., & Witjes, S. (2018). The circular economy: New or refurbished as CE 3.0?—Exploring controversies in the conceptualisation of the circular economy through a focus on history and resource value retention options. *Resources, Conservation and Recycling, 135*, 246–264.

Repo, P., & Anttonen, M. (2017, June). *Emerging consumer perspectives on circular economy*. The 13th Nordic Environmental Social Science Conference HopefulNESS, 6.-8.6.2017.

Repp, L., Hekkert, M., & Kirchherr, J. (2021). Circular economy-induced global employment shifts in apparel value chains: Job reduction in apparel production activities, job growth in reuse and recycling activities. *Resources, Conservation and Recycling, 171*, 105621.

Romero-Hernández, O., & Romero, S. (2018). Maximising the value of waste: From waste management to the circular economy. *Thunderbird International Business Review, 60*(5), 757–764.

Salmenperä, H., Pitkänen, K., Kautto, P., & Saikku, L. (2021). Critical factors for enhancing the circular economy in waste management. *Journal of Cleaner Production, 280*,. https://doi.org/10.1016/j.jclepro.2020.124339

Sandhu, K., Burton, P., & Dedekorkut-Howes, A. (2017). Between hype and veracity; privatisation of municipal solid waste management and its impacts on the informal waste sector. *Waste Management, 59*, 545–556.

Savini, F. (2021). The circular economy of waste recovery, incineration and urban reuse. *Journal of Environmental Planning and Management, 64*(12), 2114–2132. https://doi.org/10.1080/09640568.2020.1857226

Schumpeter, J. A. (1939). Business cycles: A theoretical, historical ad statistical analysis of the capitalist process. In *Mc Graw_Hill book* (p. 461).

Sharma, H. B., Vanapalli, K. R., Samal, B., Cheela, V. S., Dubey, B. K., & Bhattacharya, J. (2021). Circular economy approach in solid waste management system to achieve UN-SDGs: Solutions for post-COVID recovery. *Science of the Total Environment, 800*, 149605.

UNDP. (2020). Human development report: the new frontier-human development and the Anthropocene. UNDP. Retrieved from https://hdr.undp.org/sites/default/files/hdr2020.pdf

United Nations Environment Programme (UNEP). (2005). *Solid waste management* (Vol. I). ISBN: 92-807-2676-5.

Wang, C. C., & Lin, G. C. S. (2013). Emerging geography of technological innovation in China's ICT industry: Region, inter-firm linkages and innovative performance in a transitional economy. *Asia Pacific Viewpoint, 54*(1), 33–48. https://doi.org/10.1111/apv.12005

Washburn, B. E. (2012). Avian use of solid waste transfer stations. *Landscape and Urban Planning, 104*(3–4), 388–394.

Weitzman, M. L. (2017). Sustainability and technical progress. *In the Economics of Sustainability, 99*(1), 329–341. https://doi.org/10.4324/9781315240084-21

World Bank. (2021, June). Global economic prospects. World Bank. https://openknowledge.worldbank.org/handle/10986/35647. License: CC BY 3.0 IGO.

Yaduvanshi, N. R., Myana, R., & Krishnamurthy, S. (2016). Circular economy for sustainable development in India. *Indian Journal of Science and Technology, 9*(46), 1–9.

Yeboah-Assiamah, E. (2015). Involvement of private actors in the provision of urban sanitation services; potential challenges and precautions a conceptual paper. *Management of Environmental Quality: An International Journal, 26*(2), 270–287. https://doi.org/10.1108/MEQ-08-2014-0130

Yin, R. K. (2018). *Case study research and applications: Design and methods*. SAGE Publications.

Zemanek, J., Wozniak, A., & Malinowski, M. (2011). The role and place of solid waste transfer station in the waste management system. *Infrastruktura i Ekologia Terenów Wiejskich* (11).

Institutionalization and Enforcement of Circular Fisheries Practices in Uganda: The Military's Role in Implementing the National Fisheries Policy

Annet K. Nabatanzi-Muyimba and Joshua Mugambwa

1 Introduction

Enhancing the sustainable development process in developing nations is critical and urgent. However, the pace at which sustainable development can be realized is hinged on the ability of a government to formulate appropriate policies and its capability to implement them effectively and articulately (Abunyewah et al., 2022; Ugwuanyi & Chukwuemeka, 2013). The policy is a central and dominant organizing and shaping principle of society and every aspect of human life (Shore & Wright, 1997). Policies are artefacts, language, technologies, techniques,

A. K. Nabatanzi-Muyimba · J. Mugambwa (✉)
Faculty of Management and Public Policy, Makerere University Business School, Kampala, Uganda
e-mail: jmugambwa@mubs.ac.ug

A. K. Nabatanzi-Muyimba
e-mail: anabatanzi@mubs.ac.ug

M. O. Erdiaw-Kwasie and G. M. M. Alam (eds.), *Circular Economy Strategies and the UN Sustainable Development Goals*, Sustainable Development Goals Series,
https://doi.org/10.1007/978-981-99-3083-8_19

agents of change and instruments of power that create regulation, coherence, legitimacy and course of action (Shore, 2012). Policies emanate from government, public institutions, non-governmental organizations (NGOs), social-cultural institutions, political parties, private bodies, etc. They are formulated to address particular problems and intended to produce some known and measurable outcomes (Shore, 2012). The usual "policy cycle" rotates from agenda setting, policy formulation, policy adoption and policy implementation to policy evaluation.

Policy implementation is the process of putting policies into effect with the aim of social and economic betterment and is manifested as programmes, procedures, regulations or practices (O'Toole, 2000). Hence, administrative competence and effectiveness in policy implementation are judged based on evidence of bridging the gap between the policy's intention and actual achievements (Nadgrodkiewicz et al., 2012). Most of the time policies are often rolled out in developing nations without achieving the desired results (Cobbinah et al., 2015; Kellermann & Jones, 2013; Muhammad, 2014). Studies show that while in many developed countries, the main area of disagreement over policy occurs during formulation, in developing countries, most disagreements and failures occur during implementation (Al Mamun et al., 2023; Grindle & Thomas, 1990). Moreover, there need to be more studies on policy implementation in developing countries to draw lessons (Adebayo et al., 2010). In Uganda, implementing public policies and programmes remains a big challenge. According to Benkenstein (2011), fisheries problems in Uganda are primarily not from a lack of policies but from a lack of effective policy implementation.

Consequently, military norms and values are reshaping the fabric of policy implementation in many countries, including those in the developed world (Shore, 2012). Militarization refers to re-organizing non-military systems and institutions in line with military goals, norms and values, assigning civilian work to the military and increasing resource allocation for military purposes (Shore, 2012). The army has been involved in implementing government policies in Uganda on Presidential orders that the public service is corrupt and inefficient. The strategy of involving the military in enforcing sustainability practices in the fisheries sector makes Uganda's case special and interesting to this study. In particular, the National Fisheries Policy (NFP) is being enforced by the Uganda People's Defence Forces (UPDF). Other policy and programme implementations affected by militarization include National Agricultural Advisory Services

(NAADS), operation wealth creation (OWC) (ACODE, 2015; Muhumuza, 2016), the *construction* of the Uganda standard gauge railway (CGTN Africa, 2018; Independent, 2018), and recently the construction of national roads, schools and health centres (Bagala, 2021). This militarization, a clear deviation from public service systems and institutions, has aroused mixed feelings over the army's institutional mandate, professionalism and ethical competence in executing non-military tasks.

This chapter aims to contribute and add to existing knowledge on policy implementation and sustainable development in the developing world by examining the institutionalization and enforcement of circular fisheries practices in a developing country context. Thus, the specific objectives of the chapter are: (1) to examine the institutional framework, achievements and challenges of the national fisheries policy implementation in Uganda; (2) to examine the implementation effectiveness and effects of military enforcement of the national fisheries policy in Uganda? (3) Draw lessons and make recommendations that further both theory and practice. Empirically, the chapter examines the effectiveness and effects of the military in implementing the National Fisheries Policy (NFP) in Uganda. We adopt Morestin's (2012) framework for analysing public policy implementation using the dimensions of effects (effectiveness, unintended effects and equity); and implementation effectiveness (cost, feasibility and acceptability). Our approach to achieving the study objectives is further guided by three theoretical perspectives: institutional theory, complex adaptive systems theory and agency theory. The chapter scope concentrates on capture fisheries which are most at risk of depletion. While the Ugandan context provides clear evidence of military involvement in policy implementation in the developing world, the chapter further reviews related literature and analyses publically available documents and qualitative data collected from 12 focus group discussions (FGD) and 15 key informant interviews to achieve the specified objectives.

The chapter begins with an introduction to policy implementation practices in the developing world. The second part presents the related literature and the theoretical framework for sustainable policy implementation. The third part presents the methodology adopted. The fourth part highlights the institutional framework, achievements and challenges in the implementation of national fisheries policy in Uganda, while the fifth part reports the study findings on the militarization of NFP implementation. The sixth part covers the discussion, followed by the conclusion in the seventh section and the implications, recommendations and limitations in section eight.

2 Related Literature

2.1 Institutionalization and Enforcement of Circular Fisheries Practices

Sustainability has become a key public policy goal and an entrepreneurial imperative (as cited in de Jesus et al., 2019). Fisheries and aquaculture resources worldwide face grave sustainability threats. As a result, governments and regulatory bodies are encouraged to develop innovative solutions, adapt and change policies around circular economy principles (CE) as well as an ecosystem approach to fisheries management (EAFM) to help improve their sustainability (Fortnam, 2019; Nunoo et al., 2015; Pounds, 2021). According to de Jesus et al. (2019), CE is a transition pathway towards sustainability, which focuses on overcoming challenges of the linear economic model of resource use that follows a raw material-production-consumption-waste/disposal process. The linear model requires large quantities of resources to manufacture products sold to consumers and discarded as waste when worn out or no longer needed (Erdiaw-Kwasie et al., 2023; Piscicell & Ludden, 2016). Hence, it encourages inefficient use of raw materials, natural resources and energy, causing rapid depletion of resources, especially those that are not renewable. In particular, there is a need to ensure that actors in the fish value chain enforce and adopt circular economy principles that focus on reducing material inputs, reusing and recycling waste, and restoring and regenerating fishery resources to promote competitiveness, productivity, efficiency and sustainability (Regueiro et al., 2022). This chapter adopts Kirchherr et al. (2017) comprehensive definition of a circular economy (CE) as "an economic system that adopts business models which replace the "end-of-life" concept with reducing, alternatively reusing, recycling and recovering materials in production, distribution and consumption processes operating at the micro-level (products, companies, consumers), meso-level (industry/sector) and macro-level (city, region, nation and beyond), to achieve sustainable development through creating environmental quality, economic prosperity and social equity to the benefit of current and future generations". The vision of CE is "a cleaner and more consistent human/nature balance in which resource input and waste, emission, and energy leakage are minimised by slowing, closing, and narrowing material and energy loops" (de Jesus et al., 2019). However, the development of circular practices in the fisheries sector globally is still in its infancy, with a few developed economies embracing the principles, but many are still strategizing the practicality and applicability of the concept in their specific contexts (Pounds, 2021). The European Union and the People's Republic of China are taking the lead in adopting a circular economy as a policy priority (Kirchherr et al., 2018; Piscicell & Ludden, 2016; Repp et al., 2021). Despite its infancy, Piscicell and Ludden (2016) promote the transition to a circular economy because it eliminates waste from the industrial chain by creating closed ends through which resources can be recovered or reused to generate more value. Further, they assert that circular practices generate significant savings on production costs and lessen dependence on virgin materials and scarce resources in production, thus reducing commodity price fluctuations. The potential of a circular economy to increase or decrease employment opportunities has also been reported (Repp et al., 2021). Based on their evidence from the apparel industry, it is suggested that the increasing adoption of CE principles in Europe will likely lead to a decline in employment opportunities in non-European low-upper middle-income countries exporting apparels to EU. However, there is great potential for job creation resulting from downstream waste management, reuse and recycling.

The transition from the traditional linear economy (production-consumption-disposal model) to a circular economy is made even more difficult because it requires a lot of support in terms of better knowledge, better regulation and better funding (Jonker & Navarro, 2018), which are major challenges in the developing world. The major barriers to the transition into a circular economy are economic, political, cultural, market, regulatory and technological (Kirchherr et al., 2018; Piscicell & Ludden, 2016). For instance, cultural barriers include hesitant organizational culture that resists CE adoption; the limited willingness of firms to collaborate in the value chain; lack of consumer awareness and interest in sustainability and compliant circular products; companies continuing to operate in a linear system which is perceived to be cheaper; and non-integration of CE in company strategy, mission, vision, goals and key performance indicators. Market barriers are specified as low prices for CE compliant products/materials, limited standardization, high investment cost requirement and limited funding for circular business models. Regulatory barriers include obstructive existing laws and regulations, limited circular procurement and lack of global consensus on CE regulation, while technological barriers are characterized by a lack of technical ability to deliver high-quality remanufactured products, limited circular product designs, lack of demonstration projects and data on impacts of CE adoption. Although de Jesus and Mendonça (2018) found regulatory and technological barriers to some of the most pressing challenges in the transition to CE, Kirchherr et al. (2018) did not find them among the most important barriers. This difference may be attributed to Kirchherr et al.'s (2018) study focused on the European Union with advanced regulatory and technological capabilities. This fact also points us to the potential difference in the perceived impact of CE, its barriers and actual implementation between the developed and the developing world. Scholars have in addition highlighted the key drivers of the transition to a circular economy to include conducive public policies that provide the enabling environment, the willingness of businesses to experiment with CE, the need for government-stakeholder alliances and networks to advance the CE, disruptive circular innovations in large existing firms, adoption of common circular inspired mission, strategy and partnerships among stakeholders; circular policy entrepreneurship, creating separate organizational structures and circular start-ups outside the existing firm; leadership commitment towards sustainability targets can also drive radical circular business model innovations, and digitization as an acknowledged cause of disruptions and accelerator of the transition to CE in terms of availing data, automation, networking and enabling customer interfaces among others (Kirchherr et al., 2022; Neligan et al., 2023). Although some studies provide insights into the characteristics, capacities, strategies, drivers and barriers to the sustainability transition (de Jesus et al., 2019; de Jesus & Mendonça, 2018; Kirchherr et al., 2018, 2022), there has been little empirical work on institutionalization and enforcement of the deliberate sustainability transformation of fisheries governance and management systems (Fortnam, 2019). In Uganda, the implementation of the national fisheries policy remains a challenge and more interventions are particularly required to institutionalize and enforce fisheries practices that are more transparent, responsible, inclusive, cost-effective and environmentally compliant in order to protect and preserve the natural fisheries resources for the present and future generations (MAAIF, 2017).

Institutionalization refers to formulating, changing and strengthening the regulatory framework to embrace the new norms, values, structures and practices in the dominant governance and management system, creating feedback mechanisms and dedicating funds and personnel to support the new activities. Key strategies that are important in the institutionalization of circular fisheries management systems include building and strengthening of cross-scale relations vertically and horizontally at leadership levels, lobbying for higher level support, development of formal institutions and organizational structures, efforts to routinize human and financial resourcing, education, communication and motivation to change social behaviour positively, attitude and mindset, constituency building and mobilization, securing the support and commitment of local governments, harnessing and incorporating the traditional fishing rules into the formal institutions (Fortnam, 2019; Nunoo et al., 2015). The embedding and diffusing of new norms, values, structures and practices so that the new governance and management system persists beyond its initiation are critical to successful institutionalization (Fortnam, 2019). The literature also guides on the forces that drive or limit institutionalization (Fortnam, 2019). He articulates mainly two driving forces that include decentralization and collaborative leadership, that provide the opportunity to involve powerful actors with a mix of political, administrative and personal qualities and skills, leverage resources, develop coalitions with formal authority and control of resources, and social networks in support of developing new institutional arrangements that promote sustainable fisheries management. On the other part, the opposing forces are numerous and include the high resource intensity, the need for long-term investment and the fact that it takes long for the sustainability goals to be achieved, limited national and local government budgets, deeply entrenched social-politics and culture that make implementing new systems difficult, resistance to change and rigidity by actors and beneficiaries who seek to protect their vested interests by maintaining the status quo. Hence, sustainable fisheries policy implementation cannot operate in isolation but rather in the wider social, economic, political and ecological economy, meaning that other sector priorities and forces can derail the progress towards sustainability in the fisheries sector.

Enforcement, on the other side, is the process of managing the practical application, access and use of policy according to the defined conditions. It involves creating, categorization, management and monitoring of creating, categorizing, managing and monitoring the execution of a specific set of policy requirements for access and use of a resource, product or service. The monitoring activities in enforcement include identifying the rule breakers and prosecution activities where guilt or innocence is ascertained, or penalties are assessed and implemented when guilt is proved (Anderson, 1989). Cooperation, manpower, motivation, and financial and infrastructural commitments are needed to enforce fisheries policy effectively. Further, it calls for respect for human, labour, social and cultural rights to avoid social abuses such as forced evictions, child labour, forced labour, detention without trial, violence and other harmful practices against fishing communities in the enforcement of fisheries policy (Teh et al., 2019). While safety and healthy working conditions, certification and eco-labelling are increasingly prioritized in fisheries. Enforcement of equal and equitable participation in the national fisheries sector and applying the fishing rules equally to all participants without discrimination, harassment and intimidation are important in developing an inclusive and equitable fisheries management model. In particular, gender equity is gaining attention in fisheries because women contribute significantly to households' well-being and food security, education of their children and the wider economy (Teh et al., 2019).

2.2 *Theoretical Framework for Sustainable Policy Implementation*

The literature review reveals some theoretical approaches to achieving effective policy implementation. Institutional theory is adopted in conceptualizing how institutions are formed, how they operate and change, and how they influence behaviour of individual actors. Institutions have underlying values, norms, practices and structures that affect human behaviour and outcomes that can be achieved. Institutions are the prescriptions humans use and follow to organize all forms of repetitive and structured interactions and actions, including those within families, neighbourhoods, markets, firms, churches, private associations and governments at all scales (Ostrom, 2005). Theoretically, the institutional theory explains that formal and informal regulative, normative and cognitive pillars determine behaviour, decisions and policy implementation actions (Scott, 2006). Institutions, therefore, reduce uncertainties in policy implementation by guiding human interactions and the choice of the most efficient alternatives. Furthermore, they regulate human behaviour by restricting the pursuit of selfish interests. Institutions create the regulatory framework within which people interact, work and act in accordance with expectations of each other. Therefore, expectations are manifested through an established system of norms, values and rules that guide people to act almost unconsciously and automatically and intuitively know what implies the expected behaviour. Although they can sometimes be quite vague, frequently changing and modifying established institutional frameworks can lead to unpredictable consequences. Therefore, this study uses an institutional theory to address how and what rules restrict negative behaviour or constrain destructive actions in implementing and managing the national fisheries policy. It is enforcement, however, that helps actors respect the institutional rules and increase their commitment and adherence to the norms and structure of the fisheries governance and management system to continue receiving positive outcomes. Therefore, the military's intended role is to enforce compliance with the institutional rules as laid out in the policy and guidelines to attain and maintain sustainable fisheries resources management (Jentoft, 2004).

However, positive fisheries policy outcomes may not necessarily be acquired through a rigid formal regulatory system established and enforced by government institutions (Brundtland, 1987). Indigenous people may also generate a specific type of behaviour conditioned by close interaction with nature and adjust their actions based on their worldview. Indigenous people may have social-cultural institutions combining laws, policies and processes, operating based on consistent governance and management system, and developed concerning intrinsic values that are preconditions for effective and sustainable resource use. While the institutional theory helps to account for macro (national, regional, global) patterns in policy implementation, it has limitations in accounting for meso (sector/industry) and micro (organizational/community/individual) level differences in policy implementation. The complex adaptive systems theory (CAST) thus supports the institutional theory by accounting for adaptations, complex relationships and interconnections between the diverse range of actors, organizations and stakeholders that enable and shape the institutionalization and control of behaviour in fisheries policy implementation.

The CAST suggests that the most productive state for a system is the "edge of chaos" where there is maximum variety and creativity. Complex adaptive systems focus on the relations and interconnections of the system components to control dynamics and responses of heterogeneous nature. The CAST considers the role of adaptation, self-organization and learning in building collective competence (Pype et al., 2018). Managing fisheries resources and policy is complex, and unpredictable and outcomes vary depending on the circumstances, stakeholders, and ecological and environmental contexts. Hence, fisheries policy implementation has to adapt and be innovative in processes, structures and strategies depending on the circumstances. In particular, creative interactions and collective actions emerge from complex and continuous relations (Filipe et al., 2010). CAST further allows analysis of reasons inherent to the lower-level processes and their control (Lansing, 2003). Therefore, effective implementation of the policy requires quality relationships between actors, information sharing and the creation of feedback loops, which in turn influence the generation of knowledge and new behaviour in the fisheries management team (Pype et al., 2018).

According to Lester and Steward (2000), the agents in implementing public policies can be broadly categorized as the bureaucrats, including government ministries, administrative departments and agencies (MDAs), the Legislature (legislative bodies that formulate, monitor and evaluate implemented policies), the Courts (enforce policy through the judicial means), pressure groups (influence the guidelines and regulations in a way that benefits them) and the community and local authorities (consume or implement public programmes at the local level). Thus, policy implementation consists of rule-making, rule administration and rule-adjudication. Literature asserts that the legislative intent, the administrative capacity of the implementing bureaucracy, interest group and opposition activity, and executive support impact policy implementation. In the same regard, Lester and Steward (2000) argue that distortion is likely to occur when legislative agencies have too close relationships with administrative agencies.

3 Methodology

3.1 Data Collection and Analysis

This study adopted qualitative methods, which were useful in answering questions about experience, meaning and perspectives from the standpoint of the participants. Qualitative methods are used to learn how participants interpret the social world. Human beings construct meanings in unique ways, which are dependent on context, experiences and frames of reference (Bullock & Trombley, 2000; van Esch & van Esch, 2013). Further, Barclay et al. (2017) recommended qualitative research methods to understand social and economic dynamics and human behaviour in fisheries resources.

To achieve the objectives, qualitative data was collected from 12 focus group discussions (FGD) and 15 key informant interviews. The participants were purposively selected as guided by the local leaders. The local leaders helped to identify the residents involved in fisheries policy implementation. In some cases, participants helped to identify knowledgeable informants by snowballing. The primary informants were fishers, net makers and repairers, boat builders and repairers, gear producers, fish processors, boat owners and fish traders; civil servants including fisheries, agriculture, veterinary, accounts, production, commercial and health officers; local council leaders, sub-county chiefs and beach management units (BMU) leaders in different fishing sub-counties in Uganda. All respondents had been involved in policy implementation for not less than 5 years, which reveals adequate experience accumulated over time. The fishers, local leaders and technical officers' focus groups were organized differently so that each would actively participate and inform the study with freedom (Morgan, 2002), but also to effectively control social desirability (Patton, 2002). Efforts were taken to ensure the validity and reliability of data. The use of focus group discussions and interview triangulation and flexibility facilitated counterbalancing effects in which data and system shortcomings and strengths were smoothed out. Respondent validation strengthened trust in the results, brought new insights and motivated the refinement of findings (Bromley et al., 2003). Data was collected from fishing sub-counties from all four regions of Uganda, North, South, East and West, targeting the major lakes: Victoria, Kyoga, Kazinga channel, Kwania, Albert and Edward. Respondents were particularly selected using purposive sampling. This chapter adopted Morestin (2012) framework for analysing public policy implementation using the dimensions of effects (effectiveness, unintended effects and equity); and implementation effectiveness (cost, feasibility and acceptability).

3.2 Ethical Considerations

Informed consent was obtained from respondents to tape-record their voices. Participants were at liberty to withdraw at any stage of the study if they so wished. Audio recordings in the local language, "Luganda", were transcribed and translated into English. To ensure confidentiality, files were password protected to restrict access. In addition, to hide the identity of respondents, pseudo-names were used to identify the study participants. Further, the Makerere University Social Science Research ethics committee in Uganda approved and cleared this study under reference number MAK SS REC03.10425/AR. The research objectives and procedures that involved their recording were clarified to the respondents before their involvement. Assurance of transparency and confidentiality enabled participants to disclose the necessary information for the study freely.

4 Institutional Framework, Achievements and Challenges in NFP Implementation

4.1 Institutional Framework for Policy Implementation in Uganda

The key institutions, structures and actors in public policy implementation in Uganda are the Executive comprising the presidency and ministers, parliament, Office of the Prime Minister, judiciary, public service technocrats or bureaucrats as well as other relevant actors like interest groups and the opposition. Those responsible for policy management processes are members of the Executive Office. Structurally, the Central Government is mostly not the executants of its own policies. Local Government authorities mainly carry out policies (LG), Government Ministries, Departments and Agencies (MDAs), public corporations, firms or other contracted agencies. However, divisions and distances between such agencies render the implementation of policy complicated and uncertain.

The President who heads the Executive governs in the public interest and must have an effective means of ensuring that policy decisions are taken and executed through the ministers and civil servants, police and the military (Heywood, 1997). In a parliamentary democracy, the legislature or parliament is a site for policy-making, monitoring and evaluation. The Parliament of Uganda exercises controls over the Executive through its committees. The committee system effectively oversees policy, budget and programmes' implementation. The most important standing committees of Parliament of Uganda are those that exercise control over finances: the Public Accounts Committee, the Committee on Estimates and the Committee on Public Undertaking. Parliament allots time every week to raise questions about the Executive's functioning. The "question hour" is the forum where members can elicit information on any aspect of the performance of the Executive on policy implementation. The adjournment debates and no-confidence motions provide occasions for the entire or part of government policy implementation to be criticized. The introduction of Prime Minister (PM) question time is one of the visible signs of enhanced accountability and policy implementation monitoring in the country. It enables Members of the House to question the PM regularly. These events are covered by the press and highlight the government's performance. The role of parliament in policy implementation is guided by the 1995 Constitution of Uganda chapter 4, section 21(4) as amended.

In Uganda, a public policy becomes enforceable after parliament passes its Bill. Then the Ministry of Finance issues a certificate of financial implication (Orem & Zikusooka, 2010). A certificate of financial implication verifies that the financial implication of any new programme has been examined and that there are resources available for implementing the proposed programme.

The Office of the Prime Minister (OPM) ensures effective coordination of all government policy implementation. This is undertaken within the jurisdiction of the Policy Coordination Committee (PCC), Implementation and Coordination Steering Committee (ICSC) and the Technical Implementation Coordination Committee (TICC). The National Development Programme (NDP), the Implementation Forum and mandatory quarterly sector and local government meetings and reports further enhance policy implementation in Uganda. The Prime Minister (PM) is the driver of policy implementation where the PM has not been effective. Instead, PCC is largely pre-occupied with issues concerning the Public Sector Management Working Group (PSM-WG) and other emerging pressings of national importance. This deviation from the institutional and structural responsibility has left no room for effective coordination of public policy implementation and reporting (UNDP, 2016). The Implementation Unit (IU) and the Delivery Unit (DU) in OPM have to ensure that plans and government programmes are implemented. The OPM undertakes bi-annual and annual iterations of the Government Annual Performance Review (GAPR), which marginally includes performance reviews against UNDP objectives and strategic targets. This assessment is mainly conducted based on national budget resources released. However, the review omits critical aspects, notably the private sector's contribution, civil society organizations (CSOs) and the media (UNDP, 2016). Ggoobi (2016) observes that Uganda needs a policy-on-policy implementation to guide OPM and other structures for effectiveness and efficiency.

The judiciary is equally engaged in public policy implementation. It has the power to strike down the executive by declaring quasi-judicial and legislative actions that regard policy implementation as unconstitutional. It is also the ultimate interpreter of constitutional provisions, provides legislative and regulatory frameworks, and promotes accountability, the rule of law and the observance of human rights in policy implementation. Rule of law is particularly useful in promoting transparency, fairness, predictability and confidence in court decisions, enforceable contracts, basic security and access to justice. Another structure with a fundamental role in policy implementation is the Local Government (LG). Structurally, Central Government formulates public policy (Uganda Constitution, 1995: schedule 6), which is implemented by the different levels of governance, while Local Government Act 1997, Chapter 243, Part IV spells out the functions and powers of local government councils and one of its essential roles is that of overseeing the implementation of government and council policies. Local government implementation structures include the district, sub-county and parish authorities. However, complaints associated with this structure include slow and long procurement processes, laxity by the Ministry of Local Government to approve district procurement committees and late disbursement of funds by the Central Government, which delays policy implementation (BMAU, 2015).

On the other hand, line ministries carry out technical supervision, technical advice and monitoring of LG policy implementation in liaison with international agencies. Districts can create sectorial committees responsible for planning, supervising and monitoring policy implementation. Ministry of LG and the LG Finance Commission Act link the district and government in implementing policies. Important to note is that some local government actors make policies to enhance individual interests, which Ayeko-Kümmeth (2015) terms "Micro-hegemony". She uses the term to refer to the manipulation of political, social and economic power in policy implementation by some actors in the different levels of LG, while international agencies and development partners provide the technical assistance and resources for developing and implementing policy and new institutional arrangements (Fortnam, 2019). However, the role of the army in policy implementation is not well articulated in the institutional framework.

4.2 *Achievements and Challenges in the Implementation of the NFP in Uganda*

Uganda, located in East Africa, has no coastline and is landlocked. With about 20% of its surface area covered by water, Uganda has enormous resource potential for capture fisheries and aquaculture production. Uganda's fisheries landscape includes the five large lakes of Victoria, Kyoga, Albert, Edward, George and Kazinga Channel; 160 minor lakes, rivers such as Albert Nile, wetlands and floodplains. All of which are critical habitats, breeding and nursery grounds for fish and hence, suitable sites for fish farming. Uganda's fisheries resources are also diverse in fish species, including tilapia, Nile perch, catfish, lungfish, etc. (MAAIF, 2011).

Fisheries resources are among the most significant natural endowments of Uganda. The fisheries sub-sector contributes 12% to the agricultural gross domestic product (GDP) and 3% to the National GDP (ACSA-Uganda, 2019; NaFIRRI, 2013). While it is estimated that 72% of produced fish is consumed locally while 28% is exported (MAAIF, 2017). Therefore, the sector has potential to make significant contributions to the achievement of national development goals, such as poverty reduction and economic growth. The sector provides high-quality food and a source of protein to over 17 million people and provides direct employment and a source of income for about 300,000 people. It also supports about 5.3 million livelihoods in fisheries-dependent households (MAAIF, 2017). Despite her potential to generate 1 million metric tonnes of fish annually, Uganda only produces 460,000 metric tonnes and 100,000 metric tonnes of catch fish and framed fish, respectively (MAAIF, 2017). This poor fish production performance is attributed to unregulated illegal fishing and inadequate fisheries laws and guidelines enforcement. The loophole in enforcement has been attributed to the centralized management system characterized by inadequate human, technological, logistical and financial provisions for the institutions in charge of fisheries management, research and monitoring to carry out their activities. However, between 2000 and 2003, the Government of Uganda began transforming the fisheries industry into a sustainable one through its department of fisheries resources. The transformation was driven by major problems experienced that included depletion of fisheries resources resulting from the increased use of destructive illegal fishing gears that catch immature fish, poor post-harvest handling and processing methods leading to waste and loss, environmental damage and resource mismanagement resulting from the poor fisheries policy implementation (MAAIF, 2004).

Consequently, a number of statutory instruments were introduced to address the key challenges identified in fisheries management and policy implementation. These instruments include beach management unit (BMU) rules (2003, 2016), national fisheries policy (2004), fish act (2011), fish rules (2003, 2004) and fishing rules (2010, 2017). They were meant to provide legal empowerment of *BMUs* with sustainable fisheries planning and *management practices* in partnership with local governments; regulate fishing activities; guide the fisheries licencing process and to provide information for fish quality assurance; and issuance of fish sanitary certificates, respectively. In addition, under those guidelines, the decentralized management units were expected to involve the small-scale fishing communities and indigenous people in restoring, conserving, protecting and co-managing local aquatic resources and ecosystems (MAAIF, 2004).

In particular, the National Fisheries Policy (NFP) was intended to control illegal fishing in order to improve livelihoods and reduce poverty in fishing communities through sustainable fisheries management practices (Benkenstein, 2011). The policy's fundamental aim is ensuring increased and sustainable fish production and utilization by properly managing capture fisheries, promoting aquaculture and reducing post-harvest losses. The NFP (2004) provides guidelines for the fisheries sector in Uganda and outlines 13 strategic policy areas. The vision of the NFP is to utilize fisheries resources for socio-economic transformation optimally. The policy aims to attain sustainable fisheries, food and nutrition security, economic growth, poverty reduction, employment creation and promotion of domestic, regional and international trade. The 13 NFP strategic policy areas include: (a) sustainable fisheries management and development, (b) decentralization and community involvement in management, (c) district, sub-county and community partnership in management, (d) institutions and funding mechanisms, (e) investment in fisheries, (f) planning and policy-making, g) information dissemination, (h) environment and fisheries, (i) aquaculture, (j) post-harvest fish quality and added value, (k) fish marketing and trade, (l) human resource development and (m) research. These thirteen strategies were developed to align the fisheries policy with Uganda's Vision 2025, the Poverty Eradication Action Plan (PEAP) and the Plan for Modernisation of Agriculture (PMA). These three plans constituted the national planning framework when formulating and adopting NFP in 2004 but have since been replaced with the National Development Plan (NDP) and Vision 2040. The broader transformations of the national fisheries policy are decentralization and co-management, market liberalization and sustainable development, as in NDP III and the national poverty reduction strategic plan. The Fish Act CAP 197 operationalizes the NFP and enacts the various statutes for effective and sustainable management of fisheries resources. Hence, the ethical and accountable implementation of the fisheries policy would not only contribute to the achievement of sustainable development goals (SDG) 1 and 2, which target reducing poverty and hunger, respectively, but also to the achievement of SDG 8 on good jobs and economic growth; SDG 9 on the industry, innovation and infrastructure; SDG 12 on responsible consumption and production; SDG 13 on climate action; SDG 14 on life below water; SDG 16 on peace and justice; and SDG 17 on partnerships for the goals.

Whereas NFP was introduced in 2004 to improve the sustainable management and performance of the fisheries sector, Uganda continued to suffer a number of challenges, among which were dwindling fish stocks and fish diversity (MAAIF, 2017; Nadiope, 2010). For instance, between 2011/2012 and 2015/2016, fish stock was reduced by 85%. This problem has first and foremost been exacerbated by illegal fishing methods that catch immature fish, fishing in breeding and nursery areas that can further lead to fish depletion. Secondly, sustainable fisheries management was vested upon the Central Government department of fisheries resources, but staff were not posted. Thus, administration and management were inadequately equipped to meet the sector's needs (MAAIF, 2015). Although the objective of introducing decentralization, co-management and role sharing through BMUs was to address the discrepancies in participation among resource users and all stakeholders in planning and resource management, the participation of fishing communities was found to be either minimal or absent. A number of problems have since been attributed to such non-participatory management approaches, including lack of collaborative management leading to poor coordination and circulation of activities, non-formulation of byelaws and less enforcement of laws and regulations. Those gaps further increased the use of illegal fishing gears and methods, limited collection, dissemination, sharing and use of information leading to inadequate knowledge of the policy, laws and regulations among the fishing communities and stakeholders; lack of a continuous system of fish quality and safety assurance; poor and inadequate arbitration of fisheries conflicts and disputes; lack of adequate monitoring and supervisory mechanisms to control illegal fishing activities, among others (MAAIF, 2011). Further, insufficient funding, lack of political will, corruption and the sheer size of surveillance are some of the other major challenges in policy implementation in Uganda's fisheries sector (Benkenstein, 2011). Consequently, in 2014, BMUs reached the peak of their poor performance, and many fish factories closed due to fish scarcity, resulting in a loss of jobs and dwindling fish exports (Mugambwa et al., 2017).

Further, like other African countries, fishing communities in Uganda are among the poorest of the poor, and fish is a critical resource for their livelihood and survival. However, these fishing communities continue to pollute and use poor fishing methods that are rapidly depleting fish stocks (Lubaale, 2019). Moreover, they mainly live a rural life with very limited human and social capital and little or no financial capital to access other physical assets to enable them to adequately add value and investment in fish processing and aquaculture. This problem is coupled with other unscrupulous practices among these communities that frustrate enforcement and scare away fisheries policy implementors with insults and attacks, including deadly ones such as drowning and assassinating officers. Furthermore, unethical practices among public policy implementers, such as acceptance of brides, corruption, embezzlement of funds and negligence of duty, have largely made the implementation of the national fisheries policy ineffective. In summary, therefore, challenges of increasing encroachment on fisheries resources use of the poor fishing method and overfishing, poor landings and environmental degradation persisted, resulting into decreased fish stocks, food and nutritional insecurity, reduced exports, unemployment and ultimately increased poverty among the fishing communities (Kauta, 2016; MAAIF, 2017; Naluwairo, 2005).

4.3 Involvement of the Army in Fisheries Policy Implementation in Uganda

The government intervened in the implementation challenges by involving the military in the surveillance and enforcement of the national fisheries policy on all of Uganda's waters (ACSA-Uganda, 2019). In

2015, President Museveni provided a platform for fishers to share their concerns. In consultation with the ministers and technocrats, the President responded to the fishers concerns by giving a directive to halt the operations of BMUs. However, the move created a vacuum in policy implementation and gave freedom to the fishers to act anyhow. That vacuum prompted the President to assign the army the responsibility of managing fish landing sites as an amicable solution. The military was particularly engaged to enforce the NFP in liaison with fisheries officers in order to wipe out the illegalities (Benkenstein, 2011).

Presently, fishing activities are being monitored by the UPDF Marine Fisheries Enforcement Unit. The unit enforces the national fisheries policy through sensitizing the fishing communities on illegal fishing and acceptable hygiene at the landing sites, surveying the waters, confiscating and destroying illegal fishing equipment, and arresting culprits and presenting them for possible prosecution. Some stakeholders welcomed and embraced the army involvement in the implementation of the fisheries policy and were interested in having illegal fishing practices put to a stop (Buregeya, 2020). The *National Collaborating Centre for Healthy Public Policy* (NCCHPP) asserts that military involvement in fisheries attracted immediate compliance, enforced use of better nets, and as a result fish stocks increased. Nevertheless, NFP implementation continues to face numerous challenges including those resulting from the army's incompetence and lack of professionalism, authoritarian leadership style, corruption and brutality; lack of partnerships and collaborations with other government agencies, such as Uganda National Bureau of Standards (UNBS) and Uganda Revenue Authority (URA) in stopping the importation of substandard fishing equipment into the country, which further frustrate policy enforcement efforts (Semakula, 2018). It is against that background that an empirical study was conducted to establish the effectiveness and effects of military involvement in the fisheries policy implementation in Uganda.

5 Empirical Findings on the Effectiveness and Effects of the Military Enforcement of the National Fisheries Policy in Uganda

The results in Table 1, I below, indicate that the fisheries policy implementation by the army has had some effectiveness in reducing illegal, unreported and unregulated fishing; ensured professional monitoring, surveillance and protection of fisheries resources; apprehending and prosecution of offenders and enabled the re-opening of fisheries factories and exports. This was due to competence building for the army through training and tapping ideas from experienced fishers and experts, enforcement authority and support from the President, as well as provision of the necessary equipment, motivation and facilitation. However, there are unintended effects such as displacement and forced migration of fishers, limited consultation and sharing of useful information, skills and knowledge between the sub-county technical staff and the army, brutality and loss of lives, loss of trust in the system, bribery and army's provision of support and protection for the illegal fishers, limited cooperation between the fishers, technical team and the army, connivance and collusion, hesitancy to support the army, abuse of human rights, corruption, profiting from confiscated fish, materials and equipment, political interference and release of offenders without being tried. These unintended effects have led to unemployment, fear, hopelessness and despair among the fishing communities. Some illegal fishing continues unnoticed due to the large lake waters and lack of cooperation, illegal fish being sold in the markets and continuance of illegal fishing under the knowledge and protection of the army officers. As far as equitable policy implementation is concerned, findings indicate that the army undermines the rules, lacks social and technical skills, is corrupt and discriminative; there is limited sharing of useful information between the army and technical officers and withholding important information from the army by the community members. Hence, lack of equity has led to continued illegal fishing and selling of illegal fish.

Regarding implementation effectiveness, the findings and analysis are in Table 1 shows that the engagement of the army is an added cost to the already existing structure, moreover, the community members have abandoned rescue work to the army, and some lives are lost as a result. Furthermore, some Fishers flee islands due to harassment from the army, while many fishers have become unemployed. Consequently,

Table 1 A matrix of findings and analysis regarding Military engagement in fisheries policy implementation in Uganda

Dimension	*Supporting statements*	*Emerging themes*	*Generalized consequences on policy implementation*
Effects			
Effectiveness	"*Equally, the Police and army officers have been trained in prosecution and apprehending illegal fishermen. They use the ideas from training, consultation from experienced fishers and from experts. They sensitize the fishers at different landing sites where questions are asked and responses given. The fishers also share their skills and knowledge of the lake with fisheries officers that help to locate illegal fishers. Using this expertise, the army have been able to reduce illegal fishing. Fishers learn and understand what is expected at the lake. The fishers change behaviour and stop illegal fishing […]*" (FGD 12) "*The army has been able to reduce illegal fishing here […]*" (FGD 4)	Competence	• Creative ideas from experienced people • Improved skills and expert knowledge through training • Prosecution and apprehending illegal fishermen • Reduced illegal fishing • Reduced illegal fishing gear • Re-opening of fisheries factories • Professional monitoring of fisheries resources • Positive fishing activities and practices • Increased fish stocks
	The Fisheries Protection Unit (FPU) collected illegal fishing gear of more than Shs2b,	Law enforcement	

(continued)

Table 1 (continued)

Dimension	*Supporting statements*	*Emerging themes*	*Generalized consequences on policy implementation*
	"*The army has helped increase fish stock in quantity and size. The fisheries protection unit has helped improve compliance in terms of recommended fishing vessel and reducing illegalities which have led to re-opening of 4 fish processing factories in the country […] for example the fish processing factory in Kasensero in Kyebe Subcounty was re-opened […]*" (Interviewee 13)	Compliance	
	"*When the army came it was facilitated and has done commendable work […]. The fisheries officers could not singlehandedly monitor activities regularly because they don't have boats …motor faster boats can help the officers do the work well on the lake […]*" (Interviewee 8)	Equipment, motivation and facilitation	
	"*The army working on fisheries enforcement was a directive from the President. The army is well paid and has fast boats, informers and guns…without support, things are difficult even when we know what is required of us. We have not marked our breeding grounds due to lack of money [….]*" (Interviewee 6) "*The army is provided with food, fuel, boats, engines […] this was enough for them to do the work well […]*" (FGD 8) "*Yes, now you see what the army has been able to do which was initially impossible during the time of BMUs, they have the Presidential backing, that's authority […]*" (Interviewee 4)	Authority and support	

Dimension	*Supporting statements*	*Emerging themes*	*Generalized consequences on policy implementation*
Unintended effects	"*[…] The army does not consult us (the sub-county technical staff), we also don't interfere with their job. In a way this affects our effectiveness […]*" (Interviewee 3)	Limited Cooperation	• Increased illegalities • Unethical practices • Unemployment • Reduced fish stocks
	"*Some individuals, communities together with the marine army connive and create organised cliques. They clear certain routes which are used for illegal fishing, and transportation of illegal fish for export. Some army officers close and designate new illegal landing sites where dubious activities are carried out. When some fishermen learn that the marine army has left, they continue doing illegal fishing. People prepare local brew (Waragi), wash in the lake while the community are watching […]*" (FGD 8)	Connivance	• Fear • Loss of lives • Undersized/immature fish sold in markets • Some illegal fishing goes unnoticed due to the large lake waters and lack of cooperation • Illegal fish is sold in the market • Loss of trust in the system • Unreported illegal fishing

(continued)

Table 1 (continued)

Dimension	*Supporting statements*	*Emerging themes*	*Generalized consequences on policy implementation*
	"*60 per cent of the fishermen who used to operate at different major landing sites have migrated to other nearby Landing Sites on Buvuma and Misoma islands in Tanzania in quest for fish species […]*" (FGD 11)	Forced migration	
	"*The army does not consult or take the advice of the sub-county technical staff. The army only need inspection reports from them. The fisheries officers responded by failing to pick telephone calls from the army. The fisheries officers decided only to sit back and to wait to watch while illegal fishing is done. Which escalates illegal fishing […]*" (Interviewee 2)	Limited consultation Uncollaborative attitude Limited sharing of skills and knowledge between the sub-county technical staff and the army officers	

Dimension	*Supporting statements*	*Emerging themes*	*Generalized consequences on policy implementation*
	"*We support the army's role in fighting illegal fishing methods but the UPDF marines has veered off this purpose. When they come to our areas, they beat us with sticks. They don't segregate whether you are a fisherman or not. In Buvuma District, deaths have been recorded as a result of the brutality, beaten, handcuffed and others taken away to undisclosed destinations […]*" (FGD 3)	Brutality	
	"*What should be done is to target those who are using illegal fishing methods and those dealing in immature fish. But when they raid our villages, they don't discriminate whether you work in a shop or ride a motorcycle. Even when you have nothing to do with fishing, you are beaten, […]*" (FGD 2) "*Some leaders fear to speak out because whether you are a leader or ordinary person, fisherman or not, you are targeted fish trader/woman […]*" (Interviewee 9)	Suppression of freedom to speak	
	"*When soldiers raid landing sites or islands, they do not discriminate when inflicting torture. Even those who are not involved in the fishing business are targeted indiscriminately, get involved in human rights abuse […]*" (Interviewee 15)	Human rights abuse	

(continued)

Table 1 (continued)

Dimension	*Supporting statements*	*Emerging themes*	*Generalized consequences on policy implementation*
	"*When we take the offenders to the authorities, the authorities instead ask for bribe. So, we have lost trust in the system. For example, the salt dried fish stopped, are being sold by army men some of it ends up in markets […]*" (Interviewee 11)	Bribery Extortion Army support of the illegal fishers	
	"*The army leaders talk as if they know a lot […] we no longer respond to their calls […] they don't want advise from technical fisheries officers. They only need inspection reports from you […] we only do inspection, no enforcement […] we left that for the army […] we sit back and watch [.….]*" (Interviewee 3)	Limited cooperation	
	"*They inform the illegal fishers not to work when they will do surveillance… they work with the army [….] some army officers stop fishers on route from fishers who lack movement permit. They confiscate the fish [….] they then sell this fish and get money for their benefit […]*" (Interviewee 7)	Connivance Profiteering by army officers	

Dimension	*Supporting statements*	*Emerging themes*	*Generalized consequences on policy implementation*
	"*some community members connive and give money to the army officials who allow them to fish. Similarly, the marine army who catch illegal fish sellers confiscates the immature fish, but they sell it in truck full. They also do not take the offenders to court. The army has closed some landing sites like Sangobay but still do their deals. They create new landing sites where the illegal fishermen pay to them […]*" (Interviewee 10) "*The Local council members (LCs) know the illegal fishers but they fail to report them to the police, they have failed to even advice the army on what is necessary to curb illegalities on the lake like registering all the fishermen […]*" (Interviewee 7)	Hesitancy to support the army Limited sharing of useful information Corruption	
	"*The politicians become a stumbling block in the fight against illegal fishing. They often accuse the operations team and fabricate stories to tarnish the image of the army […]*" (Interviewee 13)	Political interference Release of offenders	
Equity	"*The marine army refuses us - fishers to go with ice on the lake. They have also banned fish smoking. Yet all these methods are for value addition and preservation of fish. This deprives us of incomes. Yet we have licenses for fishing and smoking […]*" (FGD 14)	Undermining the rules	• Illegal fishing and sell of illegal fish

(continued)

Table 1 (continued)

Dimension	*Supporting statements*	*Emerging themes*	*Generalized consequences on policy implementation*
	"*A fleet of motorcycles take immature feet along this road from Katosi. I know that but that work is for the marines [....] The army gives chance and time for the fishers to make money as they fish illegally the army is then paid for that. They make safe routes for them to carry immature fish [....] Fisheries officers know those who use monofilament (steel nets) are but are not reprimanded [...] Some army officers close and designate its new illegal landing sites where dubious activities are carried out [...]*" (Interviewee 9)	Limited sharing of useful information	
	"*Some of these officials in the fisheries protection unit do not even know the measurements of fish. Besides they ask for bribes from some people, who take the immature fish to some factories and to DR Congo [...]. These officials themselves are corrupt. They catch people with premature fish, they confiscate it and sell it to other people. Some premature fish is being caught and sold on other landing sites [...]*" (FGD 5)	Lack of technical skills Corruption	

Dimension	*Supporting statements*	*Emerging themes*	*Generalized consequences on policy implementation*
	"*It's the army who support some fishers who use kokota and illegal equipment. They are not arrested. […]*" (Interviewee 10)	Withholding information from army	
Implementation			
Cost	"*The army maintenance on the lake is an added cost. The BMU's were doing the job with the fisheries officers originally. Some of the fishing communities were doing volunteer services […]*" (FGD 6)	Army is an added cost	• Costly army maintenance • Leaving the rescue work to the army leads to loss of lives • Hopelessness • Unemployment
	"*the safety on the lake is left in the hands of the army; even when people are drowning in the lake, fishers believe it's the army to rescue them after all they are the ones benefitting [paid by government]. They do not come out to help […]*" (Interviewee 3)	Abandonment of responsibility by some community members	
	"*Many youths have no capital to buy the required fishing equipment since what they were using was destroyed by the army […]*" (Interviewee 14)	Lack of capital	

(continued)

Table 1 (continued)

Dimension	*Supporting statements*	*Emerging themes*	*Generalized consequences on policy implementation*
	"*When the army steps up its patrols of the lakes, forces fishermen to flee the islands and lake shores as revenues also drop […]*" (Interviewee 5)	Displacement/forced migration	
Feasibility	"*The army cannot be anywhere all the time; the lake waters are too wide. New Landings can be developed anytime, anywhere…the army alone has no capacity to monitor fishing activities […]*" (FGD 9)	Lack of capacity and being few in number	• Illegal fishing in isolated locations
	"*Fishers in cliques connive and get in the waters at night to do illegal fishing and get out of the water at 4:00 am, when the army is not available and most people are unable to see and catch them. […]*" (FGD 3)	Connivance	
	"*The young unacceptable fish ends up in markets but nothing is done by the market operators to arrest those dealing in illegal fish. Other fishers collude and sink the un acceptable boats (manyatta) in the lake waters. They remove the boats in the night and illegally fish […]*" (Interviewee 14)	Collusion	
Acceptability	"*The fishing community has been forced to accept the trend of fishing monitoring by the army […]*" (FGD 1)	Forced acceptability	Illegal fishing in isolated locations

Source Primary data

these acts have made the implementation of the policy costly to maintain and have caused fear and loss of lives, despair and hopelessness due to unemployment. Findings indicate further that the feasibility of implementation is low due to the lack of capacity by the military to control all the national water bodies since the marine army officers are few in number and lack modern equipment. In addition, there is connivance among the fishers on one side as well as between the army officers and the fishers, all of which lead to continuance of illegal fishing in isolated locations. Furthermore, findings show that there has been forced acceptability and compliance by the fishing community due to the power imbalances caused by the army's involvement. However, this consequently leads to isolated unreported illegal fishing activities and practices.

6 Discussion

The emerging themes guide the discussion in view of objectives of the study. First, an analysis of the policy implementation effectiveness using the army is made, and later the effects of the military involvement in the fisheries policy implementation are discussed. It should be noted that supporting literature on engagement of army in fisheries policy implementation is sparse.

6.1 Effectiveness of the Military Enforcement of the National Fisheries Policy in Uganda

The study reveals that the deployment of the army in the implementation of the national fisheries policy was costly and a duplication of the already existing structure in the implementation of the fisheries policy. Regarding implementation feasibility, the army could not single-handedly monitor all the parts of the waters, yet the fishing community abandoned their collaborative and voluntary role to stop illegal and irresponsible fishers. In power imbalance, the army has forced the fishing community to accept their position. These limitations have allowed illegal fishing to continue in isolated locations on the waters. According to Nunoo et al. (2015), there are always cost-effective traditional management systems which are very efficient in enforcing the fishing rules and regulations in the communities, and these need to be respected and cooperated with; hence, the army needs to work with community leaders in monitoring and enforcing the fishing rules. Similarly, OECD (2008) observes that policy reforms often entail costs while their benefits are less certain, more diffuse and delayed in time. Costs make the transition to a new fisheries governance and management system a complex task in which political economy considerations are of key importance (Fortnam, 2019). Sometimes reforms yield negative outcomes with a massive increase in overhead costs and a resource fall (OECD, 2008). Wu et al. (2015) note the importance of policy capacity, which refers to the possession of the competencies and capabilities important to policy implementation. Competencies are categorized into three general types of skills essential for policy success that is analytical, operational and political, while policy capabilities are assessed at the individual, organizational and system resource levels. The capacity of the army to implement the fisheries policy in Uganda was thus enabled by the funding, equipping and resourcing, training and authority accorded by the Presidential machinery.

6.2 *Effects of the Military Enforcement of the National Fisheries Policy in Uganda*

In terms of effectiveness, the army has been able to regulate illegal and unreported fishing to a large extent, regular monitoring of the lakes, apprehending and prosecuting illegal fishermen has resulted into better fish stocks and re-opening of fisheries factories, reduced illegal fishing and reduced use of illegal fishing gear. The capacity of the army to enforce the fisheries policy and achieve the policy goals is attributed to the authority and resourcing given by the Presidential machinery. The army was adequately motivated and supported with equipment such as speed boats that enabled law enforcement. In addition, they were given some training that built their competencies in weak areas. Hudson et al. (2019) found that support is a crucial success factor for implementation. It improves the perceptions, experiences and behaviour that shape the implementation process. It involves assessing existing capacity to deliver, knowing what is being done well, what needs improving and how best to build new capacity. Further capacity building should concentrate on the "how". Hence, investment in skills and competencies development, training, peer learning, information, guidance, management skills and other such interventions is necessary. This approach informed the World Health Organisation's Regional Office for Europe's health systems transformation initiative, which is focused on making change happen and putting in place the necessary infrastructure for the initiative to succeed (Hunter et al., 2018; WHO, 2018). However, combining all three functions of capacity building, including problem problem-solving, managing and regulating, especially a fusion of compliance and support within one agency, is problematic (Hudson et al., 2019). Relatedly, Muhumuza (2016) found that the army cherishes four principles: discipline, transparency, openness and patriotism, which are the foundation of quality service delivery that the citizens demand. World over, the army can promote development using discipline, integrity, passion and accountability through training. According to Stewart (2014), the twenty-first-century military is not limited to martial activity, it often takes on additional roles, including diplomatic and development functions. However, the military must not be crafted into the system as a replacement of the local government or civilian authorities but as a means of building functional institutions and collaborative systems involving multiple agencies.

6.3 Unintended Effects of the Military Enforcement of the National Fisheries Policy

Findings indicate that whereas there are reductions in illegal and unreported fishing practices, several unintended effects arise from engaging the army in fisheries policy implementation in Uganda. Participants informed that many indigenous people have been displaced and forced to migrate from their homelands and work areas, thus becoming unemployed. In addition, the limited sharing of skills and information between the sub-county technical staff and the army, reduced participation and collaboration with the community, brutality and loss of lives, loss of trust in the system, bribery, supporting the illegal fishers for personal gain, connivance, corruption, political interference, profiting from the sale of confiscated fish, materials and equipment and illegal release of offenders have created pockets of illegal fishing to continue, causing inequality and inequity, unemployment, poverty and hopelessness among the poor fishers. Hence, the original intention of NFP in improving the lives of the fishers is being undermined. These results are particularly supported by Nunoo et al. (2015) and Fortnam (2019) who report similar findings in Ghana and Philippines fisheries industries, respectively.

Previous literature has sighted similar challenges leading to policy implementation failures elsewhere. Ajulor (2018) pointed out the neglect of target beneficiaries and lack of consideration of policy environment. *Beneficiaries* are the *targeted* population who directly reap from the policy *implemented* in their communities. Consideration of the beneficiary community enhances or fails the sustainability of the implementation. One of the challenges of policy implementation is the inability to involve the target beneficiaries in the policy process (Dialoke et al., 2017). Further, Ggoobi (2016) attributed policy failure in Uganda to the neglect of the rural areas and improper management. The government imposes policies without considering whether it meets the needs of the people or not. Dialoke et al. (2017) recognized the need to involve the target beneficiary at every policy implementation stage to ensure transparency and accountability. Programmes carried out with the consent of the target beneficiary will not be sabotaged (Ajulor, 2015). Those operating at higher levels need some grasp of what actually happens at the frontline. It is further reported that a mix of bottom-up approach and co-management cannot be ignored in modern fisheries management (Fortnam, 2019; Nunoo et al, 2015). Hence, the "bottom–up" school of thought on policy implementation and the notion of the "street-level bureaucrat" whose adaptive or discretionary power can prove instrumental in determining the success or failure of a policy is still relevant (Lipsky, 1980).

The study revealed frequent corruption in fisheries policy implementation in Uganda, even when the army is engaged. These findings agree with Fortnam (2019), who found that the corrupt culture of Philippines politics, coupled with patron-client relations and the influence of political elites afford the local and central government leaders control over state resources, discretion over development priorities and law enforcement that perpetuate bribes and corruption in fisheries resource management. Furthermore, Ikechukwu and Chukwuemeka (2013) highlight forces that negatively affect policy implementation process and activities, including ineffective and corrupt political leadership under which public bureaucracy thrives, the pervasive and deep-rooted corruption within the public bureaucracy, the pressure and influence of primordial demands and values on the bureaucracy. These factors and circumstances constitute serious obstacles to effective policy implementation by public bureaucracy. Hence, corruption remains a formidable challenge amidst weak institutions, lack of political rights and an insufficiently free media. (Transparency International, 2019). Regarding political interference and patronage, Rogger (2018) found that increased legislative and executive influence increases the likelihood of a project failure, reduced project quality and increased misuse of funds. So, increased legislative oversight may lead to distortions in the technocratic process for political gain and policy failure since there is normally a trade-off between political oversight and bureaucratic autonomy.

Undermining rules fail policy monitoring, enforcement and compliance, as found in the current study. World Bank (2014) observed that undermining rules happen if powerful actors fear that the outcome may reduce their relative power now or in the future, they may attempt to block the adoption or undermine the implementation of policies that could enhance social and economic welfare. This tendency has especially significant implications for households at the bottom of the income pyramid and other marginalized groups because their bargaining power tends to be more limited. Power asymmetries can harm society and negatively impact policy effectiveness through exclusion, capture and clientelism (World Bank, 2014). Fortnam (2019) particularly found that destructive commercial fishing and mining in Lunuda Bay in the Philippines could not be stopped because the interests of the most powerful family in the province would be affected while mining companies influence local politics and decisions in their favour by bribing local officials and financing election of pro-mining candidates in addition to mayors pardoning apprehended illegal fishers in return for support during elections.

The study identified information-sharing challenges as failing fisheries policy implementation in Uganda. Nunoo et al. (2015) report fishers' ignorance of the rules and regulations due to inadequate communication of the fishing rules and regulations by leaders of community fisheries associations as well as government officials. Similarly, Weaver (2009) identifies information problems where targets lack information that would make compliance more likely as one of the key challenges in policy implementation and as a cause of all the other problems including: monitoring problems—where target compliance may be difficult or costly; resource problems in terms of lack of the resources to comply even if they want to; autonomy problems that are lack of power to make decisions that comply with policy and attitudinal problems involving hostility and mistrust of targets. While Hudson et al. (2019) identified five interacting factors that can contribute to over-optimism: complexity (underestimation of the delivery challenges); evidence base (insufficient objective, accurate and timely information on costs, time scales, benefits and risks); misunderstanding of stakeholders (optimism about the ability to align different views); behaviour and incentives (interested parties boosting their own prospects); and accountability (decision-makers seeking short-term recognition). All of these need information sharing for policy implementation effectiveness.

7 Conclusion

This chapter reveals that the legal instruments, structures and institutions necessary to enforce sustainable and responsible fishing practices are largely in place but the issue that remains is to implement them effectively and, in particular translate and communicating the national fisheries policy to the level understood by all stakeholders such as the local fishers, local leaders, fishing companies and others. Thus, there is still a need to effectively enforce the rights, responsibilities, incentives and sanctions to complete the policy cycle. Despite the rollout of the national fisheries policy in 2004 under the responsibility of the BMUs and the government fisheries department, challenges of increasing encroachment on fisheries resources, use of the poor fishing method and overfishing, poor landings and environmental degradation persisted resulting in decreased fish stocks and diversity, food and nutritional insecurity, reduced exports and closure of factories, unemployment and ultimately increased poverty among the fishing communities. The empirical results reveal that the involvement of the army in the enforcement of the national fisheries policy has to some extent, reduced illegal, *unreported* and *unregulated* (IUU) *fishing* in Uganda; improved apprehension and prosecution of illegal fishermen; increased fish stock levels resulting into re-opening of some fisheries factories and resumptions of fish exports, reduced illegal fishing and reduced illegal fishing gear hence, enhanced acceptability and compliance. However, these seemingly positive achievements have been costly while some of them have been unethically obtained and/or accompanied by unintended negative outcomes. Hence, militarization of policy implementation and generally public service delivery is not a sustainable solution. The unintended effects of the army's involvement include exclusion and displacement of some fishers, corruption, brutality and violation of human rights leading to loss of lives, hopelessness, despair, forced migration, unemployment and poverty. These unethical practices have foiled the achievement of the core objectives and institutions embedded in the NFP especially the failure to improve the quality of life of fishers and adoption of responsible fishing practices that promote resource circularity and sustainability in Uganda. The overall objective of every government is to bring about a qualitative improvement in its citizens' standard of living and promote sustainable growth and development. Realizing these noble objectives entails not only the formulation of policies but also their effective implementation (Ikechukwu & Chukwuemeka, 2013). The NFP was supposed to have led to tremendous levels of social and economic development however, the reverse has been the case. Therefore, implementing the NFP to promote responsible, circular and sustainable fisheries practices in Uganda has neither been effective with the public bureaucracy nor with the army's engagement.

8 Implications, Recommendations and Limitations

Theoretically, it is evident that fisheries policy implementation to promote circular and or sustainable fisheries practices can be enhanced by utilizing the institutional theory, agency theory and the complex adaptive systems theory (CAST). In the institutional context, laws are enforced, and the actors are empowered. Actors own the authority to participate in collective organizations and decision-making, thus, have a legitimate capacity to use agency while achieving sustainable collective goals. The institutional theory posits that formal and informal institutions determine the opportunity structure and the choice among options. The CAST contributes to sustainable fisheries policy implementation by encouraging adaptations to institutional rules and regulations in social interactions, environments and circumstances. Wide-spread adaptations and innovations can lead to the emergence of new and non-traditional means that promote fishery resources circularity and sustainability. For instance, adaptation and innovations are required in addressing and controlling climate change effects such as heavy rains that cause flooding of water banks and long periods of drought that reduce water levels, all of which affect fish breeding capacities and fish stocks; to reduce over-dependence on natural and/ or biological restocking of fisheries resources and excessive catching of fish without restocking with fish fries hence ignoring modern ways of restoring and regenerating the fisheries resources. Fish processors lack concern for reducing waste and reusing fish by-products and offcuts in processing fish feeds and reintroducing them into the water bodies; ignorance of fish food certification and eco-labelling, which are circular practices that should be promoted and enforced by the national fisheries policy.

The empirical findings imply that fisheries policy implementation involves a lot of social complexities, so the army cannot work unilaterally. As the results suggest, the army by acting unilaterally replaced the civil servants in committing unethical practices and what makes it worse is the inability of society to act on them considering the power relations. It is important to iterate the fact that African fishing is mostly indigenous done by native communities around the water bodies and mainly done on a small scale, which implies that it is sustainable. However, the growth in commercial fishing to feed the increasing population and support fish processing for export is threatening sustainability of fishery resources. Hence, the restrictive fisheries policy has become a challenge and a limitation to social and economic progress of the native fishing communities. In fact, the increased poverty among the fishing communities is not only strongly linked to decreasing fish stocks but also to the army's brutal implementation of the national fisheries policy. Further, the limited information sharing between the military enforcement unit, the fishers, local leaders and fisheries officers is the major sabotage failing the policy implementation and perpetuating fishing illegalities. The findings further elucidate the linkage between human rights violations and the minimal achievements of the military enforcement of the fisheries policy. This implies that excessive use of force in fisheries policy implementation increases social injustices, which are detrimental to the overall governance improvements and the progress towards rights-based sustainable development of the fisheries sector. The results also confirm that military enforcement of circular fisheries practices has led to forced migration and increased unemployment among small-scale fishers causing despair and hopelessness which may result in social unrest and insecurity in the communities. However, it can potentially increase jobs among commercial fishing, processing and export firms due to increasing fish stocks.

This chapter recommends reforms to strengthen institutions, systems and respect of statutory structures for public policy implementation. Instead of pushing them aside and replacing them with the army, the public bureaucracy should be deployed, motivated and empowered. Additionally, ethical behaviour and accountability be promoted in the entire public and civil service delivery. This is because unethical practices have persisted even when the policy enforcement is entrusted with the army. We thus recommend clear segregation of duties of public servants from those of the army in the implementation of policies and the specific agency roles of the army formally be institutionalized in the national fisheries policy framework. The UPDF marine fisheries enforcement unit should focus on ensuring that the fishing rules and regulations are complied with, while the fisheries officers should concentrate on inspection and assurance of quality standards. In particular, the role of the military in the fisheries sector should be more in monitoring, and surveillance and limited to providing security in the deep waters. A collaborative relationship should exist between the army, public service and other public and private agencies in policy implementation rather than competing with each other. Fisheries policy implementers should promote collective action through co-management since it has a developmental focus where both government, fishers and other stakeholders act as equal partners in sustainable fisheries management. Management by some form of partnership is likely to be an effective and sustainable policy option. This partnership arrangement would be able to yield the desired long-term returns with the potential for; greater reliability and accuracy of data and information, more suitable and effective regulation; enhanced acceptability and compliance with management measures; reduced enforcement costs; reduced conflicts; and strengthened commitment to and participation by concerned interested parties. Furthermore, the involvement of both purposeful and responsible local, political and bureaucratic leaders should be ensured.

There is a need to implement policies based on established ideal bureaucratic rules and principles. The bureaucrats need to resist primordial political pressures and demand for special favours strongly. There is a need to improve the working conditions and provide other incentives for the public bureaucrats to build their morale and commitment to work like in the case of the army or even better. Particularly, improved pay packages and equipment may significantly diminish corruption tendencies and promote the proper use of allocated funds. Information and communication technology (ICT) should also be implemented in fisheries policy. ICT should be utilized to improve access to and sharing information, tracking illegal activities, monitoring and surveillance of waters, and aiding fisheries database development and management. Gold (2017) recommends the establishment of a central unit to track the progress of policy implementation. This trend now exists in 25 countries and aims to fulfil several functions: (a) performance monitoring by tracking progress against key policy priorities by analysing a constant stream of departmental performance data; (b). problem-solving through undertaking field visits and surveys to identify delivery obstacles and flagging where additional resources may be needed to address specific problems; and (c). progress assessing through supplying heads of government with routine progress reports. While most of such units have been located at the centre of government machinery, it does not necessarily have to be the case. Other such progress track units can be established in key ministries or simply created for specific priority programmes. When well-constructed, these units can contribute to policy implementation, while weak units simply lead to additional significant problems. However, care must be exercised in generalizing the findings of this study due to limitations in its geographical coverage and small sample size while being aware that some of the conclusions have been drawn based on analysis of publicly available documents. The chapter's scope is limited to capturing fisheries. This is because the NFP (2004) which is our case study lays little emphasis on aquaculture, and it is only in 2017 that the NFP (2004) was reviewed and upgraded to National Fisheries and Aquaculture Policy (NFAP) deliberately emphasizing aquaculture to bridge the gaps in fish production from capture sources, and align the NFP to Uganda National Development Plan (NDP III) and Vision 2040. Thus, the NFAP has embedded more sustainable fisheries and aquaculture practices. The reviewed national fisheries policy intends to balance sustainable exploitation of the fisheries resources while ensuring fish availability for both present and future generations without degrading the environment (MAAIF, 2017). However, the Parliamentary bill for NFAP is not yet passed into law up to date and, thus, is not yet implementable. Hence, there is a no evidence of enforcement of the national aquaculture rules and guidelines and circular aquaculture practices. Thus, future studies may focus on them.

References

Abunyewah, M., Okyere, S. A., Frimpong, L. K., Diko, S. K., Erdiaw-Kwasie, M. O., & Boateng, V. (2022). Fire risk communication in the urban informal sector: Evidence from traditional marketplaces in Accra, Ghana. *Risk, Hazards & Crisis in Public Policy*.

ACODE. (2015). *Transformation of agriculture for wealth creation involvement of UPDF in NAADS programme and its effectiveness*. Info sheet No. 33, 2015.

ACSA-Uganda. (2019). *National fisheries and Aquaculture Policy: What you need to know*. www.acsa-ug.org

Adebayo, K., Babu, S. C., & Rhoe, V. (2010). Institutional capacity for designing and implementing agricultural and rural development policies and strategies in Nigeria. NSSP Brief 9. Abuja, Nigeria: International Food Policy Research Institute (IFPRI). http://ebrary.ifpri.org/cdm/ref/collection/p15738coll2/id/213

Ajulor, O. V. (2015). A policy analysis model of participation and change management. *Journal of, Creativity and Innovations for Sustainable Development, Research and Development Institute, 15*(2), 31–38.

Ajulor, O. V. (2018). The challenges of policy implementation in Africa and sustainable development goals. *PEOPLE: International Journal of Social Sciences, 3*(3), 1497–1518

Al Mamun, A., Islam, A. R. M., Alam, G. M., Sarker, M. N. I., Erdiaw-Kwasie, M. O., Bhandari, H., & Mallick, J. (2023). Livelihood vulnerability of char land communities to climate change and natural hazards in Bangladesh: An application of livelihood vulnerability index. *Natural Hazards, 115*(2), 1411–1437.

Anderson, L. G. (1989). Enforcement Issues in selecting fisheries management policy. *Marine Resource Economics, 6*, 261–277.

Ayeko-Kümmeth, J. (2015). *The politics of public policy decisions in local government in Uganda*. Bayreuth University.

Bagala A. (2021). *Museveni orders army to build schools, health centres*. https://www.monitor.co.ug/uganda/news/national/museveni-orders-army-to-build-schools-health-centres-3466800

Barclay, K., Michelle Voyer, M., & Mazur, N. A. (2017). The importance of qualitative social research for effective fisheries management. *Fisheries Research, 186*, 426.

Benkenstein, A. (2011). *'Troubled Waters' Sustaining Uganda's Lake Victoria Nile Perch Fishery*. The South African Institute of International Affairs (SAII), Governance of Africa's Resources Programme, Research Report 9. South Africa. https://media.africaportal.org/documents/saia_rpt_09_benkenstein_20111031.pdf

BMAU. (2015). *Public Sector Management development projects: What are the key implementation issues?* Budget Monitoring and Accountability Unit, Ministry of Finance, Planning & Economic Development, Briefing Paper, Kampala.

Bromley, H., et al. (2003). *Glossary of Qualitative Research Terms*. King's College, University of London.

Brundtland, G. H. (1987). Our common future: Report of the world commission on environment and development. United Nations, 20 March 1987.

Buregeya, D. (2020). The Uganda People's Defence Forces (UPDF) *fisheries enforcement troops have intensified operations against illegal fishing in the*

Greater Masaka region to control a fresh outbreak of transaction in immature fish. https://www.newvision.co.ug/articledetails/80120

Bullock, A., & Trombley, S. (Eds.). (2000). *The new Fontana dictionary of modern thought* (3rd ed.). HarperCollins.

CGTN Africa. (2018). *Uganda's military will learn railway construction in China*. https://news.cgtn.com/news/7955544d7a597a6333566d54/share_p.html

Cobbinah, P. B., Erdiaw-Kwasie, M. O., & Amoateng, P. (2015). Rethinking sustainable development within the framework of poverty and urbanisation in developing countries. *Environmental Development, 13*, 18–32.

de Jesus, A., & Mendonça, S. (2018). Lost in transition? Drivers and barriers in the ecoinnovation road to the circular economy. *Ecological Economics, 145*, 75–89.

de Jesus, A., Antunes, P., Santos, R., & Mendonca, S. (2019). Pathways to a circular economy: Envisioning priorities through a Delphi approach. *Journal of Cleaner Production, 228*(August), 1494–1513.

Dialoke, I., Ukah, F. O., & Maduagwuna I. V. (2017, March). Policy formulation and implementation in Nigeria: The bane of underdevelopment. *International Journal of Capacity Building in Education and Management (ICBM), 3*(2), 22–27.

Erdiaw-Kwasie, M. O., Abunyewah, M., Yusif, S., & Erdiaw-Kwasie, A. (2023). Does circular economy knowledge matter in sustainable service provision? A moderation analysis. *Journal of Cleaner Production, 383*, 135429.

Filipe, J. A., Coelho, M. F., Ferreira, M. A., & Pedro, M. I, (2010). Analyzing fisheries management through complexity and Chaos theories framework. *Journal of Mathematics and Technology, 1*(2), 5–12.

Fortnam, M. P. (2019). Forces opposing sustainability transformations: Institutionalisation of ecosystem-based approach to fisheries management. *Ecology and Society, 24*(4), 33.

Ggoobi, R. (2016). Why implementation fail in Uganda. *Africa Research Institute*. https://www.africaresearchinstitute.org/newsite/blog/why-implementation-fails-in-uganda/. Retrieved May 16, 2021.

Gold, J. (2017). *Tracking delivery global trends and warning signs in delivery units*. Institute for Government.

Grindle, M., & Thomas, J. (1990). After the decision: Implementing policy reforms in developing countries. *World Development, 18*(8), 1163–1181.

Heywood, A. (1997). *Politics*. Macmillan Press Limited.

Hudson, B., Hunter, D., & Peckham, S. (2019). Policy failure and the policy-implementation gap: Can policy support programs help? *Policy Design and Practice, 2*(1), 1–14. https://doi.org/10.1080/25741292.2018.1540378

Ikechukwu, U. B., & Chukwuemeka, E. E. O. (2013, March). The obstacles to effective policy implementation by the public bureaucracy in developing nations: The case of Nigeria. *Kuwait Chapter of Arabian Journal of Business and Management Review, 2*(7), 59–68.

Independent. (2018). UPDF ready for role in Standard Gauge Railway construction. https://www.independent.co.ug/video-updf-ready-take-part-standard-gauge-railway-construction/

Jentoft, S. (2004). Institutions in fisheries: What they are, what they do, and how they change. *Marine Policy, 28*(2), 137–149.

Jonker, J., & Navarro, N. M. (2018). *Circular City Governance: An explorative research study into current barriers and governance practices in circular city transitions across Europe*. European Investment Bank. www.europa.eu

Kauta, N. (2016). Paper presented at the Sixth Annual joint Agricultural sector review conference Commonwealth Resort.

Kellermann, A. L., & Jones, S. S. (2013). What will it take to achieve the as-yet-unfulfiled promises of health information technology. *Health Affairs, 32*(1), 63–68.

Kirchherr, J., Bauwens, T., & Ramos, T. B. (2022). *Circular disruption: Concepts, enablers and ways ahead. Business strategy and the environment*. Special Issue. Wiley.

Kirchherr, J., Piscicell, L., Bour, R., Kostense-Smit, E., Muller, J., Huibrechtse-Truijens, A., & Hekkert, M. (2018). Barriers to the circular economy: Evidence from European Union (EU). *Ecological Economics, 150*, 264–272.

Kirchherr, J., Reike, D., & Hekkert, M. (2017). Conceptualising the circular economy: Analysis of 114 definitions. *Resources, Conservation & Recycling, 127*(2017), 221–232.

Lansing, J. S. (2003). Complex adaptive systems. *Annual Review of Anthropology, 32*, 183–204.

Lester, J. P., & Stewart, J., Jr. (2000). *Public policy: An evolutionary approach* (2nd ed.). Wadsworth.

Lipsky, M. (1980). *Street-level bureaucracy: Dilemmas of the individual in public services*. Russell Sage Foundation.

Lubaale, G. (2019). Poverty in Uganda: Causes and strategies for reduction with great emphasis on ethics and ecological justice. *Sociology and Anthropology, 7*(1): 9–19.

MAAIF. (2004). *The national fisheries policy*. Ministry of Animal Industry and Fisheries.

MAAIF. (2011). *Department of fisheries resources annual report 2010/2011*. Ministry of Animal Industry and Fisheries.

MAAIF. (2015). *Agriculture sector budget execution report financial year 2013/14*. Ministry of Animal Industry and Fisheries.

MAAIF. (2017). *The national fisheries and aquaculture policy*. Ministry of Animal Industry and Fisheries.

Morestin, F. (2012). *A framework for analyzing public policies: Practical guide*. National Collaborating Centre for Healthy Public Policy. http://www.ncchpp.ca/docs/Guide_framework_analyzing_policies_En.pd

Morgan, D. L. (2002). Focus group interviewing. In J. F. Gubrium & J. A. Holstein (Eds.), *Handbook of interviewing research: Context & Method* (pp. 141–159). Sage Publications Inc.

Mugambwa, J., Nabeta, I. N., Kaberuka, W., Munene, J. C., & Rudaheranwa, N. (2017). Theorizing policy implementation: Making sense of the fish stock restoration story. *International Journal of Policy Studies, 8*(2), 45–65.

Muhammad, F. (2014). Leadership, governance and public policy implementation competencies in the broader public sector. *European Journal of Business and Management, 6*(36), 66–73.

Muhumuza, F. K. (2016, September 27). *Involvement of military personnel in agricultural extension services: Implications for uptake of emerging technologies*. Public Policy Dialogue on Agrarian Reforms and Transformation of Agriculture in Uganda, Kampala. https://umi.ac.ug/index.php/research/public-dialogues/item/924-involvement-of-military-personnel-in-agricultural-extension-services-implications-for-uptake-of-emerging-technologies

Nadgrodkiewicz, A., Nakagaki, M., & Tomicic, M. (2012). *Improving public governance: Closing the implementation gap between law and practice*. Center for International Private Enterprise and Global Integrity. https://www.cipe.org/resources/improving-public-governance-closing-implementation-gap-law-practice/

Nadiope, E. (2010). *Analysis of the implementation and enforcement of the Fish Act, CAP. 197 (2000) on Lake Victoria in Uganda* [Master of Science (Environment and Natural Resources Management)]. Makerere University.

NaFIRRI. (2013). *National Fisheries Resources Research Institute profile* (2nd ed.). National Agricultural Research Organisation.

Naluwairo, R. (2005). Promoting common property rights in fisheries management in Uganda: A review of the national fisheries policy and the proposed legislation. https://www.acode-u.org/uploadedFiles/PBP8.pdf

Neligan, A., Baumgartner, R. J., Geissdoerfer, M., & Schögg, J. (2023). Circular disruption: Digitalisation as a driver of circular economy business models. *Business Strategy and the Environment, 32*(3), 1175–1188.

Nunoo, F. K. E., Asiedu, B., Olauson, J., & Intsiful, G. (2015). Achieving sustainable fisheries management: A critical look at traditional fisheries management in the marine artisanal fisheries of Ghana, West Africa. *Journal of Energy and Natural Resources Management (JENRM), 2*(1), 15–23.

OECD. (2008). *What next? The challenges of policy implementation tertiary education for the knowledge society* (Vol. 2). ISBN 978-92-64-04652-8. https://www.oecd-ilibrary.org/docserver/9789264046535-14-en.pdf?expires=1628447903&id=id&accname=guest&checksum=60A1F4B821F7C36716C1504080CC3376

Orem, J. N., & Zikusooka, C. M. (2010). Health financing reform in Uganda: How equitable is the proposed National Health Insurance scheme? *International Journal for Equity in Health, 9*, 23–28.

Ostrom, E. (2005). *Understanding institutional diversity* (Princeton paperbacks). Princeton University Press.

O'Toole, L. J. (2000). Research on policy implementation: assessment and prospects. *Journal of Public Administration Research and Theory, 10*(2), 263–288.

Patton, M. Q. (2002). *Qualitative evaluation and research methods* (3rd ed.). Sage.

Piscicell, L., & Ludden, G. D. S. (2016). The potential of design for behaviour change to foster the transition to circular economy. In *Design Research Society, 50th Anniversary Conference* (pp. 1–16).

Pounds. A. (2021). *An introduction to circular economy principles in aquaculture*. The latest series of articles from the GAIN project, funded by the EU Horizon 2020 Fund. www.globalseafood.org

Pype, P., Mertens, F., & Helewaut, F. (2018). Healthcare teams as complex adaptive systems: Understanding team behaviour through team members' perception of interpersonal interaction. *BMC Health Services Research, 18*, 570. https://doi.org/10.1186/s12913-018-3392-3

Regueiro, L., Newton, R., Soula, M., Méndez, D., Kok, B., Little, D.C., Pastres, R., Johansen, J., & Ferreira, M. (2022). Opportunities and limitations for the introduction of circular economy principles in EU aquaculture based on the regulatory framework. *Journal of Industrial Ecology, 26*(6), 2033–2044.

Repp, L., Hekkert, M., & Kirchherr, J. (2021). Circular economy-induced global employment shifts in apparel value chain: Job reduction in apparel production activities, Job growth in reuse and recycling activities. *Resources, Conservation & Recycling, 171*(2021), 105621.

Rogger, D. (2018). *The consequences of political interference in bureaucratic decision making: Evidence from Nigeria*. Policy Research Working Paper; No. 8554. World Bank. https://openknowledge.worldbank.org/handle/10986/30244. License: CC BY 3.0 IGO.

Scott, T. (2006). *The relationship of workplace empowerment and organizational commitment among First Nations and Inuit Health Branch Nurses* [MA, University of Manitoba, Winnipeg].

Semakula, J. (2018, Friday, 31 August). UPDF wants URA, UNBS to ban Illegal fishing gear. *New Vision*.

Shore, C. (2012). Anthropology and public policy. In *The SAGE handbook of social anthropology*. Sage.

Shore, C., & Wright, S. (1997). Policy: A new field of Anthropology. In C. Shore & S. Wright (Eds.), *Anthropology of Policy: Critical Perspectives on Governance and Power*, 3–39. Routledge.

Stewart, A. (2014). *U.S. army agriculture development teams: A grassroots effort in Afghanistan supporting development and tackling insurgency*. http://www.sciencediplomacy.org/

Teh, L. C. L., Caddell, R., Allison, E. H., Finkbeiner, E. M., Kittinger, J. N., Nakamura, K., & Ota, Y. (2019). The role of human rights in implementing socially responsible seafood. *PLoS ONE, 14*(1), e0210241.

Transparency International. (2019). Sub-Saharan Africa: Undemocratic regimes undermine anti-corruption efforts. https://www.transparency.org/en/news/cpi2018-subsaharan-africa-regional-analysis

Ugwuanyi, B. I., & Chukwuemeka, E. O. (2013). The obstacles to effective policy implementation by the public bureaucracy in developing nations: The case of Nigeria. *Kuwait Chapter of Arabian Journal of Business and Management Review, 2*(7), 59–68.

UNDP. (2016). *Delivering sustainable energy in a changing climate. Strategy note on sustainable energy 2017–2021.*

Van Esch, P., & Van Esch, L. (2013). Justification of a qualitative methodology to investigate the emerging concept: The dimensions of religion as underpinning constructs for mass media social marketing campaigns. *Journal of Business Theory and Practice, 1*(2), 214–243.

Weaver, K. 2009. *Target compliance: The final frontier of policy implementation*. Brookings Institution [Google Scholar].

WHO (World Health Organisation). (2018). Leading health system transformation to the next level. Expert meeting Durham, United Kingdom, July 12–13, 2017. WHO [Google Scholar].

World Bank. (2014). *Enhancing governance for development: Why policies fail*. https://openknowledge.worldbank.org/bitstream/handle/10986/25880/9781464809507_Ch02.pdf?sequence=31&isAllowed=y

Wu, X., Ramesh, M., & Howlett, M. (2015). Policy capacity: A conceptual framework for understanding policy competences and capabilities. *Policy and Society, 34*(3–4), 165–171. https://doi.org/10.1016/j.polsoc.2015.09.001

GAYO's Sustainable Community Waste Management Model: Impacts and Lessons For Circular Futures

Kirk Bright Enu, Joshua Amponsem, Desmond Alugnoa, Betty Osei Bonsu, and Andy Bonaventure Nyamekye

1 Introduction

Sub-Saharan Africa (SSA) is the fastest-urbanizing region in the world. The population of the region is increasing at an average rate of 2.7% per annum, causing many cities and towns to grow beyond their carrying capacities (OECD/FAO, 2016). As a result, city planners and authorities are unable to keep up with the rapid change and are faced with many governance challenges (Cobbinah et al., 2021; Erdiaw-Kwasie et al., 2020). Chief among these is improper waste management which is complicated by the unprecedented rates of waste generation and changing consumption patterns (Oduro-Appiah et al., 2019). This often results in environmental pollution, public health concerns and the degradation of

K. B. Enu (✉) · J. Amponsem · D. Alugnoa · B. O. Bonsu
Green Africa Youth Organization, Accra, Ghana
e-mail: kirk@greenafricayouth.org

A. B. Nyamekye
Citigroup, New York City, USA

M. O. Erdiaw-Kwasie and G. M. M. Alam (eds.), *Circular Economy Strategies and the UN Sustainable Development Goals*, Sustainable Development Goals Series,
https://doi.org/10.1007/978-981-99-3083-8_20

ecosystems and their regulatory, provisioning and cultural services as lands and hydrological systems are turned into waste disposal sites (Cobbinah et al., 2021; Wantzen et al., 2019).

The case of Ghana is no different; waste generation volumes continue to increase and a sustainable method for final disposal has not yet been found (Debrah et al., 2022). The widespread model in many municipalities has been collecting waste from households and commercial areas with the assistance of waste management enterprises, both in the formal and informal sector, and disposing them in either designated or undesignated dumping sites (Kyere et al., 2019). For designated dumping sites, most are not engineered, neither are they properly controlled. Thus, they get full in no time and create a new chain of socio-ecological challenges, including public health exposure resulting from leachate runoff (Mudu et al., 2021). It is in view of these shortcomings that alternative waste management models, away from linear models like landfilling, have been proposed. One of these is the attempt to introduce the principles of circular economy into municipal waste management.

The adoption of the circular economy in waste management comes with many benefits. For instance, Gopinath (2020) reported that the implementation of circular economy in England led to an increase in recycling rates from 11.2% in 2000–2001 to 43.7% in 2016–2017. Potting et al. (2017) also reported that 93% of all waste generated in the Netherlands is processed for new uses with an estimated 79% of the quantity going into recycling. This, the authors attributed to the intensive application of circular economy principles in the country. Conlon et al. (2019) stated that implementation of circular economy creates an avenue for the creation of waste-based businesses which creates employment for local people. This is affirmed by Nkansah et al. (2015) whose research showed that incorporating scrap metals recycling into the manufacturing industry has the potential to create employment and subsequently reduce dependence on virgin materials for production.

City authorities in SSA are beginning to appreciate circularity, for instance, through composting and recycling with examples in Durban, Nairobi and Accra. Notwithstanding, such efforts have not made significant contribution to sustainable waste management regime (Grant, 2016). This is evidenced in how key stakeholders like waste collectors and waste pickers who are key in waste recovery, for instance, are not formerly recognized in the waste institutional framework (Oteng-Ababio et al., 2017). Moreover, earlier studies on municipal waste management and circular economy focused on major cities which are often very complex (e.g. Mudu et al., 2021; Njoroge et al., 2014; Sakijege, 2019). This leaves a gap in peri-urban communities and presents the opportunity to test an improved waste management model in a location that is not as sophisticated, despite undergoing rapid urbanization, and to institute sustainable governance mechanisms before the area saturates.

Against this backdrop, the Green Africa Youth Organization (GAYO), a youth-led non-profit which works in environmental sustainability, proposed and piloted a circular economy waste management model, the Sustainable Community Waste Management (SCWM) model, in New Edubiase in 2018 (Ghana NPAP, 2022). The SCWM model was tested through a year-long project (Sustainable Community Project), to assess its contribution to sustainable waste management. The Sustainable Community Project envisions a sustainable community as a community where an incentive-based waste value chain is implemented for municipal solid waste management. The objectives of the project were to (1) enhance stakeholder collaboration and coordination in the waste management sector; (2) demonstrate and harness economic opportunities (green jobs) in sustainable waste management; (3) create an environmentally conscious society that practises sound waste management; and (4) promote zero-waste in the municipality. In this paper, we assess the SCWM model through a review of project reports and interview of project coordinators. Through this, impacts and lessons from the project are documented to inform replication of such sustainable waste management models in other communities in SSA. The findings presented are useful for planning and waste management governance.

2 Literature Review

2.1 Origins of Circular Economy Concept

Despite its present popularity in research and policy, disagreements remain over how circular economy should be defined. This could be traced to its diverse origins, having stemmed from disciplines like economics, ecology and environmental studies (Ekins et al., 2020). The concept of circular economy has gained popularity but is more debated among scholars in engineering and the sciences as well as organizational studies (Henrysson & Nuur, 2021). Governments; non-governmental organizations; national agencies; international organizations like the European

Union; industry players; and think tanks like the Ellen McArthur Foundation are increasingly championing circular economy in contemporary times (Kirchherr & van Santen, 2019).

A circular economy is defined as "a regenerative system in which resource input and waste, emissions, and energy leakage are minimized by slowing, closing, and narrowing material and energy loops" (Camilleri, 2019, p. 2). This can be achieved through long-lasting design, maintenance, repair, reuse, remanufacturing, refurbishing and recycling. This contrasts with the widely used linear economy model which is a "take-make-dispose of" model. The circular economy model has been proposed as a viable alternative and has been the subject of development debates since the mid-1900s when calls for environmental conservation started to gain momentum (Kreienkamp, 2019).

The concept has been deeply ingrained in the global development agenda: sustainable development goal 12 (sustainable consumption and production) (Conlon et al., 2019). Two key factors are paramount in transforming economies to become more circular: balanced utilization of resources and effective waste management (Smol et al., 2020). In that vein, the circular economy model aims at closing the product value chain at all phases of the product's life span and ultimately returns waste into the product value chain (Abad-Segura et al., 2021) Different aspects of these two factors relate to different stakeholders in the waste management sector such as governance institutions, producers, consumers, waste management service providers, researchers, innovators, the media, civil society organizations and more. Salmenperä et al. (2021) explain that dialogue between these multiplicities of stakeholders, resource and knowledge-sharing as well as the illustration of the economic benefits of circular economy are equally key to the successful implementation of the model. These key factors were inculcated into the SCWM model which was tested through the Sustainable Community Project in New Edubiase.

2.2 *Circular Economy Adoption Across Regions*

Literature on the adoption of the circular economy model in waste management has varied from region to region. For example, in Europe, Smol et al. (2020) opined that for the circular economy principles to be effectively incorporated into the water and wastewater sectors, the existing 4Rs (reduce, reuse, recycle, recover) needed to be extended to include "reclaim" and "rethink". The authors explained that the integration of "reclaim" and "rethink" will ensure that all circular economy implementation in the wastewater sector will be more comprehensive, touching on the technological, organizational and societal aspects. In the production and waste management of tires through the circular economy model, Araujo-Morera et al. (2021) proposed the use of the 7Rs (reduce, reuse, recycle, redesign, renew, repair, recover). The authors expounded that for sustainable usage of the tires, they must be redesigned to reduce fuel consumption; pollutant emissions and maintain road safety. The authors also asserted that the tires ought to be manufactured from renewable sources (Russian dandelion and Guayule) to reduce the dependence on fossil fuels. The authors also proposed the reuse of alternative chemicals (fillers) in the manufacturing process whiles old and worn-out tires could be recovered and repaired for reuse. These measures will ensure that the components of the tires remain in the production cycle even after their lifespan has elapsed.

In Asia, Sri Lanka's approach to the implementation of circular economy in the waste management sector is through social entrepreneurship. For example, social enterprises in the country are empowered to upcycle discarded banana peels into waste fibres and other social enterprises also collect high-density polyethylene and low-density polyethylene plastics and upcycle them into plastic packaging materials, such as gunny bags, curtains and old sarees (Jayasinghe & Liyanage, 2018). In war-affected communities in the country, women receive second-hand clothing from the United Kingdom and redistribute them for reuse. This approach towards the adoption of circular economy principles in the country creates employment for women who are engaged in the reselling of these clothes. In so doing, materials initially considered waste are reintroduced into the production value chain. This helps to limit the amount of waste disposed into the environment (Conlon et al., 2019).

In the case of SSA, Buch et al. (2021) argued that inclusivity is key in the implementation of circular economy. This could be done through the establishment of cooperative organizations. The authors indicated that informal waste workers, such as waste collectors and waste pickers, play an important role in the waste value chain. They recover, refurbish, repair and reuse the waste in landfills to reduce the amount of waste discarded into the environment. Thus, they should be recognized by government institutions to incorporate them into the circular economy model. The authors explained that their recognition and integration into the formal waste sector would help increase waste upcycling rates. In Uganda, Joshi et al. (2019) suggested a decentralized circular economy model of plastic waste management. The authors indicated that it is more cost-effective to patronize the services of waste pickers who sort through waste at dump-sites to recover sellable materials, including glass, plastic and metals in production. This approach is more efficient in many poor regions thus ensuring that disposal, collection, remanufacture and use of materials take place in the same community. In Tanzania, Aparcana (2017) also advocated for the inclusion of the informal sector due to their immense contribution to waste collection and disposal. In Kenya, the adoption of circular economy principles touched on the complete ban of deforestation for charcoal production and the recycling of organic waste, such as coconut husks, to make charcoal briquettes. This has helped reduce deforestation and generated employment in the production and sale of these briquettes (UNEP, 2019).

Ghana also adopted a renewable energy act (Act 843) which among others, has a goal to increase the renewable energy sources in the energy mix in the country to 10% by 2030 (GoG, 2019). This has incentivized many to explore the recycling of organic wastes like coconut husk and palm kernels into charcoal briquettes (Bonsu et al., 2020). Others also convert organic waste into compost, a practice which is gaining traction (Doe et al., 2023). Plastics are also recovered by waste pickers from landfill sites and sold to recycling companies through middlemen (Oduro-Appiah et al., 2019). These plastics are then recycled into products like chairs and bowls. Some are also upcycled into sustainable fashionable products like raincoats while old clothes are redesigned and resold on the market to prevent dumping in landfills (James & Kent, 2019). Scrap metals are recovered and either reused or repurposed for other functions, such as copper wires and iron rods (Nkansah et al., 2015). These various efforts are, however, sector specific and not well coordinated.

2.3 Towards Circular Economy Models in Sub-Saharan Africa

Many circular economy models have been proposed over the years. There have been decision-making support models (e.g. Ghinea & Gavrilescu, 2010; Karmperis et al., 2013), life cycle assessment models (e.g. Gentil et al., 2010; Kulczycka et al., 2015; Winkler & Bilitewski, 2007), input–output models (e.g. Towa et al., 2020), cost–benefit models (e.g. Di Foggia & Beccarello, 2020) and others have been integrated approaches (e.g. Shamshiry et al., 2011). Most of these models have proven to be quite effective, however, they were proposed for the Global North. As a result, they may not fit the SSA context since there are geographic differences and waste compositions and volumes also differ. In developed countries, for instance, paper and cardboard make up the greatest percentage of waste (Karim & Wetterhan, 2020) but in SSA, organic waste forms the greatest component in the waste stream (Adu-Boahen et al., 2014). This suggests that a working waste management model has to be tailor-made for SSA cities.

In addition, circular economy initiatives in SSA have often been overly centralized, with workers in the informal sector poorly engaged and remunerated despite their immense contribution to waste collection and recycling (Oduro-Appiah et al., 2019). This is why Joshi et al. (2019) call for a decentralization of the waste management regime in SSA. Ghana National Plastic Action Partnership (Ghana NPAP, 2022) also advocates for effective collaboration between all stakeholders to promote good governance, behaviour change, innovation and resource mobilization for addressing plastic pollution in Ghana.

In Ghana, the Informal Service Chain model and the Extended Producer Responsibility framework were proposed by Oduro-Appiah et al. (2019) and Quartey et al. (2015), respectively. Despite the novelty of these models, they place emphasis on individual actors in the waste management sector, such as waste collectors or city planners. However, managing municipal waste in a sustainable way requires a careful consideration of the different actors and stakeholders in the waste management sector, their functions, their interrelations and how their roles coalesce in decision-making (Durán & Messina, 2019). Waste management touches on economy and ecology, and thus it is plausible that all relevant stakeholders are involved in its planning and operationalization.

3 Materials and Methods

3.1 The Sustainable Community Waste Management Model

The SCWM model adopts a comprehensive and systems approach to waste management (Fig. 1). It is premised on the argument that sound municipal waste management cannot be achieved without inclusive and active stakeholder and citizen involvement. While stakeholders often have decision-making powers, improving citizens' knowledge on waste

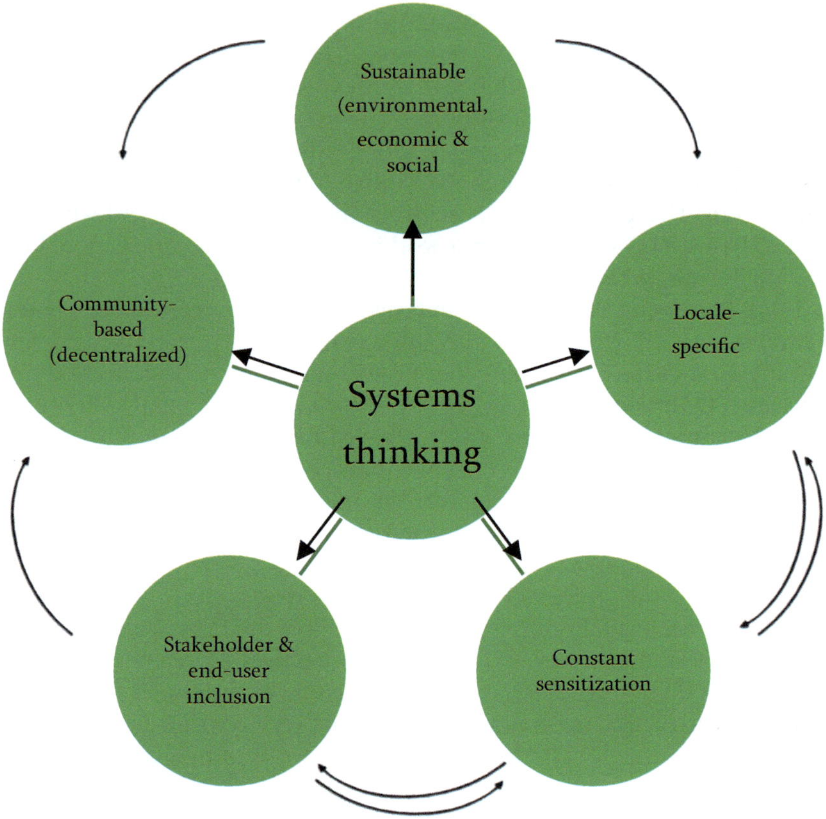

Fig. 1 Sustainable Community Waste Management framework

and the environment also promotes greater community responsibility in environmental conservation and helps make implemented interventions sustainable over the long term.

Sustainable municipal waste management is complex, traversing multiple sectors, including environmental management, health, disaster reduction and employment among others with the various stakeholders in these sectors. Systems thinking presents an approach for viewing these multiplicities of sectors and stakeholders, including end-users, and the interactions between them collectively rather than as individual parts.

This has been one of the main gaps in earlier circular economy waste management models.

The SCWM model is anchored on systems thinking, which is expected to promote sustainable waste management, providing optimized benefits for nature such as biodiversity conservation, society such as improved human health and economy such as employment and income. This may include promoting waste recovery and recycling. The sustainable waste management advanced must also be tailor-designed for the community in which it is implemented, considering the specific conditions in the community, such as the waste composition and which products could be recycled from them.

The model depends greatly on constant sensitization and engagement of both top-level stakeholders and citizens at the community level. This is expected to promote inclusivity of stakeholders from all levels and all sectors in waste management discourses, decision-making platforms and in dialogues. In practice, promoting inclusivity may include ensuring that economically vulnerable groups such as people with low skills, waste collectors and waste pickers, who are dependent on waste for their livelihoods are formally recognized, included and given room to operate in the waste management sector. This also includes ensuring that those who operate in the waste management sector, especially informal workers, are not exploited and promote their well-being. Additionally, it helps obtain feedback useful for promoting sound waste management as well as improving the knowledge of locals apart from ensuring that the waste management is not top-down.

3.2 *Case Site Description: New Edubiase*

New Edubiase is the capital of the Adansi South District in the Ashanti Region of Ghana (Fig. 2). The community has a total population of about 20,000 people, with the majority being youthful. Most residents of the town are employed in the agricultural sector and cultivate food crops like tubers and vegetables and cash crops like cocoa. A significant number of the populace are also employed in the services sector. New Edubiase also falls within the rainforest zone of Ghana and as such is characterized by semi-deciduous forests and thick vegetation. Notwithstanding, there has been significant land use change and little virgin forests remain outside the forest reserves in the outskirts of the town.

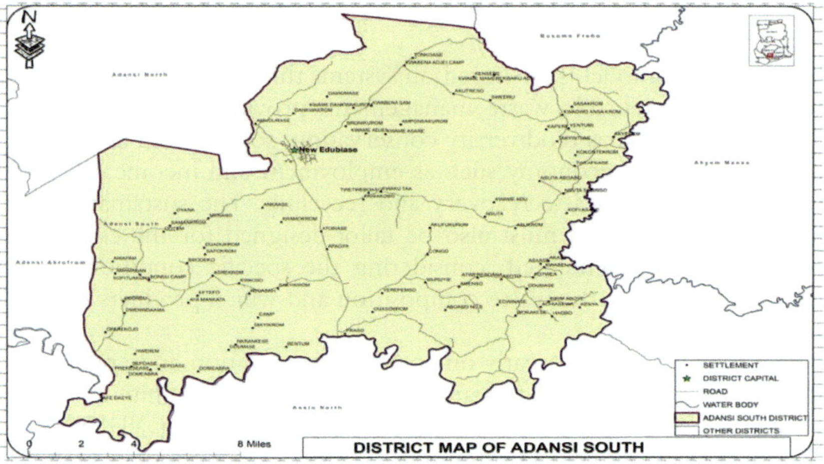

Fig. 2 Map of districts in Ashanti Region indicating the location of Adansi South District

The justification for choosing New Edubiase is that the town is one of the rapidly urbanizing areas in the Ashanti Region of Ghana and exhibits many of the characteristics of the average communities in Ghana. The town has been one of the communities sorely saddled with waste management challenges. It is asserted that only 1.6% of household wastes are properly collected and 61.8% of residents disposed of their solid and liquid waste through open space dumping, including into rivers and streams (ASD, 2014). A significant percentage also engages in open burning. While these practices lead to environmental pollution and ecosystem degradation, they also pose a multiplicity of health hazards such as high exposure to respiratory ailments and hygiene-related diseases like diarrhoea and cholera (Addo et al., 2017). A waste composition analysis was conducted prior to the implementation of the Sustainable Community Project which revealed that organic waste (44%) forms the greatest percentage of wastes in New Edubiase (Fig. 3).

3.3 Data Sources and Collection

A cross-sectional study design was adopted. Primary data was collected from project coordinators through interviews while secondary data was

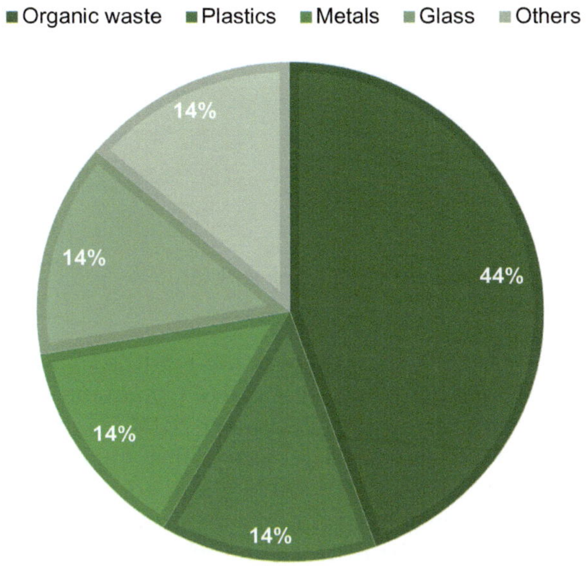

Fig. 3 Waste composition of New Edubiase before project implementation in 2018 (*Source* SCP Edubiase Final report)

obtained from project reports. The interviews were conducted virtually between April 25 and June 31, 2021. The collection of primary data was limited to project coordinators because some of the project reports were already prepared with the help of resident surveys. This way, the interviews conducted for this study helped to validate some of the results in the project reports as well as ask emerging curious questions. The sampling approach adopted then was purposive. Six project coordinators were interviewed in all in the English language.

3.4 Data Analysis

The interviews were analysed using the narrative analysis method according to Stuckey (2014)'s three-tier qualitative data analysis processes. A few key quotes from respondents were used as evidences to buttress some points that emerged. The data entry and processing were done in Microsoft Excel.

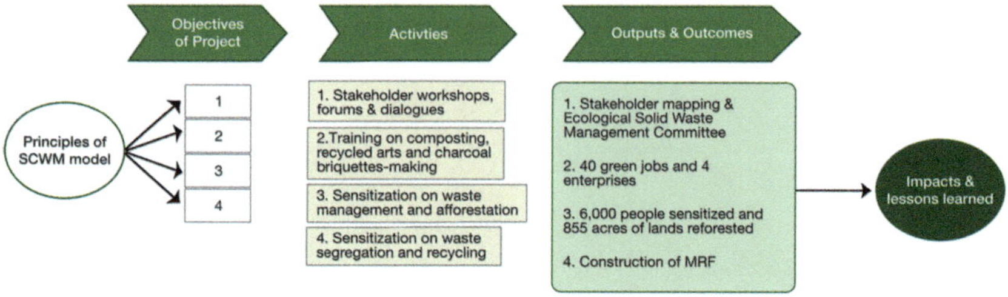

Fig. 4 Theory of change analytical framework

The theory of change framework was also used, as an evaluative framework, to assess the impact of the Sustainable Community Project, which was implemented from January to December 2018. The theory of change underlies most community-based projects either implicitly or explicitly. According to Morra-Imas et al. (2009), the theory of change is useful in that it helps in building "a commonly understood vision of the long-term goals, how they will be reached, and what will be used to measure progress along the way". By feeding the model with quantitative and qualitative data, the model helps to measure the results of an intervention against the baseline conditions that were initially encountered.

Since the SCWM model was tested through the Sustainable Community Project, the theory of change enabled us to show the link between the inputs, activities, outputs and outcomes of the project (Fig. 4). This helped to show the resulting impacts made from the project and consequently, the usefulness and potential of the SCWM model. The results were presented with a discussion of the four objectives of the Sustainable Community Project, including what was done under each, the outcomes and impacts achieved and the challenges and gaps that remain.

3.5 Study Limitation

The study's main limitations were that many of the outcomes of the SCWM model implemented through the Sustainable Community Project were not statistically tested to fully understand the full benefits and disbenefits, if any, of the model. This was because of limited financial and human resources. Nonetheless, this review has offered insights into how a circular economy model adopting a systems thinking approach can be useful for sustainable municipal waste management.

4 Results and Discussions

4.1 Enhancement of Stakeholder Collaboration and Coordination

Many stakeholders are involved in municipal waste management, and each have their interests and functions. There are those with regulatory functions, such as government agencies; those with profit orientations such as producers and businesses; those with functions in the waste collection and recycling phase such as waste service providers and recycling companies; and those who are not-for-profit like non-governmental organizations and community-based organizations. Waste management may also have strong linkages with spatial planning (Cobbinah, 2017), environmental protection, water governance (Wantzen et al., 2019), employment among others (Bai et al., 2017). This underscores why it is necessary to bring all these different stakeholders together to work hand in glove to promote sustainable municipal waste management.

As a first step to piloting of the SCWM model in New Edubiase, a stakeholder mapping was done to identify who the main actors in the waste management regime in the town are and to develop strategies for engaging them. The local government administration, Adansi South Municipal Assembly, were the first to be contacted through writing which was followed up with in-person visits. They were first to be contacted because they have the legal mandate to manage communities under their jurisdiction, including New Edubiase, and they are recognized by the local people and have their trust (Musah-Surugu et al., 2019). Contacts were established with the Environmental Health and Development Planning departments as well as assembly members in the municipality who represent various electoral areas. Through a snowball approach, other stakeholders were identified, including traditional authorities and local waste management enterprises from both the formal and informal sectors. Table 1 details all stakeholders who were identified and engaged and their mandated functions. The exercise also helped to understand each stakeholder's interests and level of influence which helped to determine how each should be strategically engaged.

By rallying these multiplicities of stakeholders, including residents, around a common vision of promoting sustainable municipal waste management, it optimized the waste management regime in New Edubiase, with each stakeholder playing their distinct functions within the framework of the SCWM model. The Ecological Solid Waste Management Committee established which was made up of officers from the local

Table 1 List of stakeholders engaged in the project and their functions

Stakeholder	*Nature of role*	*Function*
Adansi South District	Planning/ regulatory	They have political and administrative authority under the 1993 Local Government Act 462 to lead planning and development in the district
Traditional authorities	Regulatory	They are customary institutions whose key function is promoting the welfare of their community members
Assembly members	Planning	They are elected officials who act as the link between communities and the local government administration, advocating for development
Informal waste workers	Waste collection	They are made up of individuals or small enterprises who use small or low-technology equipment, often engaging in waste collection, sorting, trading and even processing
Zoomlion Limited	Waste collection	Zoomlion is a waste management company in Ghana focusing on waste collection, environmental sanitation and more recently, waste recycling
Market leaders	Waste generation	They are vendors in the marketplace who have been selected by their peers to be leaders to coordinate their activities, often relating to the specific goods they trade in and represent their interests in forums
Households	Waste generation	Households comprise of one or several persons living within the same dwelling

government administration as well as traditional leaders, formulated by-laws for governing waste management practices in the town. This also helped to promote sound waste management practices.

Despite their different vested interests and varying powers, understanding of the benefits of the SCWM model such as the prospects of green jobs, incomes and other socio-economic benefits were key incentives that won stakeholders' cooperation. This helped to show that advancing waste management in an inclusive and systems way is likely to help achieve a sustainable municipal waste management regime. On another hand, promoting inclusion in waste management will also help to promote the interest of informal waste workers who often work in poor conditions, are abused and cheated (Oduro-Appiah et al., 2019). Studies have shown that ultimately, this promotes social cohesion and has positive implications for addressing crime and insecurity as people earn a decent living (Sembiring & Nitivattananon, 2010).

Due to their key position and powers as the development and spatial planning lead in New Edubiase, the Adansi South Municipal Assembly was expected to take a leading role among the stakeholders. The Local Government Act 462 of 1993 states that waste municipal assemblies are responsible for the management of all waste generated in each district. However, it was challenging getting departments like the Environmental Health and Development Planning to take the leading role as expected despite many efforts. The factors accounting for this were many. The national building regulation of Ghana, LI 1630 of 1996, states that "A building for residential, commercial, industrial, civic or cultural use shall have a facility for refuse disposal, a standardized dustbin, and other receptacles approved by the assembly in which all the waste generated shall be stored pending final collection by the trucks to final disposal site" (GAYO, 2021, p. 13). This policy evidences the linear model (landfilling) in waste management in Ghana which informs the development of the Municipal Medium Term Development Plan for most municipalities. In this sense, the piloting of the SCWM model became a conflict to the policy orientation of the assembly, which impacted the commitment of officers to the smooth implementation of the model.

There were also some scepticisms over the green job prospects available in the waste management sector. Although this was communicated with examples implemented in other regions at several forums from the onset of the project, it was not until the materials recovery facility (MRF) (Fig. 5) was constructed and the trainings begun that these outputs like the recycled products, including plastics backpacks and raincoats, could be seen. Thus, some stakeholders were not always high on enthusiasm, especially with their respective functions to attend to, which affected the project timeline. Green jobs were a key incentive in the project, beginning with the construction of the MRF. Even though it took time engaging traditional authorities and some assembly members to secure a piece of land for the MRF, it would have been better for the project if it was delivered early on. This implies that as important as communication is, it is not enough to maintain enthusiasm, especially when advancing a waste management model that goes against the status quo. Thus, in any future replication of the SCWM model, it will be plausible for outputs like the MRF and other green jobs and products to be delivered early on for stakeholder buy-in and cooperation.

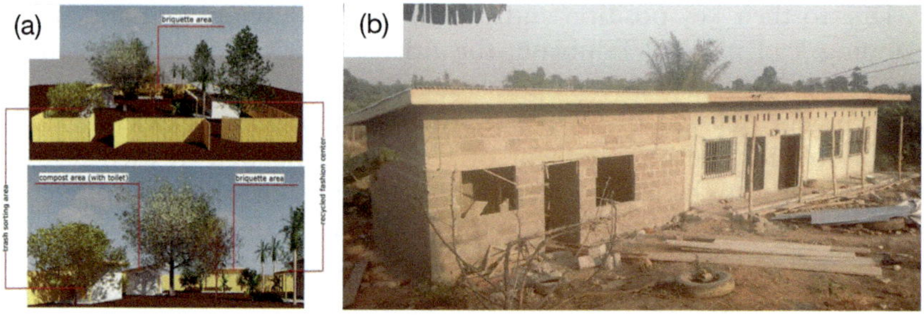

Fig. 5 Materials recovery facility **a** Plan. **b** During construction

4.2 Economic Opportunities in Sustainable Waste Management

By piloting the SCWM model, the economic potential in waste recycling was demonstrated. Under linear waste management models, economic opportunities largely exist in providing waste management services like waste collection and disposal. However, a circular economy model optimizes these and opens an array more in waste recovery, sorting and recycling.

In New Edubiase, informal waste collectors were trained to enhance their work while others were trained in waste sorting and recycling. Regarding recycling, some were trained in the recycling of plastic waste into products like backpacks, handbags, raincoats, curtains, aprons and other recycled arts products. Compost was also made from organic waste. Some of this compost was tested on vegetable farms near the MRF. The compost production received great welcome due to its potential usefulness for residents, majority of whom are farmers. The construction of the MRF also helped in changing perceptions around waste, including perceiving them as resources that can be used to make other useful products. For instance, the conversion of organic waste into compost impressed farmers in the community who find the cost of agrochemicals too expensive.

> *Some people cannot afford the government compost, so they find the compost to be very good for them, which is why we are increasing it [the production] since the project officially ended.* [Sustainable Community Project Coordinator, New Edubiase]

Some organic wastes were also used to make fuel briquettes for domestic use. Over the project duration, 500 kg of compost were produced while 20 tonnes of plastic sachet waste were recycled which were sold at an exhibition (Fig. 6a), raising GHC38,970 in revenue.

The project also had an output on green enterprise development. As part of this, there was financial literacy, sales and marketing training which were undertaken to help green entrepreneurs connect better to the markets and maximize profits. As a result, employment avenues were created, and four individuals were provided capital and technical support to start their own green business enterprises. The impact made in the enterprise development efforts were mainly providing employment avenues for individuals. In all, 40 people (26 females and 14 males) directly received employment through the project (Table 2). These beneficiaries were able to make income, including the project implementers successfully securing a €1000 deal for the delivery of over 100 pieces of recycled arts products to be sold at the Christmas Market in Köln,

Fig. 6 Recycled arts products **a** During launch and exhibition. **b** During sale at the Christmas Market in Köln, Germany

Table 2 Breakdown of direct green jobs created

	Male	*Female*
Briquette-making		6
Recycled arts	2	15
Compost-making	8	5
Project support	3	1
Total	13	27

Source SCP Edubiase Final report

Germany (Fig. 6b). The development of green business enterprises and training organized for waste workers also helped to erode some of the negative perceptions that cloud their activities such as waste work not being a decent livelihood, having poor remuneration and with people in the sector not having dignity (Coletto & Bisschop, 2017) since many community members expressed interest in enrolling in the training on making products from plastic waste, more than the project had capacity to host.

> *Before we completed the MRF, many parents and some of the stakeholders had already submitted names of their wards they wanted us to train to make the bags and aprons from the sachet waste, even though we couldn't train all of them at once. This was surprising because before, not many people wanted to work with the waste, even with Zoomlion.* [Sustainable Community Project Coordinator, New Edubiase]

Regarding gaps, although the compost yielded very positive outcomes when tested in vegetable gardens on the MRF, it needs to be taken through laboratory testing and certification is needed from the Ministry of Food and Agriculture, Ghana, before it can be made available to farmers on commercial basis despite their eagerness to use it. The project had focused only on making fuel briquettes and compost from organic waste. However, this can be expanded as organic waste can also be used to make animal feed for poultry farming (Truong et al., 2019) or used as biogas to produce energy (Atelge et al., 2020). In addition, recycled arts products like backpacks, raincoats and aprons were welcomed by the community during the exhibition; however, a market analysis is needed to understand its acceptability and to guide its uptake as a sustainable business. Furthermore, individuals trained in enterprise development within the waste value chain will need a lot of support by way of technical and financial services. This can be provided by enhancing collaboration with the Ghana Enterprises Agency (formerly known as National Board for Small-Scale Industries) which has decentralized offices in districts across Ghana.

4.3 Promotion of Sound Waste Management Practices

A key part of the Sustainable Community Project in New Edubiase was residents' sensitization on waste and the environment and proper waste disposal practices. Since the project hinged on community members' ability and willingness to segregate wastes at the source and properly dispose of them, intensive house-to-house awareness-raising campaigns and school and media campaigns were embarked on. After a series of sessions, waste bins were distributed to households and a dry run was tested. Four people were engaged in the community waste management education over the duration of the project. Six thousand people from over 1300 households were sensitized on proper waste management practices and environmental conservation. This was independent of the periodic radio and school outreaches. As outcomes, these efforts helped to significantly reduce waste volumes sent to landfills and the improper disposal of wastes on lands and into hydrological systems (Fig. 7).

> *At first, sights of heaps of waste in public spaces and the markets were very common but they are rare now. This has been one of the most visible impacts of the project*. [Programmes Manager, GAYO]

While most project outputs were implemented in a few weeks or months, the sensitization efforts lasted for about 7 months over the project duration, which the project's progression is significantly attributed to. Through the sensitization efforts and the collaborative work of the Ecological Solid Waste Management Committee, open burning was significantly curtailed while activities that pollute the environment, such as dumping of refuse into streams, were also reduced. This was because citizens became more environmentally aware, and more people behaved in a pro-environmental way which is important for the long-term sustainability of the project outcomes. Other studies by Alipour et al. (2015) using the case of Iran and Debrah et al. (2021) in a review of waste management in developing countries made similar conclusions and advocated

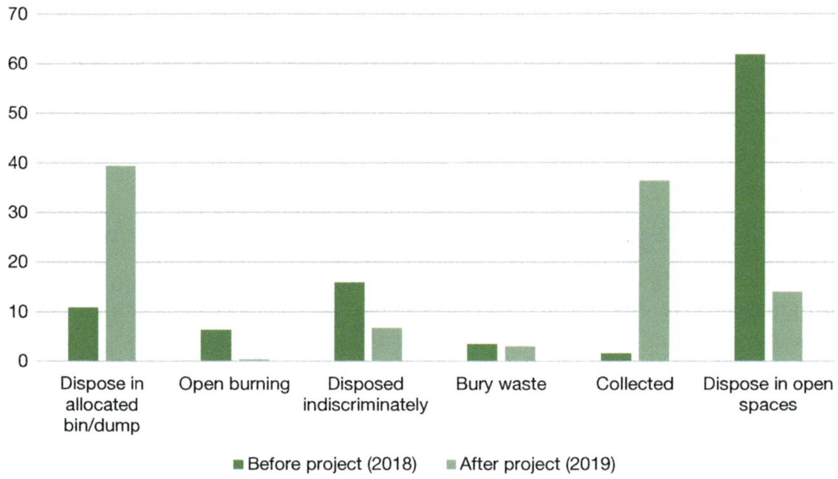

Fig. 7 Waste disposal practices (*Source* SCP Edubiase Final report)

for waste management and environmental education to be included in school curriculums. This notwithstanding, appropriate policy formulation and law enforcement are equally key to fostering and maintaining pro-environmental behavioural change in any community. The work of the Ecological Solid Waste Management Committee had helped to develop byelaws to guide waste management but enforcement was greatly missing due to human resources constraints.

In addition, some degraded forests were reforested. In all, about 855 acres of lands were reforested and planted over the duration of the project. These helped regulate the climate in the community and improve thermal comfort, conserve and foster biodiversity, improve air quality and increase the carbon sinks in the face of the global climate crisis. However, quantitative studies will need to be conducted to deduce the specific impact by way of ecosystem services delivery of these restored forests.

4.4 Promotion of Zero-Waste

In zero-waste, "We aim to send nothing to a landfill. We reduce what we need, reuse as much as we can, send little to be recycled, and compost what we cannot" (GAYO, 2021, p. 17). In the Sustainable Community

Project, zero-waste was primarily advanced through waste segregation at source and recycling. Several authors have demonstrated that most municipal solid wastes in Ghana are organics (e.g. Miezah et al., 2015; Richard et al., 2021; Seshie et al., 2020). With 67.4% of households depending on wood-based fuels as a source of energy (Bawakyillenuo et al., 2021) and 44.7% of people also engaged in agriculture (ILO, 2020), there is opportunity to advance the recycling of organic wastes into fuel briquettes and compost, respectively. Plastics, being one of the major waste components, are also recycled into recycled arts products such as raincoats and bags (GAYO, 2021).

Waste segregation at source, which was not widely practised at the time of the project onset, was introduced in New Edubiase which made recycling efforts more effective (Fig. 8). Several residents now practice waste segregation which helps enhance waste collection and recycling, which hitherto was little known in New Edubiase. These efforts also helped to improve waste collection since waste was better disposed of and waste workers had been trained on how best to go about their activities.

A major project output was the construction of the MRF. A 2-acre piece of land was procured with the assistance of the traditional authorities and some assembly members for the project. Arrangements were then

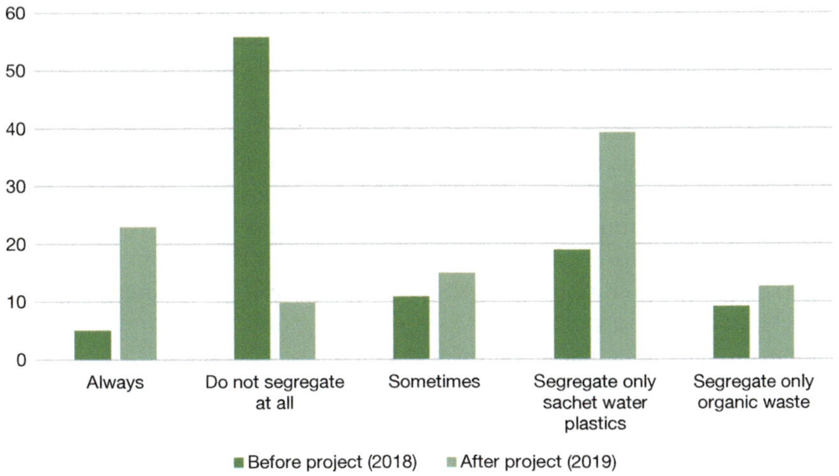

Fig. 8 Waste segregation practices (*Source* SCP Edubiase Final report)

made with some households and market women to divert their waste to the MRF rather than to landfills. The MRF had the capacity to take and process about 2 tonnes of waste each week which is why the volume of waste diverted there had to be controlled. At the facility, there were separate spaces for the processing of different wastes, be it organic or plastic. The MRF became an innovation hub for the project showing the usefulness of wastes and creating employment avenues for informal waste workers together with unemployed youth and women who were trained to recycle waste into various green products.

Despite the potential of fuel briquettes, such as helping to reduce deforestation, having better heat value and being healthy for use since they emit minimal smoke, residents were not as enthusiastic about them as they are about the compost. This needs to be studied to understand how their acceptability can be improved as better alternatives to traditional charcoal and firewood. Education on the dangers of firewood through smoke emissions also needs to be increased.

In addition, the MRF was sited quite distantly (about 6 km) from the central areas of the town as it was difficult to secure land near the town centre. Although it is plausible to have waste disposal sites in the outskirts of communities, the nature of the MRF not posing the health dangers landfills may pose and being an innovation and recycling hub also implies that it is better to have it close by so that recycled products are readily available to the market. Thus, the remote location increased the cost of transporting wastes to the MRF. This defeated the benefit of keeping waste management decentralized so that wastes are managed, included sorting and recycling, at the community level rather than kilometres away so costs are kept minimal as well as for local people to exploit the opportunities available. Also, the MRF was not easily visible to stakeholders and community members which meant that the project team had to put in extra effort to communicate the project's outputs. Perhaps, if the MRF was located close enough to the busy areas of the town, it would have helped to gain even higher cooperation from the community since evidence of the benefits of the project could easily be seen.

In the Sustainable Community Project, waste reduction was not explored as a key output as part of advancing zero-waste. However, there is opportunity to cut down on single-use food and drink containers, reuse old clothing as done in Sri Lanka and replacing especially plastic packaging with reusable bags (Conlon et al., 2019). In a broader sense, this requires policy re-orientation at the national level such as a ban on single-use plastics.

5 Conclusions and Lessons Learned

This study was advanced to review a circular economy model, the SCWM model, which was piloted in a project dubbed the Sustainable Community Project in New Edubiase, Ghana. The model was developed to enhance stakeholder collaboration in the waste management sector, demonstrate the economic opportunities in sustainable waste management, foster sound waste management practices and promote zero-waste.

From the review, we note that factors such as inclusiveness of both stakeholders and the community as end-users as well as intensive sensitization, which were key principles of the SCWM model, helped in raising environmentally conscious residents and promoting sound waste management practices. We also make the conclusion that despite the importance of communication, delivering visible and tangible outcomes of zero-waste such as green jobs and products early on help to maintain stakeholder enthusiasm and support. This is critical as the status quo of the waste management regime does not directly support zero-waste and as such that promise of better need to be evidenced early on. The economic opportunities in zero-waste such as waste sorting and recycling plastics into bags, curtains and aprons and organic waste into compost and fuel briquette help to change negative perceptions around work in the waste sector.

Specific lessons learned include the following:

1. Advancing a circular economy waste management model at the local level in Ghana is challenging because it contradicts the present linear model which relies on disposing wastes in landfills. However, adopting a systems approach and promoting inclusivity as was done in the SCWM model in New Edubiase helps to obtain stakeholder buy-in and cooperation. This helps to leverage on the expertise and resources of stakeholders, including district assemblies, assembly members, traditional authorities, formal/informal waste workers, the media, residents, among others to promote circular economy.
2. Even so, delivering key benefits of circular economy such as the green job prospects available in waste collection, sorting and recycling early on helps to give more impetus for the adoption of models

like the SCWM model and to maintain stakeholder interest and cooperation. In this sense, green jobs available in circular economy are key incentives for stakeholders and residents and the earlier they are produced, the better.

3. A circular economy waste management model, even the SCWM model, works best when it can be modified to suit each distinct locale. This is because waste compositions may differ from place to place and as such the same interventions may not be applicable for every community.
4. While waste such as plastics and organics can be recycled into various products like backpacks and compost, respectively, deciding on which ones to invest in a circular economy model ought to be informed not only by the municipal waste composition but also by which products will likely be in demand. In the case of New Edubiase for instance, the charcoal briquettes produced under the SCWM model were not as well received by the community as the compost.
5. A municipal waste management model that is sustainable ought to factor in urban environmental change and population growth especially in the city or town's context in the planning processes. This will help in the siting and allocation of enough space for the collection and transportation of waste and the development of infrastructure such as MRFs. In the experience of New Edubiase during the piloting of the SCWM model, the MRF had to be sited 6 km away from the town centre as the urban plan had not made provision for handling waste management.
6. Decentralizing municipal waste management and allowing local stakeholders to play active roles helps make waste management more effective and enable local people to harness the opportunities available in circular economy. Notwithstanding, having the support of central government, such as in instituting legislations like banning single-use plastics, can help make implementation at the local level easier.

Future research can focus on undertaking quantitative studies to fully understand the benefits of some of the outcomes of circular economy models like the SCWM model such as the tonnage of waste reduced that would have otherwise ended up in landfills and the associated cost savings as well as forests conserved because of the use of charcoal briquettes recycled from organic waste rather than traditional charcoal produced through deforestation and associated ecosystem services. Accessing the acceptability of recycled products such as fuel briquettes, compost and recycled arts products like bags, aprons, curtains and raincoats through surveys can also be studied. Furthermore, future research can explore the feasibility and cost–benefit analysis of reducing single-use plastics, including for packaging.

References

Abad-Segura, E., Batlles-delaFuente, A., González-Zamar, M.-D., & Belmonte-Ureña, L. J. (2021). Implications for sustainability of the joint application of bioeconomy and circular economy: A worldwide trend study. *Sustainability, 13*(13), 7182.

Addo, H. O., Dun-Dery, E. J., Afoakwa, E., Elizabeth, A., Ellen, A., & Rebecca, M. (2017). Correlates of domestic waste management and related health outcomes in Sunyani, Ghana: A protocol towards enhancing policy. *BMC Public Health, 17*(1), 1–10.

Adu-Boahen, K., Atampugre, G., Antwi, K., Osman, A., Osei, K., Mensah, E., & Adu-Boahen, A. (2014). Waste management practices in Ghana: Challenges and prospect, Jukwa Central Region. *International Journal of Development and Sustainability, 3*(3), 530–546.

Alipour, S., Rahmati, F. S., & Akbarbekyi, M. (2015). The role of citizenship education and participation in waste management. *Cumhuriyet Üniversitesi Fen Edebiyat Fakültesi Fen Bilimleri Dergisi, 36*(3), 3918–3935.

Aparcana, S. (2017). Approaches to formalization of the informal waste sector into municipal solid waste management systems in low-and middle-income countries: Review of barriers and success factors. *Waste Management, 61*, 593–607.

Araujo-Morera, J., Verdejo, R., López-Manchado, M. A., & Santana, M. H. (2021). Sustainable mobility: The route of tires through the circular economy model. *Waste Management, 126*, 309–322.

ASD. (2014). *Adansi South District Medium-Term Development Plan (2014–2017)*. Development Planning Office, Adansi South District. https://new-ndpc-static1.s3.amazonaws.com/CACHES/PUBLICATIONS/2016/04/04/AR_Adansi+South_2014-2017+DMTDP.pdf

Atelge, M., Krisa, D., Kumar, G., Eskicioglu, C., Nguyen, D. D., Chang, S. W., Atabani, A., Al-Muhtaseb, A. H., & Unalan, S. (2020). Biogas production from organic waste: Recent progress and perspectives. *Waste and Biomass Valorization, 11*(3), 1019–1040.

Bai, X., McPhearson, T., Cleugh, H., Nagendra, H., Tong, X., Zhu, T., & Zhu, Y.-G. (2017). Linking urbanization and the environment: Conceptual and empirical advances. *Annual Review of Environment and Resources, 42*(1), 215–240.

Bawakyillenuo, S., Crentsil, A. O., Agbelie, I. K., Danquah, S., Boakye-Danquah, E. B., & Menyeh, B. O. (2021). *The landscape of energy for cooking in Ghana: A review.*

Bonsu, B. O., Takase, M., & Mantey, J. (2020). Preparation of charcoal briquette from palm kernel shells: Case study in Ghana. *Heliyon, 6*(10), e05266.

Buch, R., Marseille, A., Williams, M., Aggarwal, R., & Sharma, A. (2021). From waste pickers to producers: An inclusive circular economy solution through development of cooperatives in waste management. *Sustainability, 13*(16), 8925.

Camilleri, M. A. (2019). The circular economy's closed loop and product service systems for sustainable development: A review and appraisal. *Sustainable Development, 27*(3), 530–536.

Cobbinah, P. B. (2017). Managing cities and resolving conflicts: Local people's attitudes towards urban planning in Kumasi, Ghana. *Land Use Policy, 68*, 222–231.

Cobbinah, P. B., Asibey, M. O., Zuneidu, M. A., & Erdiaw-Kwasie, M. O. (2021). Accommodating green spaces in cities: Perceptions and attitudes in slums. *Cities, 111*, 103094.

Cobbinah, P. B., Erdiaw-Kwasie, M., & Adams, E. A. (2021). COVID-19: Can it transform urban planning in Africa? *Cities & Health, 5*(sup1), S48–S51.

Coletto, D., & Bisschop, L. (2017). Waste pickers in the informal economy of the global South: Included or excluded? *International Journal of Sociology and Social Policy.*

Conlon, K., Jayasinghe, R., & Dasanayake, R. (2019). Circular economy: Waste-to-wealth, jobs creation, and innovation in the global south. *World Review of Science, Technology and Sustainable Development, 15*(2), 145–159.

Debrah, J. K., Teye, G. K., & Dinis, M. A. P. (2022). Barriers and challenges to waste management hindering the circular economy in Sub-Saharan Africa. *Urban Science, 6*(3), 57.

Debrah, J. K., Vidal, D. G., & Dinis, M. A. P. (2021). Raising awareness on solid waste management through formal education for sustainability: A developing countries evidence review. *Recycling, 6*(1), 6.

Di Foggia, G., & Beccarello, M. (2020). Drivers of municipal solid waste management cost based on cost models inherent to sorted and unsorted waste. *Waste Management, 114*, 202–214.

Doe, B., Aboagye, P. D., Osei-Owusu, P. K., Amoah, T., Aidoo, A., & Amponsah, N. Y. (2023). Towards circular economy and local economic development in Ghana: Insights from the coconut waste value chain. *Circular Economy and Sustainability, 3*(1), 347–372.

Durán, C. E. S., & Messina, S. (2019). Urban management model: Municipal solid waste for city sustainability. In *Municipal solid waste management*. IntechOpen.

Ekins, P., Domenech Aparisi, T., Drummond, P., Bleischwitz, R., Hughes, N., & Lotti, L. (2020). *The circular economy: What, why, how and where.*

Erdiaw-Kwasie, M. O., Abunyewah, M., Edusei, J., & Alimo, E. B. (2020). Citizen participation dilemmas in water governance: An empirical case of Kumasi, Ghana. *World Development Perspectives, 20*, 100242.

GAYO. (2021). *Zero waste strategy: Ghana*. Green Africa Youth Organization. https://greenafricayouth.com/sustainable-community-project/zerowaste/

Gentil, E. C., Damgaard, A., Hauschild, M., Finnveden, G., Eriksson, O., Thorneloe, S., Kaplan, P. O., Barlaz, M., Muller, O., & Matsui, Y. (2010). Models for waste life cycle assessment: Review of technical assumptions. *Waste Management, 30*(12), 2636–2648.

Ghana NPAP. (2022). *Plastic Action Initiative Tracker* (p. 22). Ghana National Plastic Action Partnership. https://weforum.ent.box.com/s/yu5lo0dqjhovjgwraifp0sv0gv63mgcg/file/1007782368046

Ghinea, C., & Gavrilescu, M. (2010). Decision support models for solid waste management—An overview. *Environmental Engineering and Management Journal (EEMJ), 9*(6), 869–880.

GoG. (2019). *Ghana renewable energy master plan.*

Gopinath, D. (2020). Waste management in England and the 'circular economy' model. *Geography, 105*(1), 39–46.

Grant, R. (2016). The "Urban Mine" in Accra, Ghana. *RCC Perspectives, 1*, 21–30.

Henrysson, M., & Nuur, C. (2021). The role of institutions in creating circular economy pathways for regional development. *The Journal of Environment & Development, 30*(2), 149–171.

ILO. (2020). *Sector skills strategy: Agriculture sector* (p. 5). International Labour Office. Skills and Employability Branch. https://www.ilo.org/wcmsp5/groups/public/---ed_emp/---ifp_skills/documents/publication/wcms_754214.pdf

James, A. S. J., & Kent, A. (2019). Clothing sustainability and upcycling in Ghana. *Fashion Practice, 11*(3), 375–396.

Jayasinghe, R., & Liyanage, N. (2018). *Sustainability through Eco-entrepreneurship: Analyzing the opportunities and challenges for micro and small-scale eco-enterprises in Sri Lanka.*

Joshi, C., Seay, J., & Banadda, N. (2019). A perspective on a locally managed decentralized circular economy for waste plastic in developing countries. *Environmental Progress and Sustainable Energy, 38*(1), 3–11.

Karim, M., & Wetterhan, J. T. (2020). A comparative study of solid waste management in the United States, Europe and Asia. *Environmental Engineering, 4*, 003–011.

Karmperis, A. C., Aravossis, K., Tatsiopoulos, I. P., & Sotirchos, A. (2013). Decision support models for solid waste management: Review and game-theoretic approaches. *Waste Management, 33*(5), 1290–1301.

Kirchherr, J., & van Santen, R. (2019). Research on the circular economy: A critique of the field. *Resources, Conservation and Recycling, 151*, 104480.

Kreienkamp, J. (2019, November). *The long road to Paris the history of the global climate change regime*. Global Governance Institute.

Kulczycka, J., Lelek, L., Lewandowska, A., & Zarebska, J. (2015). Life cycle assessment of municipal solid waste management—Comparison of results using different LCA models. *Polish Journal of Environmental Studies, 24*(1), 125–140.

Kyere, R., Addaney, M., & Akudugu, J. A. (2019). Decentralization and solid waste management in urbanizing Ghana: Moving beyond the status quo. In *Municipal solid waste management*. IntechOpen.

Miezah, K., Obiri-Danso, K., Kádár, Z., Fei-Baffoe, B., & Mensah, M. Y. (2015). Municipal solid waste characterization and quantification as a measure towards effective waste management in Ghana. *Waste Management, 46*, 15–27.

Morra-Imas, L. G., Morra, L. G., & Rist, R. C. (2009). *The road to results: Designing and conducting effective development evaluations*. World Bank Publications.

Mudu, P., Akua Nartey, B., Kanhai, G., Spadaro, J. V., Fobil, J., & World Health Organization. (2021). *Solid waste management and health in Accra, Ghana*.

Musah-Surugu, I. J., Ahenkan, A., & Bawole, J. N. (2019). Local government financing of climate change in Ghana: Politics of aid and central government dependency syndrome. *Journal of Asian and African Studies, 54*(5), 619–637.

Njoroge, B., Kimani, M., & Ndunge, D. (2014). Review of municipal solid waste management: A case study of Nairobi, Kenya. *International Journal of Engineering and Science, 4*(2), 16–20.

Nkansah, A., Attiogbe, F., & Kumi, E. (2015). Scrap metals' role in circular economy in Ghana, using Sunyani as a case study. *African Journal of Environmental Science and Technology, 9*(11), 793–799.

Oduro-Appiah, K., Afful, A., Kotey, V. N., & De Vries, N. (2019). Working with the informal service chain as a locally appropriate strategy for sustainable modernization of municipal solid waste management systems in lower-middle income cities: Lessons from Accra, Ghana. *Resources, 8*(1), 12.

OECD/FAO. (2016). Agriculture in Sub-Saharan Africa: Prospects and challenges for the next decade. *OECD-FAO Agricultural Outlook, 2025*(181), 1–39.

Oteng-Ababio, M., Owusu-Sekyere, E., & Amoah, S. T. (2017). Thinking globally, acting locally: Formalizing informal solid waste management practices in Ghana. *Journal of Developing Societies, 33*(1), 75–98.

Potting, J., Hekkert, M., Worrell, E., & Hanemaaijer, A. (2017). *Circular economy: Measuring innovation in the product chain* (Issue 2544). PBL Publishers.

Quartey, E. T., Tosefa, H., Danquah, K. A. B., & Obrsalova, I. (2015). Theoretical framework for plastic waste management in Ghana through extended producer responsibility: Case of sachet water waste. *International Journal of Environmental Research and Public Health, 12*(8), 9907–9919.

Richard, N., Richard, A.-A., Yeboah, A. B., & Ahwiren, A. K. (2021). Household Solid Waste Management in La-Nkwantanang-Madina Municipality of the Greater Accra Region, Ghana. *Journal of Environmental Studies [JES], 24*, 1–14.

Sakijege, T. (2019). Repercussions of improved municipal solid waste management on flood risk reduction: The Case of Dar es Salaam, Tanzania. *Journal of Geoscience and Environment Protection, 7*(09), 177.

Salmenperä, H., Pitkänen, K., Kautto, P., & Saikku, L. (2021). Critical factors for enhancing the circular economy in waste management. *Journal of Cleaner Production, 280*, 124339.

Sembiring, E., & Nitivattananon, V. (2010). Sustainable solid waste management toward an inclusive society: Integration of the informal sector. *Resources, Conservation and Recycling, 54*(11), 802–809.

Seshie, V. I., Obiri-Danso, K., & Miezah, K. (2020). Municipal solid waste characterisation and quantification as a measure towards effective waste management in the Takoradi Sub-Metro, Ghana. *Ghana Mining Journal, 20*(2), 86–98.

Shamshiry, E., Nadi, B., Bin Mokhtar, M., Komoo, I., Saadiah Hashim, H., & Yahaya, N. (2011). Integrated models for solid waste management in tourism regions: Langkawi Island, Malaysia. *Journal of Environmental and Public Health, 2011.*

Smol, M., Adam, C., & Preisner, M. (2020). Circular economy model framework in the European water and wastewater sector. *Journal of Material Cycles and Waste Management, 22*(3), 682–697.

Stuckey, H. L. (2014). The first step in data analysis: Transcribing and managing qualitative research data. *Journal of Social Health and Diabetes, 2*(01), 006–008.

Towa, E., Zeller, V., & Achten, W. M. (2020). Input-output models and waste management analysis: A critical review. *Journal of Cleaner Production, 249*, 119359.

Truong, L., Morash, D., Liu, Y., & King, A. (2019). Food waste in animal feed with a focus on use for broilers. *International Journal of Recycling of Organic Waste in Agriculture, 8*(4), 417–429.

UNEP. (2019). *Sustainability of sugarcane bagasse briquettes and charcoal value chains in Kenya*. February, 21–21.

Wantzen, K. M., Alves, C. B. M., Badiane, S. D., Bala, R., Blettler, M., Callisto, M., Cao, Y., Kolb, M., Kondolf, G. M., & Leite, M. F. (2019). Urban Stream and Wetland Restoration in the Global South—A DPSIR Analysis. *Sustainability, 11*(18), 4975.

Winkler, J., & Bilitewski, B. (2007). Comparative evaluation of life cycle assessment models for solid waste management. *Waste Management, 27*(8), 1021–1031.

Index

M. O. Erdiaw-Kwasie and G. M. M. Alam (eds.), *Circular Economy Strategies and the UN Sustainable Development Goals*, Sustainable Development Goals Series,
https://doi.org/10.1007/978-981-99-3083-8

GPSR Compliance

The European Union's (EU) General Product Safety Regulation (GPSR) is a set of rules that requires consumer products to be safe and our obligations to ensure this.

If you have any concerns about our products, you can contact us on ProductSafety@springernature.com

In case Publisher is established outside the EU, the EU authorized representative is:

Springer Nature Customer Service Center GmbH
Europaplatz 3
69115 Heidelberg, Germany

Batch number: 10726052

Printed by Printforce, the Netherlands